Analysis and Control of Nonlinear Infinite Dimensional Systems

This is volume 190 in
MATHEMATICS IN SCIENCE AND ENGINEERING
Edited by William F. Ames, *Georgia Institute of Technology*

A list of recent titles in this series appears at the end of this volume.

ANALYSIS AND CONTROL OF NONLINEAR INFINITE DIMENSIONAL SYSTEMS

Viorel Barbu
SCHOOL OF MATHEMATICS
UNIVERSITY OF IAŞI
IAŞI, ROMANIA

ACADEMIC PRESS, INC.
Harcourt Brace Jovanovich, Publishers

Boston San Diego New York
London Sydney Tokyo Toronto

This book is printed on acid-free paper. ♾

ACADEMIC PRESS, INC.
1250 Sixth Avenue, San Diego, CA 92101-4311

United Kingdom Edition published by
ACADEMIC PRESS LIMITED
24-28 Oval Road, London NW1 7DX

Library of Congress Cataloging-in-Publication Data

Barbu, Viorel.
Analysis and control of nonlinear infinite dimensional systems / Viorel Barbu.
p. cm.—(Mathematics in science and engineering; v. 189)
Includes bibliographical references and index.
ISBN 0-12-078145-X
1. Control theory. 2. Mathematical optimization. 3. Nonlinear operators. I. Title. II. Series.
QA402.3.B343 1993
003′.5—dc20 92-2851
CIP

PRINTED IN THE UNITED STATES OF AMERICA

92 93 94 95 EB 9 8 7 6 5 4 3 2 1

Contents

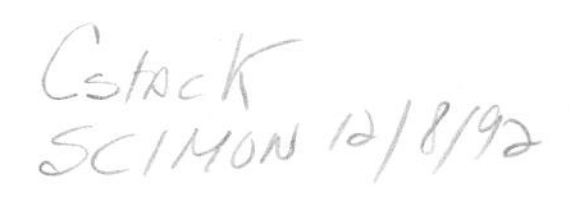

Preface

In contemporary mathematics control theory is complementarily related to analysis of differential systems, which concerns existence, uniqueness, regularity, and stability of solutions. In fact, as remarked by Lawrence Markus, control theory concerns the synthesis of systems starting from certain prescribed goals and the desired behavior of solutions. We tried to write this book in this dual perspective: analysis–synthesis having as its subject the class of nonlinear accretive control systems in Banach space. Since its inception in the 1960s the theory of nonlinear accretive (monotone) operators and of nonlinear differential equations of accretive type has occupied an important place among functional methods in the theory of nonlinear partial differential equations, along with the Leray–Schauder degree theory. Its areas of application include existence theory for nonlinear elliptic and parabolic boundary value problems and problems with free boundary.

The optimal control problems studied in this book are governed by state equations of the form $Ay = Bu + f$ and $y' + Ay = Bu + f$, where A is a nonlinear accretive (multivalued) operator in a Banach space X, B is a linear continuous operator from a controller space U to X, and u is a control parameter. Very often in applications A is an elliptic operator on an open domain of the Euclidean space with suitable boundary conditions. The cost functional is in general not differentiable, and since the state equation is nonlinear this leads to a nonsmooth and nonconvex optimization problem, which requires a specific treatment. In concrete situations such a problem reduces to a nonlinear distributed optimal control problem, and a large class of industrial optimization processes can be put into this form. In fact, the optimal control theory of nonlinear distributed parameter systems has grown in the last decade into an applied mathematical discipline with its own interest and a large spectrum of applications. However, here we shall confine ourselves to the treatment of a limited number of problems with the main emphasis on optimal control problems with free and moving boundary. Nor is this book comprehensive in any way as far as concerns theory of monotone operators and of nonlinear differential equations of accretive type in Banach spaces. The exposition was restricted to a certain body of basic results and methods along with certain significant examples in partial differential equations.

Part of the material presented in Chapter III and V appeared in a preliminary form in my 1984 Pitman Lectures Notes, *Optimal Control of Variational Inequalities*.

This book was completed while the author was Otto Szasz Visiting Professor at University of Cincinnati during the academic year 1990–1991 and the material has been used by the author for a one year graduate-level given course at University of Iaşi and University of Cincinnati.

Part of the manuscript has been read by my colleagues and former students, Professor I. Vrabie, Dr. D. Tătaru, and Dr. S. Aniţa, who contributed valuable criticism and suggestions. To Professor Gh. Moroşanu I also owe special thanks for his careful reading of the original manuscript and his constructive comments, which have led to a much better presentation.

V. Barbu
Iaşi, November 1991

Notation and Symbols

$\mathbf{R}^N$ — the N-dimensional Euclidean space

$\mathbf{R}$ — the real line $(-\infty, \infty)$

$\mathbf{R}^+ = [0, +\infty)$, $\mathbf{R}^- = (-\infty, 0]$, $\overline{\mathbf{R}} = (-\infty, +\infty]$

Ω — open subset of $\mathbf{R}^N$

$\partial\Omega$ — the boundary of Ω

$Q = \Omega \times (0, T)$, $\Sigma = \partial\Omega \times (0, T)$ where $0 < T < \infty$

$\|\cdot\|_X$ — the norm of a linear normed space X

X^*, X' — the dual of the space X

$L(X, Y)$ — the space of linear continuous operators from X to Y

∇f — the gradient of the map$f$: $X \to Y$

∂f — the subdifferential (or the Clarke gradient) of f: $X \to \mathbf{R}$

B^* — the adjoint of the operator B

$\overline{C}$ — the closure of the set C

$\operatorname{int} C$ — the interior of C

$\operatorname{conv} C$ — the convex hull of C

sign — the signum function on X: $\operatorname{sign} x = \dfrac{x}{\|x\|_X}$ if $x \neq 0$
$\operatorname{sign} 0 = \{x; \|x\| \le 1\}$

$C^k(\Omega)$ — the space of real valued functions on Ω that are continuously differentiable up to order k, $0 \le k \le \infty$

$C_0^k(\Omega)$ — the subspace of functions in $C^k(\Omega)$ with compact support in Ω

$\mathscr{D}(\Omega)$ — the space $C_0^\infty(\Omega)$

$\dfrac{dk_u}{dt^k}$, $u^{(k)}$ — the derivative of order k of u: $[a, b] \to x$

$\mathscr{D}'(\Omega)$ — the dual of $\mathscr{D}(\Omega)$, i.e., the space of distributions on Ω

$C(\overline{\Omega})$ — the space of continuous functions on $\overline{\Omega}$

$L^p(\Omega)$ — the space of p-summable functions u: $\Omega \to \mathbf{R}$ endowed with the norm $\|u\|_p = \left(\int_\Omega |u(x)|^p \, dx\right)^{1/p}$, $1 \le p < \infty$,
$\|u\|_\infty = \operatorname{ess\,sup}_{x \in \Omega} |u(x)|$ for $p = \infty$

$W^{m,p}(\Omega)$ — the Sobolev space $\{u \in L^p(\Omega);\ D^\alpha u \in L^p(\Omega),\ |\alpha| \le m\}$, $1 \le p \le \infty$

$W_0^{m,p}(\Omega)$ — the closure of $C_0^\infty(\Omega)$ in the norm of $W^{m,p}(\Omega)$

$W^{-m,q}(\Omega)$	the dual of $W_0^{m,p}(\Omega)$; $\frac{1}{p} + \frac{1}{q} = 1$.
$H^k(\Omega)$, $H_0^k(\Omega)$	the spaces of $W^{k,2}(\Omega)$ and $W_0^{k,2}(\Omega)$, respectively.
$L^p(0, T; X)$	the space of p-summable functions from (a, b) to X (Banach space) $1 \le p \le \infty$, $-\infty \le a < b \le \infty$
$\mathrm{AC}([a, b]; X)$	the space of absolutely continuous functions from $[a, b]$ to X
$\mathrm{BV}([a, b]; X)$	the space of functions with bounded variation on $[a, b]$
$W^{k,p}([a, b]; X)$	the space $\{u \in \mathrm{AC}([a, b]; X); \frac{du}{dt} \in L^p(a, b; X)\}$

Chapter 1 | Preliminaries

The aim of this chapter is to provide some basic results pertaining to geometric properties of normed spaces, semigroups of class C_0, and infinite dimensional vectorial functions defined on real intervals. Some of these results, which can be easily found in textbooks or monographs, are given without proof or with a sketch of proof only.

1.1. The Duality Mapping

Throughout this section X will be a real normed space and X^* will denote its dual. The value of a functional $x^* \in X^*$ at $x \in X$ will be denoted by either (x, x^*) or $x^*(x)$, as is convenient. The norm of X will be denoted by $\|\cdot\|$, and the norm of X^* will be denoted by $\|\cdot\|_*$. If there is no danger of confusion we omit the asterisk from the notation $\|\cdot\|_*$ and denote both the norms of X and X^* by the symbol $\|\cdot\|$.

We shall use the symbol lim or $\to$ to indicate the *strong convergence* in X and w-lim or $\rightharpoonup$ for the *weak convergence* in X. By w*-lim or $\rightharpoonup$ we shall indicate the *weak-star* convergence in X^*. The space X^* endowed with the weak star topology will be denoted by X^*_w.

Define on X the mapping $J: X \to 2^{X^*}$:

$$J(x) = \{x^* \in X^*; (x, x^*) = \|x\|^2 = \|x^*\|^2\} \qquad \forall x \in X. \tag{1.1}$$

By the Hahn–Banach theorem we know that for every $x_0 \in X$ there is some $x_0^* \in X^*$ such that $(x_0, x_0^*) = \|x_0\|$ and $\|x_0^*\| \leq 1$. Indeed, by the Hahn–Banach theorem the linear functional $f: Y \to \mathbf{R}$ defined by $f(x) = \alpha\|x_0\|$ for $x = \alpha x_0$, where $Y = \{\alpha x_0;\ \alpha \in \mathbf{R}\}$, has a linear continuous extension $x_0^* \in X^*$ on X such that $|(x_0^*, x)| \leq \|x\| \ \forall x \in X$. Hence, $(x_0^*, x_0) = \|x_0\|$ and $\|x_0^*\| \leq 1$ (in fact, $\|x_0^*\| = 1$). Clearly, $x_0^*\|x_0\| \in J(x_0)$ and so $J(x) \neq \emptyset$ for every $x \in X$.

The mapping $J: X \to X^*$ is called the *duality mapping* of the space X.

In general, the duality mapping J is multivalued. For instance, if $X = L^1(\Omega)$, where Ω is a measurable subset of $\mathbf{R}^N$ then it is readily seen that every $v \in L^\infty(\Omega)$ such that $v(x) \in \text{sign}\, u(x) \cdot \|u\|_{L^1(\Omega)}$ a.e. $x \in \Omega$ belongs to $J(u)$. (Here, we have denoted by *sign* the function $\text{sign}\, u = 1$ if $u > 0$, $\text{sign}\, u = -1$ if $u < 0$ and $\text{sign}\, 0 = [-1, 1]$.)

It turns out that the properties of the duality mapping are closely related to the nature of the spaces X and X^*, more precisely to the convexity and smoothing properties of the closed balls in X and X^*.

Recall that the space X is called *strictly convex* if the unity ball B of X is strictly convex, i.e., the boundary ∂B contains no line segments.

The space X is said to be *uniformly convex* if for each $\varepsilon > 0$, $0 < \varepsilon < 2$, there is $\delta(\varepsilon) > 0$ such that if $\|x\| = 1$, $\|y\| = 1$, and $\|x - y\| \geq \varepsilon$, then $\|x + y\| \leq 2(1 - \delta(\varepsilon))$.

Obviously, every uniformly convex space X is strictly convex. Hilbert spaces as well as the spaces $L^p(\Omega)$, $1 < p < \infty$, are uniformly convex spaces (see, e.g., G. Köthe [1]). Recall also that in virtue of the Milman theorem (see, e.g., K. Yosida [1]) every uniformly convex Banach space X is reflexive (i.e., $X = X^{**}$). Conversely, it turns out that every reflexive Banach space X can be renormed such that X and X^* become strictly convex. More precisely, one has the following important result due to E. Asplund [1].

Theorem 1.1. *Let X be a reflexive Banach space with the norm $\|\cdot\|$. Then there is an equivalent norm $\|\cdot\|_0$ on X such that X is strictly convex in this norm and X^* is strictly convex in the dual norm $\|\cdot\|_0^*$.*

Regarding the properties of the duality mapping associated with strictly or uniformly convex Banach spaces, we have

Theorem 1.2. *Let X be a Banach space. If the dual space X^* is strictly convex then the duality mapping $J: X \to X^*$ is single valued and demicontinuous, i.e., it is continuous from X to X_w^*. If the space X^* is uniformly convex then J is uniformly continuous on every bounded subset of X.*

Proof. Clearly, for every $x \in X$, $J(x)$ is a closed convex subset of X^*.

Since $J(x) \subset \partial B$, where B is the open ball of radius $\|x\|$ and center 0, we infer that if X^* is strictly convex then $J(x)$ consists of a single point. Now let $\{x_n\} \subset X$ be strongly convergent to x_0 and let x_0^* be any weak star limit point of $\{Jx_n\}$. (Since the unit ball of the dual space is w*-compact (Yosida [1], p. 137) such a x_0^* exists.) We have

$$(x_0, x_0^*) = \|x_0\|^2 \geq \|x_0^*\|^2$$

because the closed ball of radius $\|x_0\|$ in X^* is weak star closed. Hence $\|x_0\|^2 = \|x_0^*\|^2 = (x_0, x_0^*)$. In other words, $x_0^* = J(x_0)$, and so

$$J(x_n) \rightharpoonup J(x_0),$$

as claimed.

To prove the second part of the theorem let us establish first the following lemma.

Lemma 1.1. *Let X be a uniformly convex Banach space. If $x_n \rightharpoonup x$ and $\limsup_{n\to\infty} \|x_n\| \leq \|x\|$, then $x_n \to x$ as $n \to \infty$.*

Proof. By hypothesis, $(x_n, x^*) \to (x, x^*)$ for all $x \in X$, and so

$$\|x\| \leq \liminf_{n\to\infty} \|x_n\| \leq \|x\|.$$

Hence $\lim_{n\to\infty} \|x_n\| = \|x\|$. Now, we set

$$y_n = \frac{x_n}{\|x_n\|}, \qquad y = \frac{x}{\|x\|}.$$

Clearly, $y_n \rightharpoonup y$ as $n \to \infty$. Let us assume that $y_n \nrightarrow y$ and argue from this to a contradiction. Indeed, in this case we have a subsequence y_{n_k}, $\|y_{n_k} - y\| \geq \varepsilon$, and so there is $\delta > 0$ such that $\|y_{n_k} + y\| \leq 2(1 - \delta)$. Letting $n_k \to \infty$ and recalling that the norm $y \to \|y\|$ is weakly lower semicontinuous, we infer that $\|y\| \leq 1 - \delta$. The contradiction we have arrived at completes the proof of the lemma. ■

Proof of Theorem 1.2 (continued). Assume now that X^* is uniformly convex. We suppose that there exist sequences $\{u_n\}, \{v_n\}$ in X such that $\|u_n\|, \|v_n\| \leq M$, $\|u_n - v_n\| \to 0$ for $n \to \infty$, $\|J(u_n) - J(v_n)\| \geq \varepsilon > 0$ for all n, and argue from this to a contradiction. We set $x_n = u_n\|u_n\|^{-1}$, $y_n = v_n\|v_n\|^{-1}$. Clearly, we may assume without any loss of generality that $\|u_n\| \geq \alpha > 0$ and that $\|v_n\| \geq \alpha > 0$ for all n. Then, we can easily see that

$$\|x_n - y_n\| \to 0 \qquad \text{as } n \to \infty$$

and

$$(x_n, J(x_n) + J(y_n)) = \|x_n\|^2 + \|y_n\|^2 + (x_n - y_n, J(y_n))$$
$$\geq 2 - \|x_n - y_n\|.$$

Hence

$$\tfrac{1}{2}\|J(x_n) + J(y_n)\| \geq 1 - \tfrac{1}{2}\|x_n - y_n\| \qquad \forall n.$$

Inasmuch as $\|J(x_n)\| = \|J(y_n)\| = 1$ and the space X^* is uniformly convex, this implies that $\lim_{n\to\infty}(J(x_n) - J(y_n)) = 0$. On the other hand, we have

$$J(u_n) - J(v_n) = \|u_n\|(J(x_n) - J(y_n)) + \left(\|u_n\|^0 - \|v_n\|\right)J(y_n),$$

so that $\lim_{n\to\infty}(J(u_n) - J(v_n)) = 0$ strongly in X^*. This completes the proof. ■

Now let us give some examples of duality mappings.

1. $X = H$ is a Hilbert space identified with its own dual. Then $J = I$ the identity operator in H. If H is not identified with its dual H', then the duality mapping $J: H \to H'$ is the canonical isomorphism Λ of H onto H'. For instance, if $H = H_0^1(\Omega)$ and $H' = H^{-1}(\Omega)$, then $J = \Lambda$ is defined by

$$(\Lambda u, v) = \int_\Omega \nabla u \cdot \nabla v \, dx \qquad \forall u, v \in H_0^1(\Omega). \tag{1.2}$$

In other words, $J = \Lambda$ is the Laplace operator $-\Delta$ under Dirichlet boundary conditions in $\Omega \subset \mathbf{R}^N$.

2. $X = L^p(\Omega)$, where $1 < p < \infty$ and Ω is a measurable subset of $\mathbf{R}^N$. Then, the duality mapping of X is given by

$$J(u)(x) = |u(x)|^{p-2}u(x)\|u\|_{L^p(\Omega)}^{2-p} \qquad a.e.\ x \in \Omega, \quad \forall u \in L^p(\Omega). \tag{1.3}$$

Indeed, it is readily seen that

$$\int_\Omega J(u)u\,dx = \left(\int_\Omega |u|^p\,dx\right)^{2/p} = \left(\int_\Omega |J(u)|^q\,dx\right)^{2/q},$$

where $1/p + 1/q = 1$. If $X = L^1(\Omega)$, then as we will see later

$$J(u) = \{v \in L^\infty(\Omega);\ v(x) \in \operatorname{sign} u(x) \cdot \|u\|_{L^1(\Omega)} \text{ a.e. } x \in \Omega\}. \tag{1.4}$$

3. Let $X = W_0^{1,p}(\Omega)$, where $1 < p < \infty$ and Ω is a bounded and open subset of $\mathbf{R}^N$. Then,

$$J(u) = -\sum_{i=1}^{N} \frac{\partial}{\partial x_i}\left(\left|\frac{\partial u}{\partial x_i}\right|^{p-2} \frac{\partial u}{\partial x_i}\right) \qquad \|u\|_{W_0^{1,p}(\Omega)}^{2-p}. \tag{1.5}$$

In other words, $J: W_0^{1,p}(\Omega) \to W^{-1,q}(\Omega), 1/p + 1/q = 1$, is defined by

$$(J(u), v) = \sum_{i=1}^{N} \int_{\Omega} \left|\frac{\partial u}{\partial x_i}\right|^{p-2} \frac{\partial u}{\partial x_i}\frac{\partial v}{\partial x_i} \qquad dx \|u\|_{W_0^{1,p}(\Omega)}^{2-p}. \tag{1.6}$$

4. Let $X = C(\overline{\Omega})$, where $\overline{\Omega}$ is a compact subset of $\mathbf{R}^N$. Then, the duality mapping $J: C(\overline{\Omega}) \to M(\overline{\Omega})$ (the space of all bounded Radon measures on $\overline{\Omega}$) is given by (E. Sinestrari [1])

$$J(y) = \left\{\mu \|y\|_{C(\overline{\Omega})};\ \mu(y) \le \max\left\{y(x_0) \operatorname{sign} y(x_0), x_0 \in M_y\right\},\right.$$
$$\left.\mu \in M(\overline{\Omega})\right\} \qquad \forall y \in C(\overline{\Omega}),$$

where

$$M_y = \left\{x_0 \in \overline{\Omega};\ |y(x_0)| = \|y\|_{C(\overline{\Omega})}\right\}.$$

We shall see later that the duality mapping J of the space X can be equivalently defined as the subdifferential (Gâteaux differential if X^* is strictly convex) of the function $x \to \frac{1}{2}\|x\|^2$.

1.2. Compact Mappings in Banach Spaces

Let X, Y be linear normed spaces. An operator $T: X \to Y$ is called *compact* if it maps every bounded subset of X into a relatively compact subset of Y.

If D is a bounded and open subset of X and $T: D \to X$ is continuous, then for any $p \in X$ such that $p \notin F(\partial D), F = I - T$ (∂D is the boundary of D), we may define the topological degree of F in p relative to D, denoted $d(F, D, p)$ having the following properties (*the Leray–Schauder degree*):

(i) $d(I, D, p) = 1 \quad$ if $p \in D$,

$d(I, D, p) = 0 \qquad \text{if } p \notin D;$

(ii) If $d(I, D, p) \neq 0$, then the equation $Fx = p$ has at least one solution $x \in D$;

(iii) If $H(t)$ is a compact homotopy in X such that $p \notin H(t)(\partial D)$ $\forall t \in [0, T]$, then

$$d(I - H(t), D, p) \equiv \text{constant} \qquad \forall t \in [0, T]. \qquad (2.1)$$

The construction of the Leray–Schauder topological degree is a classical result, which can be found in most textbooks devoted to nonlinear equations in infinite dimensional spaces (see, for instance, J. Schwartz [1]). Here, we shall confine ourselves to deriving as a simple consequence of Leray–Schauder degree theory an important result of nonlinear analysis, the *Schauder fixed point theorem.*

Theorem 2.1 (Schauder). *Let K be a closed, bounded and convex subset of a Banach space X. Let T: $K \to K$ be a continuous and compact operator that maps K into itself. Then the equation $Tx = x$ has at least one solution $x \in K$.*

Proof. Consider first the particular case where $K = \{u \in X; \|u\| \leq R\}$. Consider the homotopy $H(t) = tT$, $0 \leq t \leq 1$, and note that $0 \notin (I - H(t))\partial K$ $\forall t \in [0, 1)$, i.e.,

$$u - tTu \neq 0 \qquad \text{for } \|u\| = R \text{ and } t \in [0, 1).$$

Hence $d(I - H(t), K, 0) \equiv$ constant, and therefore $d(I - T, K, 0) \neq 0$, as claimed.

In the general case, consider a ball B centered in origin such that $K \subset \overline{B}$ and take a compact extension $\tilde{T}$ of T on B such that $R(\tilde{T}) \subset K$ ($R(\tilde{T})$ is the range of $\tilde{T}$). (The existence of a such a map $\tilde{T}$ is a well-known result due to Djugundi, which extends the classical theorem of Tietze.) Hence, $\tilde{T}$ maps B into itself and according to the first part of the proof it has at least one fixed point $x \in B$. Since $R(\tilde{T}) \subset K$, we infer that $x \in K$ and $Tx = x$, as claimed. ■

It is clear that the following version of Schauder theorem is also true: *Let K be a compact convex subset of a Banach space X and let T: $K \to K$ be a continuous operator. Then T has at least one fixed point in K.*

The Schauder fixed point theorem along with the Leray–Schauder degree theory represent powerful instruments in the study of infinite dimensional equations with compact nonlinearities. Their applications include a large spectrum of nonlinear problems in theory of partial differential equations, ordinary and integral equations, game theory, and other fields. Here, we shall indicate only one application, to the existence of a saddle point for convex–concave functions (J. von Neumann theorem). First, we shall indicate an extension of Schauder theorem to multivalued operators (Kakutani theorem).

Recall that a multivalued mapping $T: X \to 2^Y$ (X, Y are metric spaces) is said to be *upper semicontinuous* (u.s.c.) in x if for every $\varepsilon > 0$ there is $\delta = \delta(x, \varepsilon) > 0$ such that

$$T(B(x, \delta)) \subset B(T(x), \varepsilon). \tag{2.2}$$

Here, $B(x, \delta)$ is the ball of radius δ and center x in X, whilst

$$B(T(x), \varepsilon) = \bigcup_{y \in T(x)} B(y, \varepsilon).$$

The mapping T is said to be upper semicontinuous on X if it is upper semicontinuous at every $x \in X$. It is readily seen that the graph of an u.s.c. mapping is closed. Conversely, if $R(T)$ is compact and the graph of T is closed, then T is u.s.c.

Theorem 2.2 (Kakutani). *Let K be a compact convex subset of a Banach space and let $T: K \to X$ be an upper semicontinuous mapping with convex values $T(x)$ such that $T(x) \subset K$, $\forall x \in K$. Then there is at least one $x \in K$ such that $x \in T(x)$.*

Proof. We will prove first that for every $\varepsilon > 0$ there is a continuous function $f_\varepsilon: K \to K$ such that

$$(B(x, \varepsilon) \times B(f_\varepsilon(x), \varepsilon)) \cap T \neq \emptyset \qquad \forall x \in K. \tag{2.3}$$

In other words, for every $x \in K$ there is $(x_\varepsilon, y_\varepsilon) \in K \times K$ such that $y_\varepsilon \in Tx_\varepsilon$, $\|x_\varepsilon - x\| \leq \varepsilon$, $\|y_\varepsilon - f_\varepsilon(x)\| \leq \varepsilon$.

Indeed, consider a finite cover of K with subsets of the form $B(x_i, \delta(x_i, \varepsilon)) = U_i$, $x_i \in K$, $i = 1, \ldots, m$, and take a finite subordinated partition of unity $\{\alpha_i\}_{i=1}^m$, i.e., $0 \leq \alpha_i \leq 1$, supp $\alpha_i \subset U_i$, $\Sigma_{i=1}^m \alpha_i = 1$, and the α_i are continuous. For instance, we may take

$$\alpha_i(x) = \mu_i(x) \left(\sum_{i=1}^m \mu_i(x) \right)^{-1} \qquad \forall x \in K,$$

where $\mu_i(x) = \max\{0, \delta(x_i, \varepsilon) - \|x - x_i\|\}$. Then, pick $y_i \in T(U_i)$ and define the function

$$f_\varepsilon(x) = \sum_{i=1}^{m} \alpha_i(x) y_i \qquad \forall x \in K,$$

which satisfies the condition (2.3). Moreover, $f_\varepsilon(K) \subset K$. Then, by the Schauder theorem, there is $x_\varepsilon \in K$ such that $f_\varepsilon(x_\varepsilon) = x_\varepsilon$. In other words, for every $\varepsilon > 0$ there are $(\tilde{x}_\varepsilon, \tilde{y}_\varepsilon) \in K \times K$ such that $\|\tilde{x}_\varepsilon - x_\varepsilon\| \leq \varepsilon$, $\|\tilde{y}_\varepsilon - f_\varepsilon(x_\varepsilon)\| \leq \varepsilon$ and $\tilde{y}_\varepsilon \in T\tilde{x}_\varepsilon$. Since the set K is compact, we may assume without any loss of generality that $x_\varepsilon \to x$ and $\tilde{y}_\varepsilon \to x$ for $\varepsilon \to 0$. By the upper semicontinuity property we infer that $x \in Tx$, thereby completing the proof. ■

Let U be a convex subset of X. We recall that the function $f: U \to \mathbf{R}$ is said to be convex if

$$f(\lambda x + (1 - \lambda)y) \leq \lambda f(x) + (1 - \lambda)f(y) \qquad \forall x, y \in U, 0 \leq \lambda \leq 1.$$

If $U \subset X$ and $V \subset Y$ are two convex subsets of linear spaces X and Y, respectively, the function $H: U \times V \to \mathbf{R}$ is called convex–concave if $H(u, v)$ is convex as function of u and concave as function of v. (A function g is said to be concave if $-g$ is convex.)

Theorem 2.3 (J. von Neumann). *Let X and Y be real Banach spaces and let $U \subset X$, $V \subset Y$ be compact convex subsets of X and Y, respectively. Let $H: U \times V \to \mathbf{R}$ be a continuous, convex–concave function. Then there is $(u_0, v_0) \in U \times V$ such that*

$$H(u_0, v) \leq H(u_0, v_0) \leq H(u, v_0) \qquad \forall u \in U, \forall v \in V. \tag{2.4}$$

Such a point (u_0, v_0) is called a *saddle point* of the function H and it is readily seen that (2.4) implies min–max equality:

$$H(u_0, v_0) = \min_{u \in U} \max_{v \in V} H(u, v) = \max_{v \in V} \min_{u \in U} H(u, v),$$

which plays an important role in game theory.

Proof of Theorem 2.3. Define the mappings

$$T_1 u = \{v \in V;\ H(u, v) \geq H(u, w)\ \forall w \in V\},$$
$$T_2 v = \{u \in U;\ H(u, v) \leq H(w, v)\ \forall w \in U\},$$

and set

$$T(u,v) = (T_2 v, T_1 u) \qquad \forall (u,v) \in U \times V.$$

Clearly $T: U \times V \to U \times V$ is upper semicontinuous and $T(u,v)$ is a convex set for every $(u,v) \in U \times V$. Then, by Kakutani's theorem there is at least one $(u_0, v_0) \in U \times V$ such that $(u_0, v_0) \in T(u_0, v_0)$. Clearly, this point satisfies (2.4), i.e., it is a saddle point of H. This complete the proof. ■

If the spaces X and Y are reflexive, Theorem 2.3 remains valid if U and V are merely bounded, closed, and convex subsets. This follows by the same proof using the Schauder theorem (Tichonov theorem) in locally convex spaces (see Edwards [1]).

1.3. Absolutely Continuous Functions with Values in Banach Spaces

Let X be a real (or complex) Banach space and let $[a,b]$ be a fixed interval on the real axis. A function $x:[a,b] \to X$ is said to be *finitely valued* if it is constant on each of a finite number of disjoint measurable sets $A_K \subset [a,b]$ and equal to zero on $[a,b] \setminus \cup_k A_k$. The function x is said to be *strongly measurable* on $[a,b]$ if there is a sequence $\{x_n\}$ of finite valued functions that converges strongly in X and almost everywhere on $[a,b]$ to x. The function x is said to be *Bochner integrable* if there exists a sequence $\{x_n\}$ of finitely valued functions on $[a,b]$ to X that converges almost everywhere to x and such that

$$\lim_{n \to \infty} \int_a^b \|x_n(t) - x(t)\| \, dt = 0.$$

A necessary and sufficient condition that $x:[a,b] \to X$ is Bochner integrable is that x is strongly measurable and that $\int_a^b \|x(t)\| \, dt < \infty$. The space of all Bochner integrable functions $x:[a,b] \to X$ is a Banach space with the norm

$$\|x\|_1 = \int_a^b \|x(t)\| \, dt,$$

and is denoted $L^1(a,b;X)$.

More generally, the space of all (classes) of strongly measurable functions x on $[a, b]$ to X such that

$$\|x\|_p = \left(\int_a^b \|x(t)\|^p \, dt \right)^{1/p} < \infty$$

for $1 \leq p < \infty$ and $\|x\|_\infty = \text{ess sup}_{t \in [a,b]} \|x(t)\| < \infty$, will be denoted $L^p(a, b; X)$. This is a Banach space in the norm $\| \cdot \|_p$.

If X is reflexive, then the dual of $L^p(a, b; X)$ is the space $L^q(a, b; X^*)$, where $1/p + 1/q = 1$ (see Edward's book [1]). Recall also that a function $x: [a, b] \to X$ is said to be *weakly measurable* if for any $x^* \in X^*$, the function $t \to (x(t), x^*)$ is measurable. According to the Pettis theorem, if X is separable then every weakly measurable function is strongly measurable, and so these two notions coincide.

An X-valued function x defined on $[a, b]$ is said to be *absolutely continuous* on $[a, b]$ if for each $\varepsilon > 0$ there exists $\delta(\varepsilon)$ such that $\sum_{n=1}^N \|x(t_n) - x(s_n)\| \leq \varepsilon$, whenever $\sum_{n=1}^N |t_n - s_n| \leq \delta(\varepsilon)$ and $(t_n, s_n) \cap (t_m, s_m) = \varnothing$ for $m \neq n$. Here, (t_n, s_n) is an arbitrary subinterval of (a, b).

A classical result in real analysis says that any real valued absolutely continuous function is almost everywhere differentiable and it is expressed as the indefinite integral of its derivative. It should be mentioned that this result fails for X-valued absolutely continuous functions if X is a general Banach space. Indeed, if $X = L^1(a, b)$ and $x: [a, b] \to X$ is defined by

$$x(t) = \begin{cases} 1 & \text{if } a \leq s \leq t, \\ 0 & \text{if } t < s < b, \end{cases}$$

then it is readily seen that x is absolutely continuous on $[a, b]$ but it is not differentiable in any point of the interval.

However, if the space X is reflexive we have (Komura [1]):

Theorem 3.1. *Let X be a reflexive Banach space. Then every X-valued absolutely continuous function x on $[a, b]$ is almost everywhere differentiable on $[a, b]$ and*

$$x(t) = x(a) + \int_a^t \frac{d}{ds} x(s) \, ds \qquad \forall t \in [a, b], \tag{3.1}$$

where $dx/dt: [a, b] \to X$ is derivative of x, i.e.,

$$\frac{d}{dt} x(t) = \lim_{\varepsilon \to 0} \frac{x(t + \varepsilon) - x(t)}{\varepsilon}.$$

Proof. Since x is absolutely continuous, it is continuous and so the set $\{x(t);\ t \in [a,b]\}$ is compact. Replacing X by the strong closure of the subspace generated by $\{x(t);\ t \in [a,b]\}$, we may assume without loss of generality that X is separable. Let us denote by $V(x,t)$ the total variation of x on the interval $[a,t]$, i.e.,

$$V(x,t) = \sup \sum_{i=1}^{n} \|x(t_{i=1}) - x(t_i)\|,$$

where the supremum is taken over all partitions $a = t_0 < t_1 < \cdots < t_b = b$ of $[a,b]$. Clearly, $t \to V(x,t)$ is bounded and monotonically increasing on $[a,b]$. Note also the inequality

$$\|x(t+\varepsilon) - x(t)\| \leq V(x, t+\varepsilon) - V(x,t), \qquad a \leq t < t+\varepsilon < b,$$

which yields

$$\int_a^{b-\varepsilon} \frac{\|x(t+\varepsilon) - x(t)\|}{\varepsilon}\, dt \leq V(x,b) < \infty.$$

Hence,

$$\limsup_{\varepsilon \to 0} \frac{\|x(t+\varepsilon) - x(t)\|}{\varepsilon} < \infty \qquad \text{a.e. } t \in (a,b).$$

Let $\{x_n^*\}_{n=1}^{\infty}$ be a sequence in X^* that is dense in X^*. For every n, there exists a subset I_n of $[a,b]$ such that $t \to (x(t), x_n^*)$ is differentiable at all $t \in [a,b] \setminus I_n$ and $m(I_n) = 0$ (m is the Lebesgue measure).

We set

$$I_0 = \left\{ t \in [a,b];\ \limsup_{\varepsilon \to 0} \frac{\|x(t+\varepsilon) - x(t)\|}{\varepsilon} = \infty \right\}$$

and $N = \cup_{n=0}^{\infty} I_n$. Clearly, $m(I) = 0$, and for every $t \in [a,b] \setminus \mathbf{N}$, $\|x(t+\varepsilon) - x(t)\|/\varepsilon$ is bounded for $\varepsilon \to 0$. Hence,

$$\lim_{\varepsilon \to 0} \left(\frac{x(t+\varepsilon) - x(t)}{\varepsilon}, x^* \right) = (\dot{x}(t), x^*) \tag{3.2}$$

exists for all $t \in [a,b] \setminus \mathbf{N}$ and all $x^* \in X^*$. In other words, x is weakly differentiable a.e. on $[a,b]$. The function $\dot{x}$ (the weak derivative of x) is weakly measurable and so by Pettis' theorem it is strongly measurable. By

(3.2) we infer that

$$\|\dot{x}(t)\| \le \limsup_{\varepsilon \to 0} \left\| \frac{x(t+\varepsilon) - x(t)}{\varepsilon} \right\| \qquad \forall t \in [a,b] \setminus \mathbf{N},$$

and so by the Fatou lemma $\|\dot{x}\| \in L^1(a,b)$.

On the other hand, we have

$$(x(t) - x(a), x^*) = \int_a^t (\dot{x}(s), x^*)\, ds \qquad \forall t \in [a,b],\ x^* \in X^*,$$

and therefore

$$x(t) = x(a) + \int_a^t \dot{x}(s)\, ds \qquad \forall t \in [a,b].$$

This yields

$$\lim_{\varepsilon \to 0} \frac{x(t+\varepsilon) - x(t)}{\varepsilon} = \dot{x}(t) \qquad \text{strongly in } X, \text{ a.e. } t \in (a,b).$$

Hence $\dot{x} = dx/dt$, and this completes the proof. ■

Let us denote by $\mathscr{D}(a,b)$ the space of all infinitely differentiable real valued functions on $[a,b]$ with compact support in (a,b). The space $\mathscr{D}(a,b)$ is topologized as an inductive limit of $\mathscr{D}_K(a,b)$ where K ranges over all compact subsets of (a,b) and $\mathscr{D}_K(a,b) = \{\varphi \in \mathscr{D}(a,b);\ \text{supp}\ \varphi \subset \text{K}\}$. We shall denote by $\mathscr{D}'(a,b;X)$ the space of all continuous operators from $\mathscr{D}(a,b)$ to X. An element u of $\mathscr{D}'(a,b;X)$ is called a X-valued distribution on (a,b). If $u \in \mathscr{D}'(a,b;X)$ and j is a natural number, then the equation

$$u^{(j)}(\varphi) = (-1)^j u(\varphi^{(j)}) \qquad \forall \varphi \in \mathscr{D}(a,b)$$

defines another distribution $u^{(j)}$, which is called the derivative of order j of u.

We note that every element $u \in L^1(a,b;X)$ defines uniquely the distribution (again denoted u)

$$u(\varphi) = \int_a^b u(t)\varphi(t)\, dt \qquad \forall \varphi \in \mathscr{D}(a,b), \tag{3.3}$$

and so $L^1(a,b;X)$ can be regarded as a subspace of $\mathscr{D}'(a,b;X)$. In all that follows, we shall identify a function $u \in L^1(a,b;X)$ with the distribution u defined by (3.3).

Let k be a natural number and $1 \leq p \leq \infty$. We shall denote by $W^{k,p}([a,b];X)$ the space of all X-valued distributions $u \in \mathscr{D}'(a,b;X)$ such that

$$u^{(j)} \in L^p(a,b;X) \qquad \text{for } j = 0,1,\ldots,k. \tag{3.4}$$

Here, $u^{(j)}$ is the derivative of order j of u in the sense of distributions.

We shall denote by $A^{1,p}([a,b];X)$, $1 \leq p \leq \infty$, the space of all absolutely continuous functions u from $[a,b]$ to X having the property that they are a.e. differentiable on (a,b) and $du/dt \in L^p(a,b;X)$. If the space X is reflexive, it follows by Theorem 3.1 that $u \in A^{1,p}([a,b];X)$ if and only if u is absolutely continuous on $[a,b]$ and $du/dt \in L^p(a,b;X)$.

It turns out that the space $W^{1,p}$ can be identified with $A^{1,p}$. More precisely, we have:

Theorem 3.2. *Let X be a Banach space and let $u \in L^p(a,b;X)$, $1 \leq p \leq \infty$. Then the following conditions are equivalent*:

(i) $u \in W^{1,p}([a,b];X)$;

(ii) *There is $u^0 \in A^{1,p}([a,b];X)$ such that $u(t) = u^0(t)$, a.e. $t \in (a,b)$. Moreover, $u' = du^0/dt$ a.e. in (a,b).*

Proof. For simplicity, we shall assume that $[a,b] = [0,T]$.

Let $u \in W^{1,p}([0,T];X)$, i.e., $u \in L^p(0,T;X)$ and $u' \in L^p(0,T;X)$, and define the regularization u_n of u,

$$u_n(t) = \int_0^T u(s)\rho((t-s)n)\,ds \qquad \forall t \in [0,T], \tag{3.5}$$

where $\rho \in \mathscr{D}(\mathbf{R})$ is such that $\int \rho(s)\,ds = 1$, $\rho(t) = \rho(-t)$, $\operatorname{supp} \rho \subset [-1,1]$. It is well-known that $u_n \to u$ in $L^p(0,T;X)$ for $n \to \infty$. Note also that u_n is infinitely differentiable. Let $\varphi \in \mathscr{D}(0,T)$ be arbitrary but fixed. Then, by (3.5), we see that

$$\begin{aligned}\int_0^T \frac{du_n}{dt}(t)\,\varphi(t)\,dt &= -\int_0^T u_n(t)\,\frac{d\varphi}{dt}(t)\,dt = -\int_0^T u(t)\,\frac{d\varphi_n}{dt}(t)\,dt \\ &= u'\left(\frac{d\varphi_n}{dt}\right) = \int_0^T (u')_n \varphi\,dt\end{aligned}$$

$$\text{if } \operatorname{supp} \varphi \subset \left(\frac{1}{n}, T - \frac{1}{n}\right).$$

Hence, $du_n/dt = (u')_n$ a.e. in $(1/n, T - 1/n)$. On the other hand, letting n tend to ∞ in the equation

$$u_n(t) - u_n(s) = \int_s^t \frac{du_n}{d\tau}(\tau)\, d\tau,$$

we get

$$u(t) - u(s) = \int_s^t u'\, d\tau \qquad \text{a.e. } t, s \in (0,T),$$

because $(u')_n \to u'$ in $L^p(0,T;X)$. The latter equation implies that u admits an extension to an absolutely continuous function u^0 on $[0,T]$ that satisfies the equation

$$u^0(t) - u^0(0) = \int_0^t u'(\tau)\, d\tau \qquad \forall t \in [0,T].$$

Hence, (i) $\Rightarrow$ (ii).

Conversely, assume now that $u \in A^{1,p}([0,T];X)$. Then,

$$\begin{aligned} u'(\varphi) &= -\int_0^T u(t)\varphi'(t)\, dt = -\lim_{\varepsilon \to 0} \frac{1}{\varepsilon} \int_0^T u(t) \frac{\varphi(t) - \varphi(t-\varepsilon)}{\varepsilon}\, dt \\ &= -\lim_{\varepsilon \to 0} \frac{1}{\varepsilon} \int_0^{T-\varepsilon} (u(t) - u(t+\varepsilon))\varphi(t)\, dt \\ &\quad - \lim_{\varepsilon \to 0} \frac{1}{\varepsilon} \int_{T-\varepsilon}^T u(t)\varphi(t)\, dt \\ &\quad + \lim_{\varepsilon \to 0} \frac{1}{\varepsilon} \int_0^{\varepsilon} u(t)\varphi(t-\varepsilon)\, dt \qquad \forall \varphi \in \mathscr{D}(0,T). \end{aligned}$$

Hence

$$u'(\varphi) = \int_0^T \frac{du}{dt}(t)\, \varphi(t) \qquad \forall \varphi \in \mathscr{D}(0,T).$$

This shows that $u' \in L^p(0,T;X)$ and $u' = du/dt$. This completes the proof of Theorem 3.2. ■

In particular, it follows by the proof of Theorem 3.2 that we may choose $u_1 \in A^{1,p}([a,b];X)$ such that $u_1(t) = u_1(a) + \int_a^t du_1/ds\, ds\ \forall t \in [a,b]$.

Theorem 3.3. *Let X be reflexive and let $u \in L^p(a,b;X)$, $1 < p \leq \infty$. Then the following two conditions are equivalent*:

(i) $u \in W^{1,p}([a,b];X)$:

(ii) *There is* $C > 0$ *such that*

$$\int_a^{b-h} \|u(t+h) - u(t)\|^p \, dt \le C|h|^p \qquad \forall h \in [0, b-a].$$

Proof. (i) $\Rightarrow$ (ii). By Theorem 3.1, we know that

$$u(t+h) - u(t) = \int_t^{t+h} \frac{du_1}{ds}(s)\, ds \qquad \forall t, t+h \in [a,b],$$

where $u_1 \in A^{1,p}([a,b];X)$, i.e., $du_1/dt \in L^p(a,b;X)$. This yields

$$\int_a^{b-h} \|u(t+h) - u(t)\|^p \, dt \le |h|^p \int_a^{b-h} dt \int_t^{t+h} \left\| \frac{du_1}{ds} \right\|^p ds.$$

Since

$$\lim_{h \downarrow 0} \int_a^b \frac{dt}{h} \int_t^{t+h} \left\| \frac{du_1}{ds} \right\|^p ds = \int_a^b \left\| \frac{du_1}{dt}(t) \right\|^p dt,$$

we find the estimate (ii).

(ii) $\Rightarrow$ (i). Let u_n be the regularization of u. A little calculation involving formula (3.5) reveals that $\{u_n'\}$ is bounded in $L^p(a,b;X)$. Since $u_n \to u$ in $L^p(a,b;X)$, $u_n' \to u'$ in $\mathscr{D}'(a,b;X)$, and $\{u_n'\}$ is weakly compact in $L^p(a,b;X)$, we infer that $u' \in L^p(a,b;X)$, as claimed. ∎

Remark 3.1. If $u \in W^{1,1}([a,b];X)$, then it follows as in the preceding that

$$\int_a^{b-h} \|u(t+h) - u(t)\| \, dt \le C|h| \qquad \forall h \in [0, b-a].$$

However, this inequality does not characterize the functions u in $W^{1,1}([a,b];X)$, but the functions u with bounded variation on $[a,b]$ (see H. Brézis [4]).

Let V and H be a pair of real Hilbert spaces such that $V \subset H \subset V'$ in the algebraic and topological senses. Here, V' is the dual space of V and H is identified with its own dual. Denote by $|\cdot|$ and $\|\cdot\|$ the norms of H and V, respectively, and by $(\cdot,\cdot)$ the duality between V and V'. If $v_1, v_2 \in H$, then (v_1, v_2) is the scalar product in H of v_1 and v_2.

Denote by $W([a,b];V)$ the space

$$W([a,b];V) = \{u \in L^2(a,b;V); u' \in L^2(a,b;V')\}, \tag{3.6}$$

where u' is the derivative of u in the sense of $\mathscr{D}'(a,b;V')$. By Theorem 3.2, we know that every $u \in W([a,b];V)$ can be identified with an absolutely continuous function $v^0: [a,b] \to V'$. However, we have a more precise result.

Theorem 3.4. *Let $u \in W([a,b];V)$. Then there is a continuous function $u^0: [a,b] \to H$ such that $u(t) = u^0(t)$* a.e. $t \in (a,b)$. *Moreover, if $u, v \in W([a,b];V)$ then the function $t \to (u(t), v(t))$ is absolutely continuous on $[a,b]$ and*

$$\frac{d}{dt}(u(t),v(t)) = (u'(t),v(t)) + (u(t),v'(t)) \qquad \text{a.e. } t \in (a,b). \tag{3.7}$$

Proof. Let $u, v \in W([a,b];V)$ and $\psi(t) = (u(t), v(t))$. As we have seen in Theorem 3.2, we may assume that $u, v \in AC([a,b];V')$ and

$$\lim_{\varepsilon \downarrow 0} \int_a^{b-\varepsilon} \left\| \frac{u(t+\varepsilon) - u(t)}{\varepsilon} - u'(t) \right\|_{V'}^2 dt = 0,$$

$$\lim_{\varepsilon \downarrow 0} \int_a^{b-\varepsilon} \left\| \frac{v(t+\varepsilon) - v(t)}{\varepsilon} - v'(t) \right\|_{V'}^2 dt = 0.$$

Then, we have

$$\lim_{\varepsilon \downarrow 0} \int_a^{b-\varepsilon} \left| \frac{\psi(t+\varepsilon) - \psi(t)}{\varepsilon} - (u'(t),v(t)) - (u(t),v(t)) \right| dt = 0.$$

Hence $\psi \in W^{1,1}([a,\mathrm{b}];\mathbf{R})$ and

$$\frac{d\psi}{dt}(t) = (u'(t),v(t)) + (u(t),v'(t)) \qquad \text{a.e. } t \in (a,b),$$

as claimed.

Now, in Eq. (3.7) we take $v = u$ and integrate from s to t. We get

$$\tfrac{1}{2}\left(|u(t)|^2 - |u(s)|^2\right) = \int_s^t (u'(\tau), u(\tau))\, d\tau.$$

Hence, the function $t \to |u(t)|$ is continuous. On the other hand, for every $v \in V$ the function $t \to (u(t), v)$ is continuous. Since $|u(t)|$ is bounded on $[a, b]$, this implies that for every $v \in H$ the function $t \to (u(t), v)$ is continuous, i.e., $u(t)$ is H-weakly continuous. Then, from the obvious equation

$$|u(t) - u(s)|^2 = |u(t)|^2 + |u(s)|^2 - 2(u(t), u(s)), \qquad \forall t, s \in [a, b]$$

it follows that $\lim_{s \to t} |u(t) - u(s)| = 0$, as claimed. ■

The spaces $W^{1,p}([a, b]; X)$, as well as $W([a, b]; V)$, play an important role in theory of differential equations in infinite dimensional spaces.

The following compacity result, which is a sharpening of Arzelà–Ascoli theorem, is in particular useful in this context.

Theorem 3.5 (Aubin). *Let X_0, X_1, X_2 be Banach spaces such that*

$$X_0 \subset X_1 \subset X_2, \qquad X_i \text{ reflexive for } i = 0, 1,$$

and the injection of X_0 into X_1 is compact. Let $1 < p_i < \infty$, $i = 0, 1$. Then the space

$$W = L^{p_0}(a, b; X_0) \cap W^{1, p_1}([a, b]; X_2)$$

is compactly imbedded in $L^{p_0}(a, b; X_1)$.

The proof relies on the following property of the spaces X_i (see J. L. Lions [1], p. 58). For every $\varepsilon > 0$ there exists $C_\varepsilon > 0$ such that

$$\|u\|_{X_1} \leq \varepsilon \|u\|_{X_0} + C_\varepsilon \|u\|_{X_2} \qquad \forall u \in X.$$

1.4. Linear Differential Equations in Banach Spaces

In this section, we shall review some classical results on semigroups of linear operators and the linear Cauchy problem in Banach space. For a detailed presentation of the subject we refer the reader to the books of K. Yosida [1], E. Hille and R. S. Phillips [1], and A. Pazy [2].

1.4.1. Semigroups of Class C_0

Let X be a Banach space with the norm $\|\cdot\|$. A one parameter family $S(t)$, $0 \le t < \infty$, of continuous linear operators from X to X is a *semigroup of class* C_0 on X if:

(i) $S(0) = I$;
(ii) $S(t)S(s) = S(t+s)$ for all $t, s \ge 0$ (the semigroup property);
(ii) $\lim_{t \to 0} S(t)x = x$ for every $x \in X$.

(Here, I is the unity operator in X.)

If $S(t)$ is a semigroup of class C_0 then it follows that there exist constants $\omega \in \mathbf{R}$ and $M \ge 1$ such that

$$\|S(t)\|_{L(X,X)} \le Me^{\omega t} \qquad \forall t \ge 0. \tag{4.1}$$

(Here, $L(X, X)$ is the algebra of linear continuous operators on X and $\|\cdot\|_{L(X,X)}$ is the norm in $L(X, X)$.)

In particular, if $M = 1$ and $\omega = 0$, then $S(t)$ is called a *semigroup of contractions on* X.

The linear operator A: $D(A) \subset X \to X$ defined by

$$D(A) = \left\{x \in X;\ \lim_{t \downarrow 0} \frac{S(t)x - x}{t} \text{ exists}\right\},$$

$$Ax = \lim_{t \downarrow 0} \frac{S(t)x - x}{t} \qquad \forall x \in D(A), \tag{4.2}$$

is called the *infinitesimal generator* of the semigroup $S(t)$. The set $D(A)$ is the domain of A.

Proposition 4.1. *The infinitesimal generator A is a closed, linear, and densely defined operator in X. If $x \in D(A)$, then $S(t)x \in D(A)$ for all $t \ge 0$, $S(t) \in C^1([0, \infty); X)$, and*

$$\frac{d}{dt}S(t)x = AS(t)x = S(t)Ax \qquad \forall t \ge 0. \tag{4.3}$$

Proof. We shall prove first that $D(A)$ is a dense subset of X. Let $x \in X$ be arbitrary but fixed. For every $t > 0$, define

$$x_t = t^{-1}\int_0^t S(s)x\,ds.$$

It is readily seen that $x_t \to x$ as $t \to 0$ and $x_t \in D(A)$ for all $t > 0$. More precisely, one has

$$\lim_{\varepsilon \downarrow 0} \frac{S(\varepsilon)x_t - x_t}{\varepsilon} = \frac{S(t)x - x}{t} \qquad \forall t > 0.$$

Now let $x \in D(A)$. By the semigroup property, it follows that

$$\frac{S(\varepsilon) - I}{\varepsilon} S(t)x = S(t)\frac{S(\varepsilon)x - x}{\varepsilon} \qquad \forall t \geq 0, \varepsilon > 0.$$

This yields

$$\lim_{\varepsilon \downarrow 0} \frac{S(\varepsilon) - I}{\varepsilon} S(t)x = \lim_{\varepsilon \downarrow 0} \frac{S(\varepsilon)x - x}{\varepsilon} = S(t)Ax.$$

Hence, $S(t)x \in D(A)$ and

$$\frac{d^+}{dt} S(t)x = AS(t)x = S(t)Ax \qquad \forall t \geq 0.$$

To conclude the proof of (4.3), it remains to be shown that the left derivative of $S(t)x$ exists and equals $S(t)Ax$. This follows from

$$\begin{aligned}\lim_{\varepsilon \downarrow 0} \left(\frac{S(t)x - S(t-\varepsilon)x}{\varepsilon} - S(t)Ax \right) &= \lim_{\varepsilon \downarrow 0} S(t-\varepsilon)\left(\frac{S(\varepsilon)x - x}{\varepsilon} - Ax \right) \\ &\quad + \lim_{\varepsilon \downarrow 0} S(t-\varepsilon)Ax - S(t)Ax \\ &= 0.\end{aligned}$$

Let us prove now that A is closed. For this purpose, consider a sequence $\{x_n\} \subset D(A)$ such that $x_n \to x_0$ and $Ax_n \to y_0$ in X for $n \to \infty$. Letting n tend to ∞ in the equation

$$S(t)x_n - x_n = \int_0^t S(s)Ax_n \, ds \qquad \forall t \geq 0,$$

we get

$$S(t)x_0 - x_0 = \int_0^t S(s)y_0 \, ds \qquad \forall t \geq 0,$$

which implies that $y_0 = Ax_0$, as claimed. ∎

The characterization of the infinitesimal generators of C_0-semigroups is given by the famous theorem of Hille and Yosida (Theorem 4.1), which represents the core of the whole theory of C_0-semigroups.

Theorem 4.1. *A linear operator A in a Banach space X is the infinitesimal generator of a semigroup of class C_0 if and only if*

(i) *A is closed and densely defined;*

(ii) *There exist $M > 0$ and $\omega \in \mathbf{R}$ such that*

$$\|(\lambda I - A)^{-n}\|_{L(X,X)} \leq \frac{M}{(\lambda - \omega)^n} \qquad \text{for all } \lambda > \omega, n = 1, 2 \dots . \tag{4.4}$$

Proof. Suppose that A is the infinitesimal generator of a C_0-semigroup $S(t)$ satisfying condition (4.1). Define, for $\lambda \in \mathbf{C}$, $\operatorname{Re} \lambda > \omega$,

$$R(\lambda)x = \int_0^\infty e^{-\lambda t} S(t) x \, dt \qquad \forall x \in X.$$

We have

$$\frac{d^n}{d\lambda^n} R(\lambda)x = \int_0^\infty e^{-\lambda t}(-t)^n S(t) x \, dt \qquad \forall n = 1, 2, \dots,$$

and this yields

$$\left\| \frac{d^n}{d\lambda^n} R(\lambda)x \right\|_{L(X,X)} \leq Mn!(\operatorname{Re} \lambda - \omega)^{n+1} \|x\|.$$

On the other hand, it is readily seen that $AR(\lambda)x = \lambda R(\lambda)x - x$ for all $\lambda \in \mathbf{C}$, $\operatorname{Re} \lambda > \omega$. Hence, $R(\lambda) = (\lambda I - A)^{-1}$ for $\operatorname{Re} \lambda > \omega$. Taking in account the well-known equation

$$\frac{d^n}{d\lambda^n}(\lambda I - A)^{-1} = (-1)^n n!(\lambda I - A)^{-n-1},$$

we obtain for $(\lambda I - A)^{-n}$ the estimate (4.4).

We shall prove now that if A satisfies condition (4.4), then A generates a C_0-semigroup satisfying condition (4.1). For simplicity, we shall assume that $\omega = 0$, the general case being obtained by appropriately translating the particular case we are considering.

For n natural, consider the operators

$$J_n = (I - n^{-1}A)^{-1}, \qquad A_n = n(J_n - I),$$

and note that $AJ_n x = J_n Ax = A_n x$ for all $x \in D(A)$. Moreover, A_n and J_n belong to $L(X, X)$ for all n and $\|J_n x\| \leq M\|x\|$ by virtue of condition (4.1). Hence,

$$\|A_n x\| \leq M\|Ax\| \qquad \forall x \in D(A), \tag{4.5}$$

$$\lim_{n\to\infty} J_n x = x \qquad \forall x \in \overline{D(A)} = X. \tag{4.6}$$

Since $A_n \in L(X, X)$, the approximating equation

$$\frac{du_n}{dt} = A_n u_n, \qquad t \geq 0,$$
$$u_n(0) = x, \tag{4.7}$$

has for each $x \in X$ a unique solution $u_n \in C^1([0, \infty); X)$. We set

$$S_n(t)x = u_n(t) \qquad \forall t \geq 0.$$

By uniqueness of the Cauchy problem (4.7) we see that $S_n(t)$ is a C_0-semigroup on X and A_n is the infinitesimal generator of $S_n(t)$. As a matter of fact, $S_n(t)$ is given by the exponential formula

$$S_n(t) = e^{A_n t} = e^{-nt} \sum_{k=0} \frac{(nt)^k J_n^k}{k!} \qquad \forall t \geq 0,$$

and this yields (by virtue of condition (4.4))

$$\|S_n(t)\|_{L(X,X)} \leq M \qquad \forall t > 0, \quad n = 1, 2 \ldots. \tag{4.8}$$

On the other hand, by (4.7) we have

$$\frac{d}{ds}(S_n(t-s)S_m(s)x) = S_n(t-s)(A_m - A_n)S_m(s)x \qquad \text{for } 0 \leq s \leq t.$$

This yields

$$\|S_n(t)x - S_m(t)x\|$$
$$\leq \int_0^t \|S_n(t-s)\|_{L(X,X)}\|S_m(s)\|_{L(X,X)}\|A_n x - A_m x\|\, ds$$
$$\forall m, n, \text{ and } t \geq 0,$$

and by (4.5) and (4.8) we infer that $S_n(t)x$ converges (uniformly) in t (on compacta) for every $x \in X$. We set

$$S(t)x = \lim_{n\to\infty} S_n(t)x \qquad \forall t \geq 0,\ x \in X. \tag{4.9}$$

It is readily seen that $S(t)$ is a semigroup of class C_0 on X and

$$S(t)x - x = \int_0^t S(s)Ax\,ds \qquad \forall x \in D(A),\ t \geq 0.$$

Hence, if $\tilde{A}$ is the infinitesimal generator of $S(t)$, we have $D(A) \subset D(\tilde{A})$ and $A = \tilde{A}$ on $D(A)$. On the other hand, by first part of theorem we know that $(I - n^{-1}\tilde{A})^{-1} \in L(X, X)$. Hence, $(I - n^{-1}A)^{-1} = (I - n^{-1}\tilde{A})^{-1}$ and $D(A) = D(\tilde{A})$, as desired. ■

We shall denote by e^{AT} the semigroup generated by A.

In particular, we derive from Theorem 4.1 the following characterization of infinitesimal generators of C_0-semigroups of contractions.

Corollary 4.1 (Hille–Yosida–Philips). *The linear operator A is the infinitesimal generator of a C_0-semigroup of contractions if and only if*:

(i) *A is closed and densely defined*; *and*

(ii) $\|(\lambda I - A)^{-1}\|_{L(X,X)} \leq 1/\lambda\ \forall \lambda > 0.$

We shall see later that condition (ii) characterizes the m-dissipative operators in X.

1.4.2. Analytic Semigroups

Let $S(t)\colon [0, \infty) \to L(X, X)$ be a semigroup of class C_0, and let A be its infinitesimal generator.

The semigroup $S(t)$ is said to be *differentiable* if $S(t)x \in D(A)$ for all $t > 0$. By Proposition 4.1, it follows that if $S(t)$ is differentiable then for any $x \in X$ the function $t \to S(t)x$ is infinitely differentiable and

$$\frac{d^n}{dt^n} S(t)x = \left(AS\left(\frac{t}{n}\right)\right)^n x \qquad \forall t > 0,\ x \in X. \tag{4.10}$$

Moreover, by the closed graph theorem $AS(t) \in L(X, X)$ for all $t > 0$.

The semigroup $S(t)$ is said to be *analytic* if for every $a > 0$ (equivalently, for some $a > 0$),

$$\sup_{0<t\leq a} t\|AS(t)\| < \infty. \tag{4.11}$$

It follows by (4.10) that if the semigroup $S(t)$ is analytic, then the function $t \to S(t)x$ is analytic on $(0, \infty)$. Conversely, if for every $x \in X$, $S(t)x$ is analytic on $(0, \infty)$ the semigroup $S(t)$ is differentiable and (4.11) follows from (4.10).

Theorem 4.2. *Let A be a closed and densely defined linear operator on X. Then A is the infinitesimal generator of an analytic semigroup $S(t)$, $\|S(t)\|_{L(X,X)} \leq M_0$ for all $t > 0$ if and only if there exist $\theta \in (\pi/2, \pi)$ and $M > 0$ such that*

$$\|(\lambda I - A)^{-1}\|_{L(X,X)} \leq \frac{M}{|\lambda|} \qquad \forall \lambda \in \Sigma, \tag{4.12}$$

where $\Sigma = \{\lambda \in \mathbf{C};\ \lambda \neq 0, |\arg \lambda| < \theta\}$.

Proof. Assume first that condition (4.12) is satisfied. Then, define $S(t) \in L(X, X)$ by the Dunford integral

$$S(t) = (2\pi i)^{-1} \int_\Gamma e^{\lambda t}(\lambda I - A)^{-1}\, d\lambda \qquad \forall t \geq 0, \tag{4.13}$$

where Γ is a piecewise smooth curve in Σ defined by $\Gamma = \Gamma_0 \cup \Gamma_1 \cup \Gamma_2$, $\Gamma_0 = \{re^{-i(\theta-\varepsilon)};\ 1 \leq r < \infty\}$, $\Gamma_1 = \{re^{i(\theta-\varepsilon)};\ 1 \leq r < \infty\}$, $\Gamma_2 = \{e^{i\alpha};\ -(\theta - \varepsilon) < \alpha < \theta - \varepsilon\}$.

By condition (4.12), $S(t)$ is well-defined and belongs to $L(X, X)$ for all $t \geq 0$. We have

$$\begin{aligned} AS(t)x &= (2\pi i)^{-1} \int_\Gamma \lambda e^{\lambda t}(\lambda I - A)^{-1} x\, d\lambda \\ &\quad -(2\pi i)^{-1} x \int_\Gamma e^{\lambda t}\, d\lambda \\ &= (2\pi i)^{-1} \int_\Gamma \lambda e^{\lambda t}(\lambda I - A)^{-1} x\, d\lambda \qquad \forall x \in X. \end{aligned}$$

Hence, by (4.12), we have

$$\begin{aligned}\|AS(t)x\| &\le C\left(\int_1^\infty |e^{re^{-i(\theta-\varepsilon)}t}|\,dr\right.\\&\qquad\left.+\int_1^\infty |e^{re^{i(\theta-\varepsilon)}t}|\,dr + 1\right)\|x\|\\&\le C_1 t^{-1}\|x\| \qquad \forall t>0.\end{aligned}$$

On the other hand,

$$\begin{aligned}\frac{d}{dt}S(t)x &= (2\pi i)^{-1}\int_\Gamma \lambda e^{\lambda t}(\lambda I - A)^{-1}x\,d\lambda\\&= AS(t)x, \qquad \forall t>0.\end{aligned}$$

Hence, $u(t) = S(t)x$ is the solution to the Cauchy problem

$$\begin{aligned}\frac{du}{dt} &= Au \qquad \text{in } [0,\infty),\\u(0) &= x. \end{aligned}\tag{4.14}$$

Let us prove that problem (4.14) has at most one solution $\|u(t)\| \le M$ $\forall t \ge 0$. Indeed, if u and v are two such solutions, then $w = u - v$ satisfies the equation $dw/dt = Aw$, $w(0) = 0$, and its Laplace transform $\hat{w}(\lambda)$, $\lambda > 0$, is the solution to homogeneous equation $(\lambda I - A)\hat{w}(\lambda) = 0$. Hence, $\hat{w}(\lambda) = 0$ for $\lambda > 0$, which clearly implies that $w = 0$. Then, by the uniqueness of the Cauchy problem (4.14) it follows that $S(t+s) = S(t)S(s)$ and that A is the infinitesimal generator of $S(t)$.

Conversely, if A is the infinitesimal generator of an analytic semigroups $S(t)$, then by (4.10) and (4.11) it follows that

$$\left\|\frac{d^n}{dt^n}S(t)\right\|_{L(X,X)} \le \frac{C^n}{t^n}n! \qquad \forall t>0.$$

Then, the semigroup $S(t)$ has an extension $S(z)$ defined by

$$S(z) = S(t) + \sum_{n=1}^{\infty}\frac{d^nS(t)}{dt^n}\,\frac{(z-t)^n}{n!}$$

in a sector $\{z \in \mathbf{C};\ |z-t| < \rho t/C\}$, $0 < \rho < 1$. Hence, $S(z)$ is analytic in $\Delta = \{z;\ |\arg z| < \arctan 1/C\}$. Then, we have

$$(\lambda I - A)^{-1} = \int_0^\infty e^{-\lambda t}S(t)\,dt = \int_\Gamma e^{-\lambda z}S(z)\,dz, \qquad \operatorname{Re}\lambda > 0,$$

where Γ is a piecewise smooth curve in Δ of the form encountered before. By using this formula we find estimate (4.12). ■

1.4.3. The Cauchy Problem

If A is the infinitesimal generator of a C_0-semigroup $S(t) = e^{At}$, then, as seen before, the homogeneous initial value problem (the Cauchy problem)

$$\begin{aligned} \frac{du}{dt} &= Au, \qquad t \geq 0, \\ u(0) &= x, \end{aligned} \tag{4.15}$$

has for every $x \in D(A)$ a unique solution $u \in C^1([0, \infty); X)$ given by $u(t) = e^{At}x$.

Consider now the nonhomogeneous Cauchy problem

$$\begin{aligned} \frac{du}{dt} &= Au + f, \qquad t \in [0, T], \\ u(0) &= x, \end{aligned} \tag{4.16}$$

where $f \in L^1(0, T; X)$ and $x \in X$.

The continuous function $u: [0, T] \to X$ defined by the variation of constants formula

$$u(t) = e^{At}x + \int_0^t e^{A(t-s)}f(s)\, ds \qquad \forall t \in [0, T] \tag{4.17}$$

is called *mild solution* to Cauchy problem (4.16).

A function $u: [0, T] \to X$ is said to be a *classical solution* to problem (4.15) if it is continuously differentiable on $[0, T]$, $u(t) \in D(A)$ for all $t \in [0, T]$, and it satisfies Eq. (4.16) on $[0, T]$.

It is readily seen that if A generates a C_0-semigroup, then problem (4.16) has at most one classical solution $u(t)$. Indeed, if y is a solution to

$$\frac{dy}{dt} = Ay \qquad \text{in } [0, \infty), \qquad y(0) = 0,$$

then, for $0 \leq s \leq t < \infty$, we have

$$\frac{d}{ds}(e^{A(t-s)}y(s)) = e^{A(t-s)}Ay(s) - e^{A(t-s)}Ay(s) = 0.$$

Hence $e^{A(t-s)}y(s) \equiv 0$.

Regarding the existence of classical solutions to problem (4.16), we have the following well-known theorem whose proof will be omitted.

Theorem 4.3. *Assume that A is the infinitesimal generator of a C_0-semigroup e^{At} on X and $x \in D(A)$. Assume either*

(i) $f \in C^1([0,T]; X)$, *or*

(ii) $f \in C([0,T]; X)$, $Af \in C([0,T]; X)$.

Then the function u given by formula (4.17) *is the unique classical solution to the Cauchy problem* (4.16). ■

A function $u \in W^{1,1}([0,T]; X)$ that satisfies almost everywhere Eq. (4.16) and the initial condition is called *strong solution* to problem (4.16).

Theorem 4.4. *Assume that A is the infinitesimal generator of a semigroup of class C_0. Let $x \in D(A)$. If $f \in W^{1,1}([0,T]; X)$, then the mild solution u to problem* (4.16) *is a strong solution.*

Proof. We have

$$v(t) = \int_0^t e^{A(t-s)} f(s)\, ds = \int_0^t e^{As} f(t-s)\, ds, \qquad \forall t \in [0,T], \quad (4.18)$$

and this implies that $v \in W^{1,1}([0,T]; X)$. Hence, v is a.e. differentiable on $(0,T)$ and by a little calculation we see that

$$v'(t) = \lim_{h \downarrow 0} \frac{e^{Ah} - I}{h} v(t) + \lim_{h \downarrow 0} \frac{1}{h} \int_t^{t+h} e^{A(t+h-s)} f(s)\, ds.$$

Hence, $v'(t) = Av(t) + f(t)$ a.e. $t \in (0,T)$, as claimed. ■

If the semigroup e^{At} is analytic, then the mild solution is a strong solution under weaker assumptions on f.

More precisely, we have (see A. Pazy [2]):

Theorem 4.5. *Let A be the infinitesimal generator of an analytical semigroup e^{At}. If $x \in X$ and f is Hölder continuous, i.e.,*

$$\|f(t) - f(s)\| \leq C|t-s|^{\theta} \qquad \forall t, s \in [0,T], \quad (4.19)$$

where $0 < \theta \leq 1$, *then the mild solution* u *to problem* (4.16) *is continuously differentiable on* $(0, T]$, $Au \in C([\delta, T]; H)$ *for every* $\delta > 0$, *and* u *satisfies Eq.* (4.16) *on the interval* $(0, T]$.

Moreover, if $x \in D(A)$, *then* $Au \in C([0, T]; X)$ *and* u *is a classical solution to* (4.16).

Proof. We have

$$u(t) = e^{At}x + v(t), \qquad \forall t \in [0, T],$$

where v is defined by formula (4.18). Since $e^{At}x \in C^1([0, T]; X)$ and satisfies the homogeneous equation (4.15), it suffices to prove the theorem for $u = v$, i.e., in the case $x = 0$.

We may write v as

$$\begin{aligned} v(t) &= \int_0^t e^{A(t-s)}(f(s) - f(t))\, ds + \int_0^t e^{A(t-s)} f(t)\, ds \\ &= v_1(t) + v_2(t). \end{aligned}$$

It is readily seen that $v_2(t) \in D(A)$ and $Av_2(t) = (e^{At} - I)f(t)$, $\forall t \in [0, T]$. Hence, $Av_2 \in C([0, T]; X)$.

Now, by condition (4.19) and analyticity condition (4.11) it follows that $Ae^{A(t-s)}(f(s)) - f(t)) \in L^1(0, T; X)$, and since A is closed this implies that $v_1(t) \in D(A)$ for all t and

$$Av_1(t) = \int_0^t Ae^{A(t-s)}(f(s) - f(t))\, ds.$$

Since

$$\|Ae^{A(t-s)}(f(s) - f(t))\| \leq L|t - s|^{\theta - 1} \qquad \forall t, s \in [0, T],$$

we see that $Av_1 \in C([0, T]; X)$. Hence, $Av \in C([0, T]; X)$ and

$$\frac{d}{dt} v(t) = Av(t) \qquad \forall t \in [0, T],$$

as claimed. ■

Theorem 4.6. *Let* A *be the infinitesimal generator of an analytic semigroup in a Hilbert space* H. *If* $f \in L^2(0, T; H)$ *and* $x \in D(A)$, *then the mild* *solution u to the Cauchy problem* (4.16) *belongs to* $W^{1,2}([0,T]; H)$ *and*

$$\|u'\|_{L^2(0,T;H)} + \|Au\|_{L^2(0,T;H)} \le C(\|x\|_{D(A)} + \|f\|_{L^2(0,T;H)}). \quad (4.20)$$

(Here, $\|\cdot\|_{D(A)}$ is the graph norm $\|Ax\| + \|x\|$.)

Proof. Let v be the function (4.18). We have

$$\hat{v}(\lambda) = \int_0^\infty e^{-\lambda t} v(t)\, dt = (\lambda I - A)^{-1} \hat{f}(\lambda) \qquad \text{for } \lambda = \alpha + i\xi,$$

where $\alpha \ge \rho > 0$. This yields (we extend v and f by 0 outside $[0,T]$)

$$A\hat{v}(\lambda) = \lambda(\lambda I - A)^{-1} \hat{f}(\lambda) - \hat{f}(\lambda) \qquad \forall \lambda, \operatorname{Re} \lambda = \alpha,$$

and by the Parseval formula,

$$\int_{-\infty}^{\infty} \|A\hat{v}(\lambda)\|^2\, d\xi = 2\pi \int_0^\infty e^{-2\alpha t} \|Au(t)\|^2\, dt.$$

Hence

$$\int_0^\infty e^{-2\alpha t} \|Au(t)\|^2\, dt \le C \int_{-\infty}^{\infty} \|f\|^2\, d\xi,$$

and estimate (4.20) follows. ■

1.4.4. Examples

A large class of infinite dimensional problems, including linear parabolic equations, the wave equation, symmetric hyperbolic systems, and linear functional–delay equations, can be studied in the framework of C_0-semigroups. Here, we shall present only a few standard examples of linear operators that generate C_0-semigroups.

Proposition 4.2. *Let A be linear, self-adjoint, positive operator on a real Hilbert space H. Then* $-A$ *generates an analytic semigroup of contractions on H.*

Proof. Let H be endowed with the norm $|\cdot|$ and scalar product $(\cdot,\cdot)$. Denote by $R(\lambda I + A)$ the range of $\lambda I + A$, $\lambda \ge 0$. Since

$$(Au, u) \ge 0 \qquad \forall u \in D(A),$$

it is readily seen that $R(\lambda I + A)$ is a closed subset of H. On the other hand, $R(\lambda I + A)$ is dense in H. Indeed, otherwise there would exist $y \in H$, $y \neq 0$, such that

$$(\lambda x + Ax, y) = 0 \qquad \forall x \in D(A).$$

This implies that $x \to (Ax, y)$ is continuous on H and so $y \in D(A)$. Then, for $x = y$ the previous equation gives $y = 0$. Hence $R(\lambda I + A) = H$ $\forall \lambda > 0$. On the other hand, we have

$$|(\lambda I + A)^{-1} x| \leq \frac{|x|}{\lambda} \qquad \forall \lambda > 0,\ x \in H.$$

Hence, $-A$ generates a C_0-semigroup of contractions $S(t) = e^{-At}$. For $x \in D(A)$ the function $u(t) = e^{-At}x$ is the (classical) solution to the Cauchy problem

$$\frac{du}{dt} + Au = 0 \qquad \forall t \geq 0, \qquad u(0) = x. \tag{4.21}$$

We have

$$t\left|\frac{du}{dt}\right|^2 + t\left(Au, \frac{du}{dt}\right) = 0 \qquad \forall t > 0.$$

Hence,

$$t\left|\frac{du}{dt}(t)\right|^2 + \frac{t}{2}\frac{d}{dt}(Au(t), u(t)) = 0 \qquad \forall t \geq 0. \tag{4.22}$$

On the other hand, we have (multiplying Eq. (4.21) by u)

$$\frac{1}{2}\frac{d}{dt}|u(t)|^2 + (Au(t), u(t)) = 0 \qquad \forall t \geq 0 \tag{4.23}$$

and

$$\frac{1}{2}\frac{d}{dt}|u(t+h) - u(t)|^2 + (Au(t+h) - Au(t), u(t+h) - u(t)) = 0$$
$$\forall t, h \geq 0.$$

Hence, the function $t \to |(du/dt)(t)|$ is decreasing on $(0, \infty)$. Then, integrating (4.22) on the interval $(0, t)$, we get

$$\int_0^t s\left|\frac{du}{ds}(s)\right|^2 ds \leq \int_0^t (Au(s), u(s))\, ds \qquad \forall t \geq 0,$$

whilst by (4.23)

$$\int_0^t (Au(s), u(s))\, ds + \frac{1}{2}|u(t)|^2 \le \frac{1}{2}|x|^2.$$

Hence

$$t^2\left|\frac{du}{dt}(t)\right|^2 \le |x|^2 \qquad \forall t > 0,$$

i.e.,

$$|Ae^{-At}x| \le |x|/t \qquad \forall t > 0,\ x \in H, \tag{4.24}$$

as claimed. ■

Let Ω be a bounded and open subset of $\mathbf{R}^N$ with a smooth boundary $\partial\Omega$. Then, it follows by Proposition 4.2 that the operator $A_2 : D(A_2) \subset L^2(\Omega) \to L^2(\Omega)$ defined by

$$A_2 u = -\Delta u, \qquad D(A_2) = H_0^1(\Omega) \cap H^2(\Omega), \tag{4.25}$$

generates an analytic semigroup of contractions on $L^2(\Omega)$.

More generally, consider for $1 \le p < \infty$ the operator $A_p = -\Delta$ with $D(A_p) = W_0^{1,p}(\Omega) \cap W^{2,p}(\Omega)$ for $p > 1$ and $D(A_1) = \{u \in W_0^{1,1}(\Omega);\ u \in L^1(\Omega)\}$ for $p = 1$.

Proposition 4.3. *For $1 \le p < \infty$, A_p generates an analytic semigroup of contractions in $X = L^p(\Omega)$.*

Proof. It follows that A_p is m-accretive in $L^p(\Omega)$, i.e., $R(\lambda I + A_p) = L^p(\Omega)$ for all $\lambda > 0$ and $\|(\lambda I + A_p)^{-1}\|_{L(X,X)} \le \lambda^{-1}$ $\forall \lambda > 0$ (see Proposition 3.9 in Chapter 2). Moreover,

$$D(A_1) \subset W_0^{1,q}(\Omega), \qquad \text{where } 1 < q < \frac{N}{N-1}. \tag{4.26}$$

Hence, A_p is the infinitesimal generator of a C_0-semigroup of contractions on $L^p(\Omega)$. ■

The analyticity of the semigroup $e^{-A_p t}$ follows in the case $1 \le p \le 2$ from Proposition 4.4 following and the analyticity of $e^{-A_2 t}$. The case $p > 2$ follows from verifying directly conditions of Theorem 4.2 (see Pazy [2]).

Proposition 4.4. *Let Ω be a bounded and open subset of $\mathbf{R}^N$ and let $S(t)$ be the semigroup generated on $L^1(\Omega)$ by Δ with zero Dirichlet conditions. Then $S(t)(L^1(\Omega)) \subset L^\infty(\Omega)$ for all $t > 0$ and*

$$\|S(t)u_0\|_{L^\infty(\Omega)} \le Ct^{-N/2}\|u_0\|_{L^1(\Omega)} \qquad \forall u_0 \in L^1(\Omega), t > 0, \quad (4.27)$$

where C is independent of u_0.

Interpolating between $L^1(\Omega)$ and $L^\infty(\Omega)$, it follows by inequality (4.27) that

$$\|S(t)u_0\|_{L^q(\Omega)} \le Ct^{-(N/2)(1-1/q)}\|u_0\|_{L^1(\Omega)} \qquad \forall t > 0. \qquad (4.28)$$

Proof. For $N = 1$, inequality (4.27) follows by representing $S(t)u_0$ as Fourier series with respect to the Laplace eigenfunctions.

We set $u = S(t)u_0$ and recall that

$$\frac{\partial u}{\partial t} - \Delta u = 0 \qquad \text{in } \Omega \times \mathbf{R}^+$$
$$u(0, x) = u_0(x), \qquad x \in \Omega. \qquad (4.29)$$

Without loss of generality, we may assume that u is a classical solution to (4.29).

Let k be a positive integer. We multiply Eq. (4.29) by $t^{k+2}|y|^{p-2}y$, where $p > 2$, and integrate on $\Omega \times \mathbf{R}^+$. We get

$$t^{k+2}p^{-1}\frac{d}{dt}\|u(t)\|_p^p + (p-1)t^{k+2}\int_\Omega |\nabla u(x,t)|^2|u(x,t)|^{p-2}\,dx \le 0 \qquad \forall t > 0,$$

i.e.,

$$t^{k+2}\frac{d}{dt}\|u(t)\|_p^p + 2t^{k+2}\int_\Omega |\nabla|u(x,t)|^{p/2}|^2\,dx \le 0. \qquad (4.30)$$

Then, by the Sobolev imbedding theorem,

$$\int_\Omega |\nabla|u(x,t)|^{p/2}|^2\,dx \ge C\left(\int_\Omega |u(x,t)|^{pN/(N-2)}\,dx\right)^{(N-2)/N}$$

if $N > 2$. If $N = 2$,

$$\int_\Omega |\nabla|u(x,t)|^{p/2}|^2\,dx \ge C\left(\int_\Omega |u(x,t)|^{pq}\,dx\right)^{1/q} \qquad \forall q > 1.$$

Then, substituting this in (4.30), we get

$$t^{k+2}\frac{d}{dt}\|u(t)\|_p^p + Ct^{k+2}\|u(t)\|_p^p \le 0 \qquad \forall t > 0,$$

where $\alpha = N/(N-2)$ and $\|\cdot\|_p$ is the $L^p(\Omega)$ norm. Now, we integrate from 0 to t and get

$$t^{k+2}\|u(t)\|_p^p + C\int_0^t s^{k+2}\|u(s)\|_p^p\,ds$$
$$\le (k+2)\int_0^t s^{k+1}\|u(s)\|_p^p\,ds. \tag{4.31}$$

Now we have, by the Hölder inequality,

$$\int_\Omega |u|^p\,dx = \int_\Omega |u|^\beta |u|^{p-\beta}\,dx$$
$$\le \left(\int_\Omega |u|\,dx\right)^\beta \left(\int_\Omega |u|^{(p-\beta)/(1-\beta)}dx\right)^{1-\beta},$$

where $0 < \beta < 1$ is such that $(p-\beta)(1-\beta)^{-1} = \alpha p$. This yields

$$\|u(t)\|_p^p \le \|u(t)\|_1^\beta \|u(t)\|_{\alpha p}^{p-\beta} \qquad \forall t > 0,$$

and substituting into (4.31), we get

$$t^{k+2}\|u(t)\|_p^p + C\int_0^t s^{k+2}\|u(s)\|_{\alpha p}^p\,ds$$
$$\le (k+2)\|u_0\|_1^\beta \int_0^t s^{k+1}\|u(s)\|_{\alpha p}^{p-\beta}\,ds$$
$$\le (k+2)\|u_0\|_1^\beta$$
$$\times \int_0^t \|u(s)\|_{\alpha p}^{p-\beta} s^{(k+2)(p-\beta)/p} s^{(k+1)-(k+2)(p-\beta)/p}\,ds$$
$$\le (k+2)\|u_0\|_1^\beta$$
$$\times \left(\int_0^t \|u(s)\|_{\alpha p}^p s^{k-2}\,ds\right)^{(p-\beta)/p} \left(\int_0^t s^{(k+1)p/\beta-(k+2)(p-\beta)/\beta}ds\right)^{\beta/p}$$
$$= (k+2)\|u_0\|_1^\beta \left(\int_0^t \|u(s)\|_{\alpha p}^p s^{k+2}\,ds\right)^{(p-\beta)/p}$$
$$\times (k+1-p/\beta)^{-\beta/p} t^{(k+1)\beta/p-1}.$$

Then, using the well-known inequality

$$ab \leq \frac{a^p}{p} + \frac{b^q}{q}, \qquad \frac{1}{p} + \frac{1}{q} = 1,$$

we get

$$\begin{aligned} t^{k+2}\|u(t)\|_p^p &+ C\int_0^t s^{k+2}\|u(s)\|_{\alpha p}^p\, ds \\ &\leq \tfrac{1}{2}C\int_0^t s^{k+2}\|u(s)\|_{\alpha p}^p\, ds \\ &\quad + C(k+2)^{p/\beta}\|u_0\|_1^p(k+1-p/\beta)^{-1}t^{k+1-p/\beta}. \end{aligned}$$

Finally,

$$\begin{aligned} \|u(t)\|_p &\leq C(k+2)^{1/\beta}\|u_0\|_1(k+1-p/\beta)^{-1/p}t^{(k+1)/p-1/\beta}t^{-(k+2)/p} \\ &\leq C(k+2)^{1/\beta}(k+1-p/\beta)^{-1/p}t^{-1/\beta-1/p}\|u_0\|_1, \end{aligned} \tag{4.32}$$

where k is arbitrary and $(p-\beta)(1-\beta)^{-1} = \alpha p$. For $p \to \infty$, we see that $\beta \to (\alpha - 1)/\alpha = 2/N$. Hence,

$$\|u(t)\|_{L^\infty(\Omega)} \leq C(k+2)^{N/2}t^{-N/2}\|u_0\|_{L^1(\Omega)} \qquad \forall t > 0, \tag{4.33}$$

as claimed. ■

Remark 4.1. In particular, it follows by Proposition 4.4 that for $y_0 \in L^1(\Omega)$ and $f \in L^1((0,T); L^1(\Omega))$ the boundary value problem

$$\begin{aligned} &\frac{\partial y}{\partial t} - \Delta y = f \qquad \text{in } \Omega \times (0,T), \\ &y(0) = y_0 \qquad \text{in } \Omega, \qquad y = 0 \qquad \text{in } \partial\Omega, \end{aligned}$$

has a unique mild solution $y \in C([0,T]; L^1(\Omega))$:

$$y(x,t) = \left(S(t)y_0 + \int_0^t S(t-s)f(s)\, ds\right)(x) \qquad \forall t \geq 0,\ x \in \Omega,$$

and for $q < (N+2)/N$ we have

$$\|y\|_{L^q(\Omega\times(0,T))} \leq C(\|u_0\|_{L^1(\Omega)} + \|f\|_{L^1(\Omega\times(0,T))}). \tag{4.34}$$

The Wave Equation. Consider the second order differential equation

$$\frac{d^2y}{dt^2} + Ay = f \qquad \text{in } (0, T),$$
$$y(0) = y_0, \qquad \frac{dy}{dt}(0) = y_1, \tag{4.35}$$

in a real Hilbert space H, where A is a self-adjoint positive definite operator on H and $f \in L^1((0, T); H)$.

Equation (4.35) can be written as a first order differential equation

$$\frac{dy}{dt} = z,$$
$$\frac{dz}{dt} = -Ay + f,$$
$$y(0) = y_0, \qquad z(0) = y_1, \tag{4.36}$$

on the product space $X = V \times H$, where $V = D(A^{1/2})$. (The space X is endowed with the scalar product

$$\langle (y, z), (y_1, z_1) \rangle = (A^{1/2}y, A^{1/2}y_1) + (z, z_1).)$$

It is readily seen that the operator

$$\mathscr{A} = \begin{Vmatrix} 0 & 1 \\ -A & 0 \end{Vmatrix}, \qquad D(\mathscr{A}) = D(A) \times V,$$

is m-dissipative on X, i.e., $R(I + \mathscr{A}) = X$ and $\langle \mathscr{A}(y, z), (y, z) \rangle \geq 0$ for all $(y, z) \in X$. Hence, $\mathscr{A}$ generates a C_0-semigroup on X and so for $y_0 \in D(A)$, $y_1 \in V$, and $f \in C^1([0, T]; H)$, problem (4.35) has a unique solution $y \in C^1([0, T]; V) \cap C([0, T]; D(A))$ with $dy/dt \in C^1([0, T]; H) \cap C([0, T]; V)$.

In particular, for $H = L^2(\Omega)$ and $A = -\Delta$, $D(A) = H_0^1(\Omega) \cap H^2(\Omega)$, and (4.35) reduces to the wave equation.

Chapter 2 | Nonlinear Operators of Monotone Type

In this chapter we introduce the concept of maximal monotone operator, along with the basic results of this theory and its relationship with convex analysis and the theory of generalized gradients. A separate section is devoted to nonlinear m-accretive operators. We present several examples and applications to nonlinear partial differential equations, such as nonlinear elliptic boundary value problems, the nonlinear diffusion equation, and first order quasi-linear partial differential equations.

2.1. Maximal Monotone Operators

2.1.1. Definitions and Basic Results

If X and Y are two linear spaces, we will denote by $X \times Y$ their Cartesian product. The elements of $X \times Y$ will be written as $[x, y]$ where $x \in X$ and $y \in Y$.

If A is a multivalued operator from X to Y, we may identify it with its graph in $X \times Y$:

$$\{[x, y] \in X \times Y;\ y \in Ax\}. \tag{1.1}$$

Conversely, if $A \subset X \times Y$, then we define

$$Ax = \{y \in Y;\ [x, y] \in A\}, \qquad D(A) = \{x \in X;\ Ax \neq \emptyset\}, \tag{1.2}$$

$$R(A) = \bigcup_{x \in D(A)} Ax, \qquad A^{-1} = \{[y, x];\ [x, y] \in A\}. \tag{1.3}$$

In this way here and in the following, we shall identify the operators from X to Y with their graphs in $X \times Y$ and so we shall equivalently speak of subsets of $X \times Y$ instead of operators from X to Y.

If $A, B \subset X \times Y$ and λ is a real number, we set:

$$\lambda A = \{[x, \lambda y]; [x, y] \in A\}; \tag{1.4}$$

$$A + B = \{[x, y + z]; [x, y] \in A, [x, z] \in B\}; \tag{1.5}$$

$$AB = \{[x, z]; [x, y] \in B, [y, z] \in A \text{ for some } y \in Y\}. \tag{1.6}$$

Throughout this chapter, X will be a real Banach space with dual X^*. Notations for norms, convergence, and duality pairings will be that introduced in Chapter 1, Section 1. In particular, the value of functional $x^* \in X^*$ at $x \in X$ will be denoted by either (x, x^*) or (x^*, x). If X is a Hilbert space unless otherwise stated, and we shall implicitly assume that it is identified with its own dual.

Definition 1.1. The set $A \subset X \times X^*$ (equivalently the operator $A: X \to X^*$) is said to be *monotone* if

$$(x_1 - x_2, y_1 - y_2) \geq 0 \qquad \forall [x_i, y_i] \in A, i = 1, 2. \tag{1.7}$$

A monotone set $A \subset X \times X^*$ is said to be *maximal monotone* if it is not properly contained in any other monotone subset of $X \times X^*$.

Note that if A is a single valued operator from X to X^* then A is monotone if

$$(x_1 - x_2, Ax_1 - Ax_2) \geq 0, \qquad \forall x_1, x_2 \in D(A). \tag{1.8}$$

A simple example of a monotone subset of $X \times X^*$ is the duality mapping J of X (see Section 1.1). Indeed, by definition of J,

$$\begin{aligned}(x_1 - x_2, y_1 - y_2) &= \|x_1\|^2 + \|x_2\|^2 - (x_1, y_2) - (x_2, y_1) \\ &\geq (\|x_1\| - \|x_2\|)^2 \qquad \forall [x_i, y_i] \in J.\end{aligned}$$

As a matter of fact, it turns out that J is maximal monotone in $X \times X^*$.

Now let us show that the map $\phi: L^1(\Omega) \to L^\infty(\Omega)$ defined by

$$\phi(u) = \{v \in L^\infty(\Omega); v(x) \in \operatorname{sign} u(x) \cdot \|u\|_{L^1(\Omega)} \text{ a.e. } x \in \Omega\}$$

is the duality mapping of $L^1(\Omega)$. (Here, $\operatorname{sign} u = u|u|^{-1}$ for $u \neq 0$, $\operatorname{sign} 0 = [-1, 1]$). Indeed, since $\phi \subset J$, as easily seen, it suffices to show that ϕ is maximal in $L^1(\Omega) \times L^\infty(\Omega)$. Let $(u_0, v_0) \in L^1(\Omega) \times L^\infty(\Omega)$ be such that

$$\int_\Omega (v_0 - v)(u_0 - u)\, dx \geq 0 \qquad \forall u \in L^1(\Omega), v \in \phi(u).$$

Note that the equation

$$\text{sign}\, u + \text{sign}\, u \cdot \|u\|_{L^1(\Omega)} \ni v_0 + \text{sign}\, u_0 \qquad \text{a.e. in } \Omega$$

has at least one solution u_1. Substituting in the preceding inequality yields

$$\int_\Omega (u_0 - u_1)(\text{sign}\, u_0 - \text{sign}\, u_1)\, dx \leq 0.$$

Hence, $u_0 = u_1$ and $v_0 \in \text{sign}\, u_1 \cdot \|u_1\|_{L^1(\Omega)}$, i.e., $[u_0, v_0] \in \phi$, as claimed.

Definition 1.2. Let A be a single valued operator from X to X^* with $D(A) = X$. The operator A is said to be *hemicontinuous* if, for all $x, y \in X$,

$$w - \lim_{\lambda \to 0} A(x + \lambda y) = Ax.$$

A is said to be *demicontinuous* if it is continuous from X to X^*_w, i.e.,

$$w - \lim_{x_n \to x} Ax_n = Ax.$$

A is said to be *coercive* if

$$\lim\, (y_n, x_n - x^0)\|x_n\|^{-1} = \infty \tag{1.9}$$

for some $x^0 \in X$ and all $[x_n, y_n] \in A$ such that $\lim_{n \to \infty} \|x_n\| = \infty$.

Proposition 1.1. *Let $A \subset X \times X^*$ be maximal monotone. Then:*

(i) *A is weakly–strongly closed in $X \times X^*$;* i.e., *if $[x_n, y_n] \in A$, $x_n \rightharpoonup x$ in X and $y_n \to y$ in X^*, then $[x, y] \in A$;*

(ii) *A^{-1} is maximal monotone in $X^* \times X$;*

(iii) *For each $x \in D(A)$, Ax is a closed convex subset of X^*.*

Proof. (i) From the obvious inequality

$$(x_n - u, y_n - v) \geq 0 \qquad \forall [u, v] \in A,$$

we see that $(x - u, y - v) \geq 0$ $\forall [u, v] \in A$, and since A is maximal this implies $[x, y] \in A$, as claimed.

(ii) is obvious.

(iii) By (i) it is obvious that Ax is a closed subset of X^* for each $x \in D(A)$. Now let $y_0, y_1 \in Ax$ and let $y_\lambda = \lambda y_0 + (1 - \lambda)y_1$, where

$0 < \lambda < 1$. From the inequalities

$$(x - u, y_0 - v) \geq 0, \qquad (x - u, y_1 - v) \geq 0 \qquad \forall [u, v] \in A,$$

we deduce that $(x - u, y_\lambda - v) \geq 0 \; \forall [u, v] \in A$, which implies that $[x, y_\lambda] \in A$ because A is maximal. The proof is complete. ■

It has been shown by G. Minty in the early 1960s that coercive maximal monotone operators are surjective. This important result, which implies a characterization of a maximal monotone operator A in terms of the surjectivity of $A + J$ (J is the duality mapping) is a consequence of the following existence theorem.

Theorem 1.1. *Let X be a reflexive Banach space and let A and B be two monotone sets of $X \times X^*$ such that $0 \in D(A)$, B is single valued, hemicontinuous, and coercive, i.e.,*

$$\lim_{\|x\| \to \infty} (Bx, x)/\|x\| = +\infty. \tag{1.10}$$

Then there exists $x \in K = \overline{\operatorname{conv} D(A)}$ such that

$$(u - x, Bx + v) \geq 0 \qquad \forall [u, v] \in A. \tag{1.11}$$

It should be noted that if A is maximal monotone, it follows from (1.11) that $0 \in Ax + Bx$.

We will prove Theorem 1.1 in several steps. We shall prove first the following lemma.

Lemma 1.1. *Let X be a finite dimensional Banach space and let B be a hemicontinuous monotone operator from X to X^*. Then B is continuous.*

Proof. Let us show first that B is bounded on bounded subsets. Indeed, otherwise there exists a sequence $\{x_n\} \subset X$ such that $\|Bx_n\| \to \infty$ and $x_n \to x_0$ as $n \to \infty$. We have

$$(x_n - x, Bx_n - Bx) \geq 0 \qquad \forall x \in X,$$

and therefore

$$\left(x_n - x, \frac{Bx_n}{\|Bx_n\|} - \frac{Bx}{\|Bx_n\|} \right) \geq 0 \qquad \forall x \in X.$$

Without loss of generality, we may assume that $Bx_n\|Bx_n\|^{-1} \to y_0$ as $n \to \infty$. This yields

$$(x_0 - x, y_0) \geq 0 \qquad \forall x \in X,$$

and therefore $y_0 = 0$. The contradiction we have arrived at shows that B is indeed bounded. Now, let $\{x_n\}$ be convergent to x_0 and let y_0 be a cluster point of $\{Bx_n\}$. Again by the monotonicity of B, we have

$$(x_0 - x, y_0 - Bx) \geq 0 \qquad \forall x \in X.$$

If in this inequality we take $x = tu + (1-t)x_0$, $0 \leq t \leq 1$, u arbitrary in X, we get

$$(x_0 - u, y_0 - B(tu + (1-t)x_0)) \geq 0 \qquad \forall t \in [0,1], u \in X.$$

Then, letting t tend to zero and using the hemicontinuity of B, we get

$$(x_0 - u, y_0 - Bx_0) \geq 0 \qquad \forall u \in X,$$

which clearly implies that $y_0 = Bx_0$, as claimed. ∎

Now we shall prove Theorem 1.1 in the case where X is finite dimensional.

Lemma 1.2. *Let X be a finite dimensional Banach space and let A and B be two monotone subsets of $X \times X^*$ such that $0 \in D(A)$, and B is single valued, continuous, and satisfies* (1.10). *Then there exists $x \in \overline{\operatorname{conv} D(A)}$ such that*

$$(u - x, Bx + v) \geq 0 \qquad \forall [u,v] \in A. \tag{1.12}$$

Proof. Redefining A if necessary, we may assume that $K = \overline{\operatorname{conv} D(A)}$ is bounded. Indeed, if Lemma 1.2 is true in this case, then replacing A by $A_n = \{[x,y] \in A; \|x\| \leq n\}$ we infer that for every n there exists $x_n \in K_n = K \cap \{x; \|x\| \leq n\}$ such that

$$(u - x_n, Bx_n + v) \geq 0 \qquad \forall [u,v] \in A_n. \tag{1.13}$$

This yields

$$(x_n, Bx_n)\|x_n\|^{-1} \leq \|\xi\|, \qquad \text{for some } \xi \in A0,$$

and by the coercivity condition (1.10) we see that there is $M > 0$ such that $\|x_n\| \leq M$ for all n. Now, on a subsequence, for simplicity again denoted n, we have $x_n \to x$. By (1.13) and the continuity of B, it is clear that x is a solution to (1.12), as claimed.

Let $T\colon K \to K$ be the multivalued operator defined by

$$Tx = \{y \in K; (u - y, Bx + v) \geq 0 \; \forall [u, v] \in A\}.$$

Let us show first that $Tx \neq \emptyset$, $\forall x \in K$. To this end, define the sets

$$K_{uv} = \{y \in K; (u - y, Bx + v) \geq 0\},$$

and notice that

$$Tx = \bigcap_{[u, v] \in A} K_{uv}.$$

Inasmuch as the K_{uv} are closed subsets (if nonempty) of the compact set K, to show that $\bigcap_{[u, v] \in A} K_{uv} \neq \emptyset$ it suffices to prove that every finite collection $\{K_{u_i, v_i}; i = 1, \ldots, m\}$ has a nonempty intersection. Equivalently, it suffices to show that the system

$$(u_i - y, Bx + v_i) \geq 0, \qquad i = 1, \ldots, m, \tag{1.14}$$

has a solution $y \in K$ for any set of pairs $[u_i, v_i] \in A$, $i = 1, \ldots, m$.

Consider the function $H\colon U \times U \to \mathbf{R}$,

$$H(\lambda, \mu) = \sum_{i=1}^{m} \mu_i \left(Bx + v_i, \sum_{j=1}^{m} \lambda_j u_j - u_i \right) \qquad \forall \lambda, \mu \in U, \tag{1.15}$$

where

$$U = \left\{ \lambda \in \mathbf{R}^m; \lambda = (\lambda_1, \ldots, \lambda_m), \lambda_i \geq 0, \sum_{i=1}^{m} \lambda_i = 1 \right\}.$$

The function H is continuous, convex in λ, and concave in μ. Then, according to the J. Von Neumann theorem (see Theorem 2.3 in Chapter 1) it has a saddle point $(\lambda_0, \mu_0) \in U \times U$, i.e.,

$$H(\lambda_0, \mu) \leq H(\lambda_0, \mu_0) \leq H(\lambda, \mu_0) \qquad \forall \lambda, \mu \in U. \tag{1.16}$$

On the other hand, we have

$$\begin{aligned} H(\lambda, \lambda) &= \sum_{i=1}^{m} \lambda_i \left(Bx + v_i, \sum_{j=1}^{m} \lambda_j u_j - u_i \right) \\ &= \sum_{i=1}^{m} \sum_{j=1}^{m} \lambda_i \lambda_j (v_i, u_j - u_i) + \sum_{i=1}^{m} \sum_{j=1}^{m} \lambda_i \lambda_j (Bx, u_j - u_i) \\ &\leq 0 \qquad \forall \lambda \in U, \end{aligned}$$

because $(v_i - v_j, u_i - u_j) \geq 0$ for all i, j.

Then, by (1.16), we see that

$$H(\lambda_0, \mu) \leq 0 \qquad \forall \mu \in U,$$

i.e.,

$$\sum_{i=1}^{m} \mu_i \left(Bx + v_i, \sum_{j=1}^{m} (\lambda_j)_0 u_j - u_i \right) \leq 0 \qquad \forall \mu \in U.$$

In particular, it follows that

$$\left(Bx + v_i, \sum_{j=1}^{m} (\lambda_j)_0 u_j - u_i \right) \leq 0 \qquad \forall i = 1, \ldots, m.$$

Hence, $y = \Sigma_{j=1}^{m} (\lambda_j)_0 u_j \in K$ is a solution to (1.14). We have therefore proved that T is well-defined on K and that $T(K) \subset K$. It is also clear that for every $x \in K$, Tx is a closed convex subset of X and T is upper semicontinuous on K. Indeed, since the range of T belongs to a compact set, to verify that T is upper-semicontinuous it suffices to show that T is closed in $K \times K$, i.e., if $[x_n, y_n] \in T$, $x_n \to x$ and $y_n \to y$, then $y \in Tx$. But the last property is obvious if one takes in account the definition of T. Then, applying Kakutani's fixed point theorem (Theorem 2.2 in Chapter 1) we conclude that there exists $x \in K$ such that $x \in Tx$, thereby completing the proof of Lemma 1.2. ■

Proof of Theorem 1.1. Let Λ be the family of all finite dimensional subspaces X_α of X ordered by the inclusion relation. For every $X_\alpha \in \Lambda$, denote by $j_\alpha : X_\alpha \to X$ the injection mapping of X_α into X and by $j_\alpha^* : X^* \to X_\alpha^*$ the dual mapping, i.e., the projection of X^* onto X_α^*. The operators $A_\alpha = j_\alpha^* A j_\alpha$ and $B_\alpha = j_\alpha^* B j_\alpha$ map X_α into X_α^* and are monotone in $X_\alpha \times X_\alpha^*$. Since B is hemicontinuous from X to X^* and the j_α^* are continuous from X^* to X_α^* it follows by Lemma 1.1 that B_α is continuous from X_α to X_α^*.

We may therefore apply Lemma 1.2, where $X = X_\alpha$, $A = A_\alpha$, $B = B_\alpha$, and $K = K_\alpha = \overline{\text{conv}\, D(A_\alpha)}$. Hence, for each $X_\alpha \in \Lambda$, there exists $x_\alpha \in K_\alpha$ such that

$$(u - x_\alpha, B_\alpha x_\alpha + v) \geq 0 \qquad \forall [u, v] \in A,$$

or equivalently,

$$(u - x_\alpha, Bx_\alpha + v) \geq 0 \qquad \forall [u, v] \in A_\alpha. \tag{1.17}$$

By using the coercivity condition (1.10), we deduce from (1.17) that $\{x_\alpha\}$ remains in a bounded subset of X. Since the space X is reflexive, every bounded subset of X is sequentially compact and so there exists a sequence $\{x_{\alpha_n}\} \subset \{x_\alpha\}$ such that

$$x_{\alpha_n} \rightharpoonup x \quad \text{in } X \qquad \text{as } n \to \infty. \tag{1.18}$$

Moreover, since the operator B is bounded on bounded subsets, we may assume that

$$Bx_{\alpha_n} \rightharpoonup y \quad \text{in } X^* \qquad \text{as } n \to \infty. \tag{1.19}$$

Since the closed convex subsets are weakly closed, we infer that $x \in K$. By (1.17), we see that

$$\limsup_{n\to\infty} (x_{\alpha_n}, Bx_{\alpha_n}) \le (u - x, v) + (u, y) \qquad \forall [u, v] \in A. \tag{1.20}$$

Without loss of generality, we may assume that A is maximal in the class of all monotone subsets $\tilde{A} \subset X \times X^*$ such that $D(\tilde{A}) \subset K = \overline{\text{conv}\, D(A)}$. (If not, we may extend A by Zorn's lemma to a maximal element of this class.) To complete the proof, let us show first that

$$\limsup_{n\to\infty} (x_{\alpha_n} - x, Bx_{\alpha_n}) \le 0. \tag{1.21}$$

Indeed, if this is not the case, it follows from (1.20) that

$$(u - x, v + y) \ge 0 \qquad \forall [u, v] \in A,$$

and since $x \in K$ and A is maximal in the class of monotone operators with domain in K it follows that $[x, -y] \in A$. Then, putting $u = x$ in (1.20), we obtain (1.21), which contradicts the working hypothesis.

Now, for u arbitrary but fixed in $D(A)$ consider $u_\lambda = \lambda x + (1 - \lambda)u$, $0 \le \lambda \le 1$, and notice that by virtue of the monotonicity of B we have

$$(x_{\alpha_n} - u_\lambda, Bx_{\alpha_n}) \ge (x_{\alpha_n} - u_\lambda, Bu_\lambda).$$

This yields

$$\begin{aligned}(1 - \lambda)(x_{\alpha_n} - u, Bx_{\alpha_n}) + \lambda(x_{\alpha_n} - x, Bx_{\alpha_n}) \\ \ge (1 - \lambda)(x_{\alpha_n} - u, Bu_\lambda) + \lambda(x_{\alpha_n} - x, Bu_\lambda)\end{aligned}$$

and, by (1.21) and (1.20),

$$(x - u, Bu_\lambda) \le \limsup_{n\to\infty} (x_{\alpha_n} - u, Bx_{\alpha_n}) \le (u - x, v) \qquad \forall [u, v] \in A.$$

Since B is hemicontinuous, the latter inequality yields

$$(u - x, v + Bx) \geq 0 \qquad \forall [u, v] \in A,$$

thereby completing the proof of Theorem 1.1. ■

We shall use now Theorem 1.1 to prove a fundamental result in theory of maximal monotone operators due to G. Minty and F. Browder.

Theorem 1.2. *Let X and X^* be reflexive and strictly convex. Let $A \subset X \times X^*$ be a monotone subset of $X \times X^*$ and let $J: X \to X^*$ be the duality mapping of X. Then A is maximal monotone if and only if, for any $\lambda > 0$ (equivalently, for some $\lambda > 0$), $R(A + \lambda J) = X^*$.*

Proof. *"If" part.* Assume that $R(A + \lambda J) = X^*$ for some $\lambda > 0$. We suppose that A is not maximal monotone, and argue from this to a contradiction. If A is not maximal monotone, there exists $[x_0, y_0] \in X \times X^*$ such that $[x_0, y_0] \notin A$ and

$$(x - x_0, y - y_0) \geq 0 \qquad \forall [x, y] \in A. \tag{1.22}$$

On the other hand, by hypothesis, there exists $[x_1, y_1] \in A$ such that

$$\lambda J(x_1) + y_1 = \lambda J(x_0) + y_0.$$

Substituting $[x_1, y_1]$ in place of $[x, y]$ in (1.22) yields

$$(x_1 - x_0, J(x_1) - J(x_0)) \leq 0.$$

Taking in account definition of J, we get

$$\|x_1\|^2 + \|x_0\|^2 \leq (x_1, J(x_0)) + (x_0, J(x_1)),$$

and therefore

$$(x_1, J(x_0)) = (x_0, J(x_1)) = \|x_1\|^2 = \|x_0\|^2.$$

Hence

$$J(x_0) = J(x_1),$$

and since the duality mapping J^{-1} of X^* is single valued (because X is strictly convex), we infer that $x_0 = x_1$. Hence $[x_0, y_0] = [x_1, y_1] \in A$, which contradicts the hypothesis.

"Only if" part. The space X^* being strictly convex, J is single valued and demicontinuous on X (Theorem 1.2 in Chapter 1). Let y_0 be an arbitrary element of X^* and let $\lambda > 0$. Applying Theorem 1.1, where

$$Bu = \lambda J(u) - y_0 \qquad \forall u \in X,$$

we conclude that there is $x \in X$ such that

$$(u - x, \lambda J(x) - y_0 + v) \geq 0 \qquad \forall [u, v] \in A.$$

Since A is maximal monotone, this implies that $[x, \lambda J(x) - y_0] \in A$, i.e., $y_0 \in \lambda J(x) + Ax$. Applying Theorem 1.1, we have implicitly assumed that $0 \in D(A)$. If not, we apply this theorem to $Bu = \lambda J(u + u_0) - y_0$ and $Au \overset{\text{def}}{=} A(u + u_0)$, where $u_0 \in D(A)$. Thus, the proof is complete. ∎

We shall see later that the assumption that X^* is strictly convex can be dropped in Theorem 1.2.

Now we shall use Theorem 1.1 to derive a maximality criteria for the sum $A + B$.

Corollary 1.1. *Let X be reflexive and let B be a hemicontinuous and bounded operator from X to X^*. Let $A \subset X \times X^*$ be maximal monotone. Then $A + B$ is maximal monotone.*

Proof. By Asplund's theorem (Theorem 1.1 in Chapter 1), we may take an equivalent norm in X such that X and X^* are strictly convex. It is clear that after this operation the monotonicity properties of $A, B, A + B$ as well as maximality do not change. Also, without loss of generality, we may assume that $0 \in D(A)$; otherwise, we replace A by $u \to A(u + u_0)$, where $u_0 \in D(A)$ and B by $u \to B(u + u_0)$. Let y_0 be arbitrary but fixed in X^*. Now, applying Theorem 1.1, where B is the operator $u \to Bu + J(u) - y_0$, we infer that there is an $x \in \overline{\text{conv}\, D(A)}$ such that

$$(u - x, Jx + Bx - y_0 + v) \geq 0 \qquad \forall [u, v] \in A.$$

Since A is maximal monotone, this yields

$$y_0 \in Ax + Bx + Jx,$$

as claimed. ∎

In particular, it follows by Corollary 1.1 that every monotone, hemicontinuous, and bounded operator from X to X^* is maximal monotone. We shall prove now that the boundness assumption is redundant.

Theorem 1.3. *Let X be a reflexive Banach space and let $B: X \to X^*$ be a monotone hemicontinuous operator. Then B is maximal monotone in $X \times X^*$.*

Proof. Suppose that B is not maximal monotone. Then there exists $[x_0, y_0] \in X \times X^*$ such that $y_0 \neq Bx_0$ and

$$(x_0 - u, y_0 - Bu) \geq 0 \qquad \forall u \in X. \tag{1.23}$$

For any $x \in X$, we set $u_\lambda = \lambda x_0 + (1 - \lambda)x$, $0 \leq \lambda \leq 1$, and put $u = u_\lambda$ in (1.23). We get

$$(x_0 - x, y_0 - Bu_\lambda) \geq 0 \qquad \forall \lambda \in [0,1], u \in X,$$

and, letting λ tend to 1,

$$(x_0 - x, y_0 - Bx_0) \geq 0 \qquad \forall x \in X.$$

Hence $y_0 = Bx_0$, which contradicts the hypothesis. ■

Corollary 1.2. *Let X be a reflexive Banach space and let A be a coercive maximal monotone subset of $X \times X^*$. Then A is surjective, i.e., $R(A) = X^*$.*

Proof. Let $y_0 \in X^*$ be arbitrary but fixed. Without loss of generality, we may assume that X, X^* are strictly convex, so that by Theorem 1.2 for every $\lambda > 0$ the equation

$$\lambda J(x_\lambda) + Ax_\lambda \ni y_0 \tag{1.24}$$

has a (unique) solution $x_\lambda \in D(A)$. Multiply Eq. (1.24) by $x_\lambda - x^0$, where x^0 is the element arising in the coercivity condition (1.9). We have

$$\lambda \|x_\lambda\|^2 + (x_\lambda - x^0, Ax_\lambda) = (x_\lambda - x^0, y_0) + \lambda(x_0, Jx_\lambda).$$

By (1.9), we deduce that $\{x_\lambda\}$ is bounded in X and so we may assume (taking a subsequence if necessary) that $\exists x_0 \in X$ such that

$$w - \lim_{\lambda \downarrow 0} x_\lambda = x_0.$$

Letting λ tend to zero in (1.24), we see that

$$\lim_{\lambda \downarrow 0} Ax = y_0.$$

Since, as seen earlier, maximal monotone operators are weakly–strongly closed in $X \times X^*$, we conclude that $y_0 \in Ax_0$. Hence $R(A) = X^*$, as claimed. ■

In particular, it follows by Corollary 1.2 and Theorem 1.3 that:

Corollary 1.3. *A monotone, hemicontinuous, and coercive operator B from a reflexive Banach space X to its dual X^* is surjective.* ■

Let us now pause briefly to give an immediate application of this surjectivity result to existence theory for nonlinear elliptic boundary value problems.

Let Ω be an open and bounded subset of the Euclidean space $\mathbf{R}^N$. Consider the boundary value problem

$$\sum_{|\alpha| \leq m} (-1)^{\alpha} D^{\alpha} A_{\alpha}(x, u, \ldots, D^m u) = f \qquad \text{in } \Omega, \tag{1.25}$$

$$\frac{\partial_u^j}{\partial \nu^j} = 0, \qquad j = 0, 1, \ldots, m-1, \qquad \text{in } \partial\Omega,$$

where $f \in W^{-m,q}(\Omega)$, $1/p + 1/q = 1$, is given and $A_{\alpha}: \Omega \times \mathbf{R}^k \to \mathbf{R}$ satisfy the following conditions:

(i) $A_{\alpha} = A_{\alpha}(x, \xi)$ are measurable in x, continuous in ξ, and $\exists g \in L^q(\Omega)$ such that

$$|A_{\alpha}(x, \xi)| \leq C\big(|\xi|^{p-1} + g(x)\big) \qquad \forall x \in \Omega, \xi \in \mathbf{R}^k; \tag{1.26}$$

(ii)

$$\sum_{|\alpha| \leq m} (A_{\alpha}(x, \xi) - A_{\alpha}(x, \eta))(\xi_{\alpha} - \eta_{\alpha}) \geq 0$$

for all $\xi, \eta \in \mathbf{R}^k$ and a.e. $x \in \Omega$.

Let $V = W_0^{m,p}(\Omega)$ and let $a: V \times V \to \mathbf{R}$ be the Dirichlet functional

$$a(u, v) = \sum_{|\alpha| \leq m} \int_{\Omega} D^{\alpha} v \cdot A_{\alpha}(x, u, \ldots, D^m u)\, dx, \qquad u, v \in V. \tag{1.27}$$

By (i) and (ii), it is readily seen that a is well-defined on all of $V \times V$ and

$$|a(u, v)| \leq \omega(\|u\|_{m,p}) \|v\|_{m,p} \qquad \forall u, v \in V. \tag{1.28}$$

(Here, $\|\cdot\|_{m,p}$ is the norm in $W_0^{m,p}(\Omega)$.)

Moreover,

$$a(u, u - v) - a(v, u - v) \geq 0 \qquad \forall u, v \in V. \tag{1.29}$$

The function $u \in V$ is called *weak solution* to problem (1.25) if

$$a(u, v) = (v, f) \qquad \forall v \in V. \tag{1.30}$$

(Here, $(\cdot, \cdot)$ denotes the usual pairing between V and V'.)

We will assume further that

(iii)

$$\lim_{\|u\|_{m,p} \to \infty} a(u, u)/\|u\|_{m,p} = \infty.$$

Proposition 1.2. *Under assumptions (i)–(iii), for every $f \in W^{-m,q}(\Omega)$ problem* (1.25) *has at least one weak solution $u \in W_0^{m,p}(\Omega)$.*

Proof. Define the operator $A: V \to V'$ by

$$(v, Au) = a(u, v) \qquad \forall u, v \in V.$$

The operator A is monotone and coercive. To apply Corollary 1.3, it suffices to show that A is hemicontinuous. As a matter of fact, we shall prove that A is demicontinuous. To this end, let $\{u_n\}$ be strongly convergent to u in V as $n \to \infty$. Extracting further subsequences if necessary, we may assume that

$$D^\alpha u_n \to D^\alpha u \qquad \text{a.e. in } \Omega \text{ for } |\alpha| \leq m.$$

This implies that

$$A_\alpha(x, u_n, \ldots, D^m u_n) \to A_\alpha(x, u, \ldots, D^m u) \qquad \text{a.e. } x \in \Omega, |\alpha| \leq m.$$

Taking in account estimate (1.26), this implies that

$$\lim_{n\to\infty} \int_\Omega A_\alpha(x, u_n, \ldots, D^m u_n)\, D^\alpha v\, dx$$

$$= \int_\Omega A_\alpha(x, u, \ldots, D^m u)\, D^\alpha v\, dx \qquad \forall v \in V, |\alpha| \leq m. \tag{1.31}$$

Indeed, by the Egorov theorem, for every $\varepsilon > 0$ there exists $\Omega_\varepsilon \subset \Omega$ such that $m(\Omega \setminus \Omega_\varepsilon) \leq \varepsilon$ and $A_\alpha(x, u_n, \ldots, D^m u_n) \to A_\alpha(x, u, \ldots, D^m u)$ uniformly in Ω_ε. On the other hand,

$$\left|\int_{\Omega\setminus\Omega_\varepsilon} A_\alpha(x, u_n, \ldots, D^m u_n)\, D^\alpha v\, dx\right|$$

$$\leq C\left(\int_{\Omega\setminus\Omega_\varepsilon} |D^\alpha v|^p dx\right)^{1/p}\left(\int_{\Omega\setminus\Omega_\varepsilon} (|D^\alpha u_n|^p + |g(x)|^q)\, dx\right)^{1/q}$$

$$\overset{\varepsilon\to 0}{\to} 0,$$

uniformly in n.

Now by (1.31) we conclude that $\mathrm{w}-\lim_{n\to\infty} Au_n = Au$ in V' as desired. ■

2.1.2. The Sum of Two Maximal Monotone Operators

A problem of great interest because of its implications for existence theory for partial differential equations is to know whether the sum of two maximal monotone operators is again maximal monotone. Before answering to this question, let us first establish some facts related to Yosida approximation of the maximal monotone operators.

Let us assume that X is a reflexive strictly convex Banach space with strictly convex dual X^*, and let A be maximal monotone in $X \times X^*$.

According to Corollary 1.1 and Corollary 1.2, for every $x \in X$ the equation

$$0 \in J(x_\lambda - x) + \lambda A x_\lambda \tag{1.32}$$

has a solution x_λ. Since

$$(x-u, Jx - Ju) \geq (\|x\| - \|u\|)^2 \qquad \forall x, u \in X,$$

and J^{-1} is single valued (because X is strictly convex), it is readily seen that x_λ is unique. Define

$$J_\lambda x = x_\lambda,$$
$$A_\lambda x = \lambda^{-1} J(x - x_\lambda). \tag{1.33}$$

for any $x \in X$ and $\lambda > 0$.

The operator $A_\lambda\colon X \to X^*$ is called the *Yosida approximation* of A and plays an important role in the smooth approximation of A. We collect in

Proposition 1.3 several basic properties of the operators A_λ and J_λ.

Proposition 1.3. *Let X and X^* be strictly convex and reflexive. Then:*

(i) *A_λ is single valued, monotone, bounded and demicontinuous from X to X^*;*

(ii) *$\|A_\lambda x\| \le |Ax| = \inf\{\|y\|;\ y \in Ax\}$ for every $x \in D(A)$, $\lambda > 0$;*

(iii) *$J_\lambda : X \to X$ is bounded on bounded subsets and*

$$\lim_{\lambda \to 0} J_\lambda x = x \qquad \forall x \in \overline{\operatorname{conv} D(A)}; \tag{1.34}$$

(iv) *If $\lambda_n \to 0$, $x_n \to x$, $A_{\lambda_n} x_n \rightharpoonup y$ and*

$$\limsup_{n,m \to \infty} (x_n - x_m, A_{\lambda_n} x_n - A_{\lambda_m} x_m) \le 0, \tag{1.35}$$

then $[x, y] \in A$ and $\lim_{m,n \to \infty} (x_n - x_m, A_{\lambda_n} x_n - A_{\lambda_m} x_m) = 0$;

(v) *For $\lambda \to 0$, $A_\lambda x \rightharpoonup A^0 x$ $\forall x \in D(A)$, where $A^0 x$ is the element of minimum norm in Ax. If X^* is uniformly convex, then $A_\lambda x \to A^0 x$ $\forall x \in D(A)$.*

The main ingredient of the proof is the following lemma.

Lemma 1.3. *Let X be a reflexive Banach space and let A be a maximal monotone subset of $X \times X^*$. Let $[u_n, v_n] \in A$ be such that $u_n \rightharpoonup u$, $v_n \rightharpoonup v$, and either*

$$\limsup_{n,m \to \infty} (u_n - u_m, v_n - v_m) \le 0 \tag{1.36}$$

or

$$\limsup_{n \to \infty} (u_n - u, v_n - v) \le 0.$$

Then $[u, v] \in A$ and $(u_n, v_n) \to (u, v)$ as $n \to \infty$.

Proof. Assume first that condition (1.36) holds. Since A is monotone, we have

$$\lim_{n,m \to \infty} (u_n - u_m, v_n - v_m) = 0.$$

Let $n_k \to \infty$ be such that $(u_{n_k}, v_{n_k}) \to \mu$. Then, clearly, we have

$$\mu \le (u, v) = 0.$$

Hence $\overline{\lim}_{n \to \infty} (u_n, v_n) \leq (u, v)$, while by monotonicity of A we have

$$(u_n - x, v_n - y) \geq 0 \qquad \forall [x, y] \in A,$$

and therefore

$$(u - x, v - y) \geq 0 \qquad \forall [x, y] \in A,$$

which implies $[u, v] \in A$ because A is maximal monotone. The second part of the lemma follows by the same argument. ■

Proof of Proposition 1.3. (i) We have

$$(x - y, A_\lambda x - A_\lambda y) = (J_\lambda x - J_\lambda y, A_\lambda x - A_\lambda y) + ((x - J_\lambda x) - (y - J_\lambda y), A_\lambda x - A_\lambda y),$$

and since $A_\lambda x \in AJ_\lambda x$, we infer that

$$(x - y, A_\lambda x - A_\lambda y) \geq 0$$

because A and J are monotone.

Let $[u, v] \in A$ be arbitrary but fixed. If multiply Eq. (1.32) by $J_\lambda x - u$ and use the monotonicity of A, we get

$$(J_\lambda x - u, J(J_\lambda x - x)) \leq \lambda(u - J_\lambda x, v),$$

which yields

$$\|J_\lambda x - x\|^2 \leq \|x - u\| \, \|J_\lambda x - x\| + \lambda \|x - u\| \, \|v\| + \lambda \|v\| \, \|J_\lambda x - x\|.$$

This implies that J_λ and A_λ are bounded on bounded subsets.

Now, let $x_n \to x_0$ in X. We set $u_n = J_\lambda x_n$ and $v_n = A_\lambda x_n$. By the equation

$$J(u_n - x_n) + \lambda v_n = 0,$$

it follows that

$$((u_n - x_n) - (u_m - x_m), J(u_n - x_n) - J(u_m - x_m)) + \lambda(u_n - u_m, v_n - v_m) + \lambda(x_m - x_n, v_n - v_m) = 0.$$

Since as seen previously J_λ is bounded, this yields

$$\lim_{n, m \to \infty} (u_n - u_m, v_n - v_m) \leq 0$$

and

$$\lim_{m, n \to \infty} ((u_n - x_n) - (u_m - x_m, J(u_n - x_n) - J(u_m - x_m)) = 0.$$

Now, let $n_k \to \infty$ be such that $u_{n_k} \rightharpoonup u$, $v_{n_k} \rightharpoonup v$ and $J(u_{n_k} - x_{n_k}) \rightharpoonup w$. By Lemma 1.3, it follows that $[u, v] \in A$, $[u - x_0, w] \in J$, and therefore

$$J(u - x_0) + \lambda v = 0.$$

We have therefore proven that $u = J_\lambda x_0$, $v = A_\lambda x_0$, and by the uniqueness of the limit we infer that $J_\lambda x_n \rightharpoonup J_\lambda x_0$ and $A_\lambda x_n \rightharpoonup A_\lambda x_0$, as claimed.

(ii) Let $[x, x^*] \in A$. Again, by the monotonicity of A we have

$$0 \le (x - J_\lambda x, x^* - A_\lambda x) \le \|x^*\| \, \|x - x_\lambda\| - \lambda^{-1} \|x - x_\lambda\|^2.$$

Hence,

$$\lambda \|A_\lambda x\| = \|x - x_\lambda\| \le \lambda \|x^*\| \qquad \forall x^* \in Ax,$$

which implies (ii).

(iii) Let $x \in \overline{\text{conv}\, D(A)}$ and $[u, u^*] \in A$. We have

$$(J_\lambda x - u, A_\lambda x - u^*) \ge 0,$$

and therefore

$$\|J_\lambda x - x\|^2 \le \lambda (J_\lambda x - u, u^*) + (u - x, J(J_\lambda x - x)).$$

Let $\lambda_n \to 0$ be such that $J(J_{\lambda_n} x - x) \rightharpoonup y$ in X^*. This yields

$$\overline{\lim_{\lambda \to 0}} \, \|J_\lambda x - x\|^2 \le (u - x, y).$$

Since u is arbitrary in $D(A)$, the preceding inequality extends to all $u \in \overline{\text{conv}\, D(A)}$, and in particular we may take $u = x$.

(iv) We have

$$\begin{aligned}
&(x_n - x_m, A_{\lambda_n} x_n - A_{\lambda_m} x_m) \\
&\quad = (x_n - x_m, AJ_{\lambda_n} x_n - AJ_{\lambda_m} x_m) \\
&\quad = (J_{\lambda_n} x_n - J_{\lambda_m} x_m, AJ_{\lambda_n} x_n - AJ_{\lambda_m} x_m) + \big((x_n - J_{\lambda_n} x_n) \\
&\quad - (x_m - J_{\lambda_m} x_m), A_{\lambda_n} x_n - A_{\lambda_m} x_m\big) \\
&\quad \ge \big((x_n - J_{\lambda_n} x_n) - (x_m - J_{\lambda_m} x_m), A_{\lambda_n} x_n - A_{\lambda_m} x_m\big) \\
&\quad \ge \big((x_n - J_{\lambda_n} x_n) - (x_m - J_{\lambda_m} x_m), \\
&\quad \lambda_n^{-1} J(x_n - J_{\lambda_n} x_n) - \lambda_m^{-1} J(x_m - J_{\lambda_m} x_m)\big).
\end{aligned}$$

Since $A_{\lambda_n} x_n = \lambda_n^{-1} J(J_{\lambda_n} x_n - x_n)$ and x_n remain in bounded subsets of X^* and X, respectively, we infer that

$$\lim_{m,n\to\infty} (x_n - x_m, A_{\lambda_n} x_n - A_{\lambda_m} x_m) = 0$$

and

$$\lim_{m,n\to\infty} (J_{\lambda_n} x_n - J_{\lambda_m} x_m, A_{\lambda_n} x_n - A_{\lambda_m} x_m) = 0.$$

Then, by Lemma 1.3 we conclude that $[x, y] \in A$ because

$$\lim_{n\to\infty} (J_{\lambda_n} x_n - x_n) = -\lim_{n\to\infty} \lambda_n J^{-1}(A_{\lambda_n} x_{\lambda_n}) = 0.$$

(v) Since Ax is a closed convex subset of X^*, and X^* is reflexive and strictly convex, the projection $A^0 x$ of 0 into Ax is well-defined and unique.

Now, let $x \in D(A)$ and let $\lambda_n \to 0$ be such that $A_{\lambda_n} x \rightharpoonup y$ in X^*. As seen in the proof of (iv), $y \in Ax$, and since $\|A_{\lambda_n} x\| \le \|A^0 x\|$ we infer that $y = A^0 x$. Hence, $A_\lambda x \rightharpoonup A^0 x$ for $\lambda \to 0$. If X^* is uniformly convex, then by Lemma 1.1 in Chapter 1 we conclude that $A_\lambda x \to Ax$ (strongly) in X^* as $\lambda \to 0$. ■

Proposition 1.4. *If $X = H$ is a Hilbert space, then:*

(i) $J_\lambda = (I + \lambda A)^{-1}$ *is nonexpansive in H (i.e., Lipschitz with Lipschitz constant* 1);

(ii) $\|A_\lambda x - A_\lambda y\| \le \lambda^{-1}\|x - y\|\ \forall x, y \in D(A),\ \lambda > 0$;

(iii) $\lim_{\lambda\to 0} A_\lambda x = A^0 x\ \forall x \in D(A)$.

Proof. (i) We set $x_\lambda = (I + \lambda A)^{-1} x$, $y_\lambda = (I + \lambda A)^{-1} y$ (I is the unity operator in H). We have

$$x_\lambda - y_\lambda + \lambda(Ay_\lambda - Ay_\lambda) \ni x - y. \tag{1.37}$$

Multiplying by $x_\lambda - y_\lambda$ and using monotonicity of A, we get

$$\|x_\lambda - y_\lambda\| \le \|x - y\| \qquad \forall \lambda > 0.$$

Now, multiplying Eq. (1.37) by $Ax_\lambda - Ay_\lambda$ we get (ii). Regarding (iii), it follows by Proposition 1.3(v). ■

Corollary 1.4. *Let X be a reflexive Banach space and let A be maximal monotone in $X \times X^*$. Then both $\overline{D(A)}$ and $\overline{R(A)}$ are convex.*

Proof. Without any loss of generality, we may assume that X and X^* are strictly convex. Then, as seen in Proposition 1.2, $J_\lambda x \to x$ for every $x \in \overline{\text{conv}\, D(A)}$. Since $J_\lambda x \in D(A)$ for all $\lambda > 0$ and $x \in X$, we conclude that $\overline{\text{conv}\, D(A)} = \overline{D(A)}$, as claimed. Since $R(A) = D(A^{-1})$ and A^{-1} is maximal monotone $X^* \times X$, we conclude that $\overline{R(A)}$ is also convex. ∎

We shall establish now an important property of monotone operators with nonempty interior of the domain.

Theorem 1.4. *Let A be a monotone subset of $X \times X^*$. Then A is locally bounded in any interior point of $D(A)$.*

Let us first prove the following technical lemma.

Lemma 1.4. *Let $\{x_n\} \subset X$ and $\{y_n\} \subset X^*$ be such that $x_n \to 0$ and $\|y_n\| \to \infty$ as $n \to \infty$. Let $B(0, r)$ be the closed ball $\{x;\ \|x\| \le r\}$. Then there exist $x^0 \in B(0, r)$ and $\{x_{n_k}\} \subset \{x_n\}$, $\{y_{n_k}\} \subset \{y_n\}$ such that*

$$\lim_{k \to \infty} \left(x_{n_k} - x^0, y_{n_k}\right) = -\infty. \tag{1.38}$$

Proof. Suppose that the lemma is false. Then there exists $r > 0$ such that for every $u \in \mathrm{B}(0, r)$ there exists $C_u > -\infty$ such that

$$(x_n - u, y_n) \ge C_u \qquad \forall n \in \mathbf{N}.$$

We may write $B(0, r) = \bigcup_k \{u \in B(0, r);\ (x_n - u, y_n) \ge -k\ \forall n\}$. Then, by the Hausdorff–Baire theorem (see e.g., K. Yosida [1], p. 11), we infer that there is k_0 such that

$$\text{int}\,\{u \in B(0, r);\ (x_n - u, y_n) > -k_0\ \forall n\} \ne \varnothing.$$

In other words, there are $\varepsilon > 0$, $k_0 \in \mathbf{N}$, and $u_0 \in B(0, r)$ such that

$$\{u;\ \|u - u_0\| \le \varepsilon\} \subset \{u \in B(0, r);\ (x_n - u, y_n) > -k_0\ \forall n\}.$$

Now, we have

$$(x_n - u, y_n) \ge -k_0 \qquad \text{and} \qquad (x_n + u_0, y_n) \ge C_{-u_0}.$$

Summing up, we get

$$(2x_n + u_0 - u, y_n) \ge -k_0 + C \qquad \forall u \in B(u_0, \varepsilon),$$

where $C = C_{-u_0}$. Now, we take $u = u_0 + 2x_n + w$, where $\|w\| = \varepsilon/2$. For n sufficiently large, we therefore have

$$(w, y_n) \geq -C + y_0 \qquad \forall w, \|w\| = \varepsilon/2,$$

which clearly contradicts the fact that $\|y_n\| \to \infty$ as $n \to \infty$. ∎

Proof of Theorem 1.4. Let $x_0 \in \operatorname{int} D(A)$ be arbitrary. Without loss of generality, we may assume that $x_0 = 0$. (This can be achieved by shifting the domain of A.) Let us assume that A is not locally bounded at 0. Then there exist sequences $\{x_n\} \subset X, \{y_n\} \subset X^*$ such that $[x_n, y_n] \in A$, $\|x_n\| \to 0$, and $\|y_n\| \to \infty$. According to Lemma 1.4 there exists, for every ball $B(0, r)$, $x^0 \in B(0, r)$ and $\{x_{n_k}\} \subset \{x_n\}$, $\{y_{n_k}\} \subset \{y_n\}$ such that

$$\lim_{k \to \infty} \left(x_{n_k} - x^0, y_{n_k}\right) = -\infty.$$

Let r be sufficiently small so that $B(0, r) \subset D(A)$. Then, $x^0 \in D(A)$ and by the monotonicity of A it follows that

$$\left(x_{n_k} - x^0, Ax^0\right) \to -\infty \qquad \text{as } k \to \infty.$$

The contradiction we have arrived at completes the proof. ∎

Now we are ready to prove the main result of this section, due to R. T. Rockafellar [4].

Theorem 1.5. *Let X be a reflexive Banach space and let A and B be maximal monotone subsets of $X \times X^*$ such that*

$$(\operatorname{int} D(A)) \cap D(B)) \neq \emptyset. \tag{1.39}$$

Then $A + B$ is maximal monotone in $X \times X^*$.

Proof. As in the previous cases, we may assume without loss of generality that X and X^* are strictly convex. Moreover, shifting the domains and ranges of A and B, if necessary, we may assume that $0 \in (\operatorname{int} D(A)) \cap D(B)$, $0 \in A0$, $0 \in B0$. We shall prove that $R(J + A + B) = X^*$. To this aim, consider an arbitrary element y in X^*. Since the operator B_λ is demicontinuous, bounded, and monotone, and so is J: $X \to X^*$, it follows by Corollaries 1.1 and 1.2 that for every $\lambda > 0$ the equation

$$J(x_\lambda) + Ax_\lambda + B_\lambda x_\lambda \ni y \tag{1.40}$$

has a solution $x_\lambda \in D(A)$. (Since J and J^{-1} are single valued and X, X^* are strictly convex, it follows by standard arguments involving the monotonicity of A and B that x_λ is unique.) Multiplying Eq. (1.40) by x_λ and using the obvious inequalities

$$(x_\lambda, Ax_\lambda) \geq 0, \qquad (x_\lambda, B_\lambda x_\lambda) \geq 0,$$

we infer that

$$\|x_\lambda\| \leq \|y\| \qquad \forall \lambda > 0.$$

Moreover, since $0 \in \operatorname{int} D(A)$, it follows by Theorem 1.4 that there exist the constants $\rho > 0$ and $M > 0$ such that

$$\|x^*\| \leq M \qquad \forall x^* \in Ax, \|x\| \leq \rho. \tag{1.41}$$

Multiplying Eq. (1.40) by $x_\lambda - \rho w$ and using the monotonicity of A, we get

$$(x_\lambda - \rho w, Jx_\lambda + B_\lambda x_\lambda - y) + (x_\lambda - \rho w, A(\rho w)) \leq 0 \qquad \forall \|w\| = 1.$$

By (1.41), we get

$$\|x_\lambda\|^2 - \rho(w, B_\lambda x_\lambda) \leq M(\rho + \|x_\lambda\|) + \|x_\lambda\|(\rho + \|y\|).$$

Hence,

$$\|x_\lambda\|^2 + \rho\|B_\lambda x_\lambda\| \leq \|x_\lambda\|(\rho + M + \|y\|) + M\rho \qquad \forall \lambda > 0.$$

We may conclude, therefore, that $\{B_\lambda x_\lambda\}$ and $\{y_\lambda = y - Jx_\lambda - B_\lambda x_\lambda\}$ are bounded in X^* as $\lambda \to 0$. Since X is reflexive, we may assume that on a subsequence, again denoted λ,

$$x_\lambda \rightharpoonup x_0, \qquad B_\lambda x_\lambda \rightharpoonup y_1, \qquad y_\lambda \in Ax_\lambda \rightharpoonup y_2, \qquad Jx_\lambda \rightharpoonup y_0.$$

Inasmuch as $A + J$ is monotone, we have

$$(x_\lambda - x_\mu, B_\lambda x_\lambda - B_\mu x_\mu) \leq 0 \qquad \forall \lambda, \mu > 0.$$

Then, by Proposition 1.3(iv),

$$\lim_{\lambda, \mu \to 0} (x_\lambda - x_\mu, B_\lambda x_\lambda - B_\mu x_\mu) = 0$$

and $[x_0, y_1] \in B$. Then, by Eq. (1.40), we see that

$$\lim_{\lambda, \mu \to 0} (x_\lambda - x_\mu, Jx_\lambda + y_\lambda - Jx_\lambda - y_\mu) = 0 \qquad y_\lambda \in Ax_\lambda, y_\mu \in Ax_\mu,$$

and since $J + A$ is maximal monotone it follows by Lemma 1.3 that $[x_0, y_0 + y_2] \in A + J$. Thus, letting λ tend to zero in (1.40) we see that

$$y \in J(x_0) + Ax_0 + Bx_0,$$

thereby completing the proof. ■

In particular, Theorem 1.5 leads to:

Corollary 1.5. *Let X be a reflexive Banach space, $A \subset X \times X^*$ a maximal monotone operator and let $B: X \to X^*$ be a demicontinuous monotone operator. Then $A + B$ is maximal monotone.* ■

More generally, it follows from Theorem 1.5 that if A, B are two maximal monotone sets of $X \times X^*$, and $D(B) = X^*$, then $A + B$ is maximal monotone.

We conclude this section with a result of the same type in Hilbert spaces.

Theorem 1.6. *Let $X = H$ be a Hilbert space and let A, B be maximal monotone sets in $H \times H$ such that $D(A) \cap D(B) \neq \emptyset$ and*

$$(v, A_\lambda u) \geq -C\left(\|u\|^2 + \lambda\|A_\lambda u\|^2 + \|A_\lambda u\| + 1\right) \qquad \forall [u, v] \in B. \quad (1.42)$$

Then $A + B$ is maximal monotone.

Proof. We have denoted by $A_\lambda = \lambda^{-1}(I - (I + A)^{-1})$ the Yosida approximation of A. For any $y \in H$ and $\lambda > 0$, consider the equation

$$x_\lambda + Bx_\lambda + A_\lambda x_\lambda \ni y, \quad (1.43)$$

which by Corollaries 1.4 and 1.5 has a solution (clearly unique) $x_\lambda \in D(B)$. Let $x_0 \in D(A) \cap D(B)$. Taking the scalar product of (1.43) with $x_\lambda - x_0$ and using the monotonicity of B and A_λ yields

$$(x_\lambda, x_\lambda - x_0) + (y_0, x_\lambda - x_0) + (A_\lambda x_0, x_\lambda - x_0) \leq (y, x_\lambda - x_0).$$

Since, as seen in Proposition 1.3, $\|A_\lambda x_0\| \leq |Ax_0|$ $\forall \lambda > 0$, this yields

$$\|x_\lambda\| \leq M \qquad \forall \lambda > 0.$$

Next, we multiply Eq. (1.43) by $A_\lambda x_\lambda$ and use inequality (1.42) to get, after some calculations,

$$\|A_\lambda x_\lambda\| \leq C \qquad \forall \lambda > 0.$$

Now, for a sequence $\lambda_n \to 0$, we have

$$x_{\lambda_n} \rightharpoonup x, \qquad A_{\lambda_n} x_{\lambda_n} \rightharpoonup y_1, \qquad y_{\lambda_n} \rightharpoonup y_2,$$

where $y_\lambda = y - x_\lambda - A_\lambda x_\lambda \in Bx_\lambda$.

Then, arguing as in the proof of Theorem 1.5 it follows by Proposition 1.3 that $[x, y_1] \in A$, $[x, y_2] \in B$, and this implies that $y \in x + Ax + Bx$, as claimed. ■

2.2. Generalized Gradients (Subpotential Operators)

2.2.1. *Subdifferential of a Convex Function*

Let X be a real Banach space with dual X^*. A *proper convex function* on X is a function $\varphi: X \to]-\infty, +\infty] = \overline{\mathbf{R}}$ that is not identically $+\infty$ and that satisfies the inequality

$$\varphi((1-\lambda)x + \lambda y) \leq (1-\lambda)\varphi(x) + \lambda\varphi(y) \tag{2.1}$$

for all $x, y \in X$ and all $\lambda \in [0, 1]$.

The function $\varphi: X \to]-\infty, +\infty]$ is said to be *lower semicontinuous* (l.s.c.) on X if

$$\liminf_{u \to x} \varphi(u) \geq \varphi(x) \qquad \forall x \in X,$$

or equivalently, every level subset $\{x \in X;\ \varphi(x) \leq \lambda\}$ is closed.

Since every level set of a convex function is convex and every closed convex set is weakly closed (this is an immediate consequence of Mazur's theorem, (K. Yosida [1], p. 109), we may therefore conclude that a proper convex function is lower semicontinuous if it is weakly lower semicontinuous.

Given a lower semicontinuous convex function $\varphi: X \to]-\infty, +\infty] = \overline{\mathbf{R}}$, $\varphi \not\equiv \infty$, we shall use the following notations:

$$D(\varphi) = \{x \in X;\ \varphi(x) < \infty\} \qquad \text{(the effective domain of } \varphi\text{)}, \tag{2.2}$$

$$\mathrm{Epi}(\varphi) = \{(x, \lambda) \in X \times \mathbf{R};\ \varphi(x) \leq \lambda\} \qquad \text{(the epigraph of } \varphi\text{)}. \tag{2.3}$$

It is readily seen that $\mathrm{Epi}(\varphi)$ is a closed convex subset of $X \times \mathbf{R}$, and its properties are closely related to those of the function φ.

Now, let us briefly describe some elementary properties of l.s.c., convex functions.

Proposition 2.1. *Let $\varphi: X \to \overline{\mathbf{R}}$ be a proper, l.s.c. and convex function. Then φ is bounded from below by an affine function, i.e., there are $x_0^* \in X^*$ and $\alpha \in \mathbf{R}$ such that*

$$\varphi(x) \geq (x, x_0^*) + \alpha \qquad \forall x \in X. \tag{2.4}$$

Proof. Let $E(\varphi) = \operatorname{Epi}(\varphi)$ and let $x_0 \in X$ and $r \in \mathbf{R}$ be such that $\varphi(x_0) > r$. By the classical separation theorem, there is a closed hyperplane $H = \{(x, \lambda) \in X \times \mathbf{R};\ (x_0^*, x) + \lambda = \alpha\}$ that separates $E(\varphi)$ and (x_0, r). This means that

$$(x_0^*, x) + \lambda \geq \alpha \qquad \forall x \in E(\varphi),$$

and

$$(x_0^*, x_0) + r < \alpha.$$

Hence, for $\lambda = \varphi(x)$, we have

$$(x_0^*, x) + \varphi(x) \geq (x_0^*, x_0) + r \qquad \forall x \in X,$$

which implies (2.4). ■

Proposition 2.2. *Let $\varphi: X \to \overline{\mathbf{R}}$ be proper, convex, and l.s.c. Then φ is continuous on* int $D(\varphi)$.

Proof. Let $x_0 \in \operatorname{int} D(\varphi)$. We shall prove that φ is continuous at x_0. Without loss of generality, we will assume that $x_0 = 0$ and that $\varphi(0) = 0$. Since the set $\{x: \varphi(x) > -\varepsilon\}$ is open it suffices to show that $\{x: \varphi(x) < \varepsilon\}$ is a neighborhood of the origin. We set $C = \{x \in X;\ \varphi(x) \leq \varepsilon\} \cap \{x \in X;\ \varphi(-x) \leq \varepsilon\}$. Clearly, C is a closed, balanced set of X (i.e., $\alpha x \in C$ for $|\alpha| \leq 1$ and $x \in C$). Moreover, C is absorbing, i.e., for every $x \in X$ there exists $\alpha > 0$ such that $\alpha x \in C$ (because the function $t \to \varphi(tx)$ is convex and finite in a neighborhood of the origin and therefore it is continuous). Since X is a Banach space, the preceding properties of C imply that C is a neighborhood of the origin, as claimed. ■

The function $\varphi^*: X^* \to \overline{\mathbf{R}}$ defined by

$$\varphi^*(p) = \sup\{(x, p) - \varphi(x);\ x \in X\} \tag{2.5}$$

is called the *conjugate* of φ.

Proposition 2.3. *Let $\varphi: X \to \overline{\mathbf{R}}$ be l.s.c., convex, and proper. Then φ^* is l.s.c., convex, and proper on the space X^*.*

Proof. As supremum of family of affine functions, φ^* is convex and l.s.c. Moreover, by Proposition 2.1 we see that $\varphi^* \not\equiv \infty$. ∎

Proposition 2.4. *Let X be reflexive and let φ be a l.s.c. proper convex function on X. Further assume that*

$$\lim_{\|x\| \to \infty} \varphi(x) = \infty. \tag{2.6}$$

Then there exists $x_0 \in X$ such that

$$\varphi(x_0) = \inf\{\varphi(x);\ x \in X\}.$$

Proof. Let $d = \inf\{\varphi(x);\ x \in X\}$ and let $\{x_n\} \subset X$ such that $d \leq \varphi(x_n) \leq d + 1/n$. By (2.6), we see that $\{x_n\}$ is bounded in X and since the space X is reflexive it is sequentially weakly compact. Hence, there is $\{x_{n_k}\} \subset \{x_n\}$ such that $x_{n_k} \rightharpoonup x$ as $n_k \to \infty$. Since φ is weakly semicontinuous, this implies that $\varphi(x) \leq d$. Hence $\varphi(x) = d$, as desired. ∎

Given a function f from a Banach space X to $\mathbf{R}$, the mapping $f': X \times X \to \mathbf{R}$ defined by

$$f'(x, y) = \lim_{\lambda \downarrow 0} \frac{f(x + \lambda y) - f(x)}{\lambda}, \qquad x, y \in X, \tag{2.7}$$

(if it exists) is called the *directional derivative* of f at x in direction y.

The function $f: X \to \mathbf{R}$ is said to be *Gâteaux differentiable* at $x \in X$ if there exists $\nabla f(x) \in X^*$ (the *Gâteaux differential*) such that

$$f'(x, y) = (y, \nabla f(x)) \qquad \forall y \in X. \tag{2.8}$$

If the convergence in (2.7) is uniform in y on bounded subsets, then f is said to be *Fréchet differentiable* and ∇f is called the *Fréchet differential* (derivative) of f.

Given a l.s.c., convex, proper function $\varphi: X \to \overline{\mathbf{R}}$, the mapping $\partial\varphi: X \to X^*$ defined by

$$\partial\varphi(x) = \{x^* \in X^*;\ \varphi(x) \leq \varphi(y) + (x - y, x^*), \forall y \in X\} \tag{2.9}$$

is called the *subdifferential* of φ.

In general, $\partial\varphi$ is a multivalued operator from X to X^*, and in accord with our convention we shall regard it as a subset of $X \times X^*$.

An element $x^* \in \partial\varphi(x)$ (if any) is called a *subgradient* of φ in x. We shall denote as usual by $D(\partial\varphi)$ the set of all $x \in X$ for which $\partial\varphi(x) \neq \emptyset$.

Let us pause briefly to give some simple examples.

1. $\varphi(x) = \frac{1}{2}\|x\|^2$. Then, $\partial\varphi = J$ (the duality mapping of the space X). Indeed, if $x^* \in Jx$, then $(x - y, x^*) = \|x\|^2 - (y, x^*) \geq \frac{1}{2}(\|x\|^2 - \|y\|^2)\ \forall y \in X$. Hence $x^* \in \partial\varphi(x)$.

Now, let $x^* \in \partial\varphi(x)$, i.e.,

$$\tfrac{1}{2}\left(\|x\|^2 - \|y\|^2\right) \leq (x - y, x^*) \qquad \forall y \in X. \tag{2.10}$$

We take $y = \lambda x$, $0 < \lambda < 1$, in (2.10), getting

$$(x, x^*) \geq \tfrac{1}{2}\|x\|^2(1 + \lambda).$$

Hence $(x, x^*) \geq \|x\|^2$.

If $y = \lambda x$ where $\lambda > 1$, we get that $(x, x^*) \leq \|x\|^2$. Hence, $(x, x^*) = \|x\|^2$ and $\|x^*\| \geq \|x\|$.

On the other hand, taking $y = x + \lambda u$ in (2.10), where $\lambda > 0$ and u is arbitrary in X, we get

$$(u, x^*) \leq \tfrac{1}{2}(\|x + \lambda u\|^2 - \|x\|^2),$$

which yields

$$(u, x^*) \leq \|x\|\,\|u\|.$$

Hence $\|x^*\| \leq \|x\|$. We have therefore proven that $(x^*, x) = \|x\|^2 = \|x^*\|^2$, as claimed.

2. Let K be a closed convex subset of X. The function $I_K: X \to \overline{\mathbf{R}}$ defined by

$$I_K(x) = \begin{cases} 0 & \text{if } x \in K, \\ +\infty & \text{if } x \notin K, \end{cases} \tag{2.11}$$

is called the *indicator function* of K, whilst its dual function H,

$$H_K(p) = \sup\{(p, u); u \in K\} \qquad \forall p \in X^*,$$

is called the *support function* of K.

It is readily seen that $D(\partial I_K) = K$, $\partial I_K(x) = 0$ for $x \in \operatorname{int} K$ (if nonempty) and that

$$\partial I_K(x) = N_K(x) = \{x^* \in X^*; (x - u, x^*) \geq 0\ \forall u \in K\} \qquad \forall x \in K. \tag{2.12}$$

For every $x \in K$, $N_K(x)$ is the *normal cone* at K in x. Parenthetically, note that if K_1 and K_2 are two closed convex subsets, then $K_1 \subset K_2$ if and only if

$$H_{K_1}(p) \leq H_{K_2}(p) \qquad \forall p \in X^*.$$

3. Let φ be Gâteaux differentiable at x. Then $\partial\varphi(x) = \nabla\varphi(x)$. Indeed, since φ is convex, we have

$$\varphi(x + \lambda(y - x)) \leq (1 - \lambda)\varphi(x) + \lambda\varphi(y)$$

for all $x, y \in X$ and $0 \leq \lambda \leq 1$. Hence,

$$\frac{\varphi(x + \lambda(y - x)) - \varphi(x)}{\lambda} \leq \varphi(y) - \varphi(x),$$

and letting λ tend to zero we see that $\nabla\varphi(x) \in \partial\varphi(x)$. Now, let w be an arbitrary element of $\partial\varphi(x)$. We have

$$\varphi(x) - \varphi(y) \leq (x - y, w) \qquad \forall y \in X.$$

Equivalently,

$$\frac{\varphi(x + \lambda y) - \varphi(x)}{\lambda} \geq (y, w) \qquad \forall\lambda > 0,\ y \in X,$$

and this implies that $(y, \nabla\varphi(x) - w) \geq 0$ for all $y \in X$. Hence, $w = \nabla\varphi(x)$.

By the definition of $\partial\varphi$ it is obvious that

$$\varphi(x) = \inf\{\varphi(u);\ u \in X\} \qquad \text{iff } 0 \in \partial\varphi(x).$$

There is a close relationship between $\partial\varphi$ and $\partial\varphi^*$. More precisely, we have:

Proposition 2.5. *Let X be a reflexive Banach space and let $\varphi\colon X \to \overline{\mathbf{R}}$ be a l.s.c., convex, proper function. Then the following conditions are equivalent:*

(i) $x^* \in \partial\varphi(x)$;

(ii) $\varphi(x) + \varphi^*(x^*) = (x, x^*)$;

(iii) $x \in \partial\varphi^*(x^*)$.

In particular, $\partial\varphi^* = (\partial\varphi)^{-1}$ *and* $(\varphi^*)^* = \varphi$.

Proof. By definition of φ^*, we see that

$$\varphi^*(x^*) \geq (x, x^*) - \varphi(x) \qquad \forall x \in X,$$

with equality if and only if $0 \in \partial_x((x, x^*) - \varphi(x))$. Hence, (i) and (ii) are equivalent. Now, if (ii) holds, then x^* is a minimum point for the function $\varphi^*(p) - (x, p)$ and so $x \in \partial\varphi^*(x^*)$. Hence, (ii) $\Rightarrow$ (iii). Since conditions (i) and (ii) are equivalent for φ^*, we may equivalently express (iii) as

$$\varphi^*(x) + (\varphi^*)^*(x) = (x, x^*).$$

Thus, to prove (ii) it suffices to show that $(\varphi^*)^* = \varphi$. It is readily seen that $(\varphi^*)^* = \varphi^{**} \leq \varphi$. We suppose now that there exists $x_0 \in X$ such that $\varphi^{**}(x_0) > \varphi(x_0)$, and we will argue from this to a contradiction. We have, therefore, $(x_0, \varphi^{**}(x_0)) \notin \mathrm{Epi}(\varphi)$ and so by the separation theorem it follows that there are $x_0^* \in X^*$ and $\alpha \in \mathbf{R}$ such that $(x_0^*, x_0) + \alpha\varphi^{**}(x_0) > \sup\{(x_0^*, x) + \alpha\lambda;\ (x, \lambda) \in \mathrm{Epi}(\varphi)\}$. After some calculation involving this inequality, it follows that $\alpha < 0$. Then dividing this inequality by $-\alpha$, we get

$$\begin{aligned}
-\left(x_0^*, \frac{x_0}{\alpha}\right) - \varphi^{**}(x_0) &> \sup\left\{\left(x_0^*, -\frac{x}{\alpha}\right) - \lambda;\ (x, \lambda) \in \mathrm{Epi}(\varphi)\right\} \\
&= \sup\left\{\left(-\frac{x_0^*}{\alpha}, x\right) - \varphi(x);\ x \in D(\varphi)\right\} \\
&= \varphi^*\left(-\frac{x_0^*}{\alpha}\right),
\end{aligned}$$

which clearly contradicts the definition of φ^{**}. ■

Theorem 2.1. *Let X be a real Banach space and let $\varphi: X \to \overline{\mathbf{R}}$ be a l.s.c. proper convex function. Then $\partial\varphi$ is a maximal monotone subset of $X \times X^*$.*

Proof. It is readily seen that $\partial\varphi$ is monotone in $X \times X^*$. To prove that $\partial\varphi$ is maximal monotone, we shall assume for simplicity that X is reflexive and refer the reader to Rockafellar's work [2] for the general case.

Continuing, we fix $y \in X^*$ and consider the equation

$$Jx + \partial\varphi(x) \ni y. \tag{2.13}$$

Let $f: X \to \overline{\mathbf{R}}$ be the convex, l.s.c. function defined by

$$f(x) = \tfrac{1}{2}\|x\|^2 + \varphi(x) - (x, y).$$

By Proposition 2.1, we see that $\lim_{\|x\| \to \infty} f(x) = +\infty$, and so by Proposition 2.4 we conclude that there exists $x_0 \in X$ such that

$$f(x_0) = \inf\{f(x); x \in X\}.$$

This yields

$$\tfrac{1}{2}\|x_0\|^2 + \varphi(x_0) - (x_0, y) \le \tfrac{1}{2}\|x\|^2 + \varphi(x) - (x, y) \qquad \forall x \in X,$$

i.e.,

$$\begin{aligned} \varphi(x_0) - \varphi(x) &\le (x_0 - x, y) + \tfrac{1}{2}\left(\|x\|^2 - \|x_0\|^2\right) \\ &\le (x_0 - x, y) + (x - x_0, Jx) \qquad \forall x \in X. \end{aligned}$$

In the latter inequality we take $x = tx_0 + (1 - t)u$, $0 < t < 1$, where u is an arbitrary element of X. We get

$$\varphi(x_0) - \varphi(u) \le (x_0 - u, y) + (u - x_0, w_t),$$

where $w_t \in J(tx_0 + (1 - t)u)$.

For $t \to 1$, $w_t \rightharpoonup w \in J(x_0)$ because J is maximal monotone and so it is strongly–weakly closed in $X \times X^*$. Hence,

$$\varphi(x_0) - \varphi(u) \le (x_0 - u, y - w) \qquad \forall u \in X,$$

and this inequality shows that $y - w \in \partial\varphi(x_0)$, i.e., x_0 is a solution to Eq. (2.13). We have therefore proven that $R(J + \partial\varphi) = X^*$, thereby completing the proof. ■

Proposition 2.6. *Let $\varphi: X \to \overline{\mathbf{R}}$ be a l.s.c. convex and proper function. Then $D(\partial\varphi)$ is a dense subset of $D(\varphi)$.*

Proof. Let x be any element of $D(\varphi)$ and let $x_\lambda = J_\lambda x$ be the solution to the equation (see (1.32))

$$J(x_\lambda - x) + \lambda\partial\varphi(x_\lambda) \ni 0.$$

Multiplying this equation by $x_\lambda - x$, we get

$$\|x_\lambda - x\|^2 + \lambda(\varphi(x_\lambda) - \varphi(x)) \le 0 \qquad \forall \lambda > 0.$$

Since by Proposition 2.1 φ is bounded from below by an affine function and $\varphi(x) < \infty$, this yields

$$\lim_{\lambda \to 0} x_\lambda = x.$$

As $x_\lambda \in D(\partial\varphi)$ and x is arbitrary in $D(\varphi)$, we conclude that $\overline{D(\varphi)} = \overline{D(\partial\varphi)}$, as claimed. ■

Proposition 2.7. *Let φ be a l.s.c., proper, convex function on X. Then* int $D(\varphi) \subset D(\partial\varphi)$.

Proof. Let $x_0 \in \operatorname{int} D(\varphi)$ and let $V = B(x_0, r) = \{x; \|x - x_0\| < r\}$ be such that $V \subset D(\varphi)$. We know by Proposition 2.2 that φ is continuous on V and this implies that the set $D = \{(x, \lambda) \in V \times \mathbf{R};\ \varphi(x) < \lambda\}$ is an open convex set of $X \times \mathbf{R}$. Thus, there is a closed hyperplane, $H = \{(x, \lambda) \in X \times \mathbf{R};\ (x, x_0^*) + \lambda = \alpha\}$, which separates $(x_0, \varphi(x_0))$ from $\overline{D}$. Hence, $(x_0, x_0^*) + \varphi(x_0) < \alpha$ and

$$(x, x_0^*) + \lambda \geq \alpha \qquad \forall (x, \lambda) \in \overline{D}.$$

This yields

$$\varphi(x_0) - \varphi(x) < -(x_0 - x, x_0^*) \qquad \forall x \in V.$$

But, for every $u \in X$, there exists $0 < \lambda < 1$ such that $x = \lambda x_0 + (1 - \lambda)u \in V$. Substituting this x in the preceding inequality and using the convexity of φ, we obtain

$$\varphi(x_0) \leq \varphi(u) + (x_0 - u, x_0^*) \qquad \forall u \in X.$$

Hence, $x_0 \in D(\partial\varphi)$ and $x_0^* \in \partial\varphi(x_0)$. ■

For every $\lambda > 0$, define the function

$$\varphi_\lambda(x) = \inf\left\{\frac{\|x - u\|^2}{2\lambda} + \varphi(u);\ u \in X\right\} \qquad \forall x \in X, \tag{2.14}$$

where $\varphi: X \to \overline{\mathbf{R}}$ is a l.s.c. proper convex function. By Propositions 2.1 and 2.4 it follows that $\varphi_\lambda(x)$ is well-defined for all $x \in X$ and the infimum defining it is attained (if the space X is reflexive). This implies by a straightforward argument that φ_λ is convex and l.s.c. on X. (Since φ_λ is everywhere defined, we conclude by Proposition 2.2. That φ_λ is continuous.)

The function φ_λ is called the *regularization* of φ, for reasons that will become clear in the following theorem.

Theorem 2.2. *Let X be a reflexive and strictly convex Banach space with strictly convex dual. Let $\varphi: X \to \overline{\mathbf{R}}$ be a l.s.c. convex, proper function and let $A = \partial\varphi \subset X \times X^*$. Then the function φ_λ is convex, continuous, Gâteaux differentiable, and $\nabla\varphi_\lambda = A_\lambda$ for all $\lambda > 0$. Moreover:*

$$\varphi_\lambda(x) = \frac{\|x - J_\lambda x\|^2}{2\lambda} + \varphi(J_\lambda x) \qquad \forall \lambda > 0,\ x \in X; \tag{2.15}$$

$$\lim_{\lambda \to 0} \varphi_\lambda(x) = \varphi(x) \qquad \forall x \in X; \tag{2.16}$$

$$\varphi(J_\lambda x) \le \varphi_\lambda(x) \le \varphi(x) \qquad \forall \lambda > 0,\ x \in X. \tag{2.17}$$

If X is a Hilbert space (not necessarily identified with its dual) then φ_λ is Fréchet differentiable on X.

Proof. We observe that the subdifferential of the function $u \to \|x - u\|^2/2\lambda + \varphi(u)$ is just the operator $u \to \lambda^{-1}J(u - x) + \partial\varphi(u)$ (see Theorem 2.3 following). This implies that every solution x_λ of the equation

$$\lambda^{-1}J(u - x) + \partial\varphi(u) \ni 0$$

is a minimum point of the function $u \to 1/(2\lambda)\|x - u\|^2 + \varphi(u)$. Recalling that $x_\lambda = J_\lambda x$, we obtain (2.15). Regarding inequality (2.17), it is an immediate consequence of (2.14). To prove (2.16), assume first that $x \in D(\varphi)$. Then as seen in Proposition 1.3, $\lim_{\lambda \to 0} J_\lambda x = x$, and by (2.17) and the lower semicontinuity of φ we infer that

$$\varphi(x) \le \liminf_{\lambda \to 0} \varphi(J_\lambda x) \le \liminf_{\lambda \to 0} \varphi_\lambda(x) \le \varphi(x).$$

If $x \notin D(\varphi)$, i.e., $\varphi(x) = +\infty$, then $\lim_{\lambda \to 0} \varphi_\lambda(x) = +\infty$ because otherwise would exist $\{\lambda_n\} \to 0$ and $C > 0$ such that

$$\varphi_{\lambda_n}(x) \le C \qquad \forall n.$$

Then, by (2.15), we see that $\lim_{n \to \infty} J_{\lambda_n} x = x$, and again by (2.17) and the lower semicontinuity of φ we conclude that $\varphi(x) \le C$, which is absurd.

To conclude the proof, it remains to show that φ_λ is Gâteaux differentiable and $\nabla\varphi_\lambda = A_\lambda$. By (2.15), it follows that

$$\begin{aligned}\varphi_\lambda(y) - \varphi_\lambda(x) &\le (J_\lambda(y) - J_\lambda(x), A_\lambda y)\\ &\quad + \frac{1}{2\lambda}\left(\|y - J_\lambda(y)\|^2 - \|x - J_\lambda(x)\|^2\right)\\ &= (y - x, A_\lambda y) + (J_\lambda(y) - y, A_\lambda y) + (x - J_\lambda(x), A_\lambda y)\\ &\quad + \frac{1}{2\lambda}\left(\|y - J_\lambda(y)\|^2 - \|x - J_\lambda(x)\|^2\right)\\ &\le (y - x, A_\lambda y).\end{aligned}$$

Hence,

$$0 \le \varphi_\lambda(y) - \varphi_\lambda(x) - (y - x, A_\lambda x) \le (y - x, A_\lambda y - A_\lambda x) \tag{2.18}$$

for all $x, y \in X$ and $\lambda > 0$. The latter inequality clearly implies that

$$\lim_{t \downarrow 0} \frac{\varphi_\lambda(x + tu) - \varphi_\lambda(x)}{t} = (u, A_\lambda x) \qquad \forall u, x \in X,$$

because A_λ is demicontinuous. Hence, φ_λ is Gâteaux differentiable and $\nabla\varphi_\lambda = (\partial\varphi)_\lambda = A_\lambda$.

Now, assume that X is a Hilbert space. Then by the monotonicity of A we see that

$$\|J_\lambda(x) - J_\lambda(y)\| \le \|x - y\| \qquad \forall x, y \in X,\ \lambda > 0.$$

This implies that $A_\lambda : X \to X$ is Lipschitz with Lipschitz constant $2/\lambda$. (As a matter of fact, it follows that it is $1/\lambda$.) Then, by the inequality (2.18), we see that

$$|\varphi_\lambda(x) - \varphi_\lambda(y) - (x - y, A_\lambda x)| \le \frac{2}{\lambda}\|x - y\|^2 \qquad \forall x, y \in X,$$

and this shows that φ_λ is Fréchet differentiable. ■

Let us consider the particular case where $\varphi = I_K$, K a closed convex subset of X, and X is a Hilbert space. Then

$$(I_K)_\lambda(x) = \frac{\|x - P_K x\|^2}{2\lambda} \qquad \forall x \in X,\ \lambda > 0, \tag{2.19}$$

where $P_K x$ is the projection of x on K. (Since K is closed and convex, $P_K x$ is uniquely defined.) Moreover, as seen in the preceding, we have

$$P_K = J_\lambda = (I + \lambda A)^{-1} \qquad \forall \lambda > 0. \tag{2.20}$$

A problem of great interest in convex optimization as well as for calculus with convex functions is to determine whether given two l.s.c., convex, proper functions f and g on X, $\partial(f+g) = \partial f + \partial g$. The following theorem due to R. T. Rockafellar [3] gives a general answer to this question.

Theorem 2.3. *Let X be a Banach space and let $f: X \to \overline{\mathbf{R}}$ and $g: X \to \overline{\mathbf{R}}$ be two l.s.c., convex, proper functions such that $D(f) \cap \operatorname{int} D(g) \neq \emptyset$. Then*

$$\partial(f+g) = \partial f + \partial g. \tag{2.21}$$

Proof. If the space X is reflexive, (2.21) is an immediate consequence of Theorem 1.5. Indeed, as seen in Proposition 2.7, $\operatorname{int} D(\partial g) = \operatorname{int} D(g)$ and so $D(\partial f) \cap \operatorname{int} D(\partial g) \neq \emptyset$. Then, by Theorem 1.5, $\partial f + \partial g$ is maximal monotone in $X \times X^*$. On the other hand, it is readily seen that $\partial f + \partial g \subset \partial(f+g)$. Hence, $\partial f + \partial g = \partial(f+g)$.

In the general case, Theorem 2.3 follows by a separation argument we will present subsequently.

Since the relation $\partial f + \partial g \subset \partial(f+g)$ is obvious, let us prove that $\partial(f+g) \subset \partial f + \partial g$. To this end, consider $x_0 \in D(\partial f) \cap D(\partial g)$ and $w \in \partial(f+g)(x_0)$, arbitrary but fixed. We shall prove that $w = w_1 + w_2$, where $w_1 \in \partial f(x_0)$ and $w_2 \in \partial g(x_0)$. Replacing the functions f and g by $x \to f(x + x_0) - f(x_0) - (x, z_1)$ and $x \to g(x + x_0) - g(x_0) - (x, z_2)$, respectively, where $w = z_1 + z_2$, we may assume that $x_0 = 0$, $w = 0$, and $f(0) = g(0) = 0$. Hence, we should prove that $0 \in \partial f(0) + \partial g(0)$. Consider the sets E_i, $i = 1, 2$, defined by $E_1 = \{(x, \lambda) \in X \times \mathbf{R}; f(x) \leq \lambda\}$, $E_2 = \{(x, \lambda) \in X \times \mathbf{R}; g(x) \leq -\lambda\}$. Inasmuch as $0 \in \partial(f+g)(0)$, we have

$$0 = (f+g)(0) = \inf\{(f+g)(x); x \in X\},$$

and therefore $E_1 \cap \operatorname{int} E_2 = \emptyset$. Then, by the separation theorem there exists a closed hyperplane that separates the sets E_1 and E_2. In other words, there are $w \in X^*$ and $\alpha \in \mathbf{R}$ such that

$$\begin{aligned} (w, x) + \alpha\lambda &\leq 0 \qquad \forall (x, \lambda) \in E_1, \\ (w, x) + \alpha\lambda &\geq 0 \qquad \forall (x, \lambda) \in E_2. \end{aligned} \tag{2.22}$$

Let us observe that the hyperplane is not vertical, i.e., $\alpha \neq 0$. Indeed, if $\alpha = 0$, then this would imply that the hyperplane $(w, x) = 0$ separates the sets $D(f)$ and $D(g)$ in the space X, which is not possible because $D(f) \cap \operatorname{int} D(g) \neq \emptyset$. Hence $\alpha \neq 0$, and to specify the situation we shall assume that $\alpha > 0$. Then, by (2.22) we see that

$$g(x) \leq -\lambda \leq (w, x) \leq -\alpha f(x) \qquad \forall x \in X,$$

and therefore $(1/\alpha)w \in \partial f(0)$, $-(1/\alpha)w \in \partial g(0)$, i.e., $0 \in \partial f(0) + \partial g(0)$, as claimed. ■

Theorem 2.4. *Let $X = H$ be a real Hilbert space and let A be a maximal monotone subset of $H \times H$. Let $\varphi: H \to \overline{\mathbf{R}}$ be a l.s.c., convex, proper function such that $D(A) \cap D(\partial\varphi) \neq \emptyset$ and, for some $h \in H$,*

$$\varphi\big((I + \lambda A)^{-1}(x + \lambda h)\big) \leq \varphi(x) + C\lambda(1 + \varphi(x)), \qquad \forall x \in D(\varphi),\ \lambda > 0. \tag{2.23}$$

Then $A + \partial\varphi$ is maximal monotone and

$$\overline{D(A + \partial\varphi)} = \overline{D(A)} \cap \overline{D(\varphi)}.$$

Proof. We shall proceed as in the proof of Theorem 1.6. Let y be arbitrary but fixed in H. Then, for every $\lambda > 0$, the equation

$$x_\lambda + A_\lambda x_\lambda + \partial\varphi(x_\lambda) \ni y$$

has a unique solution $x_\lambda \in D(\partial\varphi)$. We multiply the preceding equation by $x - J_\lambda(x_\lambda + \lambda h)$ and use condition (2.23). This yields

$$\|A_\lambda x_\lambda\|^2 + (A_\lambda x_\lambda, J_\lambda(x_\lambda) - J_\lambda(x_\lambda + \lambda h)) \leq C\lambda(\|y\| + \|h\| + \|x_\lambda\| + \varphi(x_\lambda) + 1),$$

where $J_\lambda = (I + \lambda A)^{-1}$. We get

$$\|A_\lambda x_\lambda\|^2 \leq C\lambda(\|y\| + \|h\| + \|x_\lambda\| + \varphi(x_\lambda) + 1).$$

On the other hand, multiplying the preceding equation by $x_\lambda - x_0$, where $x_0 \in D(A) \cap D(\partial\varphi)$, we get

$$\|x_\lambda\|^2 + \varphi(x_\lambda) \leq C\big(\|A_\lambda x_0\|^2 + \varphi(x_0) + 1\big).$$

Hence, $\{A_\lambda x_\lambda\}$ and $\{x_\lambda\}$ are bounded in H. Then, as seen in the proofs of Theorems 1.5 and 1.6, this implies that $x_\lambda \rightharpoonup x$, where x is the solution to the equation

$$x + \partial\varphi(x) + Ax \ni y.$$

Now, let us prove that

$$\overline{D(A)} \cap \overline{D(\varphi)} \subset \overline{D(A) \cap D(\varphi)} \subset \overline{D(A) \cap D(\partial\varphi)}.$$

Let $u \in \overline{D(A)} \cap \overline{D(\varphi)}$ be arbitrary but fixed and let h be as in condition (2.23). Clearly, there is a sequence $\{u_\lambda\} \subset D(\varphi)$ such that $u_\lambda + \lambda h \in D(\varphi)$ and $u_\lambda \to u$ as $\lambda \to 0$. Let $v_\lambda = J_\lambda(u_\lambda + \lambda h) \in D(A) \cap D(\varphi)$ (by condition (2.23)). We have

$$\|v_\lambda - u\| \leq \|J_\lambda(u_\lambda + \lambda h) - J_\lambda u\| + \|u - J_\lambda u\| \to 0 \qquad \text{as } \lambda \to 0,$$

because $u \in \overline{D(A)}$ (see Proposition 1.3). Hence,

$$\overline{D(A)} \cap \overline{D(\varphi)} \subset \overline{D(A) \cap D(\varphi)}.$$

Now, let u be arbitrary in $D(A) \cap D(\varphi)$ and let $x_\lambda \in D(A) \cap D(\partial\varphi)$ be the solution to

$$x_\lambda + \lambda(Ax_\lambda + \partial\varphi(x_\lambda)) \ni u.$$

By the definition of $\partial\varphi$, we have

$$\begin{aligned} \varphi(x_\lambda) - \varphi(u) &\leq (u - x_\lambda - \lambda Ax_\lambda, x_\lambda - u) \\ &\leq -\|u - x_\lambda\|^2 + \lambda\|A^0 u\| \, \|u - x_\lambda\| \qquad \forall \lambda > 0. \end{aligned}$$

Hence $x_\lambda \to u$ for $\lambda \to 0$, and so $D(A) \cap D(\varphi) \subset \overline{D(A) \cap D(\partial\varphi)}$, as claimed. This completes the proof. ■

Remark 2.1. In particular, condition (2.23) holds if

$$(A_\lambda(x + \lambda h), y) \geq -C(1 + \varphi(x)) \qquad \forall \lambda > 0,$$

for some $h \in H$, and all $[x, y] \in \partial\varphi$.

We conclude this section with an explicit formula for $\partial\varphi$ in term of the directional derivative.

Proposition 2.8. *Let X be a Banach space and let $\varphi: X \to \overline{\mathbf{R}}$ be a l.s.c., convex, proper function on X. Then for all $x_0 \in D(\partial\varphi)$,*

$$\partial\varphi(x_0) = \{x_0^* \in X^*; \ \varphi'(x_0, u) \geq (u, x_0^*) \ \forall u \in X\}. \tag{2.24}$$

Proof. Let $x_0^* \in \partial\varphi(x_0)$. Then, by the definition of $\partial\varphi$,

$$\varphi(x_0) - \varphi(x_0 + tu) \leq -t(u, x_0^*) \qquad \forall u \in X, t > 0,$$

which yields

$$\varphi'(x_0, u) \geq (u, x_0^*) \qquad \forall u \in X.$$

Assume not that $(u, x_0^*) \leq \varphi'(x_0, u)$ $\forall u \in X$. Since φ is convex, the function $t \to (\varphi(x_0 + tu) - \varphi(x_0))/t$ is monotonically increasing and so we have

$$(u, x_0^*) \leq t^{-1}(\varphi(x_0 + tu) - \varphi(x_0)) \qquad \forall u \in X, t > 0.$$

Hence $x_0^* \in \partial\varphi(x_0)$, and the proof is complete. ■

Formula (2.24) can be taken as definition of the subdifferential $\partial\varphi$, and we shall see later that it may be used to define generalized gradients of certain nonconvex functions.

It turns out that if φ is continuous at x, then

$$\varphi'(x_0, u) = \sup\{(u, x_0^*);\ x_0^* \in \partial\varphi(x_0)\}, \qquad u \in X. \tag{2.25}$$

2.2.2. Examples of Subdifferential Mappings

There is a general characterization of maximal monotone operators that are subdifferentials of l.s.c. convex functions due to R. T. Rockafellar [1]. A set $A \subset X \times X^*$ is said to be *cyclically monotone* if

$$(x_0 - x_1, x_0^*) + \cdots + (x_{n-1} - x_n, x_{n-1}^*) + (x_n - x_0, x_0^*) \geq 0 \tag{2.26}$$

for all $[x_i, x_i^*] \in A$, $i = 0, 1, \ldots, n$. A is said to be *maximal cyclically monotone* if it is cyclically monotone and has no cyclically monotone extensions in $X \times X^*$. It turns out that the class of subdifferential mappings coincides with that of maximal cyclically monotone operators. More precisely, one has:

Theorem 2.5. *Let X be a real Banach space and let $A \subset X \times X^*$. The set A is the subdifferential of a l.s.c. convex, proper function from X to $\mathbf{R}$ if and only if A is maximal cyclically monotone.*

We leave to the reader the proof of this theorem and we shall concentrate on some significant examples of subdifferential mappings.

1. *Maximal monotone sets (graphs) in* $\mathbf{R} \times \mathbf{R}$. Every maximal monotone set (graph) of $\mathbf{R} \times \mathbf{R}$ is the subdifferential of a l.s.c. convex proper function on $\mathbf{R}$. Indeed, let β be a maximal monotone set in $\mathbf{R} \times \mathbf{R}$ and let $\beta^0: \mathbf{R} \to \mathbf{R}$ be the function defined by

$$\beta^0(r) = \{y \in \beta(r); |y| = \inf\{|z|; z \in \beta(r)\}\} \qquad \forall r \in \mathbf{R}.$$

We know that $\overline{D(\beta)} = [a, b]$, where $-\infty \leq a \leq b \leq \infty$. The function β^0 is monotonically increasing and so the integral

$$j(r) = \int_{r_0}^{r} \beta^0(u)\, du \qquad \forall r \in \mathbf{R}, \tag{2.27}$$

where $r_0 \in D(\beta)$, is well-defined (unambiguously a real number or $+\infty$). Clearly, the function j is continuous on (a, b) and convex on $\mathbf{R}$. Moreover, $\liminf_{r \to b} j(r) \geq j(b)$ and $\liminf_{r \to a} j(r) \geq j(a)$. Finally,

$$j(r) - j(t) = \int_{t}^{r} \beta^0(u)\, du \leq v(r - t) \qquad \forall [r, v] \in \beta,\ t \in \mathbf{R}.$$

Hence $\beta = \partial j$, where j is the l.s.c. convex function defined by (2.27).

2. *Self-adjoint operators*. Let H be a real Hilbert space with scalar product $(\cdot, \cdot)$ and norm $|\cdot|$, and let A be a linear self-adjoint positive operator on H. Then, $A = \partial\varphi$ where

$$\varphi(x) = \begin{cases} \frac{1}{2}|A^{1/2}x|^2, & x \in D(A^{1/2}), \\ +\infty, & \text{otherwise}. \end{cases} \tag{2.28}$$

Conversely, any linear, densely defined operator that is the subdifferential of a l.s.c. convex function on H is self-adjoint.

To prove these assertions, we note first that any self-adjoint positive operator A in a Hilbert space is maximal monotone. Indeed, it is readily seen that the range of the operator $I + A$ is simultaneously closed and dense in H. On the other hand, if $\varphi: H \to \overline{\mathbf{R}}$ is the function defined by (2.28), then clearly it is convex, l.s.c., and

$$\varphi(x) - \varphi(u) = \tfrac{1}{2}(|A^{1/2}x|^2 - |A^{1/2}u|^2) \leq (Ax, x - u)$$
$$\forall x \in D(A),\ u \in D(A^{1/2}).$$

Hence $A \subset \partial\varphi$, and since A is maximal monotone we conclude that $A = \partial\varphi$.

Now, let A be a linear, densely defined operator on H of the form $A = \partial\psi$, where $\psi: H \to \overline{\mathbf{R}}$ is a l.s.c. convex function. By Theorem 2.2, we know that $A_\lambda = \nabla\psi_\lambda$, where $A_\lambda = \lambda^{-1}(I - (I + \lambda A)^{-1})$. This yields

$$\frac{d}{dt}\psi_\lambda(tu) = t(A_\lambda u, u) \qquad \forall u \in H, t \in [0,1],$$

and therefore $\psi_\lambda(u) = \frac{1}{2}(A_\lambda u, u)$ for all $u \in H$ and $\lambda > 0$. Calculating the Fréchet derivative of ψ_λ, we see that

$$\nabla\psi_\lambda = A_\lambda = \tfrac{1}{2}(A_\lambda + A_\lambda^*).$$

Hence $A_\lambda = A_\lambda^*$, and letting $\lambda \to 0$ this implies that $A = A^*$, as claimed.

More generally, if A is a linear continuous, symmetric operator from a Hilbert space V to its dual V^* (not identified with V), then $A = \partial\varphi$, where $\varphi: V \to \mathbf{R}$ is the function

$$\varphi(u) = \tfrac{1}{2}(Au, u) \qquad \forall u \in V.$$

Conversely, every linear continuous operator $A: V \to V'$ of the form $\partial\varphi$ is symmetric.

In particular, if Ω is a bounded and open domain of $\mathbf{R}^N$ with a sufficiently smooth boundary (of class C^2, for instance), then the operator $A: D(A) \subset L^2(\Omega) \to L^2(\Omega)$ defined by

$$Ay = -\Delta y \quad \forall y \in D(A), \qquad D(A) = H_0^1(\Omega) \cap H^2(\Omega),$$

is self-adjoint and $A = \partial\varphi$ where $\varphi: L^2(\Omega) \to \mathbf{R}$ is given by

$$\varphi(y) = \begin{cases} \frac{1}{2}\int_\Omega |\nabla y|^2\,dx & \text{if } y \in H_0^1(\Omega), \\ +\infty & \text{otherwise}. \end{cases}$$

This result remains true for a nonsmooth bounded open domain if it is convex (see Grisvard [1]).

3. *Convex integrands*. Let Ω be a measurable subset of the Euclidean space $\mathbf{R}^N$ and let $L^p(\Omega)$, $1 \le p < \infty$, be the space of all p-summable functions on Ω. We set $L_m^p(\Omega) = (L^p(\Omega))^m$.

The function $g: \Omega \times \mathbf{R}^m \to \overline{\mathbf{R}}$ is said to be a *normal convex integrand* if the following conditions hold:

(i) For almost all $x \in \Omega$, the function $g(x, \cdot): \mathbf{R}^m \to \overline{\mathbf{R}}$ is convex, l.s.c., and not identically $+\infty$;

(ii) g is $\mathscr{L} \times \mathscr{B}$ measurable on $\Omega \times \mathbf{R}^m$; that is, it is measurable with respect to the σ-algebra of subsets of $\Omega \times \mathbf{R}^m$ generated by products of Lebesgue measurable subsets of Ω and Borel subsets of $\mathbf{R}^m$.

We note that if g is convex in y and $\operatorname{int} D(g(x, \cdot)) \neq \emptyset$ for every $x \in \Omega$, then condition (ii) holds if and only if $g = g(x, y)$ is measurable in x for every $y \in \mathbf{R}^m$ (Rockafellar [5]).

A special case of a $\mathscr{L} \times \mathscr{B}$ measurable integrand is the *Caratheodory integrand.* Namely, one has:

Lemma 2.1. *Let $g = g(x, y): \Omega \times \mathbf{R}^m \to \mathbf{R}$ be continuous in y for every $x \in \Omega$ and measurable in x for every y. Then g is measurable.*

Proof. Let $\{z_i\}_{i=1}^{\infty}$ be a dense subset of $\mathbf{R}^N$ and let $\lambda \in \mathbf{R}$ arbitrary but fixed. Inasmuch as g is continuous in y, it is clear that $g(x, y) \leq \lambda$ if and only if for every n there exists z_i such that $\|z_i - y\| \leq 1/n$ and $g(x_1, z_i) \leq \lambda + 1/n$. Denote by Ω_{in} the set $\{x \in \Omega; g(x, z_i) \leq \lambda + 1/n\}$ and put $Y_{in} = \{y \in \mathbf{R}^m; \|y - z_i\| \leq 1/n\}$. Since

$$\{(x, y) \in \Omega \times \mathbf{R}^m; g(x, y) \leq \lambda\} = \bigcap_{n=1}^{\infty} \bigcup_{i=1}^{\infty} \Omega_{in} \times Y_{in},$$

we infer that g is $\mathscr{L} \times \mathscr{B}$ measurable, as desired. ■

Let assume in addition to conditions (i), (ii) the following:

(iii) There are $\alpha \in L_m^q(\Omega)$, $1/p + 1/q = 1$, and $\beta \in L^1(\Omega)$ such that

$$g(x, y) \geq (\alpha(x), y) + \beta(x) \qquad \text{a.e. } x \in \Omega,\ y \in \mathbf{R}^m, \tag{2.29}$$

where $(\cdot, \cdot)$ is the usual scalar product in $\mathbf{R}^m$;

(iv) There is $y_0 \in L_m^p$ such that $g(x, y_0) \in L^1(\Omega)$.

Let us remark that if g is independent of x, then conditions (iii) and (iv) automatically hold by virtue of Proposition 2.1.

Define on the space $X = L_m^p(\Omega)$ the function $I_g: X \to \overline{\mathbf{R}}$,

$$I_g(y) = \begin{cases} \int_\Omega g(x, y(x))\, dx & \text{if } g(x, y) \in L^1(\Omega), \\ +\infty & \text{otherwise}. \end{cases} \tag{2.30}$$

Proposition 2.9. *Let g satisfy assumptions (i)–(iv). Then the function I_g is convex, lower semicontinuous, and proper. Moreover,*

$$\partial I_g(y) = \{w \in L_m^q(\Omega); w(x) \in \partial g(x, y(x)) \text{ a.e. } x \in \Omega\}. \tag{2.31}$$

Here, ∂g is the subdifferential of the function $y \to g(x, y)$.

Proof. Let us show that I_g is well-defined (unambiguously a real number or $+\infty$) for every $y \in L^q_m(\Omega)$. Note first that for every Lebesgue measurable function $y: \Omega \to \mathbf{R}^m$ the function $x \to g(x, y(x))$ is Lebesgue measurable on Ω. For a fixed $\lambda \in \mathbf{R}$, we set

$$E = \{(x, y) \in \Omega \times \mathbf{R}^m;\ g(x, y) \leq \lambda\}.$$

Let us denote by $\mathscr{S}$ the class of all sets $\tilde{E} \subset \Omega \times \mathbf{R}^m$ having the property that the set $\{x \in \Omega;\ (x, y(x)) \in \tilde{E}\}$ is Lebesgue measurable. Obviously, $\tilde{E}$ contains every set of the form $T \times D$, where T is a measurable subset of Ω and D is an open subset of $\mathbf{R}^m$. Since $\mathscr{S}$ is a σ-algebra, it follows that it contains the σ-algebra generated by the products of Lebesgue measurable subsets of Ω and Borel subsets of $\mathbf{R}^m$. Hence $E \in \mathscr{S}$, and therefore $g(x, y(x))$ is Lebesgue measurable and so I_g is well-defined. By assumption (i), it follows that I_g is convex, whilst by (iv) we see that $I_g \not\equiv +\infty$. Let $\{y_n\} \subset L^p_m(\Omega)$ be strongly convergent to y. Then there is $\{y_{n_k}\} \subset \{y_n\}$ such that

$$y_{n_k}(x) \to y(x) \quad \text{a.e. } x \in \Omega \qquad \text{for } n_k \to \infty.$$

Then, by assumption (iii) and Fatou's lemma, it follows that

$$\liminf_{n_k \to \infty} \int_\Omega \big(g(x, y_{n_k}(x)) - (\alpha(x), y_{n_k}(x)) - \beta(x)\big)\, dx$$

$$\geq \int_\Omega (g(x, y(x)) - (\alpha(x), y(x)) - \beta(x))\, dx,$$

and therefore

$$\liminf_{n_k \to \infty} I_g(y_{n_k}) \geq I_g(y).$$

Clearly, this implies that $\liminf_{n \to \infty} I_g(y_n) \geq I_g(y)$, i.e., I_g is l.s.c. on X.

Let us now prove (2.31). It is easily seen that every $w \in L^q_m(\Omega)$ such that $w(x) \in \partial g(x, y(x))$ belongs to $\partial I_g(y)$. Now let $w \in \partial I_g(y)$, i.e.,

$$\int_\Omega (g(x, y(x)) - g(x, u(x)))\, dx \leq \int_\Omega (w(x), y(x) - u(x))\, dx$$

$$\forall u \in L^p_m(\Omega).$$

Let D be an arbitrary measurable subset of Ω and let $u \in L^p_m(\Omega)$ be defined by

$$u(x) = \begin{cases} y_0 & \text{for } x \in D, \\ y(x) & \text{for } x \in \Omega \setminus \overline{D}, \end{cases}$$

where y_0 is arbitrary in $\mathbf{R}^m$. Substituting in the previous inequality, we get

$$\int_D (g(x, y(x)) - g(x, y_0) - (w(x), y(x) - y_0))\, dx \le 0.$$

Since D is arbitrary this implies, a.e. $x \in \Omega$,

$$g(x, y(x)) \le g(x, y_0) + (w(x), y(x) - y_0) \qquad \forall y_0 \in \mathbf{R}^m.$$

Hence, $w(x) \in \partial g(x, y(x))$ a.e. $x \in \Omega$, as claimed. ■

The case $p = \infty$ is more difficult since the elements of $\partial I_g(y) \subset (L^1_m(\Omega))^*$ are measures on Ω. However, we refer the reader to Rockafellar [5] for the complete description of ∂I_g in this case.

Now, let us consider the special case where

$$g(x, y) = I_k(y) = \begin{cases} 0 & \text{if } y \in K, \\ +\infty & \text{if } y \notin K, \end{cases}$$

K being a closed, convex subset of $\mathbf{R}^m$. Then, I_g is the indicator function of the closed convex subset $\mathscr{K}$ of $L^p_m(\Omega)$ defined by

$$\mathscr{K} = \{y \in L^p_m(\Omega);\ y(x) \in K \text{ a.e. } x \in \Omega\},$$

and so by formula (2.31) we see that the normal cone $N_{\mathscr{K}} \subset L^q_m(\Omega)$ to $\mathscr{K}$ is defined by

$$N_{\mathscr{K}}(y) = \{w \in L^q_m(\Omega);\ w(x) \in N_K(y(x)) \text{ a.e. } x \in \Omega\}, \tag{2.32}$$

where $N_K(y) = \{z \in \mathbf{R}^m;\ (z, y - u) \ge 0\ \forall u \in K\}$ is the normal cone at K in $y \in K$.

In particular, if $m = 1$ and $K = [a, b]$, then

$$\begin{aligned} N_{\mathscr{K}}(y) = \{w \in L^q(\Omega);\ & w(x) = 0 \text{ a.e. in } [x \in \Omega;\ a < y(x) < b], \\ & w(x) \ge 0 \text{ a.e. in } [x \in \Omega;\ y(x) = b],\ w(x) \le 0 \text{ a.e.} \\ & \text{in } [x \in \Omega;\ y(x) = a]\}. \end{aligned} \tag{2.33}$$

Let us take now $K = \{y \in \mathbf{R}^m; \|y\| \le \rho\}$. Then,

$$N_K(y) = \begin{cases} 0 & \text{if } \|y\| < \rho, \\ \bigcup_{\lambda > 0} \lambda y & \text{if } \|y\| = \rho, \end{cases}$$

and so $N_{\mathscr{K}}$ is given by

$$\begin{aligned} N_{\mathscr{K}}(y) = \{ & w \in L^q_m(\Omega);\ w(x) = 0 \text{ a.e. in } [x \in \Omega;\ \|y(x)\| < \rho], \\ & w(x) = \lambda(x) y(x) \text{ a.e. in } [x \in \Omega;\ \|y(x)\| = \rho], \text{ where} \\ & \lambda \in L^q_m(\Omega),\ \lambda(x) \ge 0 \text{ a.e. } x \in \Omega\}. \end{aligned}$$

We propose that the reader calculate the normal cone $N_{\mathscr{C}}$ to the set

$$\mathscr{C} = \left\{ y \in L^2(\Omega);\ a \le y(x) \le b \text{ a.e. } \left[x \in \Omega;\ \int_\Omega y(x)\, dx = 1 \right] \right\}$$

and prove that

$$\begin{aligned} N_{\mathscr{C}}(y) = \{ & z = w + \lambda;\ \lambda \in \mathbf{R},\ w \in L^2(\Omega),\ w(x) = 0 \\ & \text{a.e. in } [a < y(x) < b],\ w(x) \ge 0 \text{ a.e. in } [x;\ y(x) = b], \\ & w(x) \le 0 \text{ a.e. in } [\ x;\ y(x) = a]\}. \end{aligned}$$

4. *Semilinear elliptic operators in* $L^2(\Omega)$. Let Ω be an open, bounded subset of $\mathbf{R}^N$, and let $g: \mathbf{R} \to \overline{\mathbf{R}}$ be a lower semicontinuous, convex, proper function such that $0 \in D(\partial g)$.

Define the function $\varphi: L^2(\Omega) \to \overline{\mathbf{R}}$ by

$$\varphi(y) = \begin{cases} \int_\Omega \left(\frac{1}{2} |\nabla y|^2 + g(y) \right) dx & \text{if } y \in H^1_0(\Omega) \text{ and } g(y) \in L^1(\Omega), \\ +\infty & \text{otherwise}. \end{cases} \tag{2.34}$$

Proposition 2.10. *The function φ is convex, l.s.c. and $\not\equiv +\infty$. Moreover, if the boundary $\partial\Omega$ is sufficiently smooth (for instance, of class C^2) or if Ω is convex, then*

$$\begin{aligned} \partial\varphi(y) = \{ & w \in L^2(\Omega);\ y \in H^1_0(\Omega) \cap H^2(\Omega), \\ & w(x) + \Delta y(x) \in \partial g(y(x)) \text{ a.e. } x \in \Omega\}. \end{aligned} \tag{2.35}$$

Proof. It is readily seen that φ is convex and $\not\equiv +\infty$. Let $\{y_n\} \subset L^2(\Omega)$ be strongly convergent to y as $n \to \infty$. As seen earlier,

$$\liminf_{n \to \infty} \int_\Omega g(y_n)\, dx \ge \int_\Omega g(y)\, dx,$$

and it is also clear that

$$\liminf_{n \to \infty} \int_\Omega |\nabla y_n|^2 \, dx \geq \int_\Omega |\nabla y|^2 \, dx.$$

Hence, $\liminf_{n \to \infty} \varphi(y_n) \geq \varphi(y)$.

Let us denote by $\Gamma \subset L^2(\Omega) \times L^2(\Omega)$ the operator defined by the second part of (2.35), i.e.,

$$\Gamma = \{[y, w] \in (H_0^1(\Omega) \cap H^2(\Omega)) \times L^2(\Omega);$$
$$w(x) \in -\Delta y(x) + \partial g(y(x)) \text{ a.e. } x \in \Omega\}.$$

Since the inclusion $\Gamma \subset \partial \varphi$ is obvious, it suffices to show that Γ is maximal monotone in $L^2(\Omega)$. To this end, observe that $\Gamma = A + B$, where $Ay = -\Delta y \ \ \forall y \in D(A) = H_0^1(\Omega) \cap H^2(\Omega)$, and $By = \{v \in L^2(\Omega);\ v(x) \in \partial g(y(x)) \text{ a.e. } x \in \Omega\}$. As seen earlier, the operators A and B are maximal monotone in $L^2(\Omega) \times L^2(\Omega)$. Replacing B by $y \to By - y_0$, where $y_0 \in B0$, we may assume without loss of generality that $0 \in B(0)$. On the other hand, it is readily seen that $(B_\lambda u)(x) = \beta_\lambda(u(x))$ a.e. $x \in \Omega$ for all $u \in L^2(\Omega)$, where $\beta = \partial g$ and $\beta_\lambda = \lambda^{-1}(1 - (1 + \lambda\beta)^{-1})$, is the Yosida approximation of β. We have

$$(Au, B_\lambda u) = -\int_\Omega \Delta u \, \beta_\lambda(u) \, dx = \int_\Omega \beta_\lambda'(u) |\nabla u|^2 \, dx \geq 0$$
$$\forall u \in H_0^1(\Omega) \cap H^2(\Omega),$$

because $\beta_\lambda' \geq 0$. Then applying Theorem 1.6 (or 2.4) we may conclude that $\Gamma = A + B$ is maximal monotone. ∎

Remark 2.2. Since $A + B$ is coercive, it follows from Corollary 1.2 that $R(A + B) = L^2(\Omega)$. Hence, for every $f \in L^2(\Omega)$, the Dirichlet problem

$$\begin{aligned} -\Delta y + \beta(y) &\ni f && \text{a.e. in } \Omega, \\ y &= 0 && \text{in } \partial\Omega, \end{aligned} \tag{2.36}$$

has a unique solution $y \in H_0^1(\Omega) \cap H^2(\Omega)$.

In the special case where $\beta \subset \mathbf{R} \times \mathbf{R}$ is given by

$$\beta(r) = \begin{cases} 0 & \text{if } r > 0, \\ \mathbf{R}^- & \text{if } r = 0, \end{cases}$$

problem (2.36) reduces to the *obstacle problem*

$$\begin{aligned}
-\Delta y &= f \qquad \text{a.e. in } [y > 0],\\
-\Delta y &\geq f, \quad y \geq 0 \qquad \text{a.e. in } \Omega,\\
y &= 0 \qquad \text{in } \partial\Omega.
\end{aligned} \tag{2.36$'$}$$

This is an *elliptic variational inequality*, which will be discussed in some detail later.

We note that the solution y to (2.36) is the limit in $H_0^1(\Omega)$ of the solutions y_ε to the approximating problem

$$\begin{aligned}
-\Delta y + \beta_\varepsilon(y) &= f \qquad \text{in } \Omega,\\
y &= 0 \qquad \text{in } \partial\Omega.
\end{aligned} \tag{2.37}$$

Indeed, multiplying (2.37) by y_ε and Δy_ε, respectively, we get

$$\|y_\varepsilon\|^2_{H_0^1(\Omega)} + \|\Delta y_\varepsilon\|^2_{L^2(\Omega)} \leq C \qquad \forall \varepsilon > 0,$$

and therefore $\{y_\varepsilon\}$ is bounded in $H_0^1(\Omega) \cap H^2(\Omega)$. This yields

$$\int_\Omega |\nabla(y_\varepsilon - y_\lambda)|^2\, dx + \int_\Omega (\beta_\varepsilon(y_\varepsilon) - \beta_\lambda(y_\lambda))(y_\varepsilon - y_\lambda)\, dx = 0,$$

and therefore

$$\int_\Omega |\nabla(y_\varepsilon - y_\lambda)|^2\, dx + \int_\Omega (\beta_\varepsilon(y_\varepsilon) - \beta_\lambda(y_\lambda))(\varepsilon\beta_\varepsilon(y_\varepsilon) - \lambda\beta_\lambda(y_\lambda))\, dx \leq 0,$$

because $\beta_\varepsilon(y) \in \beta((1 + \varepsilon\beta)^{-1}y)$ and β is monotone. Hence, $\{y_\varepsilon\}$ is Cauchy in $H_0^1(\Omega)$, and so $y = \lim_{\varepsilon \to 0} y_\varepsilon$ exists in $H_0^1(\Omega)$. This clearly also implies that

$$\begin{aligned}
\Delta y_\varepsilon &\to \Delta y \qquad \text{weakly in } L^2(\Omega),\\
y_\varepsilon &\to y \qquad \text{weakly in } H^2(\Omega),\\
\beta_\varepsilon(y_\varepsilon) &\to g \qquad \text{weakly in } L^2(\Omega).
\end{aligned}$$

Now, by Proposition 1.3(iv), we see that $g(x) \in \beta(y(x))$ a.e. $x \in \Omega$, and so y is the solution to problem (2.36).

5. *Nonlinear boundary Neumann conditions*. Let Ω be a bounded and open subset of $\mathbf{R}^N$ with the boundary $\partial\Omega$ of class C^2. Let $j: \mathbf{R} \to \overline{\mathbf{R}}$ be a l.s.c., proper, convex function and let $\beta = \partial j$. Define the function

$\varphi: L^2(\Omega) \to \overline{\mathbf{R}}$ by

$$\varphi(u) = \begin{cases} \frac{1}{2}\int_\Omega |\nabla u|^2\,dx + \int_{\partial\Omega} j(u)\,dx & \text{if } u \in H^1(\Omega) \\ & \text{and } j(u) \in L^1(\partial\Omega), \\ +\infty & \text{otherwise.} \end{cases} \tag{2.38}$$

Since for every $u \in H^1(\Omega)$ the trace of u on $\partial\Omega$ is well-defined and belongs to $H^{1/2}(\partial\Omega) \subset L^2(\partial\Omega)$, formula (2.38) makes sense. Moreover, arguing as in the previous example it follows that φ is convex and l.s.c. on $L^2(\Omega)$. Regarding its subdifferential $\partial\varphi \subset L^2(\Omega) \times L^2(\Omega)$, it is completely described in Proposition 2.11, due to H. Brézis [2, 3].

Proposition 2.11. *We have*

$$\partial\varphi(u) = -\Delta u \qquad \forall u \in D(\partial\varphi), \tag{2.39}$$

where

$$D(\partial\varphi) = \left\{u \in H^2(\Omega);\ -\frac{\partial u}{\partial \nu} \in \beta(u) \text{ a.e. in } \partial\Omega\right\}$$

and $\partial/\partial\nu$ is the outward normal derivative to $\partial\Omega$. Moreover, there are some positive constants C_1, C_2 such that

$$\|u\|_{H^2(\Omega)} \le C_1\|u - \Delta u\|_{L^2(\Omega)} + C_2 \qquad \forall u \in D(\partial\varphi). \tag{2.40}$$

Proof. Let $A: L^2(\Omega) \to L^2(\Omega)$ be the operator defined by

$$Au = -\Delta u, \qquad u \in D(A),$$

$$D(A) = \left\{u \in H^2(\Omega);\ -\frac{\partial u}{\partial \nu} \in \beta(u) \text{ a.e. in } \partial\Omega\right\}.$$

Note that A is well-defined since for every $u \in H^2(\Omega)$, $\partial u/\partial\nu \in H^{1/2}(\partial\Omega)$. It is easily seen that $A \subset \partial\varphi$. Indeed, by Green's formula,

$$\begin{aligned}\int_\Omega Au(u - v)\,dx &= \int_\Omega \nabla u(\nabla u - \nabla v)\,dx + \int_{\partial\Omega} \beta(u)(u - v)\,dx \\ &\ge \frac{1}{2}\int_\Omega |\nabla u|^2\,dx + \int_{\partial\Omega} j(u)\,dx \\ &\quad - \frac{1}{2}\int_\Omega |\nabla v|^2\,dx - \int_{\partial\Omega} j(v)\,dx\end{aligned}$$

for all $u \in D(A)$ and $v \in H^1(\Omega)$. Hence,

$$(Au, u - v) \geq \varphi(u) - \varphi(v) \qquad \forall u \in D(A), v \in L^2(\Omega).$$

(Here, $(\cdot, \cdot)$ is the usual scalar product in $L^2(\Omega)$.) Thus, to show that $A = \partial\varphi$, it suffices to prove that A is maximal monotone in $L^2(\Omega) \times L^2(\Omega)$, i.e., $R(I + A) = L^2(\Omega)$. Toward this aim, we fix $f \in L^2(\Omega)$ and consider the equation $u + Au = f$, i.e.,

$$\begin{aligned} u - \Delta u &= f \qquad \text{in } \Omega, \\ \frac{\partial u}{\partial \nu} + \beta(u) &\ni 0 \qquad \text{in } \partial\Omega. \end{aligned} \tag{2.41}$$

We approximate (2.41) by

$$\begin{aligned} u - \Delta u &= f \qquad \text{in } \Omega, \\ \frac{\partial u}{\partial \nu} + \beta_\lambda(u) &= 0 \qquad \text{in } \partial\Omega, \end{aligned} \tag{2.41}'$$

where $\beta_\lambda = \lambda^{-1}(1 - (1 + \lambda\beta)^{-1})$, $\lambda > 0$. Recall that β_λ is Lipschitz with Lipschitz constant $1/\lambda$ and $\beta_\lambda(u) \to \beta^0(u)$ $\forall u \in D(\beta)$, for $\lambda \to 0$.

Let us show first that Eq. (2.41)′ has a unique solution $u_\lambda \in H^2(\Omega)$. Indeed, consider the operator $u \xrightarrow{T} v|_{\partial\Omega}$ from $L^2(\partial\Omega)$ to $L^2(\partial\Omega)$, where $v \in H^1(\Omega)$ is the solution to linear boundary value problem

$$v - \Delta v = f \;\text{ in } \Omega, \qquad v + \lambda \frac{\partial v}{\partial \nu} = (1 + \lambda\beta)^{-1} u \;\text{ in } \partial\Omega. \tag{2.42}$$

(The existence of v is an immediate consequence of the Lax–Millgram lemma.) Moreover, by Green's formula we see that

$$\begin{aligned} &\|v - \bar{v}\|^2_{L^2(\Omega)} + \int_\Omega |\nabla(v - \bar{v})|^2\, dx + \frac{1}{\lambda}\int_{\partial\Omega} (v - \bar{v})^2\, d\sigma_x \\ &\quad \leq \frac{1}{\lambda}\int_{\partial\Omega} \left((1 + \lambda\beta)^{-1}u - (1 + \lambda\beta)^{-1}\bar{u}\right)(v - \bar{v})\, d\sigma_x, \end{aligned}$$

where $\{v, u\}$ and $\{\bar{v}, \bar{u}\}$ satisfy (2.42). Since

$$|(1 + \lambda\beta)^{-1}x - (1 + \lambda\beta)^{-1}y| \leq |x - y| \qquad \forall x, y \in \mathbf{R}, \lambda > 0,$$

we infer that

$$\|v - \bar{v}\|^2_{H^1(\Omega)} + \frac{1}{2\lambda}\|Tu - T\bar{u}\|_{L^2(\partial\Omega)} \leq \frac{1}{2\lambda}\|u - \bar{u}\|^2_{L^2(\partial\Omega)}.$$

Since by the trace theorem the map $v \to v|_{\partial\Omega}$ is continuous from $H^1(\Omega)$ into $H^{1/2}(\partial\Omega) \subset L^2(\partial\Omega)$, we have

$$\|v - \bar{v}\|_{H^1(\Omega)} \geq C\|Tu - T\bar{u}\|_{L^2(\partial\Omega)},$$

and so the map T is a contraction of $L^2(\partial\Omega)$. Applying the Banach fixed point theorem, we therefore conclude that there exists $u \in L^2(\partial\Omega)$ such that $Tu = u$, and so problem (2.41)[1] has a unique solution $u_\lambda \in H^1(\Omega)$. We have

$$\begin{aligned} u_\lambda - \Delta u_\lambda &= f \qquad \text{in } \Omega, \\ \frac{\partial u_\lambda}{\partial \nu} &= -\beta_\lambda(u_\lambda) \qquad \text{in } \partial\Omega. \end{aligned} \tag{2.43}$$

Since $\beta_\lambda(u_\lambda) \in H^1(\Omega)$ (because β_λ is Lipschitz) and so its trace to $\partial\Omega$ belongs to $H^{1/2}(\partial\Omega)$, we conclude by the classical regularity theory for the linear Neumann problem (see, e.g., Lions and Magenes [1], Brézis [7]) that $u_\lambda \in H^2(\Omega)$.

Let us postpone for the time being the proof of the following estimate:

$$\|u_\lambda\|_{H^2(\Omega)} \leq C(1 + \|f\|_{L^2(\Omega)}) \qquad \forall \lambda > 0, \tag{2.44}$$

where C is independent of λ and f.

Now, to obtain existence in problem (2.41), we pass to limit $\lambda \to 0$ in (2.43). Inasmuch as the mapping $u \to (u|_{\partial\Omega}, \partial u/\partial\nu|_{\partial\Omega})$ in continuous from $H^2(\Omega)$ to $H^{3/2}(\partial\Omega) \times H^{1/2}(\partial\Omega)$ and the injection of $H^2(\Omega)$ into $H^1(\Omega) \subset L^2(\Omega)$ is compact, we may assume, selecting a subsequence if necessary, that, for $\lambda \to 0$,

$$\begin{aligned} u_\lambda &\rightharpoonup u \qquad \text{in } H^2(\Omega), \\ u_\lambda &\to u \qquad \text{in } H^1(\Omega), \\ u_\lambda|_{\partial\Omega} &\to u|_{\partial\Omega} \qquad \text{in } H^{1/2}(\partial\Omega) \subset L^2(\partial\Omega), \\ \frac{\partial u_\lambda}{\partial\nu} &\rightharpoonup \frac{\partial u}{\partial\nu} \qquad \text{in } H^{1/2}(\partial\Omega) \subset L^2(\partial\Omega). \end{aligned} \tag{2.45}$$

Moreover, since by (2.44) $\{\beta_\lambda(u_\lambda)\}$ is bounded in $L^2(\partial\Omega)$, we may assume that, for $\lambda \to 0$,

$$\beta_\lambda(u_\lambda) \rightharpoonup g \qquad \text{in } L^2(\partial\Omega). \tag{2.46}$$

It is clear by (2.43), (2.45), and (2.46) that

$$u - \Delta u = f \qquad \text{in } \Omega,$$
$$\frac{\partial u}{\partial \nu} + g \ni 0 \qquad \text{a.e. in } \partial\Omega.$$

Let us show that $g(x) \in \beta(u(x))$ a.e. $x \in \Omega$. Indeed, the operator $\tilde{\beta} \subset L^2(\partial\Omega) \times L^2(\partial\Omega)$ defined by

$$\tilde{\beta} = \{[u, v] \in L^2(\partial\Omega) \times L^2(\partial\Omega); v(x) \in \beta(u(x)) \text{ a.e. } x \in \partial\Omega\}$$

is obviously maximal monotone, and

$$\tilde{\beta}_\lambda(u)(x) = \beta_\lambda(u(x)), \qquad \left((I + \lambda\tilde{\beta})^{-1} u\right)(x) = (1 + \lambda\beta)^{-1} u(x)$$
$$\text{a.e. } x \in \Omega.$$

Since, by (2.46), $\tilde{\beta}_\lambda(u_\lambda) \rightharpoonup g$, $(I + \lambda\tilde{\beta})^{-1} u_\lambda \to u$, and $\tilde{\beta}_\lambda(u_\lambda) \in \tilde{\beta}((I + \lambda\tilde{\beta})^{-1} u_\lambda)$, we conclude that $g \in \tilde{\beta}(u)$ (because $\tilde{\beta}$ is strongly–weakly closed). We have therefore proved that u is a solution to Eq. (2.41), and since f is arbitrary in $L^2(\Omega)$ we infer that $A = \partial\varphi$. Finally, letting λ tend to zero in the estimate (2.44), we obtain (2.40), as claimed. ∎

Proof of estimate (2.44). Multiplying Eq. (2.43) by $u_\lambda - u_0$, where $u_0 \in D(\beta)$ is a constant, we get after some calculation involving Green's lemma that

$$\int_\Omega \left(u_\lambda^2 + |\nabla u_\lambda|^2\right) dx \le C\left(\int_\Omega f^2\, dx + 1\right).$$

(We shall denote by C several positive constants independent of λ and f.) Hence,

$$\|u_\lambda\|_{H^1(\Omega)} \le C(\|f\|_{L^2(\Omega)} + 1) \qquad \forall \lambda > 0. \tag{2.47}$$

If Ω' is an open subset of Ω such that $\overline{\Omega'} \subset \Omega$, then we choose $\rho \in C_0^\infty(\Omega)$ such that $\rho = 1$ in $\overline{\Omega'}$. We set $v = \rho u_\lambda$ and note that

$$v - \Delta v = \rho f - u_\lambda \Delta\rho - 2\nabla\rho \cdot \nabla u_\lambda \qquad \text{in } \Omega. \tag{2.48}$$

Since v has compact support in Ω we may assume that $v \in H^2(\mathbf{R}^N)$, and Eq. (2.48) extends to all of $\mathbf{R}^N$. Then, taking the Fourier transform and using Parseval's formula, we get

$$\|v\|_{H^2(\mathbf{R}^N)} \le C(\|f\|_{L^2(\Omega)} + \|u_\lambda\|_{H^1(\Omega)}),$$

and therefore, by (2.47),

$$\|u_\lambda\|_{H^2(\Omega')} \leq C(\|f\|_{L^2(\Omega)} + 1) \qquad \forall\lambda > 0, \tag{2.49}$$

where C is independent of $\Omega' \subset\subset \Omega$.

To obtain H^2-estimates near the boundary $\partial\Omega$, let $x_0 \in \partial\Omega$, U be a neighborhood of x_0, and $\varphi: U \to Q$ be such that $\varphi \in C^2(U)$, $\varphi^{-1} \in C^2(Q)$, $\varphi^{-1}(Q_+) = \Omega \cap U$, and $\varphi^{-1}(Q_0) = \partial\Omega \cap U$, where $Q = \{y \in \mathbf{R}^N; \|y'\| < 1, |y_n| < 1\}$, $Q_+ = \{y \in Q; 0 < y_n < 1\}$, $Q_0 = \{y \in Q; y_n = 0\}$, and $y = (y', y_N) \in \mathbf{R}^N$. (Since $\partial\Omega$ is of class C^2 such a pair (U, φ) always exists.) Now we will "transport" Eq. (2.48) from $U \cap \Omega$ on Q_+ using the local coordinate φ. We set

$$w(y) = u_\lambda(\psi(y)) \qquad \forall y \in Q_+, \qquad \psi = \varphi^{-1},$$

and observe that w satisfies on Q_+ the boundary value problem

$$\begin{aligned} w - \sum_{k,j=1}^{N} \frac{\partial}{\partial y_j}\left(a_{kj}(y)\frac{\partial w}{\partial y_k}\right) + \sum_{j=1}^{N} b_j(y)\frac{\partial w}{\partial y_j} & \\ + C(y)w = g(y) \qquad \text{in } Q_+, & \\ \frac{\partial w}{\partial n} + \beta_\lambda(w) = 0 \qquad \text{in } Q_0, & \end{aligned} \tag{2.50}$$

where $g(y) = f(\psi(y))$,

$$a_{kj}(y) = \sum_{1=1}^{N} \frac{\partial\varphi_k}{\partial x_1}\frac{\partial\varphi_j}{\partial x_1} \qquad \forall y \in Q_+, \qquad \varphi = (\varphi_1, \ldots, \varphi_N),$$

and

$$\frac{\partial w}{\partial n} = \sum_{i,j=1}^{N} \frac{\partial w}{\partial y_j}\frac{\partial\varphi_j}{\partial x_i}\cos(\nu, x_i)$$

(ν is the outward normal derivative to $\partial\Omega$). Since $\varphi_N(x) = 0$ is the equation of the surface $\partial\Omega \cap U$ we may assume that $\partial\varphi_N/\partial x_j = -\cos(\nu, x_j)$, and so

$$\frac{\partial w}{\partial n} = -\sum_{j=1}^{N} \frac{\partial w}{\partial y_j} a_{jN} \qquad \text{in } Q_0.$$

Assuming for a while that $f \in C^1(\overline{\Omega})$, we see that $x = \partial w/\partial y_i$, $1 \le i \le N - 1$, satisfies the equation

$$z - \sum_{k,j=1}^{N} \frac{\partial}{\partial y_j}\left(a_{kj}\frac{\partial z}{\partial y_k}\right) + \sum_{j=1}^{N}\left(\tilde{b}_j \frac{\partial z}{\partial y_j} + \tilde{c}_j \frac{\partial w}{\partial y_j}\right) + C(y)z + C'(y)w$$
$$= \frac{\partial}{\partial y_i} g(y) \qquad \text{in } Q_+,$$
$$\frac{\partial z}{\partial n} = -\beta_\lambda'(u_\lambda)z + \sum_{j=1}^{N} \frac{\partial w}{\partial y_j}\frac{\partial a_{jN}}{\partial y_i} \qquad \text{in } Q_0. \tag{2.51}$$

Now, let $\varphi \in C_0^\infty(\overline{Q}_+)$ be such that $\rho(y) = 0$ for $\|y'\| \ge 2/3$, $2/3 < y_n < 1$, and $\rho(y) = 1$ for $\|y'\| < 1/2$ and $0 \le y_N \le 1/2$. Multiplying Eq. (2.51) by $\rho^2 z$ and integrating on Q_+, we get

$$\int_{Q_+} \rho^2 z^2\,dy + \sum_{k,j=1}^{N} \int_{Q_+} a_{kj}(y)\frac{\partial z}{\partial y_k}\frac{\partial}{\partial y_j}(\rho^2 z)\,dy + \int_{Q_0} \rho^2 \beta_\lambda'(u_\lambda) z^2\,dy$$
$$= \sum_{j=1}^{N} \int_{Q_0} \rho^2 \frac{\partial w}{\partial y_j}\frac{\partial a_{jN}}{\partial y_i}\,dy + \int_{Q_+} \frac{\partial}{\partial y_i} g(y) z(y)\,dy$$
$$- \sum_{j=1}^{N} \int_{Q_+} \rho^2\left(\tilde{b}_j \frac{\partial z}{\partial y_j} + \tilde{c}_j \frac{\partial w}{\partial y_j}\right) z\,dy$$
$$- \int_{Q_+} (Cz + C'w) z\rho^2\,dy.$$

Taking into account that

$$\sum_{k,j=1}^{N} a_{kj}(y)\xi_k \xi_j \ge \omega\|\xi\|^2 \qquad \forall y \in Q_+,\ \xi \in \mathbf{R}^N,$$

we find after some calculation that

$$\sum_{j=1}^{N} \int_{Q_+} \rho^2(y)\left(\frac{\partial z}{\partial y_j}\right)^2 dy \le C\left(\|g\|^2_{L^2(Q_+)} + \|w\|^2_{H^1(Q_+)} + 1\right).$$

Hence,

$$\left\|\rho \frac{\partial^2 w}{\partial y_i\,\partial y_j}\right\|_{L^2(Q_+)} \le C(\|f\|_{L^2(\Omega)} + \|u_\lambda\|_{H^1(\Omega)} + 1)$$
$$\text{for } i = 1, 2, \ldots, N-1,\ j = 1, \ldots, N.$$

Since $a_{NN}(y) \geq w_0 > 0$ for all $y \in Q_+$, we see by Eq. (2.50) and the last estimate that

$$\left\| \frac{\partial^2 w}{\partial y_N^2} \right\|_{L^2(Q_+)} \leq C(\|f\|_{L^2(\Omega)} + \|u_\lambda\|_{H^1(\Omega)} + 1).$$

Hence,

$$\| \rho w\|_{H^2(Q_+)} \leq C(\|f\|_{L^2(\Omega)} + 1).$$

Equivalently,

$$\|(\rho \cdot \varphi)u_\lambda\|_{H^2(U \cap \Omega)} \leq C(\|f\|_{L^2(\Omega)} + 1) \qquad \forall \lambda > 0.$$

Hence, there is a neighborhood $U' \subset U$ such that

$$\|u_\lambda\|_{H^2(U' \cap \Omega)} \leq C(\|f\|_{L^2(\Omega)} + 1) \qquad \forall \lambda > 0. \tag{2.52}$$

Now taking a finite partition of unity subordinated to a such a cover $\{U'\}$ of $\partial\Omega$ and using the local estimates (2.49) and (2.52), we get (2.44). This completes the proof of Proposition 2.11.

We have incidentally proved that for every $f \in L^2(\Omega)$, the boundary value problem (2.41) has a unique solution $u \in H^2(\Omega)$. If $\beta \subset \mathbf{R} \times \mathbf{R}$ is the graph

$$\beta(0) = \mathbf{R}^N \qquad \beta(r) \neq \varnothing \quad \text{for } r \neq 0,$$

then (2.41) reduces to the classical Dirichlet problem. If

$$\beta(r) = \begin{cases} 0 & \text{if } r > 0, \\ [-\infty, 0] & \text{if } r = 0, \end{cases} \tag{2.53}$$

then problem (2.41) can be equivalently written as

$$\begin{aligned} &y - \Delta y = f \qquad \text{in } \Omega, \\ &y \frac{\partial y}{\partial \nu} = 0, \quad y \geq 0, \quad \frac{\partial y}{\partial \nu} \geq 0 \qquad \text{in } \partial\Omega. \end{aligned} \tag{2.54}$$

This is the celebrated *Signorini's problem*, which arises in elasticity in connection with the mathematical description of friction problems. This is a problem of unilateral type and the subset Γ_0 that separates $\{x \in \partial\Omega;\ y > 0\}$ from $\{x \in \partial\Omega;\ \partial y/\partial \nu > 0\}$ is a *free boundary* and it is one of the unknowns of the problem.

For other unilateral problems of physical significance that can be written in the form (2.41), we refer to the book of Duvaut and Lions [1].

Remark 2.3. Proposition 2.11 and its corollaries remain valid if Ω *is an open, bounded, and convex subset of* $\mathbf{R}^N$. The idea, developed by P. Grisvard in his book [1], is to approximate such a domain Ω by smooth domain Ω_ε, to use the estimate (2.44) (which is valid on every Ω_ε with a constant C independent of ε), and to pass to limit. It is useful to note that the constant C in estimate (2.44) is independent of β.

6. *The nonlinear diffusion operator.* Let Ω be a bounded and open subset of $\mathbf{R}^N$ with a sufficiently smooth boundary $\partial\Omega$. Denote as usual by $H_0^1(\Omega)$ the Sobolev space of all $u \in H^1(\Omega)$ having null trace on $\partial\Omega$ and by $H^{-1}(\Omega)$ the dual of $H_0^1(\Omega)$. Note that $H^{-1}(\Omega)$ is a Hilbert space with the scalar product

$$\langle u, v \rangle = (J^{-1}u, v) \qquad \forall u, v \in H^{-1}(\Omega),$$

where $J = -\Delta$ is the canonical isomorphism (duality mapping) of $H_0^1(\Omega)$ onto $H^{-1}(\Omega)$ and $(\cdot,\cdot)$ is the pairing between $H_0^1(\Omega)$ and $H^{-1}(\Omega)$.

Let $j\colon \mathbf{R} \to \overline{\mathbf{R}}$ be a l.s.c., convex, proper function and let $\beta = \partial j$. Define the function $\varphi\colon H^{-1}(\Omega) \to \overline{\mathbf{R}}$ by

$$\varphi(u) = \begin{cases} \int_\Omega j(u(x))\,dx & \text{if } u \in L^1(\Omega) \text{ and } j(u) \in L^1(\Omega), \\ +\infty & \text{otherwise}. \end{cases} \tag{2.55}$$

Proposition 2.12. *Let us assume further that*

$$\lim_{|r|\to\infty} j(r)/|r| = +\infty. \tag{2.56}$$

Then the function φ *is convex and lower semicontinuous on* $H^{-1}(\Omega)$. *Moreover,*

$$\begin{aligned} \partial\varphi = \big\{[u,w] \in (H^{-1}(\Omega) \cap L^1(\Omega)) \times H^{-1}(\Omega);\, w = -\Delta v, v \in H_0^1(\Omega), \\ v(x) \in \beta(u(x)) \text{ a.e. } x \in \Omega\big\}. \end{aligned} \tag{2.57}$$

Proof. Obviously, φ is convex. To prove that φ l.s.c., consider a sequence $\{u_n\} \subset H^{-1}(\Omega) \cap L^1(\Omega)$ such that $u_n \to u$ in $H^{-1}(\Omega)$ and $\varphi(u_n) \le \lambda$, i.e., $\int_\Omega j(u_n)\,dx \le \lambda\ \forall n$. We must prove that $\int_\Omega j(u)\,dx \le \lambda$. We have already seen in the proof of Proposition 2.9 that the function $u \to \int_\Omega j(u)\,dx$ is lower semicontinuous on $L^1(\Omega)$. Since this function is convex, it is weakly lower semicontinuous in $L^1(\Omega)$ and so it suffices to show that $\{u_n\}$ is weakly compact in $L^1(\Omega)$. According to Dunford–Pettis criterion (see e.g.,

Edwards [1], p. 270), we must prove that $\{u_n\}$ is bounded in $L^1(\Omega)$ and the integrals $\int |u_n|\,dx$ are uniformly absolutely continuous, i.e., for every $\varepsilon > 0$ there is $\delta(\varepsilon)$ such that $\int_E |u_n(x)|\,dx \le \varepsilon$ if $m(E) \le \delta(\varepsilon)$ (E is a measurable set of Ω). By condition (2.56), for every $p > 0$ there exists $R(p) > 0$ such that $j(r) \ge p|r|$ if $|r| \ge R(p)$. This clearly implies that $\int_\Omega |u_n(x)|\,dx \le C$.

Moreover, for every measurable subset E of Ω, we have

$$\int_E |u_n(x)|\,dx \le \int_{E\cap\{|u_n|\ge R(p)\}} |u_n(x)|\,dx + \int_{E\cap\{|u_n|<R(p)\}} |u_n(x)|\,dx \le \frac{1}{p}\int_\Omega |u_n(x)|\,dx + R(p)m(E) \le \varepsilon,$$

if we choose $p > (2\varepsilon)^{-1}\sup \int_\Omega |u_n(x)|\,dx$ and $m(E) \le \varepsilon/2R(p)$. Hence, $\{u_n\}$ is weakly compact in $L^1(\Omega)$.

To prove (2.57), consider the operator $A \subset H^{-1}(\Omega) \times H^{-1}(\Omega)$ defined by

$$Au = \{-\Delta v;\ v \in H_0^1(\Omega),\ v(x) \in \beta(u(x)) \text{ a.e. } x \in \Omega\},$$

where $D(A) = \{u \in H^{-1}(\Omega) \cap L^1(\Omega);\ \exists v \in H_0^1(\Omega),\ v(x) \in \beta(u(x))$ a.e. $x \in \Omega\}$. To prove that $A = \partial\varphi$, we will show separately that $A \subset \partial\varphi$ and that A is maximal monotone. Let us show first that A is maximal monotone in $H^{-1}(\Omega) \times H^{-1}(\Omega)$. Let f be arbitrary but fixed in $H^{-1}(\Omega)$. We must show that there exist $u \in H^{-1}(\Omega) \cap L^1(\Omega)$ and $v \in H_0^1(\Omega)$ such that

$$u - \Delta v = f \quad \text{in } \Omega, \qquad v(x) \in \gamma(u(x)) \quad \text{a.e. } x \in \Omega;$$

equivalently,

$$u - \Delta v = f \quad \text{in } \Omega, \qquad u(x) \in \beta(v(x)) \quad \text{a.e. } x \in \Omega,\ u \in H^{-1}(\Omega) \cap L^1(\Omega), \qquad v \in H_0^1(\Omega), \tag{2.58}$$

where $\gamma = \beta^{-1}$.

Consider the approximate equation

$$\gamma_\lambda(v) - \Delta v = f \quad \text{in } \Omega, \qquad v = 0 \quad \text{in } \partial\Omega, \tag{2.59}$$

where $\gamma_\lambda = \lambda^{-1}(1 - (1 + \lambda\gamma)^{-1})$, $\lambda > 0$. It is readily seen that (2.59) has a unique solution $v_\lambda \in H_0^1(\Omega)$. Indeed, since $-\Delta$ is maximal monotone from $H_0^1(\Omega)$ to $H^{-1}(\Omega)$ and $v \to \gamma_\lambda(v)$ is monotone and continuous from

$H_0^1(\Omega)$ to $H^{-1}(\Omega)$ (in fact, from $L^2(\Omega)$ to itself) we infer by Corollary 1.1 that $v \to \gamma_\lambda(v) - \Delta v$ is maximal monotone in $H_0^1(\Omega) \times H^{-1}(\Omega)$, and by Corollary 1.2 that it is surjective. Let $v_0 \in D(\gamma)$. Multiplying Eq. (2.59) by $v_\lambda - v_0$, we get

$$\int_\Omega |\nabla v_\lambda|^2\, dx + \int_\Omega \gamma(v_0)(v_\lambda - v_0)\, dx \le (v_\lambda - v_0, f).$$

Hence, $\{v_\lambda\}$ is bounded in $H_0^1(\Omega)$. Then, on a subsequence, again denoted λ, we have

$$v_\lambda \rightharpoonup v \quad \text{in } H_0^1(\Omega), \qquad v_\lambda \to v \quad \text{in } L^2(\Omega).$$

Thus, extracting further subsequences, we may assume that

$$\begin{aligned} v_\lambda(x) &\to v(x) \qquad \text{a.e. } x \in \Omega, \\ (1 + \lambda\gamma)^{-1} v_\lambda(x) &\to v(x) \qquad \text{a.e. } x \in \Omega, \end{aligned} \tag{2.60}$$

because by condition (2.56) it follows that $D(\gamma) = R(\beta) = \mathbf{R}$ (β is coercive) and so $\lim_{\lambda \to 0}(1 + \lambda\gamma)^{-1} r = r$ for all $r \in \mathbf{R}$ (Proposition 1.3).

We set $g_\lambda = \gamma_\lambda(v_\lambda)$. Then, letting λ tend to zero in (2.59), we see that $g_\lambda \to u$ in $H^{-1}(\Omega)$ and

$$u - \Delta v = f \quad \text{in } \Omega, \qquad v \in H_0^1(\Omega).$$

It remains to be shown that $u \in L^1(\Omega)$ and $u(x) \in \gamma(v(x))$ a.e. $x \in \Omega$.

Multiplying Eq. (2.59) by v_λ, we see that

$$\int_\Omega g_\lambda v_\lambda\, dx \le C \qquad \forall \lambda > 0.$$

On the other hand, for some $u_0 \in D(j)$ we have $j(g_\lambda(x)) \le j(u_0) + (g_\lambda(x) - u_0)v$ $\forall v \in \beta(g_\lambda(x))$. This yields

$$\int_\Omega j(g_\lambda(x))\, dx \le C \qquad \forall \lambda > 0,$$

because $(1 + \lambda\gamma)^{-1} v_\lambda \in \beta(g_\lambda)$.

As seen before, this implies that $\{g_\lambda\}$ is weakly compact in $L^1(\Omega)$. Hence, $u \in L^1(\Omega)$ and

$$g_\lambda \rightharpoonup u \quad \text{in } L^1(\Omega) \qquad \text{for } \lambda \to 0. \tag{2.61}$$

On the other hand, by (2.60) it follows by virtue of the Egorov theorem that for every $\varepsilon > 0$ there exists a measurable subset $E_\varepsilon \subset \Omega$ such that

$m(\Omega \setminus E_\varepsilon) \le \varepsilon$, $\{(1+\lambda\gamma)^{-1}v_\lambda\}$ is bounded in $L^\infty(E_\varepsilon)$, and

$$(1+\lambda\gamma)^{-1}v_\lambda \to v \quad \text{uniformly in } E_\varepsilon \qquad \text{as } \lambda \to 0. \tag{2.62}$$

Recalling that $g_\lambda(x) \in \gamma((1+\lambda\gamma)^{-1}v_\lambda(x))$ and that the operator $\tilde{\gamma} = \{[u, v] \in L^1(E_\varepsilon) \times L^\infty(E_\varepsilon);\ v(x) \in \gamma(u(x)) \text{ a.e. } x \in E_\varepsilon\}$ is maximal monotone in $L^1(E_\varepsilon) \times L^\infty(E_\varepsilon)$, we infer, by (2.61) and (2.62), that $[u, v] \in \tilde{\gamma}$, i.e., $v(x) \in \gamma(u(x))$ a.e. $x \in E_\varepsilon$. Since ε is arbitrary, we infer that $v(x) \in \gamma(u(x))$ a.e. $x \in \Omega$, as desired.

To prove that $A \subset \partial\varphi$, we shall use the following lemma, which is a special case to a general result due to Brézis and Browder [1].

Lemma 2.2. *Let Ω be an open subset of $\mathbf{R}^N$. If $w \in H^{-1}(\Omega) \cap L^1(\Omega)$ and $u \in H_0^1(\Omega)$ are such that*

$$w(x)u(x) \ge -|h(x)| \qquad \text{a.e. } x \in \Omega, \tag{2.63}$$

for some $h \in L^1(\Omega)$, then $wu \in L^1(\Omega)$ and

$$w(u) = \int_\Omega w(x)u(x)\,dx. \tag{2.64}$$

(Here, $w(u)$ is the value of functional $w \in H^{-1}(\Omega)$ at $u \in H_0^1(\Omega)$.)

Proof. The proof relies on an approximation result for the functions of $H_0^1(\Omega)$ due to Hedberg [1].

Let $u \in H_0^1(\Omega)$. Then there exists a sequence $\{u_n\} \subset H_0^1(\Omega) \cap L^\infty(\Omega)$ such that $\operatorname{supp} u_n$ is a compact subset of Ω, $u_n \to u$ in $H_0^1(\Omega)$, and

$$|u_n(x)| \le \inf(n, |u(x)|), \qquad u_n(x)u(x) \ge 0 \qquad \text{a.e. } x \in \Omega. \tag{2.65}$$

Then,

$$w(u_n) = \int_\Omega w(x)u_n(x)\,dx \qquad \forall n. \tag{2.66}$$

On the other hand, by (2.63) we have

$$wu_n + |h|\frac{u_n}{u} = (wu + |h|)\frac{u_n}{u} \ge 0 \qquad \text{a.e. in } \Omega,$$

and so by the Fatou lemma $wu + |h| \in L^1(\Omega)$ and

$$\liminf_{n\to\infty} \int_\Omega \left(wu_n + |h|\frac{u_n}{u}\right) dx \ge \int_\Omega (wu + |h|)\,dx$$

because, on a subsequence, $u_n(x) \to u(x)$ a.e. $x \in \Omega$.

We have therefore proved that $wu \in L^1(\Omega)$ and

$$\liminf_{n \to \infty} \int_\Omega wu_n \, dx \geq \int_\Omega wu \, dx.$$

On the other hand, $wu_n \to wu$ a.e. in Ω and, by (2.65), $|wu_n| \leq |wu|$ a.e. in Ω. Then, by the Lebesgue dominated convergence theorem, $wu_n \to wu$ in $L^1(\Omega)$, and letting $n \to \infty$ in (2.66) we get (2.64), as desired. ■

Now, to conclude the proof of Proposition 2.12, consider an arbitrary element $[u, -\Delta v] \in A$, i.e., $u \in H^{-1}(\Omega) \cap L^1(\Omega)$, $v \in H_0^1(\Omega)$, $v(x) \in \beta(u(x))$ a.e. $x \in \Omega$. We have

$$\langle Au, u - \bar{u} \rangle = (v, u - \bar{u}) \qquad \forall \bar{u} \in H^{-1}(\Omega) \cap L^1(\Omega).$$

Since $v(x)(u(x) - \bar{u}(x)) \geq j(u(x)) - j(\bar{u}(x))$ a.e. $x \in \Omega$, it follows by Lemma 2.2 that

$$\langle Au, u - \bar{u} \rangle = (v, u - \bar{u}) = \int_\Omega v(x)(u(x) - \bar{u}(x)) \, dx$$

$$\geq \int_\Omega j(u(x)) \, dx - \int_\Omega j(u(x)) \, dx \qquad \forall \bar{u} \in D(\varphi).$$

Hence,

$$\langle Au, u - \bar{u} \rangle \geq \varphi(u) - \varphi(\bar{u}) \qquad \forall \bar{u} \in H^{-1}(\Omega),$$

thereby completing the proof. ■

Remark 2.4. Condition (2.56) is equivalent to $R(\beta) = \mathbf{R}$ and β^{-1} is bounded. Indeed, by the definition of $\partial j = \beta$, we have

$$j(r) \leq j(r_0) + y(r - r_0) \qquad \forall r \in \mathbf{R},$$

and (2.56) implies that β is coercive and so β^{-1} is everywhere defined and bounded on bounded subsets. Conversely, if β^{-1} is bounded on bounded subsets, then for every $z \in \mathbf{R}$, $|z| \leq \rho$ there is $v \in D(\beta)$ and $M > 0$ such that $[v, z] \in \beta$ and $|v| < M$. Then, by the inequality

$$j(r) - j(v) \geq z(r - v) \qquad \forall r \in \mathbf{R},$$

we get $\rho|r| \leq j(r) + M_\rho \ \forall r \in \mathbf{R}$. Hence $\lim_{|\mathrm{r}| \to \infty} \mathrm{j(r)}/|\mathrm{r}| = \infty$, as claimed. It should be noted that this result remains true for a general l.s.c. convex function j on a Banach space X.

2.2.3. Generalized Gradients

Here we shall present the Clarke generalized gradient along with some of its most important properties. Throughout this subsection, X is a real Banach space with the norm denoted $\|\cdot\|$, the dual X^*, and the duality pairing $(\cdot, \cdot)$.

A function $f: X \to \mathbf{R}$ is called *locally Lipschitz* if for every $\rho > 0$ there exists $L_\rho > 0$ such that

$$|f(x) - f(y)| \leq L_\rho \|x - y\| \qquad \forall \|x\|, \|y\| \leq \rho.$$

Given a locally Lipschitz function $f: X \to \mathbf{R}$, the function $f^0: X \times X \to \mathbf{R}$ defined by

$$f^0(x, v) = \limsup_{\substack{y \to x \\ \lambda \downarrow 0}} \frac{f(y + \lambda v) - f(y)}{\lambda} \qquad \forall x, v \in X, \tag{2.67}$$

is called the *directional derivative of f*. It is easily seen that f^0 is finite, positively homogeneous, and subadditive in v, i.e.,

$$f^0(x, \lambda v) = \lambda f^0(x, v), \qquad f^0(x, v_1 + v_2) \leq f^0(x, v_1) + f^0(x, v_2).$$

Then, by the Hahn–Banach theorem, there exists a linear functional on X, denoted η, such that

$$\eta(v) \leq f^0(x, v) \qquad \forall v \in X. \tag{2.68}$$

Since $|f^0(x, v)| \leq L\|v\|$ $\forall v \in X$, it follows that η is continuous on X. Hence, $\eta \in X^*$ and

$$(v, \eta) \leq f^0(x, v) \qquad \forall v \in X. \tag{2.69}$$

By definition, *the generalized gradient* (Clarke's gradient) of f at x, denoted $\partial f(x)$ is the set of all $\eta \in X^*$ satisfying (2.69). In other words,

$$\partial f(x) = \{\eta \in X^*; (v, \eta) \leq f^0(x, v)\ \forall v \in X\}. \tag{2.70}$$

Clearly, this definition extends to all functions f that are Lipschitz in a neighborhood of x. Note that if x is a local minimum point or maximum point of f, then $0 \in \partial f(x)$.

Let us observe that if f is locally Lipschitz and Gâteaux differentiable at x, then $\nabla f(x) \in \partial f(x)$. Indeed, we have

$$(v, \nabla f(x)) = f'(x, v) \leq f^0(x, v) \qquad \forall v \in X.$$

If f is continuously differentiable in a neighborhood of x, then $\nabla f(x) = \partial f(x)$ because in this case $f'(x,v) = f^0(x,v)\ \forall v \in X$.

On the other hand, if f is convex (and locally Lipschitz), then ∂f reduces to the subdifferential of f defined in Section 2.2 (see (2.9)). Indeed we have, by (2.67),

$$f^0(x,v) = \lim_{\varepsilon \downarrow 0} \sup_{\|y-x\| \le \varepsilon\delta} \sup_{0<\lambda<\varepsilon} \lambda^{-1}(f(y+\lambda v) - f(y)),$$

where δ is positive and arbitrary. Since the function $\lambda \to \lambda^{-1}(f(y+\lambda v) - f(y))$ is decreasing (because f is convex), we infer that

$$\begin{aligned} f^0(x,v) &= \lim_{\varepsilon \downarrow 0} \sup_{\|y-x\| \le \delta\varepsilon} \varepsilon^{-1}(f(y+\varepsilon v) - f(y)) \\ &\le \lim_{\varepsilon \downarrow 0} \varepsilon^{-1}(f(x+\varepsilon v) - f(x)) + 2L\delta. \end{aligned}$$

Hence $f^0(x,v) \le f'(x,v)$. Since the opposite inequality is obvious, we have that $f^0(x,v) = f'(x,v)$, and by Proposition 2.8 we conclude that $\partial f(x)$ coincides with the subdifferential of f at x.

Proposition 2.13. *For every $x \in X$, $\partial f(x)$ is convex and weak star compact in X^*. Moreover,*

$$f^0(x,v) = \max\{(v,\eta);\ \eta \in \partial f(x)\} \qquad \forall x, v \in X, \tag{2.71}$$

and the map $\partial f\colon X \to 2^{X^*}$ is weak star upper semicontinuous, i.e., if $x_n \to x$ and $y_n \in \partial f(x_n)$ is weak star convergent to y in X^*, then $y \in \partial f(x)$.

Proof. By definition it is obvious that $\partial f(x)$ is convex, closed, and bounded in X^*. Hence, by the Alaoglu theorem, it is weak star compact. To prove (2.71), we will assume that $f^0(x,v_0) > \max\{(v_0,\eta);\ \eta \in \partial f(x)\}$ for some $v_0 \in X$ and will argue from this to a contradiction. Indeed, by the Hahn–Banach theorem there is $\zeta \in X^*$ such that

$$(v,\zeta) \le f^0(x,v) \quad \forall v \in X, \qquad \text{and} \qquad (v_0,\zeta) = f^0(x,v_0).$$

Hence, $\zeta \in \partial f(x)$ and $f^0(x,v_0) > (v_0,\zeta)$, which leads to a contradiction. We shall prove now that f^0 is upper semicontinuous. Let $x_n \to x$ and $v_n \to v$. For every n, there exists $h_n \in X$ and $\lambda_n \in (0,1)$ such that $\|h_n\| + \lambda_n \le 1/n$ and

$$f^0(x_n,v_n) \le (f(x_n + h_n + \lambda_n v_n) - f(x_n + h_n))\lambda_n^{-1} + \frac{1}{n}.$$

This yields

$$\limsup_{n\to\infty} f^0(x_n, v_n) \le f^0(x, v),$$

as claimed. Now, let $x_n \to x$, $\eta_n \rightharpoonup \eta$ be such that $\eta_n \in \partial f(x_n)$. We have

$$(v, \eta_n) \le f^0(x_n, v) \qquad \forall v \in X,$$

and letting $n \to \infty$ we get $(v, \eta) \le f^0(x, v)\ \forall v \in X$. Hence, $\eta \in \partial f(x)$. The proof is complete. ■

In the case $X = \mathbf{R}^N$, $\partial f(x)$ can be equivalently defined as the convex hull of all limit points of $\nabla f(y)$ for $y \to x$. The function f being locally Lipschitz it is almost everywhere differentiable (Rademacher's theorem). Let us denote by Γ_f the set of all points $x \in \mathbf{R}^N$ where f is not differentiable. Denote by m the Lebesgue measure in $\mathbf{R}^N$.

Theorem 2.6. *Let x be arbitrary in $\mathbf{R}^N$ and let $S \subset \mathbf{R}^N$ be such that $m(S) = 0$. Then if f is locally Lipschitz in a neighborhood of x, we have*

$$\partial f(x) = \overline{\text{conv}}\{\lim \nabla f(x_n);\ x_n \to x, x_n \notin S \cup \Gamma_f\}. \tag{2.72}$$

Proof. Since $\nabla f(x_n) \in \partial f(x_n)$, it follows by Proposition 2.13 that

$$\lim\{\nabla f(x_n);\ x_n \to x, x_n \notin S \cup \Gamma_f\} \subset \partial f(x),$$

and since $\partial f(x)$ is convex, we see that

$$E = \text{conv}\{\lim \nabla f(x_n);\ x_n \to x,\ x_n \notin S \cup \Gamma_f\} \subset \partial f(x).$$

Let us denote by H the support function of E, i.e., $H(p) = \sup\{(p, u);\ u \in E\}$. To prove that $\partial f(x) \subset E$, it suffices, according to Proposition 2.13, to show that

$$H(v) \ge f^0(x, v) \qquad \forall v \in X.$$

Let $v \neq 0$ and $\varepsilon > 0$. We will prove the inequality

$$f^0(x, v) - \varepsilon \le \limsup\{(\nabla f(y), v);\ y \to x,\ y \notin S \cup \Gamma_f\}. \tag{2.73}$$

We denote by μ the right hand side of (2.73). There is $\delta > 0$ such that $(\nabla f(y), v) \le \mu + \varepsilon$ for all $y \notin S \cup \Gamma_f$, $\|y - x\| \le \delta$. We have

$$f(y + tv) - f(y) = \int_0^t (\nabla f(y + sv), v)\, ds,$$

and since by the Fubini theorem the Lebesgue measure of the set $\{y + tv;\ 0 < t < \delta/2\|v\|\} \cap (S \cup \Gamma_f)$ is zero, we infer that $f(y + tv) - f(y) \le (\mu + \varepsilon)t$ for $0 < t < \delta/2\|v\|$ and $\|x - y\| \le \delta$. This implies (2.73), as desired. ∎

Let $g: \mathbf{R} \to \mathbf{R}$ be a measurable and locally bounded function. Then, its indefinite integral

$$f(y) = \int_a^y g(x)\,dx, \qquad y \in \mathbf{R},$$

is locally Lipschitz on $\mathbf{R}$, and by Proposition 2.13 it follows that

$$\partial f(y) = \bigcap_{\delta > 0} \bigcap_{m(N) = 0} \overline{\operatorname{conv}}\, g(B_\delta(y) \setminus N) \qquad \forall y \in \mathbf{R}, \tag{2.74}$$

where $B_\delta(y) = \{x \in \mathbf{R};\ |x - y| < \delta\}$ and m is the Lebesgue measure.

The multivalued map f arising in the right hand side of (2.74) is well-known from the theory of Filipov solutions for differential equations with discontinuous right hand side. It can be equivalently represented as

$$\tilde{f}(y) = [\overline{m}(g(y)), \overline{M}(g(y))] \qquad \forall y \in \mathbf{R}, \tag{2.75}$$

where

$$\overline{m}(g(y)) = \lim_{\delta \to 0} \operatorname*{ess\,inf}_{u \in [y - \delta, y + \delta]} g(u),$$

$$\overline{M}(g(y)) = \lim_{\delta \to 0} \operatorname*{ess\,sup}_{u \in [y - \delta, y + \delta]} g(u).$$

If C is a closed subset of X, denote by $d_C(x)$ the distance from x to C, i.e.,

$$d_C(x) = \inf\{\|x - u\|;\ u \in C\}.$$

Obviously, d_C is Lipschitz with Lipschitz constant 1 (i.e., d_C is nonexpansive).

By definition, the vector $v \in X$ is said to be tangent at C in x if $d_C^0(x, v) = 0$. The set of all tangent vectors at C in x will be denoted by $T_C(x)$, and it is called the *tangent cone at C in x*. (By properties of d^0, it is readily seen that $T_C(x)$ is a closed cone of X.)

The normal cone at C in x is by definition the set

$$N_C(x) = \{\eta \in X^*;\ (v, \eta) \le 0\ \forall v \in T_C(x)\}. \tag{2.76}$$

Proposition 2.14. *Let f and g be locally Lipschitz functions on the Banach space X. Then*

$$\partial(f+g)(x) \subset \partial f(x) + \partial g(x) \qquad \forall x \in X. \tag{2.77}$$

If C is a closed subset of X and if f attains its infimum on C in x, then

$$0 \in \partial f(x) + N_C(x). \tag{2.78}$$

Proof. By definition of the directional derivative, we have

$$(f+g)^0(x,v) \leq f^0(x,v) + g^0(x,v) \qquad \forall v \in X,$$

and since, by Proposition 2.13, $f^0(x,\cdot)$ is the support function of $f(x)$, i.e., $f^0(x,\cdot) = H_{\partial f(x)}$, we have

$$H_{\partial(f+g)(x)}(v) \leq H_{\partial f(x)}(v) + H_{\partial g(x)}(v) \leq H_{\partial f(x)+\partial g(x)}(v) \qquad \forall v \in X,$$

and this implies (2.77).

Now, let $x \in C$ be such that $f(x) = \inf\{f(u);\ u \in C\}$, and let S be a closed ball centered at x such that f is Lipschitz with constant L on S. Replacing if necessary C by $C \cap S$ (the sets C and $C \cap S$ have the same normal cone at x) we may assume that $C \subset S$. Consider the function $g(y) = f(y) + \mu d_C(y)$, where $\mu \geq L$. We have

$$g(x) \leq g(y) \qquad \forall y \in S,$$

and therefore $0 \in \partial g(x)$. Then, by (2.77),

$$0 \in \partial f(x) + \mu \partial d_C(x) \subset \partial f(x) + N_C(x),$$

because $\partial d_C(x) \subset N_C(x)$. The proof is complete. ■

Let us show now that if C is convex, then $N_C(x)$ coincides with the normal cone at C in x in the sense of convex analysis, i.e.,

$$N_C(x) = \{\eta \in X^*;\ (x-u,\eta) \geq 0 \qquad \forall u \in C\}.$$

Indeed, if $(x-u,\eta) \geq 0$ for all $u \in C$, then x is a minimum point on C for the function $u \to (x-u,\eta\}$ and so, by formula (2.78),

$$0 \in -\eta + N_C(x).$$

Conversely, if $\eta \in N_C(x)$, then $(\eta, v) \leq 0$ for all $v \in X$ such that $d_C^0(x,v) = 0$. On the other hand, since the function d_C is convex, we have

$d^0 = d'$, and so

$$d^0(x, u - x) = d'(x, u - x) = \lim_{\lambda \downarrow 0} \frac{1}{\lambda}(d(1 - \lambda)x + \lambda u) - d(x)) \le 0 \qquad \forall u \in C.$$

Hence, $(\eta, x - u) \ge 0 \ \forall u \in C$, as claimed.

Now, let V be a real separable Hilbert space and let V^* be its dual (not identified with V). Let $\{e_n\}_{n=1}^{\infty}$ be an orthonormal basis in V and let X_n be the finite dimensional space generated by $\{e_1, \dots, e_n\}$. Denote by $P_n : V \to X_n$ the projection of X into X_n, i.e.,

$$P_n u = \sum_{i=1}^{n} (u, e_i)e_i .$$

Let $\Lambda_n : \mathbf{R}^n \to X_n$ be the operator

$$\Lambda_n(\tau) = \sum_{i=1}^{n} \tau_i e_i , \qquad \tau = (\tau_1, \dots, \tau_n).$$

If $f: V \to \mathbf{R}$ is a locally Lipschitz function, denote by $f^\varepsilon: V \to \mathbf{R}$ the function

$$f^\varepsilon(u) = \int_{\mathbf{R}^n} f(P_n u - \varepsilon \Lambda_n \tau) \rho_n(\tau)\, d\tau \qquad \forall u \in V, \tag{2.79}$$

where $n = [\varepsilon^{-1}]$ and $\rho_n \in C_0^\infty(\mathbf{R}^n)$ is a mollifier in $\mathbf{R}^n$, i.e., $\rho_n(\theta) = 0$ for $\|\theta\|_n > 1$, $\rho_n \ge 0$, $\int_{\mathbf{R}^n} \rho_n(\theta)\, d\theta = 1$, and $\rho_n(\theta) = \rho_n(-\theta) \ \forall \theta \in \mathbf{R}^n$.

Proposition 2.15. *The function f^ε is Fréchet differentiable and*

$$\lim_{\varepsilon \to 0} f^\varepsilon(u) = f(u) \qquad \forall u \in V. \tag{2.80}$$

If $\{u_\varepsilon\}$ is strongly convergent to u in V as $\varepsilon \to 0$ and

$$\nabla f^\varepsilon(u_\varepsilon) \rightharpoonup \xi \qquad \text{in } V^*,$$

then $\xi \in \partial f(u)$.

Proof. Clearly, we may rewrite f^ε as

$$f^\varepsilon(u) = \varepsilon^{-n} \int_{\mathbf{R}^n} f(\Lambda_n \theta) \rho_n\big((\Lambda_n^{-1} P_n u - \theta)\varepsilon^{-1}\big)\, d\theta,$$

from which we see that f^ε is Fréchet differentiable and ∇f^ε is continuous (i.e., $f^\varepsilon \in C^1(V)$). On the other hand, we have

$$f^\varepsilon(u) - f(u) = \int_{\mathbf{R}^n} (f(P_n u - \varepsilon \Lambda_n \tau) - f(u)) \rho_n(\tau)\, d\tau,$$

which yields

$$|f^\varepsilon(u) - f(u)| \leq L(\|u - P_n u\| + \varepsilon) \qquad \forall \varepsilon > 0,$$

and this implies (2.80).

Now, if $\{u_\varepsilon\}$ is as before, by the mean value theorem we have

$$\begin{aligned}
&\lambda^{-1}(f^\varepsilon(u_\varepsilon + \lambda z) - f^\varepsilon(u_\varepsilon)) \\
&\quad = \lambda^{-1} \int_{\mathbf{R}^n} (f(P_n u_\varepsilon + P_n z - \varepsilon \Lambda_n \tau) - f(P_n u_\varepsilon - \varepsilon \Lambda_n \tau)) \rho_n(\tau)\, d\tau.
\end{aligned}$$

Hence, by the Beppo–Levi theorem (or by the Fatou lemma),

$$(\nabla f^\varepsilon(u_\varepsilon), z) \leq \int_{\mathbf{R}^n} f^0(P_n u_\varepsilon - \varepsilon \Lambda_n \tau, P_n z) \rho_n(\tau)\, d\tau.$$

Since the function f^0 is upper semicontinuous in $V \times V$ and $P_n u_\varepsilon - \varepsilon \Lambda_n \tau \to u$, $P_n z \to z$ for $\varepsilon \to 0$ ($n \to \infty$), we have

$$\limsup_{n \to \infty} \int_{\mathbf{R}^n} f^0(P_n u_\varepsilon - \varepsilon \Lambda_n \tau, P_n z) \rho_n(\tau)\, d\tau \leq f^0(u, z).$$

Hence, $(\xi, z) \leq f^0(u, z)$ for all $x \in V$ and so $\xi \in \partial f(u)$, as claimed. ∎

It must be said that this approximating result can be extended to separable Banach spaces with Schauder basis.

Corollary 2.1. *If u and v are two distinct points of V, then there is a point z on the segment $[u, v]$ and $\eta \in \partial f(z)$ such that*

$$f(u) - f(v) = (\eta, u - v).$$

Proof. Let f^ε constructed as before. By the classical mean value theorem, there exists z_ε such that $z_\varepsilon = \lambda_\varepsilon u + (1 - \lambda_\varepsilon)v$, $0 \leq \lambda_\varepsilon \leq 1$, and

$$f^\varepsilon(u) - f^\varepsilon(v) = (\nabla f^\varepsilon(z_\varepsilon), u - v).$$

Extracting a subsequence, we may assume that

$$z_\varepsilon \to z \quad \text{and} \quad \nabla f^\varepsilon(z_\varepsilon) \rightharpoonup \eta \qquad \text{as } \varepsilon \to 0.$$

Then, by Proposition 2.15, it follows that

$$f(u) - f(v) = (\eta, u - v),$$

as desired. ■

Consider now on the space $X = L^p(\Omega)$, $1 \le p < \infty$, $\Omega \subset \mathbf{R}^N$, the function

$$f(y) = \int_\Omega g(x, y(x))\,dx \qquad \forall y \in L^p(\Omega),$$

where $g: \Omega \times \mathbf{R} \to \mathbf{R}$ is measurable in x and Lipschitz in y, i.e.,

$$|g(x,y) - g(x,z)| \le \alpha(x)|y - z| \qquad \forall y, z \in \mathbf{R},\ x \in \Omega,$$

where $\alpha \in L^q(\Omega)$, $1/p + 1/q = 1$, and $g(x,0) \in L^1(\Omega)$.

Clearly, f is Lipschitz in $L^p(\Omega)$ and

$$\partial f(y) \subset \{w \in L^q(\Omega);\, w(x) \in \partial g(x, y(x)) \text{ a.e. } x \in \Omega\}. \tag{2.81}$$

Indeed, we have

$$\begin{aligned} f^0(y,v) &= \limsup_{\substack{z \to y \\ \lambda \downarrow 0}} \int_\Omega \frac{g(x, z(x) + \lambda v(x)) - g(x, z(x))}{\lambda}\,dx \\ &\le \int_\Omega g^0(x, y, v)\,dx \qquad \forall v \in L^p(\Omega). \end{aligned}$$

Hence, if $\eta \in \partial f(y)$, we have

$$\int_\Omega ((\eta(x), v(x)) - g^0(x, y(x), v(x)))\,dx \le 0 \qquad \forall v \in L^p(\Omega),$$

and this yields, by standard arguments,

$$(\eta(x), v) \le g^0(x, y(x), v) \qquad \forall v \in \mathbf{R} \quad \text{a.e. } x \in \Omega,$$

as claimed.

This concept of generalized gradient extends to functions $f: X \to [-\infty, +\infty]$. If $E(f)$ is the epigraph of such a function, i.e.,

$$E(f) = \{(x, \lambda) \in X \times \mathbf{R};\, f(x) \le \lambda\},$$

and if $-\infty < f(x) < +\infty$, then, by definition,

$$\partial f(x) = \{\eta \in X^*; (\eta, -1) \in N_{E(f)}(x, f(x))\}. \tag{2.82}$$

The set $\partial f(x) \subset X^*$ is weak star closed in X^* (if nonempty). In particular, if $f = I_C$ is the indicator function of a closed subset of X, then

$$\partial I_C(x) = N_C(x) \qquad \forall x \in C.$$

Indeed, $\eta \in \partial I_C(x)$ if and only if

$$(\eta, -1) \in N_{E(I_C)}(x, 0) = N_{C \times [0,\infty]}(x, 0) = N_C(x) \times]-\infty, 0].$$

Hence, $\eta \in \partial I_C(x)$ if and only if $\eta \in N_C(x)$. This definition agrees with that given for locally Lipschitz functions. Indeed, it can be proved that if f is locally Lipschitz on X, then $\eta \in \partial f(x)$ if and only if $(\eta, -1) \in N_{E(f)}(x, f(x))$ (see, e.g., F. Clarke [2]).

We shall not pursue further the study of ∂f in this general context and refer the reader to F. Clarke [2] and R. T. Rockafellar [7, 8]. We mention, however, the following extension of Proposition 2.14.

Proposition 2.16. *Let $f_i: X \to \mathbf{R}$, $i = 1, 2$, be finite in x and let f_2 be Lipschitz in a neighborhood of x. Then one has*

$$\partial(f_1 + f_2)(x) \subset \partial f_1(x) + \partial f_2(x). \tag{2.83}$$

■

2.3. Accretive Operators in Banach Spaces

2.3.1. *Definition and Basic Properties*

Throughout this subsection, X will be a real Banach space with the norm $\|\cdot\|$, X^* will be its dual space and $(\cdot, \cdot)$ the pairing between X and X^*. We will denote as usual by $J: X \to X^*$ the duality mapping of the space X.

Definition 3.1. A subset A of $X \times X$ (equivalently, a multivalued operator from X to X) is called *accretive* if for every pair $[x_1, y_1], [x_2, y_2] \in A$, there is $w \in J(x_1 - x_2)$ such that

$$(y_1 - y_2, w) \geq 0. \tag{3.1}$$

An accretive set is said to be *maximal accretive* if it is not properly contained in any accretive subset of $X \times X$.

An accretive set A is said to be m-*accretive* if

$$R(I + A) = X. \tag{3.2}$$

Here, we have denoted by I the unity operator in X; when there will be no danger of confusion we shall simply write 1 instead of I.

A subset A is called *dissipative* (*resp*., *maximal dissipative*, m-*dissipative*) is $-A$ is accretive (resp., maximal accretive, m-accretive).

Finally, A is said to be ω-*accretive* (ω-m-*accretive*), where $\omega \in \mathbf{R}$, if $A + \omega I$ is accretive (resp., m-accretive).

The accretiveness of A can be equivalently expressed as

$$\|x_1 - x_2\| \le \|x_1 - x_2 + \lambda(y_1 - y_2)\| \qquad \forall \lambda > 0, [x_i, y_i] \in A, i = 1,2, \tag{3.3}$$

using the following lemma (Kato's lemma).

Lemma 3.1. *Let $x, y \in X$. Then there exists $w \in J(x)$ such that $(y, w) \ge 0$ if and only if*

$$\|x\| \le \|x + \lambda y\| \qquad \forall \lambda > 0. \tag{3.4}$$

Proof. Let x and y in X be such that $(y, w) \ge 0$ for some $w \in J(x)$. Then, by definition of J, we have

$$\|x\|^2 = (x, w) \le (x + \lambda y, w) \le \|x + \lambda y\| \cdot \|w\| = \|x + \lambda y\| \cdot \|w\| \qquad \forall \lambda > 0,$$

and (3.4) follows.

Suppose now that (3.4) holds. For $\lambda > 0$, let w_λ be an arbitrary element of $J(x + \lambda y)$. Without loss of generality, we may assume that $x \ne 0$. Then, $w_\lambda \ne 0$ for λ small. We set $f_\lambda = w_\lambda \|w_\lambda\|^{-1}$. Since $\{f_\lambda\}_{\lambda > 0}$ is weak star compact in X^*, there exists a generalized sequence, again denoted λ, such that $f_\lambda \rightharpoonup f$ in X^*. On the other hand, it follows from the inequality

$$\|x\| \le \|x + \lambda y\| = (x + \lambda y, f_\lambda) \le \|x\| + \lambda(y, f_\lambda)$$

that

$$(y, f_\lambda) \ge 0 \qquad \forall \lambda > 0.$$

Hence, $(y, f) \geq 0$ and $\|x\| \leq (x, f)$. Since $\|f\| \leq 1$, this implies that $\|x\| = (x, f)$, $\|f\| = 1$, and therefore $w = f\|x\| \in J(x)$, $(y, w) \geq 0$, as claimed. ■

Proposition 3.1. *A subset A of $X \times X$ is accretive if and only if inequality* (3.3) *holds for all $\lambda > 0$ and all $[x_i, y_i] \in A$, $i = 1, 2$.* ■

Proposition 3.1 is an immediate consequence of Lemma 3.1. In particular, it follows that A is ω-accretive iff

$$\|x_1 - x_2 + \lambda(y_1 - y_2)\| \geq (1 - \lambda\omega)\|x_1 - x_2\| \quad \text{for } 0 < \lambda < 1/\omega \text{ and } [x_i, y_i] \in A, i = 1, 2. \tag{3.5}$$

Hence, if A is accretive then the operator $(I + \lambda A)^{-1}$ is nonexpansive on $R(I + \lambda A)$, i.e.,

$$\|(I + \lambda A)^{-1}x - (I + \lambda A)^{-1}y\| \leq \|x - y\| \quad \forall \lambda > 0,\ x, y \in R(I + \lambda A).$$

If A is ω-accretive, then it follows by (3.5) that $(I + \lambda A)^{-1}$ is Lipschitz with Lipschitz constant $1/(1 - \lambda\omega)$ on $R(I + \lambda A)$, $0 < \lambda < 1/\omega$. Let us define the operators J_λ and A_λ:

$$J_\lambda x = (I + \lambda A)^{-1}x, \qquad x \in R(I + \lambda A); \tag{3.6}$$

$$A_\lambda x = \lambda^{-1}(x - J_\lambda x), \qquad x \in R(I + \lambda A). \tag{3.7}$$

The operator A_λ is called *the Yosida approximation of A*, and in the special case when $X = H$ is a Hilbert space it has been studied in Section 2.2 (see Proposition 2.4).

In Proposition 3.2 following, we collect some elementary properties of J_λ and A_λ.

Proposition 3.2. *Let A be ω-accretive in $X \times X$. Then:*

(a) $\|J_\lambda x - J_\lambda y\| \leq (1 - \lambda\omega)^{-1}\|x - y\|$ $\forall \lambda \in (0, 1/\omega)$, $x, y \in R(I + \lambda A)$;

(b) A_λ *is ω-accretive and Lipschitz with Lipschitz constant* $2/(1 - \lambda\omega)$ *in* $R(I + \lambda A)$, $0 < \lambda < 1/\omega$;

(c) $A_\lambda x \in AJ_\lambda x$ $\forall x \in R(I + \lambda A)$, $0 < \lambda < 1/\omega$;

(d) $(1 - \lambda\omega)\|A_\lambda x\| \leq |Ax| = \inf\{\|y\|;\ y \in Ax\}$;

(e) $\lim_{\lambda \to 0} J_\lambda x = x$ $\forall x \in \overline{D(A)} \cap_{0 < \lambda < 1/\omega} R(I + \lambda A)$.

Proof. (a) and (b) are immediate consequences of inequality (3.5).

(c) Let $x \in R(I + \lambda A)$. Then, $A_\lambda x \in \lambda^{-1}((I + \lambda A)J_\lambda x - J_\lambda x) \in AJ_\lambda x$.

(d) For $x \in D(A) \cap R(I + \lambda A)$, we have $A_\lambda x = \lambda^{-1}(J_\lambda(I + \lambda A)x - J_\lambda x)$ and, therefore, $\|A_\lambda x\| \leq |Ax|(1 - \lambda\omega)^{-1}$ $\forall x \in D(A)$.

(e) For every $x \in D(A) \cap R(I + \lambda A)$, we have

$$\|J_\lambda x - x\| = \lambda\|A_\lambda x\| \leq \frac{\lambda}{1 - \lambda\omega}|Ax| \qquad \forall \lambda \in \left(0, \frac{1}{\omega}\right).$$

Hence $\lim_{\lambda \to 0} J_\lambda x = x$. Clearly, this extends to all of $\overline{D(A)} \cap_{0<\lambda<1/\omega} R(I + \lambda A)$, as claimed. ∎

In the following we shall confine ourselves to the study of accretive subsets, the extensions to the ω-accretive sets being immediate.

Proposition 3.3. *An accretive set $A \subset X \times X$ is* m-*accretive if and only if $R(I + \lambda A) = X$ for all (equivalently, for some) $\lambda > 0$.*

Proof. Let A be m-accretive and let $y \in X$, $\lambda > 0$ be arbitrary but fixed. Then, the equation

$$x + \lambda Ax \ni y \tag{3.8}$$

may be written as

$$x = (I + A)^{-1}\left(\frac{y}{\lambda} + \left(1 - \frac{1}{\lambda}\right)x\right).$$

Then, by the contraction principle, we infer that the equation has solution for $1/2 < \lambda < +\infty$.

Now, fix $\lambda_0 > 1/2$ and write the preceding equation as

$$x = (I + \lambda_0 A)^{-1}\left(\left(1 - \frac{\lambda_0}{\lambda}\right)x + \frac{\lambda_0}{\lambda}y\right). \tag{3.9}$$

Since $(I + \lambda_0 A)^{-1}$ is nonexpansive, this equation has a solution for $\lambda \in (\lambda_0/2, \infty)$. Repeating this argument, we conclude that $R(I + \lambda A) = X$ for all $\lambda > \lambda_0/2^n$, $n = 1, \ldots,$ and so $R(I + \lambda A) = X$ for $\lambda > 0$.

Assume now that $R(I + \lambda_0 A) = X$ for some $\lambda_0 > 0$. Then, if we set Eq. (3.8) into the form (3.9), we conclude as before that $R(I + \lambda A) = X$ for all $\lambda \in (\lambda_0/2, \infty)$ and so $R(I + \lambda) = X$ for all $\lambda > 0$, as claimed. ∎

Combining Proposition 3.2 and 3.3 we conclude that $A \subset X \times X$ is m-accretive if and only if for all $\lambda > 0$ the operator $(I + \lambda A)^{-1}$ is nonexpansive on all of X.

Similarly, A is ω-m-accretive if and only if, for all $0 < \lambda < 1/\omega$,

$$\|(I + \lambda A)^{-1}x - (I + \lambda A)^{-1}y\| \leq \frac{1}{1 - \lambda\omega}\|x - y\| \qquad \forall x, y \in X. \tag{3.10}$$

By Theorem 1.2, if $X = H$ is a Hilbert space, then A is m-accretive if and only if it is maximal accretive.

A subset $A \subset X \times X$ is said to be *demiclosed* if it is in closed $X \times X_w$, i.e., if $x_n \to x$, $y_n \rightharpoonup y$, and $[x_n, y_n] \in A$, then $[x, y] \in A$ (recall that $\rightharpoonup$ denotes weak convergence). A is said to be *closed* if $x_n \to x$, $y_n \to y$, and $[x_n, y_n] \in A$ for all n imply that $[x, y] \in A$.

Proposition 3.4. *Let A be an* m-*accretive set of $X \times X$. Then A is closed and if $\lambda_n \in \mathbf{R}$, $x_n \in X$ are such that $\lambda_n \to 0$ and*

$$x_n \to x, \quad A_{\lambda_n}x_n \to y \qquad \text{for } n \to \infty, \tag{3.11}$$

then $[x, y] \in A$. If X^* is uniformly convex then A is demiclosed, and if

$$x_n \to x, \quad A_{\lambda_n}x \rightharpoonup y \qquad \text{for } n \to \infty, \tag{3.12}$$

then $[x, y] \in A$.

Proof. Let $x_n \to x$, $y_n \to y$, $[x_n, y_n] \in A$. Since A is accretive, we have

$$\|x_n - u\| \leq \|x_n + \lambda y_n - (u + \lambda v)\| \qquad \forall [u, v] \in A,\ \lambda > 0.$$

Hence,

$$\|x - u\| \leq \|x + \lambda y - (u + \lambda v)\| \qquad \forall [u, v] \in A,\ \lambda > 0.$$

Now, A being m-accretive, there is $[u, v] \in A$ such that $u + \lambda v = x + \lambda y$. Substituting in the latter inequality, we see that $x = u$ and $y = v \in Ax$, as claimed. Now, if λ_n, x_n satisfy condition (3.11), then $\{A_{\lambda_n}x_n\}$ is bounded and so $J_{\lambda_n}x_n - x_n \to 0$. Since $A_{\lambda_n}x_n \in AJ_{\lambda_n}x_n$, $J_{\lambda_n}x_n \to x$, and A is closed, we have that $[x, y] \in A$. We shall assume now that X^* is uniformly convex. Let x_n, y_n be such that $x_n \to x$, $y_n \rightharpoonup y$, $[x_n, y_n] \in A$. Since A is accretive, we have

$$(y_n - v, J(x_n - u)) \geq 0 \qquad \forall [u, v] \in A,\, n \in \mathbf{N}^*.$$

On the other hand, recalling that J is continuous on X (Theorem 1.2, Chapter 1) we may pass to limit $n \to \infty$ to obtain

$$(y - v, J(x - u)) \geq 0 \qquad \forall [u, v] \in A.$$

Now, if we take $[u, v] \in A$ such that $u + v = x + y$, we see that $y = v$ and $x = u$. Hence $[x, y] \in A$, and so A is demiclosed. The final part of Proposition 3.4 is an immediate consequence of this property, remembering that $A_{\lambda_n} x_n \in AJ_{\lambda_n} x_n$. ■

Note that an m-accretive set of $X \times X$ is maximal accretive. Indeed, if $[x, y] \in X \times X$ is such that

$$\|x - u\| \leq \|x + \lambda y - (u + \lambda v)\| \qquad \forall [u, v] \in A,\, \lambda > 0,$$

then choosing $[u, v] \in A$ such that $u + \lambda v = x + \lambda y$, we see that $x = u$ and so $v = y \in Ax$.

In particular, if X^* is uniformly convex, then for every $x \in D(A)$ we have

$$Ax = \{y \in X^*;\, (y - v, J(x - u)) \geq 0\ \forall [u, v] \in A\}.$$

Hence, Ax is a closed convex subset of X. Denote by $A^0 x$ the element of minimum norm on Ax, i.e., the projection of origin into Ax. Since the space X is reflexive, $A^0 x \neq \varnothing$ for every $x \in D(A)$. The set $A^0 \subset A$ is called the *minimal section* of A. If the space X is strictly convex, then A^0 is single valued.

Proposition 3.5. *Let X and X^* be uniformly convex and let A be an m-accretive set of $X \times X$. Then:*

(i) $A_\lambda x \to A^0 x\ \forall x \in D(A)$ *for* $\lambda \to 0$;
(ii) $\overline{D(A)}$ *is a convex set of* X.

Proof. (i) Let $x \in D(A)$. As seen in Proposition 3.2, $\|A_\lambda x\| \leq |Ax| = \|A^0 x\|\ \forall \lambda > 0$. Now, let $\lambda_n \to 0$ be such that $A_{\lambda_n} x \rightharpoonup y$. By Proposition 3.4 we know that $y \in Ax$, and therefore

$$\lim_{n \to \infty} \|A_{\lambda_n} x\| = \|y\| = \|A^0 x\|.$$

Since the space X is uniformly convex, this implies that $A_{\lambda_n}x \to y = A^0 x$ (Lemma 1.1, Chapter 1). Hence, $A_\lambda x \to A^0 x$ for $\lambda \to 0$.

(ii) Let $x_1, x_2 \in D(A)$, and $0 \le \alpha \le 1$. We set $x_\alpha = \alpha x_1 + (1-\alpha)x_2$. Then, as is easily verified,

$$\|J_\lambda(x_\alpha) - x_1\| \le \|x_\alpha - x_1\| + \lambda|Ax_1| \qquad \forall \lambda > 0,$$

$$\|J_\lambda(x_\alpha) - x_2\| \le \|x_\alpha - x_2\| + \lambda|Ax_2| \qquad \forall \lambda > 0,$$

and since the space X is uniformly convex, these imply by a standard device that

$$\|J_\lambda(x_\alpha) - x_\alpha\| \le \delta(\lambda) \qquad \forall \lambda > 0,$$

where $\lim_{\lambda \to 0} \delta(\lambda) = 0$. Hence, $x_\alpha \in \overline{D(A)}$. ∎

We know that a linear operator $A: X \to X$ is m-dissipative if and only if it is the infinitesimal generator of a C_0-contraction semigroup in X (Theorem 4.2, Chapter 1). Regarding linear m-accretive (equivalently, m-dissipative operators), we note the following density result.

Proposition 3.6. *Let X be a Banach space. Then any* m-*accretive linear operator $A: X \to X$ is densely defined, i.e.,* $\overline{D(A)} = X$.

Proof. Let $y \in X$ be arbitrary but fixed. For every $\lambda > 0$, the equation $x_\lambda + \lambda A x_\lambda = y$ has a unique solution $x_\lambda \in D(A)$. We know that $\|x_\lambda\| \le \|y\|$ for all $\lambda > 0$ and so, on a subsequence $\lambda_n \to 0$,

$$x_{\lambda_n} \rightharpoonup x, \quad \lambda_n A x_{\lambda_n} \rightharpoonup y - x \qquad \text{in } X.$$

Since A is closed, its graph in $X \times X$ is weakly closed (because it is a linear subspace of $X \times X$) and so $\lambda_n x_{\lambda_n} \to 0$, $A(\lambda_n x_{\lambda_n}) \rightharpoonup y - x$ imply that $y - x = 0$. Hence,

$$(1 + \lambda_n A)^{-1} y \rightharpoonup y.$$

We have therefore proven that $y \in \overline{D(A)}$ (recall that the weak closure of $D(A)$ coincides with the strong closure). This completes the proof. ∎

We conclude this section by introducing another convenient way to define the accretiveness. Toward this aim, denote by $[\cdot, \cdot]_s$ the directional derivative of the function $x \to \|x\|$, i.e. (see (2.7)),

$$[x, y]_s = \lim_{\lambda \downarrow 0} \frac{\|x + \lambda y\| - \|x\|}{\lambda}, \qquad x, y \in X. \tag{3.13}$$

Since the function $\lambda \to \|x + \lambda y\|$ is convex, we may define, equivalently, $[\cdot, \cdot]_s$ as

$$[x, y]_s = \inf_{\lambda > 0} \frac{\|x + \lambda y\| - \|x\|}{\lambda} \qquad \forall x, y \in X. \tag{3.14}$$

Let us now briefly list some properties of the bracket $[\cdot, \cdot]_s$.

Proposition 3.7. *Let X be a Banach space. We have*:

(i) $[\cdot, \cdot]_s : X \times X \to \mathbf{R}$ *is upper semicontinuous*;
(ii) $[\alpha x, \beta y]_s = |\beta|[x, y]_s$ *if* $\alpha, \beta > 0$, $x, y \in X$;
(iii) $[x, \alpha x + y]_s = \alpha\|x\| + [x, y]_s$ *if* $\alpha \in \mathbf{R}$, $x \in X$;
(iv) $|[x, y]_s| \leq \|y\|$, $[x, y + z]_s \leq [x, y]_s + [x, y]_s$ $\forall x, y, z \in X$;
(v) $[x, y]_s = \max\{(y, x^*); x^* \in \Phi(x)\}$ $\forall x, y \in X$,

where

$$\Phi(x) = \{x^* \in X^*; (x, x^*) = \|x\|, \|x^*\| = 1\} \qquad \text{if } x \neq 0,$$
$$\Phi(0) = \{x^* \in X^*; \|x^*\| \leq 1\}.$$

Proof. (i) has been proven in Proposition 2.13, whilst (ii), (iii), and (iv) are immediate consequences of the definition. To prove (v), we note first that

$$\Phi(x) = \partial(\|x\|) \qquad \forall x \in X,$$

and apply Proposition 2.13. ■

Now coming back to the definition of accretiveness, we see that condition (3.3) can be equivalently written as

$$[x_1 - x_2, y_1 - y_2]_s \geq 0 \qquad \forall [x_i, y_i] \in A, i = 1, 2. \tag{3.15}$$

Similarly, the condition (3.5) is equivalent to

$$[x_1 - x_2, y_1 - y_2]_s \geq -\omega\|x_1 - x_2\| \qquad \forall [x_i, y_i] \in A, i = 1, 2. \tag{3.16}$$

Summarizing, we may see that a subset A of $X \times X$ is ω-accretive if one of the following equivalent conditions hold:

(i) If $[x_1, y_1], [x_2, y_2] \in A$, then there is $w \in J(x_1 - x_2)$ such that

$$(y_1 - y_2, w) \geq -\omega \|x_1 - x_2\|;$$

(ii) $\|x_1 - x_2 + \lambda(y_1 - y_2)\| \geq (1 - \lambda\omega)\|x_1 - x_2\|$ for $0 < \lambda < 1/\omega$ and all $[x_i, y_i] \in A$, $i = 1, 2$;

(iii) $[x_1 - x_2, y_1 - y_2]_s \geq -\omega\|x_1 - x_2\|$ $\forall [x_i, y_i] \in A$, $i = 1, 2$.

In applications it is more convenient to use condition (iii) to verify the accretiveness.

We know that if X is a Hilbert space, then a continuous accretive operator is m-accretive (Theorem 1.3). This result was extended by R. Martin [1] to general Banach spaces. More generally, we have the following result established by the author in [1] (see also G. F. Webb [1]).

Theorem 3.1. *Let X be a real Banach space, A an* m-*accretive set of $X \times X$ and let $B: X \to X$ be a continuous,* m-*accretive operator with $D(B) = X$. Then $A + B$ is* m-*accretive.*

For the proof of this theorem, we refer to the author's book [2]. A direct proof making use of sharp estimates and convergence results for a discrete scheme of the form

$$x_{n+1} - x_n + h_{n+1}(x_{n+1} + Ax_{n+1} + Bx_{n+1}) \ni \theta_{n+1},$$

along with several generalizations of this theorem can be found in the work of Crandall and Pazy [4].

Other m-accretivity criteria for the sum $A + B$ of two m-accretive operators $A, B \subset X \times X$ can be obtained approximating the equation $x + Ax + Bx \ni y$ by

$$x + Ax + B_\lambda x \ni y,$$

where B_λ is the Yosida approximation of B.

We shall illustrate the method on the following example.

Proposition 3.8. *Let X be a Banach space with uniformly convex dual X^* and let A and B two* m-*accretive sets in $X \times X$ such that $D(A) \cap D(B) \neq \emptyset$ and*

$$(Au, J(B_\lambda u)) \geq 0 \qquad \forall \lambda > 0, u \in D(A). \tag{3.17}$$

Then $A + B$ is m-*accretive.*

Proof. Let $f \in X$ and $\lambda > 0$ be arbitrary but fixed. We shall approximate the equation

$$u + Au + Bu \ni f \tag{3.18}$$

by

$$u + Au + B_\lambda u \ni f, \qquad \lambda > 0, \tag{3.19}$$

where B_λ is the Yosida approximation B, i.e., $B_\lambda = \lambda^{-1}(I - (I + \lambda B)^{-1})$. We may write Eq. (3.19) as

$$u = \left(1 + \frac{\lambda}{1+\lambda} A\right)^{-1} \left(\frac{\lambda f}{1+\lambda} + \frac{(I+\lambda B)^{-1} u}{1+\lambda}\right),$$

which by the Banach fixed point theorem has a unique solution $u_\lambda \in D(A)$ (because $(I + \lambda B)^{-1}$ and $(I + \lambda A)^{-1}$ are nonexpansive).

Now, we multiply the equation

$$u_\lambda + Au_\lambda + B_\lambda u_\lambda \ni f \tag{3.20}$$

by $J(B_\lambda u_\lambda)$ and use condition (3.17) to get

$$\|B_\lambda u_\lambda\| \le \|f\| + \|u_\lambda\| \qquad \forall \lambda > 0.$$

On the other hand, multiplying Eq. (3.20) by $J(u_\lambda - u_0)$, where $u_0 \in D(A) \cap D(B)$, we get

$$\begin{aligned} \|u_\lambda - u_0\| &\le \|u_0\| + \|f\| + \|\xi_0\| + \|B_\lambda u_0\| \\ &\le \|u_0\| + \|f\| + \|\xi_0\| + |Bu_0| \qquad \forall \lambda > 0, \end{aligned}$$

where $\xi_0 \in Au_0$.

Hence,

$$\|u_\lambda\| + \|B_\lambda u_\lambda\| \le C \qquad \forall \lambda > 0. \tag{3.21}$$

Now we multiply the equation (in the sense of duality between X and X^*)

$$u_\lambda - u_\mu + Au_\lambda - Au_\mu + B_\lambda u_\lambda - B_\mu u_\mu \ni 0$$

by $J(u_\lambda - u_\mu)$. Since A is accretive, we have

$$\|u_\lambda - u_\mu\|^2 + \big(B_\lambda u_\lambda - B_\mu u_\mu, J(u_\lambda - u_\mu)\big) \le 0 \qquad \forall \lambda, \mu > 0. \tag{3.22}$$

On the other hand,

$$(B_\lambda u_\lambda - B_\mu u_\mu, J(u_\lambda - u_\mu)) \geq \Big(B_\lambda u_\lambda - B_\mu u_\mu, J(u_\lambda - u_\mu) - J\big((I+\lambda B)^{-1}u_\lambda - (I+\mu B)^{-1}u_\mu\big)\Big)$$

because B is accretive and $B_\lambda u \in B(I + \lambda B)^{-1}u)$. Since J is uniformly continuous on bounded subsets and by (3.21)

$$\|u_\lambda - (I+\lambda B)^{-1}u_\lambda\| + \|u_\mu - (I+\mu B)^{-1}u_\mu\| \leq C(\lambda + \mu),$$

this implies that $\{u_\lambda\}$ is a Cauchy sequence and so $u = \lim_{\lambda \to 0} u_\lambda$ exists. Extracting further subsequences, we may assume that

$$B_\lambda y \rightharpoonup y, \qquad f - B_\lambda u_\lambda - u_\lambda \rightharpoonup z.$$

Then, by Proposition 3.4, we see that $y \in Bu$, $z \in Au$, and so u is a solution (obviously unique) to Eq. (3.18). ■

2.3.2. *Some Examples*

Throughout this section, Ω is a bounded and open subset of $\mathbf{R}^N$ with a smooth boundary, denoted $\partial\Omega$.

1. *Sublinear elliptic operators in $L^p(\Omega)$.* Let β be a maximal monotone graph in $\mathbf{R} \times \mathbf{R}$ such that $0 \in D(\beta)$. Let $\tilde{\beta} \subset L^p(\Omega) \times L^p(\Omega)$, $1 \leq p < \infty$, be the operator defined by

$$\tilde{\beta}(u) = \{v \in L^p(\Omega); v(x) \in \beta(u(x)) \text{ a.e. } x \in \Omega\},$$
$$D(\tilde{\beta}) = \{u \in L^p(\Omega); \exists v \in L^p(\Omega) \text{ such that } v(x) \in \beta(u(x)), \text{a.e. } x \in \Omega\}. \tag{3.23}$$

It is easily seen that $\tilde{\beta}$ is m-accretive in $L^p(\Omega) \times L^p(\Omega)$ and

$$\big((I + \lambda\tilde{\beta})^{-1}u\big)(x) = (1 + \lambda\beta)^{-1}u(x) \qquad \text{a.e. } x \in \Omega,\ \lambda > 0,$$
$$(\tilde{\beta}_\lambda u)(x) = \beta_\lambda(u(x)) \qquad \text{a.e. } x \in \Omega,\ \lambda > 0,\ u \in L^p(\Omega).$$

Proposition 3.9. *Let $B: L^p(\Omega) \to L^p(\Omega)$ be the operator defined by*

$$\begin{aligned} Bu &= -\Delta u + \tilde{\beta}(u) \qquad \forall u \in D(B), \\ D(B) &= W_0^{1,p}(\Omega) \cap W^{2,p}(\Omega) \cap D(\tilde{\beta}) \qquad \text{if } 1 < p < \infty, \\ D(B) &= \{u \in W_0^{1,1}(\Omega);\ \Delta u \in L^1(\Omega)\} \cap D(\tilde{\beta})\} \qquad \text{if } p = 1. \end{aligned} \tag{3.24}$$

Then B is m-*accretive and surjective.*

We note that for $p = 2$, this result has been proved in Proposition 2.6.

Proof. Let us show first that B is accretive. If $u_1, u_2 \in D(B)$ and $v_1 \in Bu_1$, $v_2 \in Bu_2$, $1 < p < \infty$, we have, by Green's formula,

$$\begin{aligned} &\|u_1 - u_2\|_{L^p(\Omega)}^{p-2}(v_1 - v_2, J(u_1 - u_2)) \\ &\quad = -\int_\Omega \Delta(u_1 - u_2)\,|u_1 - u_2|^{p-2}(u_1 - u_2)\,dx \\ &\qquad + \int_\Omega (\beta(u_1) - \beta(u_2))(u_1 - u_2)|u_1 - u_2|^{p-2}\,dx \geq 0 \end{aligned}$$

because β is monotone (recall that $J(u)(x) = |u(x)|^{p-2}u(x)\|u\|_{L^p(\Omega)}^{2-p}$ is the duality mapping of the space $L^p(\Omega)$). In the case $p = 1$, consider the function $\gamma_\varepsilon: \mathbf{R} \to \mathbf{R}$ defined by

$$\gamma_\varepsilon(r) = \begin{cases} 1 & \text{for } r > \varepsilon, \\ \varepsilon^{-1} r & \text{for } -\varepsilon < r \leq \varepsilon, \\ -1 & \text{for } r < -\varepsilon. \end{cases} \tag{3.25}$$

The function γ_ε is a smooth approximation of the signum function,

$$\operatorname{sign} r = \begin{cases} 1 & \text{for } r > 0, \\ [-1, 1] & \text{for } r = 0, \\ -1 & \text{for } r < 0. \end{cases}$$

If $[u_i, v_i] \in B$, $i = 1, 2$, then we have

$$\begin{aligned} \int_\Omega (v_1 - v_2)\,\gamma_\varepsilon(u_1 - u_2)\,dx &= \int_\Omega |\nabla(u_1 - u_2)|^2 \gamma_\varepsilon'(u_1 - u_2)\,dx \\ &\quad + \int_\Omega (\beta(u_1) - \beta(u_2))\gamma_\varepsilon(u_1 - u_2)\,dx \\ &\geq 0 \qquad \forall \varepsilon > 0. \end{aligned}$$

For $\varepsilon \to 0$, $\gamma_\varepsilon(u_1 - u_2) \to g$ in $L^\infty(\Omega)$, where $g \in J(u)\|u\|_{L^1(\Omega)}^{-1}$, $u = u_1 - u_2$, i.e., $g(x) \in \operatorname{sign} u(x)$ a.e. $x \in \Omega$. Hence, B is accretive.

We shall prove that B is m-accretive, considering separately the cases $1 < p < \infty$ and $p = 1$.

Case 1. $1 < p < \infty$. Let us denote by A_p the operator $-\Delta$ with the domain $D(A_p) = W_0^{1,p}(\Omega) \cap W^{2,p}(\Omega)$. We have already seen that A_p is accretive. Moreover, by a classical result of Agmon *et al.* [1], $R(I + A_p) = L^p(\Omega)$ and

$$\|u\|_{W^{2,p}(\Omega) \cap W_0^{1,p}(\Omega)} \leq C\|A_p u\|_{L^p(\Omega)} \qquad \forall u \in D(A_p). \tag{3.26}$$

Hence, A_p is m-accretive. Let us prove now that $R(I + A_p + \tilde{\beta}) = L^p(\Omega)$. Replacing, if necessary, the graph β by $u \to \beta(u) - v_0$, where $v_0 \in \beta(0)$, we may assume that $0 \in \tilde{\beta}(0)$ and so $\tilde{\beta}_\lambda(0) = 0$. Then, by Green's formula,

$$\begin{aligned}\left(A_p u, J(\tilde{\beta}_\lambda u)\right) &= -\|\tilde{\beta}_\lambda(u)\|_{L^p(\Omega)}^{2-p} \int_\Omega \Delta u |\beta_\lambda(u)|^{p-2}\beta_\lambda(u)\, dx \\ &= \|\tilde{\beta}_\lambda(u)\|_{L^p(\Omega)}^{2-p} \int_\Omega |\nabla u|^2 \frac{d}{du}|\beta_\lambda(u)|^{p-2}\beta_\lambda(u)\, dx \geq 0 \\ &\qquad \forall \lambda > 0, \end{aligned} \tag{3.27}$$

and by Proposition 3.7 we conclude that $R(I + A_p + \tilde{\beta}) = L^p(\Omega)$, as claimed.

To prove surjectivity of $A_p + \tilde{\beta}$, consider the equation

$$\varepsilon u + A_p u + \tilde{\beta}(u) \ni f, \qquad \varepsilon > 0, f \in L^p(\Omega), \tag{3.28}$$

which as seen before has a unique solution u_ε, and

$$u_\varepsilon = \lim_{\lambda \to 0} u_\lambda^\varepsilon \qquad \text{in } L^p(\Omega),$$

where u_λ^ε is the solution to approximating equation $\varepsilon u + A_p u + \tilde{\beta}_\lambda(u) \ni f$. By (3.27), it follows that

$$\|A_p u_\lambda^\varepsilon\|_{L^p(\Omega)} \leq C,$$

where C is independent of ε and λ. Hence,

$$\|A_p u_\varepsilon\|_{L^p(\Omega)} \leq C \qquad \forall \varepsilon > 0,$$

which by the estimate (3.26) implies that $\{u_\varepsilon\}$ is bounded in $W_0^{1,p}(\Omega) \cap W^{2,p}(\Omega)$. Selecting a subsequence, for simplicity again denoted u_ε, we may assume that

$$\begin{aligned}
u_\varepsilon &\to u && \text{weakly in } W^{2,p}(\Omega), \text{ strongly in } L^p(\Omega),\\
A_p u_\varepsilon &\to A_p u && \text{weakly in } L^p(\Omega),\\
\tilde{\beta}_\varepsilon(u_\varepsilon) &\to g && \text{weakly in } L^p(\Omega),\\
\varepsilon u_\varepsilon &\to 0 && \text{strongly in } L^p(\Omega).
\end{aligned}$$

Since by Proposition 3.4 we know that $g \in \tilde{\beta}(u)$, we infer that u is the solution to the equation $A_p u + \tilde{\beta}(u) \ni f$, i.e., $u \in W^{2,p}(\Omega)$ and

$$\begin{aligned}
-\Delta u + \beta(u) &\ni f && \text{a.e. in } \Omega,\\
u &= 0 && \text{in } \partial\Omega.
\end{aligned} \tag{3.29}$$

Case 2. $p = 1$. We shall prove directly that $R(A_1 + \tilde{\beta}) = L^1(\Omega)$, i.e., for $f \in L^1(\Omega)$, Eq. (3.29) has a solution $u \in D(A_1) = \{u \in W_0^{1,1}(\Omega);\ \Delta u \in L^1(\Omega)\}$.

We fix f in $L^1(\Omega)$ and consider $\{f_n\} \subset L^2(\Omega)$ such that $f_n \to f$ in $L^1(\Omega)$. As seen before, the problem

$$\begin{aligned}
-\Delta u_n + \beta(u_n) &\ni f_n && \text{in } \Omega,\\
u_n &= 0 && \text{in } \partial\Omega,
\end{aligned} \tag{3.30}$$

has a unique solution $u_n \in H_0^1(\Omega) \cap H^2(\Omega)$. Let $v_n(x) = f_n(x) + \Delta u_n(x) \in \beta(u_n(x))$. By (3.30) we see that

$$\int_\Omega |v_n(x) - v_m(x)|\,dx \le \int_\Omega |f_n(x) - f_m(x)|\,dx,$$

because β is monotone and

$$\int_\Omega \Delta u\,\theta\,dx \le 0$$

for all $u \in H_0^1(\Omega) \cap H^2(\Omega)$ and some $\theta \in L^\infty(\Omega)$ such that $\theta(x) \in \operatorname{sign} u(x)$ a.e. $x \in \Omega$. Hence,

$$\begin{aligned}
v_n &\to v && \text{strongly in } L^1(\Omega),\\
\Delta u_n &\to \xi && \text{strongly in } L^1(\Omega).
\end{aligned} \tag{3.31}$$

Now, let $h_i \in L^p(\Omega)$, $i = 0, 1, \ldots, N$, $p > N$. Then, by a well-known result due to G. Stampacchia [1] (see also Dautray and Lions [1], p. 462), the boundary value problem

$$
\begin{aligned}
-\Delta \varphi &= h_0 + \sum_{i=1}^{N} \frac{\partial h_i}{\partial x_i} \qquad \text{in } \Omega, \\
\varphi &= 0 \qquad \text{in } \partial\Omega,
\end{aligned} \tag{3.32}
$$

has a unique weak solution $\varphi \in H_0^1(\Omega) \cap L^\infty(\Omega)$ and

$$\|\varphi\|_{L^\infty(\Omega)} \le C \sum_{i=0}^{N} \|h_i\|_{L^p(\Omega)}, \qquad h_i \in L^p(\Omega). \tag{3.33}$$

This means that

$$\int_\Omega \nabla\varphi \cdot \nabla\psi \, dx = \int_\Omega h_0 \psi - \sum_{i=1}^{N} \int_\Omega h_i \frac{\partial \psi}{\partial x_i} \, dx \qquad \forall \psi \in H_0^1(\Omega). \tag{3.34}$$

Substituting $\psi = u_n$ in (3.34), we get

$$-\int_\Omega \varphi \, \Delta u_n \, dx = \int_\Omega \nabla\varphi \cdot \nabla u_n \, dx = \int_\Omega h_0 u_n \, dx - \sum_{i=1}^{N} \int_\Omega h_i \frac{\partial u_n}{\partial x_i} \, dx,$$

and therefore, by (3.33),

$$\left| \int_\Omega h_0 u_n \, dx - \sum_{i=1}^{N} h_i \frac{\partial u_n}{\partial x_i} \, dx \right| \le C \|\Delta u_n\|_{L^1(\Omega)} \sum_{i=0}^{N} \|h_i\|_{L^p(\Omega)}.$$

Since $\{h_i\}_{i=0}^N \subset (L^p(\Omega))^{N+1}$ are arbitrary, we conclude that $\{(u_n, \partial u_n / \partial x_1, \dots, \partial u_n / \partial x_n)\}_{n=1}^\infty$ is bounded in $(L^q(\Omega))^{N+1}$, $1/q + 1/p = 1$. Hence,

$$\|u_n\|_{W_0^{1,q}(\Omega)} \le C \|\Delta u_n\|_{L^1(\Omega)}, \tag{3.35}$$

where $1 < q = p/(p-1) < N/(N-1)$. Hence, $\{u_n\}$ is bounded in $W_0^{1,q}(\Omega)$ and compact in $L^1(\Omega)$. Then, extracting a further subsequence if necessary, we may assume that

$$u_n \to u \qquad \text{weakly in } W_0^{1,q}(\Omega), \text{ strongly in } L^1(\Omega). \tag{3.36}$$

Then, by (3.31), it follows that $\xi = \Delta u$, and since the operator $\tilde{\beta}$ is closed in $L^1(\Omega) \times L^1(\Omega)$, we see by (3.31) and (3.36) that $v(x) \in \beta(u(x))$ a.e. $x \in \Omega$ and $u \in W_0^{1,q}(\Omega)$. Hence $R(B) = L^1(\Omega)$, and in particular B is m-accretive. ■

We have proved, therefore, the following existence result for the semilinear elliptic problem in $L^1(\Omega)$.

Corollary 3.1. *For every $f \in L^1(\Omega)$, the boundary value problem*

$$-\Delta u + \beta(u) \ni f \qquad \text{a.e. in } \Omega,$$
$$u = 0 \qquad \text{in } \partial\Omega, \tag{3.37}$$

has a unique solution $u \in W_0^{1,q}(\Omega)$ with $\Delta u \in L^1(\Omega)$, where $1 \le q < N/(N-1)$. Moreover, the following estimate holds:

$$\|u\|_{W_0^{1,q}(\Omega)} \le C\|f\|_{L^1(\Omega)} \qquad \forall f \in L^1(\Omega). \tag{3.38}$$

In particular, $D(A_1) \subset W_0^{1,q}(\Omega)$ and

$$\|u\|_{W_0^{1,q}(\Omega)} \le C\|\Delta u\|_{L^1(\Omega)} \qquad \forall u \in D(A_1). \qquad \blacksquare$$

Remark 3.1. It is clear from the previous proof that Proposition 2.8 and Corollary 3.1 remain true for linear second order elliptic operators A on Ω.

2. *The nonlinear diffusion operator in* $L^1(\Omega)$. In the space $X = L^1(\Omega)$, define the operator

$$Au = -\Delta\beta(u) \qquad \forall u \in D(A),$$
$$D(A) = \{u \in L^1(\Omega);\ \beta(u) \in W_0^{1,1}(\Omega),\ \Delta\beta(u) \in L^1(\Omega)\}, \tag{3.39}$$

where β is a maximal monotone graph in $\mathbf{R} \times \mathbf{R}$ such that $0 \in \beta(0)$ and Ω is an open bounded subset of $\mathbf{R}^N$ with smooth boundary.

Proposition 3.10. *The operator A is* m-*accretive in* $L^1(\Omega) \times L^1(\Omega)$.

Proof. Let $u, v \in D(A)$ and let γ be a smooth monotone approximation of sign (see (3.25)). Then, we have

$$\int_\Omega (Au - Av)\gamma(\beta(u) - \beta(v))\,dx$$
$$= \int_\Omega |\nabla(\beta(u) - \beta(v))|^2\gamma'(\beta u) - \beta(v))\,dx \ge 0.$$

Letting $\gamma \to$ sign, we get

$$\int_\Omega (Au - Av)\xi\,dx \ge 0,$$

where $\xi(x) \in \text{sign}(\beta(u(x)) - \beta(v(x))) = \text{sign}(u(x) - v(x))$ a.e. $x \in \Omega$. Hence, A is accretive.

Let us prove now that $R(I + A) = L^1(\Omega)$. For $f \in L^1(\Omega)$, the equation

$$u + Au = f$$

can be equivalently written as

$$\beta^{-1}(v) - \Delta v = f \quad \text{in } \Omega, \qquad v \in W_0^{1,1}(\Omega),\ \Delta v \in L^1(\Omega). \quad (3.40)$$

But according to Proposition 3.9, (Corollary 3.1), Eq. (3.40) has a solution $v \in W_0^{1,q}(\Omega)$, $\Delta v \in L^1(\Omega)$. This completes the proof. ∎

3. *Quasilinear partial differential equations of first order.* Here, we shall study the first order partial differential operator

$$Au = \sum_{i=1}^{N} \frac{\partial}{\partial x_i} a_i(u(x)), \qquad x \in \mathbf{R}^N, \quad (3.41)$$

in the space $X = L^1(\mathbf{R}^N)$. We shall use the notation $a = (a_1, a_2, \ldots, a_N)$, $\varphi_x = (\varphi_{x_1}, \ldots, \varphi_{x_N})$, $a(u)_x = \sum_{i=1}^{N}(\partial/\partial x_i)\, a_i(u(x)) = \operatorname{div} a(u)$.

The function $a: \mathbf{R} \to \mathbf{R}^N$ is assumed to be continuous.

We will define A as the closure of the operator $A_0 \subset L^1(\Omega) \times L^1(\Omega)$ defined in the following.

Definition 3.2. $A_0 = \{[u, v] \in L^1(\mathbf{R}^N) \times L^1(\mathbf{R}^N);\ a(u) \in (L^1(\mathbf{R}^N))^N$ and

$$\int_{\mathbf{R}^N} \text{sign}_0(u(x) - k)\,((a(u(x)) - a(k), \varphi_x(x)) + v(x)\varphi(x))\, dx = 0 \quad (3.42)$$

for all $\varphi \in C_0(\mathbf{R}^N)$ such that $\varphi \geq 0$, and all $k \in \mathbf{R}\}$. Here, $\text{sign}_0 r = r/|r|$ for $r \neq 0$, $\text{sign}_0 0 = 0$.

It is readily seen that if $a \in C^1(\mathbf{R})$ and $u \in C_0^1(\mathbf{R}^N)$ then $u \in D(A_0)$ and $A_0 u = a(u)_x$. Indeed, if ρ is a smooth approximation of $r \to |r|$, then we have

$$\begin{aligned}
&\int_{\mathbf{R}^N} \rho'(u(x) - k) a(u)_x \varphi\, dx \\
&\quad = \int_{\mathbf{R}^N} dx \left(\int_k^{u(x)} \rho'(s - k) a'(s)\, ds \right)_x \varphi(x)\, dx \\
&\quad = -\int_{\mathbf{R}^N} dx \left(\left(\int_k^{u(x)} \rho'(s - k) a'(s)\, ds \right), \varphi_x(x) \right),
\end{aligned}$$

where $a' = (a'_1, a'_2, \ldots, a'_N)$ is the derivative of a. Now, letting ρ' tend to sign_0, we get

$$\int_{\mathbf{R}^N} \text{sign}_0(u(x) - k)\,(a(u(x)) - a(k), \varphi_x(x)) + a(u(x))_x \varphi(x))\,dx = 0$$

for all $\varphi \in C_0(\mathbf{R}^N)$. Hence, $u \in D(A_0)$ and $A_0 u = (a(u))_x$.

Conversely, if $u \in D(A_0) \cap L^\infty(\mathbf{R}^N)$ and $v \in A_0 u$, then using the inequality (3.42) with $k = \|u\|_{L^\infty(\mathbf{R}^N)} + 1$ and $k = -(\|u\|_{L^\infty(\mathbf{R}^N)} + 1)$, we get

$$\int_{\mathbf{R}^N} ((a(u(x)) - a(k), \varphi_x(x)) + v(x)\varphi(x))\,dx = 0 \qquad \forall \varphi \in C_0^\infty(\mathbf{R}^N).$$

Hence, $v = (a(u))_x$ in the sense of distributions on $\mathbf{R}^N$.

Let A be the closure of A_0 in $L^1(\mathbf{R}^N) \times L^1(\mathbf{R}^N)$, i.e., $A = \{[u, v] \in L^1(\mathbf{R}^N) \times L^1(\mathbf{R}^N);\ \exists [u_n, v_n] \in A_0,\ u_n \to u,\ v_n \to v \text{ in } L^1(\mathbf{R}^N)\}$.

Proposition 3.11. *Let* $a\colon \mathbf{R} \to \mathbf{R}^N$ *be continuous and* $\limsup_{r \to 0} \|a(r)\|/|r| < \infty$. *Then* A *is* m-*accretive.*

We shall prove this important result in several steps.

Lemma 3.2. *A is accretive in $L^1(\mathbf{R}^N) \times L^1(\mathbf{R}^N)$.*

Proof. Let $[u, v]$ and $[\bar{u}, \bar{v}]$ be two arbitrary elements of A_0. By Definition 3.2 we have, for $k = \bar{u}(y)$, $f(x) = \psi(x, y)$ ($\psi \in C_0^\infty(\mathbf{R}^N \times \mathbf{R}^N)$, $\psi \geq 0$),

$$\int_{\mathbf{R}^N \times \mathbf{R}^N} \text{sign}_0(u(x) - \bar{u}(y))\,((a(u(x)) - a(\bar{u}(y)), \psi_x(x, y)) + v(x)\psi(x, y))\,dx\,dy \geq 0. \tag{3.43}$$

Now, it is clear that we can interchange u and $\bar{u}$, v and $\bar{v}$, x and y to obtain, by adding to (3.43) the resulting inequality,

$$\int_{\mathbf{R}^N \times \mathbf{R}^N} \text{sign}_0(u(x) - \bar{u}(y))\,((a(u(x)) - a(\bar{u}(y)), \psi_x(x, y) + \psi_y(x, y)) + (v(x) - \bar{v}(y))\psi(x, y))\,dx\,dy \geq 0 \tag{3.44}$$

for all $\psi \in C_0^\infty(\mathbf{R}^N \times \mathbf{R}^N)$, $\psi \geq 0$.

Now, we take $\psi(x, y) = (1/\varepsilon^n)\,\varphi(x + y)\rho((x - y)/\varepsilon)$, where $\varphi \in C_0^\infty(\mathbf{R}^N)$, $\varphi \geq 0$, and $\rho \in C_0(\mathbf{R}^N)$ is such that $\text{supp}\,\rho \subset \{y;\ \|y\| \leq 1\}$, $\int \rho(y)\,dy = 1$, $\rho(y) = \rho(-y)\ \forall y \in \mathbf{R}^N$.

Substituting in (3.44), we get

$$\int_{\mathbf{R}^N\times\mathbf{R}^N} \operatorname{sign}_0(u(y+\varepsilon z)-\bar{u}(y))\,(2(a(u(y+\varepsilon z)) - a(\bar{u}(y)), \nabla\varphi(y+\varepsilon z)) + (v(y+\varepsilon z) - \bar{v}(y))\,\varphi(y+\varepsilon z))\,\rho(z)\,dz \geq 0. \tag{3.45}$$

Now, letting ε tend to zero in (3.45), we get

$$\int_{\mathbf{R}^N} \theta(y)(v(y)-\bar{v}(y)\,dy + 2\int_{\mathbf{R}^N} \theta(y)(a(u(y)) - a(\bar{u}(y)), \nabla\varphi(y))\,dy \geq 0, \tag{3.46}$$

where $\theta(y) \in \operatorname{sign}(u(y)-\bar{u}(y))$ a.e. $y \in \mathbf{R}^N$. Hence, for every $\varphi \in C_0^\infty(\mathbf{R}^N)$, $\varphi \geq 0$, there exists $\theta \in J(u-\bar{u})$ such that (3.46) holds (J is the duality mapping of the space $L^1(\Omega)$). If in (3.46) we take $\varphi = \alpha(\varepsilon\|y\|^2)$, where $\alpha \in C_0^\infty(\mathbf{R})$, $\alpha \geq 0$, and $\alpha(r) = 1$ for $|r| \leq 1$, and let $\varepsilon \to 0$, we get

$$\int_{\mathbf{R}^N} \theta(y)(v(y)-\bar{v}(y))\,dy \geq 0$$

for some $\theta \in J(u-\bar{u})$. Hence, A_0 is accretive and hence so is A. ■

In order to prove that A is m-accretive, we will show that the range of $I + A_0$ is dense in $L^1(\mathbf{R}^N)$, i.e., that the equation $u + a(u)_x = f$ has a solution (in the generalized sense) for a sufficiently large class of functions f. We shall approximate this equation by the family of elliptic equations

$$u + a(u)_x - \varepsilon\,\Delta u = f \qquad \text{in } \mathbf{R}^N. \tag{3.47}$$

Lemma 3.3. *Let $a \in C^1$, a' bounded, and let $\varepsilon > 0$. Then for each $f \in L^2(\mathbf{R}^N)$, Eq. (3.47) has a solution $u \in H^2(\mathbf{R}^N)$.*

Proof. Denote by Λ the operator

$$\Lambda = -\Delta, \qquad D(\Lambda) = H^2(\mathbf{R}^N)$$

and $Bu = -a(u)_x$ $\forall u \in D(B) = H^1(\mathbf{R}^N)$. The operator $T = (I + \varepsilon\Lambda)^{-1}B$ is obviously continuous and bounded from $H^1(\mathbf{R}^N)$ to $H^2(\mathbf{R}^N)$, and therefore it is compact in $H^1(\mathbf{R}^N)$.

For a given $f \in L^2(\mathbf{R}^N)$, Eq. (3.47) is equivalent to

$$u = Tu + (I + \varepsilon\Lambda)^{-1} f. \tag{3.48}$$

Let $D = \{u \in H^1(\mathbf{R}^N);\ \|u\|^2_{L^2(\mathbf{R}^N)} + \varepsilon\|\nabla u\|^2_{L^2(\mathbf{R}^N)} < R^2\}$, where $R = \|f\|_{L^2(\mathbf{R}^N)} + 1$. We note that

$$(I + \varepsilon\Lambda)^{-1} f \notin (I - tT)(\partial D), \qquad 0 \le t \le 1. \tag{3.49}$$

Indeed, we suppose that there is $u \in \partial D$ and $t \in [0,1]$ such that

$$u - \varepsilon\,\Delta u + ta(u)_x = f \qquad \text{in } \mathbf{R}^N,$$

and we will argue from this to a contradiction. Multiplying the last equation by u and integrating on $\mathbf{R}^N$, we get

$$\|u\|^2_{L^2(\mathbf{R}^N)} + \varepsilon\|\nabla u\|^2_{L^2(\mathbf{R}^N)} + t\int_{\mathbf{R}^N} a(u)_x u\,dx = \int_{\mathbf{R}^N} fu\,dx.$$

On the other hand, we have

$$\int_{\mathbf{R}^N} a(u)_x u\,dx = -\int_{\mathbf{R}^N} (a(u), u_x)\,dx = -\int_{\mathbf{R}^N} \operatorname{div} b(u)\,dx = 0,$$

where $b(u) = \int_0^u a(s)\,ds$. Hence,

$$\|u\|^2_{L^2(\mathbf{R}^N)} + \varepsilon\|\nabla u\|^2_{L^2(\mathbf{R}^N)} \le \|f\|_{L^2(\mathbf{R}^N)}\|u\|_{L^2(\mathbf{R}^N)} \le (R-1)R < R^2,$$

and so $u \notin \partial D$.

Let us denote by $d(I - tT, D, (I + \varepsilon\Lambda)^{-1} f)$ the topological degree of the map $I - tT$ relative to D at the point $(I + \varepsilon\Lambda)^{-1} f$. By (3.49), it follows that

$$d\big(I - tT, D, (I + \varepsilon\Lambda)^{-1} f\big) = d\big(I, D, (I + \varepsilon\Lambda)^{-1} f\big)$$

for all $0 \le t \le 1$. Hence,

$$d\big(I - T, D, (I + \varepsilon\Lambda)^{-1} f\big) = d\big(I, D, (I + \varepsilon\Lambda)^{-1} f\big) = 1$$

because $(I + \varepsilon\Lambda)^{-1} f \in D$. Hence, Eq. (3.48) has at least one solution $u \in D(\Lambda) = H^2(\mathbf{R}^N)$. The proof of Lemma 3.3 is complete. ■

Lemma 3.4. *Under assumptions of Lemma* 3.3, *if* $f \in L^p(\mathbf{R}^N)$, $1 \le p \le \infty$, *then* $u \in L^p(\mathbf{R}^N)$ *and*

$$\|u\|_{L^p(\mathbf{R}^N)} \le \|f\|_{L^p(\mathbf{R}^N)}. \tag{3.50}$$

Proof. We shall treat first the case $1 < p < \infty$. Let $\alpha_n : \mathbf{R} \to \mathbf{R}$ be defined by

$$\alpha_n(r) = \begin{cases} |r|^{p-2}r & \text{if } |r| \le n, \\ n^{p-2}r & \text{if } r > n, \\ n^{p-2}r & \text{if } r < -n. \end{cases}$$

If multiply Eq. (3.47) by $\alpha_n(u) \in L^2(\mathbf{R}^N)$ and integrate on $\mathbf{R}^N$, we get

$$\int_{\mathbf{R}^N} \alpha_n(u) u \, dx \le \int_{\mathbf{R}^N} f \alpha_n(u) \, dx \tag{3.51}$$

because, as previously seen,

$$\int_{\mathbf{R}^N} a(u)_x \alpha_n(u) \, dx = \int_{\mathbf{R}^N} dx \left(\int_0^{u(x)} a'(s) \alpha_n(s) \, ds \right)_x dx = 0,$$

whilst

$$-\int_{\mathbf{R}^N} \Delta u \ \alpha_n(u) \, dx = \int_{\mathbf{R}^N} \alpha_n'(u) |\nabla u|^2 \, dx \ge 0$$

because α_n is monotonically increasing. Note the relation

$$\alpha_n(r) r \ge |\alpha_n(r)|^q \qquad \forall r \in \mathbf{R} \qquad 1/p + 1/q = 1.$$

Then, using the Hölder inequality in (3.51), we get

$$\int_{\mathbf{R}^N} |\alpha_n(u)|^q \, dx \le \left(\int_{\mathbf{R}^N} |f|^p \, dx \right)^{1/p} \left(\int_{\mathbf{R}^N} |\alpha_n(u)|^q \, dx \right)^{1/q}.$$

Hence,

$$\int_{[|u(x)| \le n]} |u(x)|^p \, dx \le \|f\|_{L^p(\mathbf{R}^N)}^p,$$

which clearly implies that $u \in L^p(\mathbf{R}^N)$ and (3.50) holds. In the case $p = 1$, we multiply Eq. (3.47) by $\delta_n(u)$, where

$$\delta_n(r) = \begin{cases} n\,r & \text{if } |r| \le n^{-1}, \\ 1 & \text{if } r > n^{-1}, \\ -1 & \text{if } r < -n^{-1}. \end{cases}$$

Note that $\delta_n(u) \in L^2(\mathbf{R}^N)$ because $m\{x \in \mathbf{R}^N;\ |u(x)| > n^{-1}\} \leq n^2\|u\|^2_{L^2(\mathbf{R}^N)}$. Then, arguing as before, we get

$$\int_{[|u(x)| \geq n^{-1}]} |u(x)|\,dx \leq \int_{\mathbf{R}^u} \delta_n(u)\,dx \leq \int_{\mathbf{R}^N} |f|\,|\delta_n(u)|\,dx$$

$$\leq n \int_{[|u| \leq n^{-1}]} |f|\,|u|\,dx$$

$$+ \int_{[|u| > n^{-1}]} |f|\,dx \leq \|f\|_{L^1(\mathbf{R}^N)}.$$

Then, letting $n \to \infty$ we get (3.50), as desired.

In the case $p = \infty$, put $M = \|f\|_{L^\infty(\mathbf{R}^N)}$. Then, we have

$$u - M + a(u)_x - \varepsilon\Delta(u - M) = f - M \leq 0 \qquad \text{a.e. in } \mathbf{R}^N.$$

Multiplying this by $(u - M)^+$ (which, as is well-known, belongs to $H^1(\mathbf{R}^N)$), we get $\int_{\mathbf{R}^N}((u - M)^+)^2\,dx \leq 0$ because

$$\int_{\mathbf{R}^N} a(u)_x (u - M)^+\,dx = 0,$$

$$-\int_{\mathbf{R}^N} \Delta(u - M)\,(u - M)^+\,dx = \int_{\mathbf{R}^N} |\nabla(u - M)^+|^2\,dx \geq 0.$$

Hence, $u(x) \leq M$ a.e. $x \in \mathbf{R}^N$. Now, we multiply the equation

$$u + M + (a(u))_x - \varepsilon\Delta(u + M) = f + M \geq 0$$

by $(u + M)^-$ and get as before that $(u + M)^- = 0$ a.e. in $\mathbf{R}^N$. Hence, $u \in L^\infty(\mathbf{R}^N)$ and

$$|u(x)| \leq \|f\|_{L^\infty(\mathbf{R}^N)} \qquad \text{a.e. } x \in \mathbf{R}^N,$$

as desired. ∎

Lemma 3.5. *Under assumptions of Lemma* 3.3, *let* $f, g \in L^2(\mathbf{R}^N) \cap L^1(\mathbf{R}^N)$ *and let* $u, v \in H^2(\mathbf{R}^N) \cap L^1(\mathbf{R}^N)$ *be the corresponding solutions to Eq.* (3.47). *Then we have*

$$\|(u - v)^+\|_{L^1(\mathbf{R}^N)} \leq \|(f - g)^+\|_{L^1(\mathbf{R}^N)}, \tag{3.52}$$

$$\|u - v\|_{L^1(\mathbf{R}^N)} \leq \|(f - g)\|_{L^1(\mathbf{R}^N)}. \tag{3.53}$$

Proof. Since (3.53) is an immediate consequence of (3.52) we shall confine ourselves to the latter estimate. If multiply the equation

$$u - v + (a(u) - a(v))_x - \varepsilon \Delta(u - v) = f - g$$

by $\xi(x) \in \text{sign}(u - v)^+$ and integrate on $\mathbf{R}^N$, we get

$$\int_{\mathbf{R}^N} (u - v)^+ \, dx + \int_{\mathbf{R}^N} (a(u) - a(v))_x \xi(x) \, dx \leq \int_{\mathbf{R}^N} (f - g)^+ \, dx.$$

Now,

$$\int_{\mathbf{R}^N} (a(u) - a(v))_x \xi(x) \, dx = \int_{[u(x) > v(x)]} (a(u(x)) - a(v(x)))_x \, dx = 0$$

by the divergence theorem because

$$a(u) = a(v) \qquad \text{in } \partial\{x; u(x) > v(x)\}.$$

Hence, $\|(u - v)^+\|_{L^1(\mathbf{R}^N)} \leq \|(f - g)^+\|_{L^1(\mathbf{R}^N)}$, as claimed. ■

Proof of Proposition 3.11. Let us show first that $L^1(\mathbf{R}^N) \cap L^\infty(\mathbf{R}^N) \subset R(I + A_0)$. To this end, consider a sequence $\{a_\varepsilon\}$ of C^1 functions such that $a_\varepsilon(0) = 0$, $a_\varepsilon \overset{\varepsilon \to 0}{\to} a$ uniformly on compacta. For $f \in L^1(\mathbf{R}^N) \cap L^\infty(\mathbf{R}^N)$, let $u_\varepsilon \in H^2(\mathbf{R}^N) \cap L^1(\mathbf{R}^N) \cap L^\infty(\mathbf{R}^N)$ be the solution to Eq. (3.47). Note the estimates

$$\|u_\varepsilon\|_{L^1(\mathbf{R}^N)} \leq \|f\|_{L^1(\mathbf{R}^N)} \qquad \|u_\varepsilon\|_{L^\infty(\mathbf{R}^N)} \leq \|f\|_{L^\infty(\mathbf{R}^N)}.$$

Moreover, applying Lemma 3.5 to the functions $u = u_\varepsilon(x)$ and $v = v_\varepsilon(x + y)$, we get the estimate

$$\int_{\mathbf{R}^N} |u_\varepsilon(x + y) - u_\varepsilon(x)| \, dx \leq \int_{\mathbf{R}^N} |f(x + y) - f(x)| \, dx$$

$$\forall y \in \mathbf{R}^N.$$

According to a well-known compactness criterion these estimates imply that $\{u_\varepsilon\}$ is compact in $L^1_{\text{loc}}(\mathbf{R}^N)$. Hence, there is a subsequence, which for simplicity will again be denoted u_ε, such that

$$u_\varepsilon \to u \qquad \text{in every } L^1(B_R), \forall R > 0,$$
$$u_\varepsilon(x) \to u(x) \qquad \text{a.e. } x \in \mathbf{R}^N, \tag{3.54}$$

where $B_R = \{x; \|x\| \leq R\}$. We shall show that $u + A_0 u = f$.

Let $\varphi \in C_0^\infty(\mathbf{R}^N)$, $\varphi \geq 0$, and let $\alpha \in C^2(\mathbf{R})$ be such that $\alpha'' \geq 0$. Multiply the equation satisfied by u_ε by $\alpha'(u_\varepsilon)\varphi$, and integrate on $\mathbf{R}^N$. Integration by parts yields

$$\int_{\mathbf{R}^N} (u_\varepsilon \alpha'(u_\varepsilon)\varphi - (\alpha'(u_\varepsilon)a_\varepsilon(u_\varepsilon) - \alpha'(k)a_\varepsilon(k))\varphi_x$$

$$+ \varphi_x \int_k^{u_\varepsilon(x)} \alpha''(s)a_\varepsilon(s)\, ds + \varepsilon(\alpha''(u_\varepsilon)|\nabla u_\varepsilon|^2\varphi - \alpha'(u_\varepsilon)\,\Delta\varphi)\, dx$$

$$= \int_{\mathbf{R}^N} f\alpha'(u_\varepsilon)\varphi\, dx.$$

Letting ε tend to zero and using (3.54), we get

$$\int_{R^N} (u\alpha'(u)\varphi - (\alpha'(u)a(u) - \alpha'(k)a(k))\varphi_x$$

$$+ \varphi_x \int_k^u \alpha''(s)a(s)\, ds - dx \leq \int_{\mathbf{R}^N} f\alpha'(u)\, dx$$

for all $\varphi \in C_0^\infty(\mathbf{R}^N)$, $\varphi \geq 0$, $k \in \mathbf{R}$, and $\alpha \in C^2(\mathbf{R})$, $\alpha'' \geq 0$. Let $\alpha(s) = \theta_\varepsilon(s - k)$, where

$$\theta_\varepsilon(r) = \int_0^r \gamma_\varepsilon(s)\, ds, \qquad r \in \mathbf{R}$$

and γ_ε is given by (3.25). Letting ε tend to zero, this yields

$$\int_{\mathbf{R}^N} \operatorname{sign}_0(u - k)\,[u\varphi - (a(u) - a(k))\varphi_x - f\varphi]\, dx \leq 0.$$

On the other hand, since $\limsup_{|r|\to 0} a(r)/|r| < \infty$, $a(u) \in L^1(\mathbf{R}^N)$. We have therefore shown that $f \in u + A_0 u$. Now, let $f \in L^1(\mathbf{R}^N)$, and let $f_n \in L^1(\mathbf{R}^N) \cap L^\infty(\mathbf{R}^N)$ be such that $f_n \to f$ in $L^1(\mathbf{R}^N)$ for $n \to \infty$. Let $u_n \in D(A_0)$ be the solution to the equation $u + A_0 u \ni f_n$. Since A_0 is accretive in $L^1(\mathbf{R}^N) \times L^1(\mathbf{R}^N)$, we see that $\{u_n\}$ is convergent in $L^1(\mathbf{R}^N)$. Hence, there is $u \in L^1(\mathbf{R}^N)$ such that

$$u_n \to u, \quad v_n + u_n \to f \qquad \text{in } L^1(\mathbf{R}^N), \qquad v_n \in A_0 u_n.$$

This implies that $f \in u + Au$, thereby completing the proof. ■

In particular, we have proved that for every $f \in L^1(\mathbf{R}^N)$ the first order partial differential equation

$$u - \sum_{i=1}^{N} \frac{\partial}{\partial x_i} a_i(u) = f \qquad \text{in } \mathbf{R}^N \tag{3.55}$$

has a unique generalized solution $u \in L^1(\mathbf{R}^N)$, and the map $f \to u$ is Lipschitz in $L^1(\mathbf{R}^N)$.

Bibliographical Notes and Remarks

Section 1. Theorem 1.1 is due to G. Minty [2], Theorems 1.2 and 1.3 were originally given by G. Minty [1] in Hilbert space and extended to reflexive Banach spaces by F. Browder [1, 2]. Theorems 1.4 and 1.5 are due to R. T. Rockafellar [3, 4], but the proofs given here are due to P. M. Fitzpatrick [1] and to H. Brézis *et al.* [1], respectively. For other significant results in the theory of monotone operators, we refer the reader to the survey of F. Browder [2].

Section 2. The results of Section 2.2.1 are essentially due to J. Moreau [2] and to R. T. Rockafellar [1]. Theorem 2.2 has been established in Hilbert space by H. Brézis [2, 4] (see also J. Moreau [1]) and in this form arises in the author's book [2]. Theorem 2.4 is due to H. Brézis [4]. For other results on convex functions and their subdifferentials, we refer to the monographs of R. T. Rockafellar [1], H. Brézis [4], and V. Barbu and T. Precupanu [1].

The concept of generalized gradient developed in Section 2.2.3 along with its basic properties are due to F. Clarke [1, 2] (see also R. T. Rockafellar [7, 8]). Proposition 2.15 has been established in the author's book [7].

Section 3. The general theory of m-accretive operators has been developed in the works of T. Kato [3] and M. G. Crandall and A. Pazy [1, 2] in connection with theory of semigroups of nonlinear contractions. Proposition 3.9 is due to H. Brézis and W. Strauss [1] and Proposition 3.11 to M. G. Crandall [1]. Other examples and applications to partial differential equations can be found in author's book [2] as well as in the monographs of R. Martin [2] and D. Zeidler [1].

Chapter 3 | Controlled Elliptic Variational Inequalities

This chapter is concerned with optimal control problems governed by variational inequalities of elliptic type and semilinear elliptic equations. The main emphasis is put on first order necessary conditions of optimality obtained by an approximating regularizing process.

Since the optimal control problems governed by nonlinear elliptic equations, and in particular by variational inequalities, are nonconvex and nonsmooth the standard methods to derive first order necessary conditions of optimality are usually inapplicable in this situation. The method we shall use here is to approximate the given problem by a family of smooth optimization problems containing an adapted penalty term and to pass to limit in the corresponding optimality conditions.

We shall discuss in detail several controlled free boundary problems to which the general theory is applied, such as the obstacle problem and the Signorini problem.

3.1. Elliptic Variational Inequalities. Existence Theory

3.1.1. Abstract Elliptic Variational Inequalities

Let X be a reflexive Banach space with the dual X^* and let $A: X \to X^*$ be a monotone operator (linear or nonlinear). Let $\varphi: X \to \overline{\mathbf{R}}$ be a lower semicontinuous convex function on X, $\varphi \not\equiv +\infty$. If f is a given element of X, consider the following problem.

Find $y \in X$ such that

$$(Ay, y - z) + \varphi(y) - \varphi(z) \leq (y - z, f) \qquad \forall z \in X. \tag{1.1}$$

This is an *abstract elliptic variational* inequality associated with the operator A and convex function φ, and can be equivalently expressed as

$$Ay + \partial\varphi(y) \ni f, \tag{1.2}$$

where $\partial\varphi \subset X \times X^*$ is the subdifferential of φ. In the special case where $\varphi = I_K$ is the indicator function of a closed convex subset K of X, i.e.,

$$I_K(x) = \begin{cases} 0 & \text{if } x \in K, \\ +\infty & \text{otherwise,} \end{cases}$$

problem (1.1) becomes:

Find $y \in K$ such that

$$(Ay, y - z) \leq (y - z, f) \qquad \forall z \in K. \tag{1.3}$$

It is useful to notice that if the operator A is itself a subdifferential $\partial\psi$ of a continuous convex function $\psi: X \to \mathbf{R}$, then the variational inequality (1.1) is equivalent to the minimization problem (the Dirichlet principle)

$$\min\{\psi(z) + \varphi(z) - (z, f);\ z \in X\} \tag{1.4}$$

or, in the case of problem (1.3),

$$\min\{\psi(z) - (z, f);\ z \in K\}. \tag{1.5}$$

As far as concerns existence in problem (1.1), we note first the following result.

Theorem 1.1. *Let $A: X \to X^*$ be a monotone, demicontinuous operator and let $\varphi: X \to \overline{\mathbf{R}}$ be a lower semicontinuous, proper, convex function. Assume that there exists $y_0 \in D(\varphi)$ such that*

$$\lim_{\|y\| \to \infty} ((Ay, y - y_0) + \varphi(y))/\|y\| = +\infty. \tag{1.6}$$

Then problem (1.1) *has at least one solution. Moreover, the set of solutions is bounded, convex, and closed in X. If the operator A is strictly monotone, i.e., $(Au - Av, u - v) = 0 \Leftrightarrow u = v$, then the solution is unique.*

Proof. By Theorem 1.5 in Chapter 2, the operator $A + \partial\varphi$ is maximal monotone in $X \times X^*$. Since by condition (1.6) it is also coercive, we conclude (see Corollary 1.2 in Chapter 2) that is surjective. Hence, Eq. (1.2) (equivalently, (1.1)) has at least one solution.

Since the set of all solutions y to (1.1) is $(A + \partial\varphi)^{-1}(f)$, we infer that this set is closed and convex (Proposition 1.1 in Chapter 2). By the coercivity condition (1.6), it is also bounded. Finally, if A (or more generally, if $A + \partial\varphi$) is strictly monotone, then $(A + \partial\varphi)^{-1}f$ consists of a single element.

■

In the special case $\varphi = I_K$, we have:

Corollary 1.1. *Let $A: X \to X^*$ be a monotone demicontinuous operator and let K be a closed convex subset of X. Assume either that there is $y_0 \in K$ such that*

$$\lim_{\|y\| \to \infty} (Ay, y - y_0)/\|y\| = +\infty, \tag{1.7}$$

or that K is bounded. Then problem (1.3) *has at least one solution. The set of all solutions is bounded, convex, and closed. If A is strictly monotone, then the solution to* (1.3) *is unique.* ■

To be more specific we shall assume in the following that $X = V$ is a Hilbert space, $X^* = V'$, and

$$V \subset H \subset V' \tag{1.8}$$

algebraically and topologically, where H is a real Hilbert space identified with its own dual. The norms of V and H will be denoted by $\|\cdot\|$ and $|\cdot|$, respectively. For $v \in V$ and $v' \in V'$ we denote by (v, v') the value of v' in v; if $v, v' \in H$, this is the scalar product in H of v and v'. The norm in V' will be denoted by $\|\cdot\|_*$.

Let $A \in L(V, V')$ be a linear continuous operator from V to V' such that, for some $\omega > 0$,

$$(Av, v) \geq \omega\|v\|^2 \qquad \forall v \in V.$$

Very often the operator A is defined by the equation

$$(u, Av) = a(u, v) \qquad \forall u, v \in V, \tag{1.9}$$

where $a: V \times V \to \mathbf{R}$ is a bilinear continuous functional on $V \times V$ such that

$$a(v, v) \geq \omega\|v\|^2 \qquad \forall v \in V. \tag{1.10}$$

In terms of a, the variational inequality (1.1) on V becomes

$$a(y, y - z) + \varphi(y) - \varphi(z) \leq (y - z, f) \qquad \forall z \in V, \qquad (1.11)$$

and,

$$y \in K, \qquad a(y, y - z) \leq (y - z, f) \qquad \forall z \in K. \qquad (1.12)$$

As we shall see later, in applications V is usually a Sobolev space on an open subset Ω of $\mathbf{R}^N$, $H = L^2(\Omega)$, and A is an elliptic differential operator on Ω with appropriate homogeneous boundary value conditions. The set K incorporates various unilateral conditions on the domain Ω or on its boundary $\partial\Omega$.

By Theorem 2.1 of Chapter 2 we have the following existence result for problem (1.11).

Corollary 1.2. *Let $a: V \times V \to \mathbf{R}$ be a bilinear continuous functional satisfying condition* (1.10) *and let $\varphi: V \to \overline{\mathbf{R}}$ be a* l.s.c., *convex proper function. Then, for every $f \in V'$, problem* (1.11) *has a unique solution $y \in V$. The map $f \to y$ is Lipschitz from V' to V.* ■

Similarly for problem (1.12):

Corollary 1.3. *Let $a: V \times V \to \mathbf{R}$ be a bilinear continuous functional satisfying condition* (1.10) *and let K be a closed convex subset of V. Then, for every $f \in V'$, problem* (1.12) *has a unique solution y. The map $f \to y$ is Lipschitz from V' to V.* ■

A problem of great interest when studying Eq. (1.11) is whether $Ay \in H$. To answer this problem, we define the operator $A_H: H \to H$,

$$A_H y = Ay \qquad \text{for } y \in D(A_H) = \{u \in V;\ Au \in H\}. \qquad (1.13)$$

The operator A_H is positive definite on H and $R(I + A_H) = H$ (I is the unit operator in H). (Indeed, by Theorem 1.3 in Chapter 2 the operator $I + A$ is surjective from V to V'.) Hence, A_H is maximal monotone in $H \times H$.

Theorem 1.2. *Under the assumptions of Corollary* 1.2, *suppose in addition that there exists $h \in H$ and $C \in \mathbf{R}$ such that*

$$\varphi\big((I + \lambda A_H)^{-1}(y + \lambda h)\big) \leq \varphi(y) + C\lambda \qquad \forall \lambda > 0,\ y \in V. \quad (1.14)$$

Then, if $f \in H$, *the solution* y *to* (1.11) *belongs to* $D(A_H)$ *and*

$$|Ay| \leq C(I + |f|). \tag{1.15}$$

Proof. Let $A_\lambda \in L(H, H)$ be the Yosida approximation of A_H, i.e.,

$$A_\lambda = \lambda^{-1}\left(I - (I + \lambda A_H)^{-1}\right), \qquad \lambda > 0.$$

Let $y \in V$ be the solution to (1.11). If in (1.11) we set $z = (I + \lambda A_H)^{-1}(y + \lambda h)$ and use condition (1.14), we get

$$(Ay, A_\lambda y) - \left(Ay, (I + \lambda A_H)^{-1} h\right) \leq (A_\lambda y, f) - \left((I + \lambda A_H)^{-1} h, f\right).$$

Since $(Ay, A_\lambda y) \geq |A_\lambda y|^2$ for all $\lambda > 0$ and $y \in V$, we get

$$|A_\lambda y|^2 \leq |A_\lambda y|\,|h| + |A_\lambda y|\,|f| + |f|\,|h| \qquad \forall \lambda > 0.$$

(Here we have assumed that A is symmetric; the general case follows by Theorem 2.4 in Chap. 2.) We get the estimate

$$|A_\lambda y| \leq C(1 + |f|) \qquad \forall \lambda > 0,$$

where C is independent of λ and f. This implies that $y \in D(A_H)$ and estimate (1.15) holds. ■

Corollary 1.4. *In Corollary* 1.3, *assume in addition that* $f \in H$ *and*

$$(I + \lambda A_H)^{-1}(y + \lambda h) \in K \qquad \text{for some } h \in H \text{ and all } \lambda > 0. \tag{1.16}$$

Then the solution y *to variational inequality* (1.3) *belongs to* $D(A_H)$, *and the following estimate holds*:

$$|Ay| \leq C(I + |f|) \qquad \forall f \in H. \quad ■ \tag{1.17}$$

3.1.2. The Obstacle Problem

Throughout this section, Ω is an open and bounded subset of the Euclidean space $\mathbf{R}^N$ with a smooth boundary $\partial\Omega$. In fact, we shall assume that $\partial\Omega$ is of class C^2. However, if Ω is convex this regularity condition on $\partial\Omega$ is no longer necessary.

Let $V = H^1(\Omega)$, $H = L^2(\Omega)$, and $A: V \to V'$ be defined by

$$(z, Ay) = a(y, z) = \sum_{i=1}^{N} \int_\Omega a_{ij}(x) y_{x_i}(x) z_{x_j}(x)\, dx$$
$$+ \int_\Omega a_0(x) y(x) z(x)\, dx + \frac{\alpha_1}{\alpha_2} \int_{\partial\Omega} y(x) z(x)\, d\sigma_x$$
$$\forall y, z \in V, \quad (1.18)$$

where α_1, α_2 are two nonnegative constants such that $\alpha_1 + \alpha_2 > 0$. If $\alpha_2 = 0$, we take $V = H_0^1(\Omega)$ and $A: H_0^1(\Omega) \to H^{-1}(\Omega)$ defined by

$$(z, Ay) = a(y, z) = \sum_{i=1}^{N} \int_\Omega a_{ij}(x) y_{x_i}(x) z_{x_j}(x)\, dx$$
$$+ \int_\Omega a_0(x) y(x) z(x)\, dx \qquad \forall y, z \in H_0^1(\Omega). \quad (1.19)$$

Here, $a_0, a_{ij} \in L^\infty(\Omega)$ for all $i, j = 1, \ldots, N$, $a_{ij} = a_{ji}$, and

$$a_0(x) \geq 0, \qquad \sum_{i,j=1}^{N} a_{ij}(x) \xi_i \xi_j \geq \omega \|\xi\|_N^2 \qquad \forall \xi \in \mathbf{R}^N,\ x \in \Omega, \quad (1.20)$$

where ω is some positive constant and $\|\cdot\|_N$ is the Euclidean norm in $\mathbf{R}^N$.

If $\alpha_1 = 0$, we shall assume that $a_0(x) \geq \rho > 0$ a.e. $x \in \Omega$.

The reader will recognize of course in the operator defined by (1.18) the second order elliptic operator

$$A_0 y = -\sum_{i,j=1}^{N} (a_{ij} y_{x_i})_{x_j} + a_0 y \quad (1.21)$$

with the boundary value conditions

$$\alpha_1 y + \alpha_2 \frac{\partial y}{\partial \nu} = 0 \qquad \text{in } \partial\Omega, \quad (1.22)$$

where $\partial/\partial\nu$ is the outward normal derivative,

$$\frac{\partial}{\partial \nu} y = \sum_{i,j=1}^{N} a_{ij} y_{x_i} \cos(\nu, x_i). \quad (1.23)$$

Similarly, the operator A defined by (1.19) is the differential operator (1.21) with Dirichlet homogeneous conditions: $y = 0$ in $\partial\Omega$.

Let $\psi \in H^2(\Omega)$ be a given function and let K be the closed convex subset of $V = H^1(\Omega)$ defined by

$$K = \{y \in V;\ y(x) \geq \psi(x) \text{ a.e. } x \in \Omega\}. \tag{1.24}$$

Note that $K \neq \emptyset$ because $\psi^+ = \max(\psi, 0) \in K$. If $V = H_0^1(\Omega)$, we shall assume that $\psi(x) \leq 0$ a.e. $x \in \partial\Omega$, which will imply as before that $K \neq \emptyset$.

Let $f \in V'$. Then, by Corollary 1.3, the variational inequality

$$a(y, y - z) \leq (y - z, f) \qquad \forall z \in K \tag{1.25}$$

has a unique solution $y \in K$.

Formally, y is the solution to the following boundary value problem known in the literature as the *obstacle problem*,

$$A_0 y = f \qquad \text{in } \Omega^+ = \{x \in \Omega;\ y(x) > \psi(x)\},$$
$$A_0 y \geq f, \quad y \geq \psi \qquad \text{in } \Omega,$$
$$y = \psi \qquad \text{in } \Omega \setminus \Omega^+, \ \frac{\partial y}{\partial \nu} = \frac{\partial \psi}{\partial \nu} \qquad \text{in } \partial\Omega^+ \setminus \partial\Omega, \tag{1.26}$$

$$\alpha_1 y + \alpha_2 \frac{\partial y}{\partial \nu} = 0 \qquad \text{in } \partial\Omega. \tag{1.27}$$

Indeed, if $\psi \in C(\overline{\Omega})$ and y is a sufficiently smooth solution, then Ω^+ is an open subset of Ω and so for every $\alpha \in C_0^\infty(\Omega^+)$ there is $\rho > 0$ such that $y \pm \rho\alpha \geq \psi$ on Ω, i.e., $y \pm \rho\alpha \in K$. Then if take $z = y \pm \rho\alpha$ in (1.25), we see that

$$\sum_{i,j=1}^{N} \int_\Omega a_{ij} y_{x_i} \alpha_{x_j}\, dx + \int_\Omega a_0 y \alpha\, dx = (f, \alpha) \qquad \forall \alpha \in C_0(\Omega^+).$$

Hence, $A_0 y = f$ in $\mathscr{D}'(\Omega^+)$.

Now, if we take $z = y + \alpha$, where $\alpha \in H^1(\Omega)$ and $\alpha \geq 0$ in Ω, we get

$$\sum_{i,j=1}^{N} \int_\Omega a_{ij} y_{x_i} \alpha_{x_j}\, dx + \int_\Omega a_0 y \alpha\, dx \geq (f, \alpha),$$

and therefore $A_0 y \geq f$ in $\mathscr{D}'(\Omega)$.

The boundary conditions (1.27) are obviously incorporated into the definition of the operator A if $\alpha_2 = 0$. If $\alpha_2 > 0$, then the boundary conditions (1.27) follow from the inequality (1.25) if $\alpha_1 \psi + \alpha_2\, \partial\psi/\partial\nu \leq 0$ a.e. in $\partial\Omega$ (see Theorem 1.3 following). As for the equation $\partial y/\partial\nu = \partial\psi/\partial\nu$ in $\partial\Omega^+$, this is a transmission property that is implied by the conditions $y \geq \psi$ in Ω and $y = \psi$ in $\partial\Omega^+$, if y is smooth enough.

In problem (1.26), (1.27), the surface $\partial\Omega^+ \setminus \partial\Omega = S$, which separates the domains Ω^+ and $\Omega \setminus \overline{\Omega}^+$ is not known *a priori* and is called the *free boundary*. In classical terms, this problem can be reformulated as follows: Find the free boundary S and the function y that satisfy the system

$$A_0 y = f \qquad \text{in } \Omega^+,$$

$$y = \psi \qquad \text{in } \Omega \setminus \Omega^+, \qquad \frac{\partial y}{\partial \nu} = \frac{\partial \psi}{\partial \nu} \qquad \text{in } S,$$

$$\alpha_1 y + \alpha_2 \frac{\partial y}{\partial \nu} = 0 \qquad \text{in } \partial\Omega.$$

In the variational formulation (1.25), the free boundary S does not appear explicitly but the unknown function y satisfies a nonlinear equation. Once y is known, the free boundary S can be found as the boundary of the coincidence set $\{x \in \Omega;\ y(x) = \psi(x)\}$.

There exists an extensive literature on regularity properties of the solution to the obstacle problem and of the free boundary. We mention in this context the earlier work of Brézis and Stampacchia [1] and the books of Kinderlehrer and Stampacchia [1] and A. Friedman [1], which contain complete references on the subject. Here, we shall present only a partial result.

Proposition 1.1. *Assume that $a_{ij} \in C^1(\overline{\Omega})$, $a_0 \in L^\infty(\Omega)$, and that conditions* (1.20) *hold. Further, assume that $\psi \in H^2(\Omega)$ and*

$$\alpha_1 \psi + \alpha_2 \frac{\partial \psi}{\partial \nu} \le 0 \qquad \text{a.e. in } \partial\Omega. \tag{1.28}$$

Then for every $f \in L^2(\Omega)$ the solution y to variational inequality (1.25) *belongs to $H^2(\Omega)$ and satisfies the complementarity system*

$$\begin{aligned} (A_0 y(x) - f(x))(y(x) - \psi(x)) = 0 \qquad & \text{a.e. } x \in \Omega, y(x) \ge \psi(x), \\ A_0 y(x) \ge f(x) \qquad & \text{a.e. } x \in \Omega, \end{aligned} \tag{1.29}$$

along with boundary value conditions

$$\alpha_1 y(x) + \alpha_2 \frac{\partial y}{\partial \nu}(x) = 0 \qquad \text{a.e. } x \in \partial\Omega. \tag{1.30}$$

Moreover, there exists a positive constant C independent of f such that

$$\|y\|_{H^2(\Omega)} \le C(\|f\|_{L^2(\Omega)} + 1). \tag{1.31}$$

Proof. We shall apply Corollary 1.4, where $H = L^2(\Omega)$, $V = H^1(\Omega)$ (respectively, $V = H_0^1(\Omega)$ if $\alpha_2 = 0$), A is defined by (1.18) (respectively, (1.19)), and K is given by (1.24).

Clearly, the operator $A_H: L^2(\Omega) \to L^2(\Omega)$ is defined in this case by

$$(A_H y)(x) = (A_0 y)(x) \qquad \text{a.e. } x \in \Omega,\ y \in D(A_H),$$
$$D(A_H) = \left\{ y \in H^2(\Omega);\ \alpha_1 y + \alpha_2 \frac{\partial y}{\partial \nu} = 0 \text{ a.e. in } \partial\Omega \right\}.$$

We shall verify condition (1.16) with $h = A_0 \psi$. To this end, consider for $\lambda > 0$ the boundary value problem

$$w + \lambda A_0 w = y + \lambda A_0 \psi \qquad \text{in } \Omega,$$
$$\alpha_1 w + \alpha_2 \frac{\partial w}{\partial \nu} = 0 \qquad \text{in } \partial\Omega,$$

which has a unique solution $w \in D(A_H)$. Multiplying this equation by $(w - \psi)^- \in H^1(\Omega)$ and integrating on Ω, we get, via Green's formula,

$$\int_\Omega |(w - \psi)^-|^2\, dx + \lambda a((w - \psi)^-, (w - \psi)^-)$$
$$- \frac{\lambda}{\alpha_2} \int_{\partial\Omega} \left(\alpha_1 \psi + \alpha_2 \frac{\partial \psi}{\partial \nu} \right) (w - \psi)^-\, d\sigma$$
$$= - \int_\Omega (y - \psi)(w - \psi)^-\, dx \leq 0.$$

Hence, $(w - \psi)^- = 0$ a.e. in Ω and so $w \in K$ as claimed. Then, by Corollary 1.4, we infer that $y \in D(A_H)$ and

$$\|A_H y\|_{L^2(\Omega)} \leq C(\|f\|_{L^2(\Omega)} + 1),$$

and since $\partial\Omega$ is sufficiently smooth (or Ω convex) this implies (1.31).
Now, if $y \in D(A_H)$, we have

$$a(y, z) = \int_\Omega A_0 y(x) z(x)\, dx \qquad \forall z \in H^1(\Omega),$$

and so by (1.25) we see that

$$\int_\Omega (A_0 y(x) - f(x))(y(x) - z(x))\, dx \leq 0 \qquad \forall z \in K. \quad (1.32)$$

The last inequality clearly can be extended by density to all $z \in K_0$, where

$$K_0 = \{u \in L^2(\Omega);\ u(x) \geq \psi(x) \text{ a.e. } x \in \Omega\}. \tag{1.33}$$

If in (1.32) we take $z = \psi + \alpha$, where α is any positive $L^2(\Omega)$ function, we get

$$(A_0 y)(x) - f(x) \geq 0 \qquad \text{a.e. } x \in \Omega.$$

Then, for $z = \psi$, (1.32) yields

$$(y(x) - \psi(x))(A_0 y)(x) - f(x) = 0 \qquad \text{a.e. } x \in \Omega,$$

which completes the proof. ■

We note that under assumptions of Theorem 1.3 the obstacle problem can be equivalently written as

$$A_H y + \partial I_{K_0}(y) \ni f, \tag{1.34}$$

where

$$\partial I_{K_0}(y) = \left\{ v \in L^2(\Omega);\ \int_\Omega v(x)(y(x) - z(x))\,dx \geq 0\ \forall z \in K_0 \right\}$$

or, equivalently,

$$\partial I_{K_0}(y) = \{v \in L^2(\Omega);\ v(x) \in \beta(y(x) - \psi(x)) \text{ a.e. } x \in \Omega\},$$

where $\beta: \mathbf{R} \to 2^{\mathbf{R}}$ is the maximal monotone graph,

$$\beta(r) = \begin{cases} 0 & \text{if } r > 0, \\ \mathbf{R}^- & \text{if } r = 0, \\ \varnothing & \text{if } r < 0. \end{cases} \tag{1.35}$$

Hence, under the conditions of Theorem 1.3, we may equivalently write the variational inequality (1.25) as

$$\begin{aligned} (A_0 y)(x) + \beta(y(x) - \psi(x)) &\ni f(x) \qquad \text{a.e. } x \in \Omega, \\ \alpha_1 y + \alpha_2 \frac{\partial y}{\partial \nu} &= 0 \qquad \text{a.e. in } \partial\Omega, \end{aligned} \tag{1.36}$$

and as seen in Section 2.2, Chapter 2, it is equivalent to the minimization problem

$$\min\left\{\tfrac{1}{2}a(y,y) + \int_\Omega j(y(x) - \psi(x))\,dx - \int_\Omega f(x)y(x)\,dx;\ y \in L^2(\Omega)\right\}, \tag{1.36$'$}$$

where $j: \mathbf{R} \to \overline{\mathbf{R}}$ is defined by

$$j(r) = \begin{cases} 0 & \text{if } r \geq 0 \\ +\infty & \text{otherwise.} \end{cases} \tag{1.37}$$

As seen elsewhere, Eq. (1.36) can be approximated by the smooth boundary value problem

$$\begin{aligned} A_0 y + \beta_\varepsilon(y - \psi) &= f && \text{a.e. in } \Omega, \\ \alpha_1 y + \alpha_2 \frac{\partial y}{\partial \nu} &= 0 && \text{a.e. in } \partial\Omega, \end{aligned} \tag{1.38}$$

where $\beta_\varepsilon(r) = -(1/\varepsilon)r^-$ (β_ε is the Yosida approximation of β).

In this context, we have a more general result. Let β be a maximal monotone graph in $\mathbf{R} \times \mathbf{R}$, and let $\psi \in H^2(\Omega)$. Let $\beta_\varepsilon = \varepsilon^{-1}(1 - (1 + \varepsilon\beta)^{-1})$ be the Yosida approximation of β. Then, for each $f \in L^2(\Omega)$, the boundary value problem

$$\begin{aligned} A_0 y + \beta_\varepsilon(y - \psi) &= f && \text{in } \Omega, \\ \alpha_1 y + \alpha_2 \frac{\partial y}{\partial \nu} &= 0 && \text{in } \partial\Omega \end{aligned} \tag{1.39}$$

has a unique solution $y_\varepsilon^f \in H^2(\Omega)$. (Problem (1.39) can be written as $A_H y + \beta_\varepsilon(y - \psi) = f$, where $y \to \beta_\varepsilon(y - \psi)$ is monotone and continuous in $L^2(\Omega)$.)

Proposition 1.2. *Assume that*

$$\left(\alpha_1 \psi + \alpha_2 \frac{\partial \psi}{\partial \nu}\right)\beta_\varepsilon(-\psi) \geq 0 \qquad \text{a.e. in } \partial\Omega. \tag{1.40}$$

Then

$$y_\varepsilon^f \to y^f \qquad \textit{strongly in } H^1(\Omega), \textit{ weakly in } H^2(\Omega), \tag{1.41}$$

where y_ε^f is the solution to boundary value problem (1.39).*Moreover, if* $f_\varepsilon \to f$ *weakly in* $L^2(\Omega)$, *then*

$$y_\varepsilon^{f_\varepsilon} \to y^f \qquad \textit{weakly in } H^2(\Omega), \textit{ strongly in } H^1(\Omega). \tag{1.42}$$

Proof. Let $y_\varepsilon = y_\varepsilon^{f_\varepsilon}$. Multiplying Eq. (1.39) by $\beta_\varepsilon(y_\varepsilon - \psi)$ and integrating on Ω, we get, by Green's formula,

$$\int_\Omega |\beta_\varepsilon(y_\varepsilon - \psi)|^2\,dx + \int_\Omega A_0(y_\varepsilon - \psi)\beta_\varepsilon(y_\varepsilon - \psi)\,dx$$

$$= \int_\Omega (f - A_0\psi)\beta_\varepsilon(y_\varepsilon - \psi)\,dx.$$

Hence

$$\int_\Omega |\beta_\varepsilon(y_\varepsilon - \psi)|^2\,dx + a(y_\varepsilon - \psi, \beta_\varepsilon(y_\varepsilon - \psi))$$

$$+ \int_{\partial\Omega} \frac{\partial}{\partial\nu}(y_\varepsilon - \psi)\beta_\varepsilon(y_\varepsilon - \psi)\,d\sigma = \int_\Omega (f - A_0\psi)\beta_\varepsilon(y_\varepsilon - \psi)\,dx.$$

Inasmuch as $(\partial/\partial\nu)(y_\varepsilon - \psi)\,\beta_\varepsilon(y_\varepsilon - \psi) = -(\alpha_1/\alpha_2)(y_\varepsilon - \psi)\,\beta_\varepsilon(y_\varepsilon - \psi) - (\partial\psi/\partial\nu + (\alpha_1/\alpha_2)\psi)\,\beta_\varepsilon(-\psi) \le 0$ in $\partial\Omega$, we infer that

$$\{\beta_\varepsilon(y_\varepsilon - \psi)\} \qquad \text{is bounded in } L^2(\Omega)$$

and $\{A_0 y_\varepsilon\}$ is bounded in $L^2(\Omega)$. We may conclude, therefore, that $\{y_\varepsilon\}$ is bounded in $H^2(\Omega)$ and on a subsequence, again denoted $\{y_\varepsilon\}$, we have

$$y_\varepsilon \to y \qquad \text{weakly in } H^2(\Omega), \text{ strongly in } H^1(\Omega), \tag{1.43}$$

$$\beta_\varepsilon(y_\varepsilon - \psi) \to \xi \qquad \text{weakly in } L^2(\Omega), \tag{1.44}$$

$$A_H y_\varepsilon = A_0 y_\varepsilon \to A_H y \qquad \text{weakly in } L^2(\Omega). \tag{1.45}$$

Clearly, we have

$$A_0 y + \xi = f \qquad \text{a.e. in } \Omega$$

$$\alpha_2 \frac{\partial y}{\partial \nu} + \alpha_1 y = 0 \qquad \text{a.e. in } \partial\Omega.$$

On the other hand, if denote by $B \subset L^2(\Omega) \times L^2(\Omega)$ the operator $By = \{w \in L^2(\Omega);\ w(x) \in \beta(y(x) - \psi(x))$ a.e. in $\Omega\}$, then B is maximal monotone and its Yosida approximation B_ε is given by

$$B_\varepsilon(y) = \beta_\varepsilon(y - \psi) \qquad \text{a.e. in } \Omega.$$

Then, by (1.44) and Proposition 1.3 in Chapter 2, we deduce that $\xi \in By$, i.e., $\xi(x) \in \beta(y(x) - \psi(x))$. Hence $y = y^f$, thereby completing the proof of Proposition 1.2. ■

In particular, Condition (1.40) is satisfied if β is given by (1.35) and so Proposition 1.1 can be viewed as a particular case of Proposition 1.2.

A simple physical model for the obstacle problem is that of an elastic membrane that occupies a plane domain Ω and is limited from below by a rigid obstacle ψ whilst it is under the pressure of a vertical force field of density f. We assume that the membrane is clamped along the boundary $\partial\Omega$ (see Fig. 1.1). It is well-known from linear elasticity that when there is no obstacle the vertical displacement $y = y(x)$, $x \in \Omega$, of the membrane satisfies the Laplace–Poisson equation. In the presence of the obstacle $y = \psi(x)$, the deflection $y = y(x)$ of the membrane satisfies the system (1.26). More precisely, we have

$$\begin{aligned} -\Delta y = f \quad & \text{in } \{x \in \Omega;\, y(x) > \psi(x)\}, \\ -\Delta y \geq f, \quad y \geq \psi \quad & \text{in } \Omega, \\ y = 0 \quad & \text{in } \partial\Omega. \end{aligned} \tag{1.46}$$

The contact region $\{x \in \Omega;\, y(x) = \psi(x)\}$ is not known *a priori* and its boundary is the free boundary of the problem.

Let us consider now the case of two parallel elastic membranes loaded by forces f_i, $i = 1, 2$, that act from opposite directions (see Fig. 1.2).The variational inequality characterizing the equilibrium solution y is (see, e.g.,

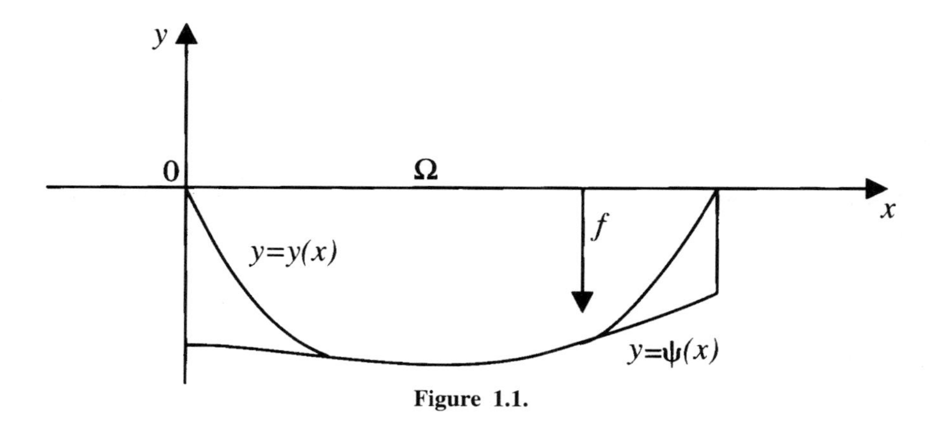

Figure 1.1.

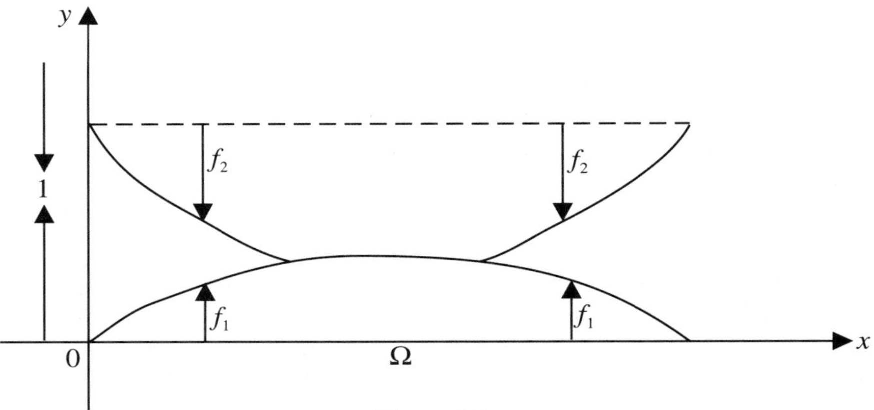

Figure 1.2.

Kikuchi and Oden [1])

$$\mu_1 \int_\Omega \nabla y_1 \cdot \nabla(y_1 - z_1)\, dx + \mu_2 \int_\Omega \nabla y_2 \cdot \nabla(y_2 - z_2)\, dx$$
$$\leq \int_\Omega f_1(y_1 - z_1)\, dx + \int_\Omega f_2(y_2 - z_2)\, dx \qquad \forall (z_1, z_2) \in K, \quad (1.47)$$

where $f_1 \geq 0$, $f_2 \leq 0$, and

$$K = \{(y_1, y_2) \in H_0^1(\Omega) \times H_0^1(\Omega);\ y_1 - y_2 \leq l \text{ a.e. in } \Omega\}. \quad (1.48)$$

Here, μ_1, μ_2 are positive constants, l is the distance between the initial positions of the unloaded membranes, and $y_1(x_1, x_2) \geq 0$, $y_2(x_1, x_2) \leq 0$ are the deflection of the membranes 1 and 2 in $(x_1, x_2) = x$.

This problem is of the form (1.25), where $H = L^2(\Omega) \times L^2(\Omega)$, $V = H_0^1(\Omega) \times H_0^1(\Omega)$, K is defined by (1.48), $f = (f_1, f_2)$, and

$$a(y, z) = \mu_1 \int_\Omega \nabla y_1 \cdot \nabla z_1\, dx + \mu_2 \int_\Omega \nabla y_2 \cdot \nabla z_2\, dx$$
$$\text{for } y = (y_1, y_2),\ z = (z_1, z_2) \in V \times V.$$

Formally, the solution $y = (y_1, y_2)$ to problem (1.47) is the solution to the free boundary problem

$$\begin{aligned}
&-\mu_1 \Delta y_1 = f_1, \quad -\mu_2 \Delta y_2 = f_2 \quad \text{in } \{x;\ y_1(x) - y_2(x) < l\},\\
&y_1 - y_2 \leq l \quad \text{in } \Omega,\\
&-\mu_1 \Delta y_1 \leq f_1, \quad -\mu_2 \Delta y_2 \geq f_2 \quad \text{in } \Omega,\\
&y_1 = 0, \quad y_2 = 0 \quad \text{in } \partial\Omega.
\end{aligned}$$

The free boundary of this problem is the boundary of the contact set $\{x; y_1(x) - y_2(x) = l\}$.

An important success of the theory of elliptic variational inequalities has been the discovery made by C. Baiocchi [1] that the mathematical model of the water flow through an isotropic homogeneous rectangular dam can be described as an obstacle problem of the type just presented. Let us now briefly described this problem.

Denote by $D = (0, a) \times (0, b)$ the dam and by D_0 the wetted region (see Fig. 1.3).The boundary S that separates the wetted region D_0 from the dry region $D_1 = D \setminus \overline{D}_0$ is unknown and it is a free boundary.

Let z be the piezometric head and let $p(x_1, x_2)$ be the unknown pressure at the point $(x_1, x_2) \in D$. We have $z = p + x_2$ in D and, by the D'Arcy law (we normalize the coefficients),

$$\Delta z = 0 \qquad \text{in } D_0. \tag{1.49}$$

Note also that z satisfies the obvious boundary conditions

$$z = h_1 \quad \text{in } AF, \qquad z = x_2 \quad \text{in } S \cup GC, \qquad z = h_2 \quad \text{in } BC,$$
$$\frac{\partial z}{\partial x_2} = 0 \quad \text{in } AB, \qquad \frac{\partial z}{\partial \nu} = 0 \quad \text{in } S, \tag{1.50}$$

where $\partial / \partial \nu$ is the normal derivative to S.

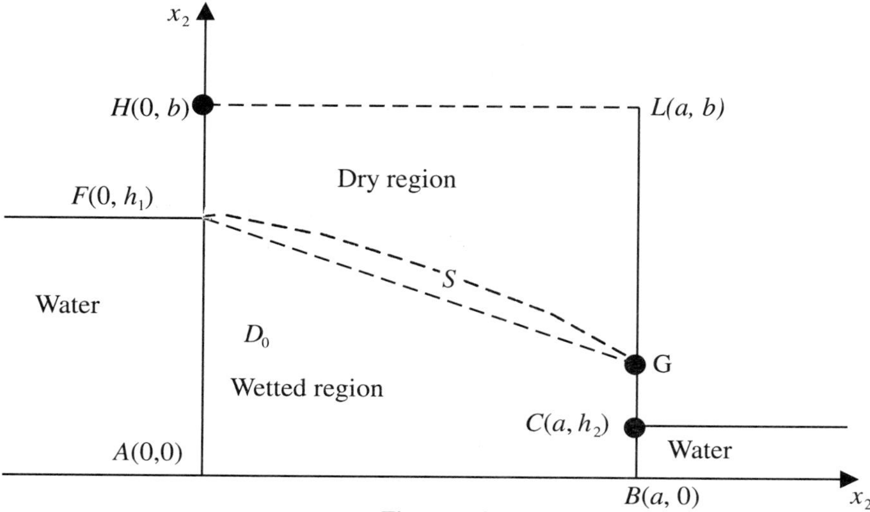

Figure 1.3.

Introduce the function

$$\tilde{z}(x_1, x_2) = \begin{cases} z(x_1, x_2) & (x_1, x_2) \in \overline{D}_0, \\ x_2 & (x_1, x_2) \in \overline{D} \setminus \overline{D}_0, \end{cases}$$

and consider the *Baiocchi transformation*

$$y(x_1, x_2) = \int_{x_2}^{b} (\tilde{z}(x_1, s) - s)\, ds \qquad \forall (x_1, x_2) \in \overline{D}.$$

We have:

Lemma 1.1. *The function y satisfies the equation*

$$\Delta y = \chi_{D_0} \qquad \text{in } \mathscr{D}'(D) \tag{1.51}$$

and the conditions

$$y > 0 \quad \text{in } D_0 \qquad y = 0 \quad \text{in } \overline{D} \setminus D_0, \qquad y = g \quad \text{in } \partial D, \tag{1.52}$$

where

$$g = 0 \qquad \text{in } FH \cup HL \cup LC,$$

$$g(x_1, 0) = \frac{1}{2} h_1^2 - \frac{x_1}{2a}(h_1^2 - h_2^2) = \alpha x_1 + \beta,$$

$$g = \frac{1}{2}(x_2 - h_2)^2 \qquad \textit{in } CB, \qquad g = \frac{1}{2}(x_2 - h_1)^2 \qquad \textit{in } AF.$$

We have denoted by χ_{D_0} the characteristic function of D_0.

Proof. We shall assume that $y \in H^1(D)$ and that the free boundary S is sufficiently smooth. Then, if $x_2 = \sigma(x_1)$ is the equation of S, we have, for every test function $\varphi \in C_0^\infty(D)$,

$$\begin{aligned}(\Delta y)(\varphi) &= \int_D y\, \Delta\varphi\, dx = -\int_D (y_{x_1}\varphi_{x_1} + y_{x_2}\varphi_{x_2})\, dx_1\, dx_2 \\ &= -\int_{D_0} \varphi_{x_1}(x_1, x_2)\left(\int_{x_2}^{\sigma(x_1)} z_{x_1}(x_1, s)\, ds \right. \\ &\qquad + \varphi_{x_2}(x_1, x_2)(x_2 - z(x_1, x_2))\, dx_1\, dx_2\end{aligned}$$

$$= -\int_{D_0} \left(\varphi(x_1, x_2) \int_{x_2}^{\sigma(x_1)} z_{x_1}(x_1, s)\, ds \right)_{x_1}$$

$$+ \varphi(x_1, x_2)(x_2 - z(x_1, x_2))_{x_2}\, dx_1\, dx_2$$

$$+ \int_{D_0} \varphi(x_1, x_2) \left(\int_{x_2}^{\sigma(x_1)} z_{x_1 x_1}(x_1, s)\, ds \right.$$

$$\left. + \sigma'(x_1) z_{x_1}(x_1, \sigma(x_1)) + \big(1 - z_{x_2}(x_1, x_2)\big) \right) dx_1\, dx_2$$

$$= \int_{D_0} \varphi(x_1, x_2)(\sigma'(x_1) z_{x_1}(x_1, \sigma(x_1))$$

$$- z_{x_2}(x_1, \sigma(x_1)) + 1)\, dx_1\, dx_2$$

because $\Delta z = 0$ in $D_0 = \{(x_1, x_2) \in D;\ x_2 < \sigma(x_1)\}$. On the other hand, we have, by (1.50),

$$0 = \frac{\partial z}{\partial \nu}\bigg|_S = \big(z_{x_2}(x_1, \sigma(x_1)) - \sigma'(x_1) z_{x_1}(x_1, \sigma(x_1))\big)\big(1 + (\sigma'(x_1))^2\big) \qquad \forall x_1 \in [0, a].$$

Hence,

$$\int_\Omega y\, \Delta\varphi\, dx = \int_{D_0} \varphi(x_1, x_2)\, dx = \chi_{D_0}(\varphi) \qquad \forall \varphi \in C_0^\infty(D),$$

and Eq. (1.51) follows.

Conditions (1.52) follow by the definition of y and by (1.50). ■

By Lemma 1.1, we may view y as the solution to the obstacle problem

$$\begin{aligned} -\Delta y \geq -1, \quad y \geq 0 \quad & \text{in } D, \\ \Delta y = 1 \quad & \text{in } \{x \in D;\ y(x) > 0\}, \\ y = g \quad & \text{in } \partial D, \end{aligned} \tag{1.53}$$

or, in the variational form,

$$\int_\Omega \nabla y \cdot \nabla(y - v)\, dx + \int_D (y - v)\, dx \leq 0, \qquad \forall v \in K, \tag{1.54}$$

where $K = \{v \in H^1(D);\ v = g \text{ in } \partial D,\ v \geq 0 \text{ in } D\}$.

By Corollary 1.3, we conclude that problem (1.54) (and consequently the dam problem (1.53)) has a unique solution $y \in K$. The free boundary S can be found solving the equation $y(x_1, x_2) = 0$. For sharp regularity properties of y and S, we refer to the book of A. Friedman [1].

3.1.3. An Elasto-Plastic Problem

Let Ω be an open domain of $\mathbf{R}^N$ and let $a: H_0^1(\Omega) \times H_0^1(\Omega) \to \mathbf{R}$ be defined by

$$a(y, z) = \int_\Omega \nabla y \cdot \nabla z \, dx \qquad \forall y, z \in H_0^1(\Omega). \tag{1.55}$$

Introduce the set

$$K = \{y \in H_0^1(\Omega); \|\nabla y(x)\|_N \leq 1 \text{ a.e. } x \in \Omega\}, \tag{1.56}$$

where $\|\cdot\|_N$ is the Euclidean norm in $\mathbf{R}^N$, and consider the variational inequality

$$y \in K, \qquad a(y, y - z) \leq (y - z, f) \qquad \forall z \in K, \tag{1.57}$$

where $f \in H^{-1}(\Omega)$.

By Corollary 1.3, this problem has a unique solution y.

If y is sufficiently smooth, then it follows as for the obstacle problem that

$$\begin{aligned} -\Delta y &= f && \text{in } \{x \in \Omega; \|\nabla y(x)\|_N < 1\} = \Omega_e, \\ \|\nabla y\|_N &= 1 && \text{in } \Omega_p = \Omega \setminus \Omega_e, \\ y &= 0 && \text{in } \partial\Omega. \end{aligned} \tag{1.58}$$

The interpretation of problem (1.58) is as follows: The domain can be decomposed in two parts Ω_e (the elastic zone) and Ω_p (the plastic zone). In Ω_e, y satisfies the classical equation of elasticity whilst in Ω_p, $\|\nabla y(x)\|_N = 1$; the surface S that separates the elastic and plastic zones is a free boundary, which is not known *a priori* and is one of the unknowns of the problem. This models the elasto-plastic torsion of a cylindrical bar of cross-section Ω that is subject to an increasing torque. The state y represents in this case the stress potential in Ω.

As noted earlier, (1.57) is equivalent to the minimization problem

$$\min\{\tfrac{1}{2}a(z, z) - (z, f); z \in K\}. \tag{1.59}$$

If $f \in L^2(\Omega)$ and $\partial\Omega$ is sufficiently smooth, then the solution y to (1.57) belongs to $H^2(\Omega)$ (Brézis and Stampacchia [1]).

It is useful to point out that

$$y = \lim_{\varepsilon \to 0} y_\varepsilon \qquad \text{in } L^2(\Omega),$$

where $y_\varepsilon \in H_0^1(\Omega)$ is the solution to the boundary value problem

$$\begin{aligned} -\Delta y - \operatorname{div} \partial h_\varepsilon(\nabla y) &= f \quad \text{in } \Omega, \\ y &= 0 \quad \text{in } \partial\Omega, \end{aligned}$$

where $h_\varepsilon : \mathbf{R}^N \to \mathbf{R}^N$ is defined by

$$h_\varepsilon(u) = \begin{cases} 0 & \text{if } \|u\|_N < 1, \\ \dfrac{(\|u\|_N - 1)^2}{2\varepsilon} & \text{if } \|u\|_N \geq 1, \end{cases}$$

and $\partial h_\varepsilon : \mathbf{R}^N \to \mathbf{R}^N$ is its differential, i.e.,

$$\partial h_\varepsilon(u) = \begin{cases} 0 & \text{if } \|u\|_N < 1, \\ \dfrac{(u\|u\|_N - 1)}{\varepsilon\|u\|_N} & \text{if } \|u\|_N \geq 1. \end{cases}$$

Let us calculate starting from (1.59) the solution to problem (1.57) in the case where $\Omega = (0, 1)$. If make the substitution

$$w(x) = \int_0^x z(s)\, ds, \qquad x \in (0, 1),$$

then problem (1.59) becomes

$$\inf\left\{\tfrac{1}{2}\int_0^1 w^2(x)\, dx - \int_0^1 w(x)\int_x^1 f(s)\, ds\, dx;\ w \in U\right\}, \tag{1.60}$$

where $U = \{u \in L^2(0, 1);\ |u(x)| \leq 1 \text{ a.e. } x \in (0, 1)\}$. Hence, the solution w to (1.60) satisfies the equation

$$w - \int_x^1 f(s)\, ds + N_U(w) \ni 0,$$

where N_U is the normal cone to U. Hence,

$$y'(x) = w(x) = \begin{cases} \displaystyle\int_x^1 f(s)\, ds & \text{if } \left|\displaystyle\int_x^1 f(s)\, ds\right| \leq 1, \\ 1 & \text{if } \displaystyle\int_x^1 f(s)\, ds > 1, \\ -1 & \text{if } \displaystyle\int_x^1 f(s)\, ds < -1. \end{cases}$$

3.1.4. *Elliptic Problems with Unilateral Conditions at the Boundary*

Consider in $\Omega \subset \mathbf{R}^N$ the boundary value problem

$$\begin{aligned} cy - \Delta y &= f \qquad \text{in } \Omega, \\ \frac{\partial y}{\partial \nu} + \beta(y) &\ni g \qquad \text{in } \Gamma_1, \\ y &= 0 \qquad \text{in } \Gamma_2, \end{aligned} \tag{1.61}$$

where Γ_1 and Γ_2 are two open, smooth, and disjoint parts of $\partial\Omega$, $\overline{\Gamma}_1 \cup \overline{\Gamma}_2 = \partial\Omega$, $f \in L^2(\Omega)$, $g \in L^2(\Gamma_1)$, c is a positive constant and β is a maximal monotone graph in $\mathbf{R} \times \mathbf{R}$.

Let $j: \mathbf{R} \to \overline{\mathbf{R}}$ be a lower semicontinuous, convex function such that $\partial j = \beta$. We set

$$V = \{y \in H^1(\Omega);\ y = 0 \text{ in } \Gamma_2\}$$

and define the operator $A: V \to V'$ by

$$(Ay, z) = a(y, z) = \int_\Omega \nabla y \cdot \nabla z\, dx + c\int_\Omega yz\, dx \qquad \forall y, z \in V. \tag{1.62}$$

Let $\varphi: V \to \overline{\mathbf{R}}$ be the function

$$\varphi(z) = \int_{\Gamma_1} j(z)\, dz \qquad \forall z \in V,$$

and let $f_0 \in V'$ be defined by

$$(f_0, z) = \int_\Omega f(x)z(x)\, dx + \int_{\Gamma_1} g(x)z(x)\, dx \qquad \forall z \in V.$$

By Corollary 1.2, the variational inequality

$$a(y, y - z) + \varphi(y) - \varphi(z) \le (f_0, y - z) \qquad \forall z \in V, \tag{1.63}$$

has a unique solution $y \in V$.

Problem (1.63) can be equivalently written as

$$\min\{\tfrac{1}{2}a(z, z) + \varphi(z) - (f_0, z);\ z \in V\}. \tag{1.64}$$

The solution y to (1.63) (equivalently, (1.64)) can be viewed as generalized solution to problem (1.61). Indeed, if in (1.63) we take $z = y - \alpha$, where $\alpha \in C_0^\infty(\Omega)$, we get

$$\int_\Omega (cy\alpha + \nabla y \cdot \nabla \alpha)\, dx = \int_\Omega f\alpha\, dx.$$

Hence, $cy - \Delta y = f$ in $\mathscr{D}'(\Omega)$. Now, multiplying this by $y - z$, where $z \in V$, and integrating on Ω, we get

$$c \int_\Omega y(y-z)\,dx + \int_\Omega \nabla y \cdot \nabla(y-z)\,dx$$
$$= \int_\Omega f(y-z)\,dx + \int_{\Gamma_1} \frac{\partial y}{\partial \nu}(y-z)\,dx.$$

More precisely, we have

$$a(y, y-z) - \left\langle \frac{\partial y}{\partial \nu}, y - z \right\rangle = (f, y-z),$$

where $\langle \cdot, \cdot \rangle$ is the pairing between $H^{1/2}(\Gamma_1)$ and $H^{-1/2}(\Gamma_1)$ (if $y \in H^1(\Omega)$ and $\Delta y \in L^2(\Omega)$, then $y \in H^{1/2}(\Gamma_1)$ and $\partial y/\partial \nu \in H^{-1/2}(\Gamma_1)$ (see Lions and Magenes [1]).

Then, by (1.63), we see that

$$\int_{\Gamma_1} (j(y) - j(z))\,dx \le \left\langle g - \frac{\partial y}{\partial \nu}, y - z \right\rangle \qquad \forall z \in V.$$

Hence, if $g - \partial y/\partial \nu \in L^2(\Gamma_1)$, we may conclude that $g - \partial y/\partial \nu \in \beta(y)$ a.e. in Γ_1. Otherwise, this simply means that

$$g - \frac{\partial y}{\partial \nu} \in \partial \tilde{\varphi}(y),$$

where $\partial \tilde{\varphi}: H^{1/2}(\Gamma_1) \to H^{-1/2}(\Gamma_1)$ is the subdifferential of the function $\tilde{\varphi}: H^{1/2}(\Gamma_1) \to \overline{\mathbf{R}}$ defined by

$$\tilde{\varphi}(v) = \int_{\Gamma_1} j(v)\,dx \qquad \forall v \in H^{1/2}(\Gamma_1).$$

In the special case where $\Gamma_1 = \partial\Omega$ and $g \equiv 0$, then as seen in Section 2.2, Chapter 2 (Proposition 2.11), $y \in H^2(\Omega)$ and

$$\|y\|_{H^2(\Omega)} \le C(I + \|f\|_{L^2(\Omega)}) \qquad \forall f \in L^2(\Omega), \tag{1.65}$$

where C is independent of f.

Moreover, we have

$$y = \lim_{\varepsilon \to 0} y_\varepsilon \qquad \text{weakly in } H^2(\Omega), \text{ strongly in } H^1(\Omega), \tag{1.66}$$

where $y_\varepsilon \in H^2(\Omega)$ is the solution to the approximating problem

$$cy_\varepsilon - \Delta y_\varepsilon = f \quad \text{in } \Omega,$$
$$\frac{\partial y_\varepsilon}{\partial \nu} + \beta_\varepsilon(y_\varepsilon) = 0 \quad \text{in } \partial\Omega, \tag{1.67}$$

and

$$\beta_\varepsilon(r) = \frac{1}{\varepsilon}\left(r - (1 + \varepsilon\beta)^{-1} r\right) \qquad \forall r \in \mathbf{R},\ \varepsilon > 0.$$

In this case, we have a more precise result. Namely:

Proposition 1.3. *Let $f_\varepsilon \to f$ weakly in $L^2(\Omega)$. Then the solution $y_\varepsilon \in H^2(\Omega)$ to problem (1.67) is weakly convergent in $H^2(\Omega)$ and strongly convergent in $H^1(\Omega)$ to the solution y^f to problem (1.61).*

Proof. As seen in the proof of Proposition 2.11, Chapter 2, we have for y_ε an estimate of the form (1.65), i.e.,

$$\|y_\varepsilon\|_{H^2(\Omega)} \le C(1 + \|f_\varepsilon\|_{L^2(\Omega)}) \qquad \forall \varepsilon > 0,$$

where C is independent of β_ε (i.e., of ε). Hence on a subsequence, again denoted ε, we have

$$y_\varepsilon \to y \qquad \text{weakly in } H^2(\Omega), \text{ strongly in } H^1(\Omega),$$
$$\frac{\partial y_\varepsilon}{\partial \nu} \to \frac{\partial y}{\partial \nu} \qquad \text{weakly in } H^{1/2}(\partial\Omega), \text{ strongly in } L^2(\Omega),$$
$$\beta_\varepsilon(y_\varepsilon) \to \xi \qquad \text{strongly in } L^2(\partial\Omega),$$
$$y_\varepsilon \to y \qquad \text{strongly in } L^2(\partial\Omega). \tag{1.68}$$

Arguing as in the proof of Proposition 1.2, we see that $\xi \in \beta(y)$ a.e. in $\partial\Omega$. Then, letting ε tend to zero in (1.68), we see that y is the solution to (1.61). ∎

We shall consider now some particular cases. If $j(r) = g_0|r|$, $r \in \mathbf{R}$, where g_0 is some positive constant, then

$$\beta(r) = g_0 \operatorname{sign} r = \begin{cases} g_0 \dfrac{r}{|r|} & \text{if } r \neq 0, \\ [-g_0, g_0] & \text{if } r = 0, \end{cases}$$

and so problem (1.61) becomes

$$\begin{aligned} cy - \Delta y &= f \qquad \text{in } \Omega, \\ \frac{\partial y}{\partial \nu} + g_0 \operatorname{sign} y &\ni 0 \qquad \text{in } \partial\Omega. \end{aligned} \tag{1.69}$$

Equivalently,

$$\begin{aligned} &cy - \Delta y = f \qquad \text{in } \Omega, \\ &\left|\frac{\partial y}{\partial \nu}\right| \le g_0, \quad y \frac{\partial y}{\partial \nu} + g_0|y| = 0 \qquad \text{a.e. in } \partial\Omega. \end{aligned} \tag{1.70}$$

The boundary conditions can be rewritten as

$$y \begin{cases} = 0 & \text{if } |\frac{\partial y}{\partial \nu}| < g_0, \\ \le 0 & \text{if } \frac{\partial y}{\partial \nu} = g_0, \\ \ge 0 & \text{if } \frac{\partial y}{\partial \nu} = -g_0, \end{cases} \qquad \text{a.e. in } \partial\Omega.$$

Hence, there are *a priori* two regions Γ^1 and Γ^2 on $\partial\Omega$ where $|\partial y/\partial \nu| < g_0$ and $|\partial y/\partial \nu| = g_0$, respectively. However, Γ^1 and Γ^2 are not known, so problem (1.69) is in fact a free boundary problem and as seen before it has a unique solution $y \in H^2(\Omega)$. Problem (1.70) models the equilibrium configuration of an elastic body Ω that is in unilateral contact with friction on $\partial\Omega$ (see Duvaut and Lions [1], Chapter IV).

The Signorini Problem. Consider now problem (1.61) in the special case where $\Gamma_1 = \partial\Omega$, $\Gamma_2 = \emptyset$, $g \equiv 0$, and

$$\beta(r) = \begin{cases} 0 & r > 0, \\ ]-\infty, 0] & r = 0, \\ \emptyset & r < 0, \end{cases} \qquad \forall r \in \mathbf{R}. \tag{1.71}$$

i.e.

$$cy - \Delta y = f \qquad \text{a.e. in } \Omega, y \ge 0, \quad \frac{\partial y}{\partial \nu} \ge 0, \quad y\frac{\partial y}{\partial \nu} = 0 \qquad \text{a.e. in } \partial\Omega. \tag{1.72}$$

This is the famous problem of Signorini, which describes the conceptual model of an elastic body Ω that is in contact with a rigid support body and is subject to volume forces f. These forces produce a deformation of Ω and a displacement on $\partial\Omega$, with the normal component negative or zero.

Other unilateral problems of the form (1.61) arise in fluid mechanics with semipermeable boundaries, climatization problems or in thermostat control of heat flow, and we refer the reader to the previously cited book of Duvaut and Lions [1].

In mechanics, one often meets problems in which constitutive laws are given by nonmonotone multivalued mappings that lead to problems of the following type (see P. D. Panagiotopoulos [1]):

$$Ay + \gamma(y) \ni f \qquad \text{in } \Omega,$$

where $A: V \to V'$ is defined by (1.18) and γ is the generalized gradient (in the sense of Clarke) of a locally Lipschitz function $\phi: V \to \mathbf{R}$. In particular, we may take

$$\phi(y) = \int_\Omega dx \int_0^{y(x)} j(\xi)\, d\xi,$$

where $j \in L^\infty_{\mathrm{loc}}(\mathbf{R})$ is such that $|j(\xi)| \leq C(1 + |\xi|^{p-1})$, $\xi \in \mathbf{R}$, and so $\partial\phi(y) \subset \{w \in L^q(\Omega);\ w(x) \in [\tilde{m}(j(y(s))), \tilde{M}(j(y(x)))]$ a.e. $x \in \Omega\}$ (see (2.75) in Chapter 2).

This is a *hemivariational inequality*, and the existence theory developed in the preceding partially extends to this class of nonlinear problems (see the book [1] by Panagiotopoulos, and the references given there).

3.2. Optimal Control of Elliptic Variational Inequalities

In this section, we shall discuss several optimal control problems governed by semilinear elliptic equations and in particular that governed by elliptic variational inequalities and problems with free boundary.

The most important objective of such a treatment is to derive a set of first order conditions for optimality (maximum principle) that is able to give complete information on the optimal control. Since the optimal control problems governed by nonlinear equations are nonsmooth and nonconvex, the standard methods of deriving necessary conditions of optimality are inapplicable here. The method is in brief the following: One approximates the given problem by a family of smooth optimization problems and afterwards tends to the limit in the corresponding optimality equations.

An attractive feature of this approach, which we shall illustrate on some model problems, is that it allows the treatment of optimal control problems governed by a large class of nonlinear equations, even of nonmonotone type and with general cost criteria.

3.2.1. General Formulation of Optimal Control Problems

Let V and H be a pair of real Hilbert spaces such that V is a dense subset of H and

$$V \subset H \subset V'$$

algebraically and topologically. We have denoted by V' the dual of V and the notation is that of Section 1.1. Thus $(\cdot,\cdot)$ is the pairing between V and V' (and the scalar product of H) and $\|\cdot\|$, $|\cdot|$ are the norms in V and H, respectively.

Consider the equation

$$Ay + \partial\varphi(y) \ni Bu + f, \tag{2.1}$$

where $A: V \to V'$ is a linear continuous operator satisfying the coercivity condition

$$(Av, v) \geq \omega\|v\|^2 \quad \forall v \in V \qquad \text{for some } \omega > 0, \tag{2.2}$$

$\varphi: V \to \overline{\mathbf{R}}$ is a lower semicontinuous, convex function, $\partial\varphi: V \to V'$ is the subdifferential of φ, $B \in L(U, V')$, and f is a given element of V'. Here, U is another real Hilbert space with the scalar product denoted $\langle \cdot,\cdot \rangle$ and norm $|\cdot|_U$ (the controller space).

As seen in Section 3.1 a large class of nonlinear elliptic problems, including problems with free boundary and unilateral conditions at the boundary, can be written in this form.

The parameter u is called the *control* and the corresponding solution y is the *state* of the system. Equation (2.1) itself will be referred to as the *state system* or *control system*.

The optimal control problem we shall study in this chapter can be put in the following general form:

(P) *Minimize the function*

$$g(y) + h(u) \tag{2.3}$$

on all $y \in V$ *and* $u \in U$ *satisfying the state system* (2.1).

Here, $g: H \to \mathbf{R}$ and $h: U \to \overline{\mathbf{R}}$ are given functions that satisfy the following conditions:

(i) *g is Lipschitz on bounded subsets of H* (i.e., *g is locally Lipschitz*) *and bounded from below by an affine function*, i.e.,

$$g(y) \geq (\alpha, y) + C \qquad \forall y \in H, \tag{2.4}$$

where $\alpha \in H$ *and C is a real constraint*;

(ii) *h is convex, lower semicontinuous, and*

$$\lim_{|u|_U \to \infty} h(u)/|u|_U = +\infty; \tag{2.5}$$

(iii) *B is completely continuous from U to* V'.

The last hypothesis is in particular satisfied if the injection of V into H is compact.

Roughly speaking, the object of control theory for system (2.1) is the adjustment of the control parameter u, subject to certain restrictions, such that the corresponding state y has some specified properties or to achieve some goals, which very often are expressed as minimization problems of the form (P). For instance, we might pick one known state y^0 and seek to find u in a certain closed convex subset $U_0 \subset U$ so that $y = y^0$. Then the least square approach leads us to a problem of the type (P), where

$$g(y) = \tfrac{1}{2}|y - y^0|^2 \qquad \text{and} \qquad h(u) = \begin{cases} 0 & \text{if } u \in U_0, \\ +\infty & \text{elsewhere}. \end{cases}$$

Other control problems such as that of finding the control u such that the free boundary of Eq. (2.1) (if this equation is a problem with free boundary) is as close as possible to a given surface S, though more complicated, also admit an adequate formulation in terms of (P). This will be discussed further in a later section.

Now let us briefly discuss existence in problem (P). A pair $(y^*, u^*) \in V \times U$ for which the infimum in (P) is attained is called *optimal pair* and the control u^* is referred as *optimal control.*

Proposition 2.1. *Under assumptions* (i)–(iii) *problem* (P) *has at least one optimal pair.*

Proof. For every $u \in U$ we shall denote by $y^u \in U$ the solution to Eq. (2.1). Note first that the map $u \to y^u$ is weakly–strongly continuous from U to V. Indeed, if $\{u_n\} \subset U$ is weakly convergent to u in U, by assumption (iii) we see that

$$Bu_n \to Bu \qquad \text{strongly in } V'$$

whilst by (2.1) and (2.2) we have

$$\omega \| y^{u_n} - y^{u_m} \|^2 \le \| Bu_n - Bu_m \|_* \| y^{u_n} - y^{u_m} \|$$

because $\partial\varphi$ is monotone in $V \times V'$. Hence,

$$\lim_{n \to \infty} y^{u_n} = y \qquad \text{exists in the strong topology of } V.$$

Now letting n tend to $+\infty$ in

$$Ay^{u_n} + \partial\varphi(y^{u_n}) \in Bu_n + f,$$

we conclude that $y = y^u$, as claimed.

Now let $d = \inf\{g(y^u) + h(u); u \in U\}$. By assumptions (i), (ii), it follows that $d > -\infty$. Now, let $\{u_n\} \subset U$ be such that

$$d \le g(y_n) + h(u_n) \le d + n^{-1} \qquad \forall n,$$

where $y_n = y^{u_n}$. By (i) and (ii) we see that $\{u_n\}$ is bounded in U, and so on a subsequence (for simplicity again denoted $\{u_n\}$) we have

$$u_n \to u^* \qquad \text{weakly in } U,$$
$$y_n \to y^* \qquad \text{strongly in } V,$$

because B is completely continuous. Since as seen in the preceding $y^* = y^{u^*}$ and $g(y^*) + h(u^*) = d$ (because g is continuous and h is weakly lower semicontinuous), we infer that (y^*, u^*) is an optimal pair for problem (P). ■

From the previous proof it is clear that as far as existence is concerned, hypotheses (i)–(iii) are too strong. For instance, it suffices to assume that g is merely continuous from V to $\mathbf{R}$. Also, if g is convex, assumption (iii) is no longer necessary because g is, in this case, weakly lower semicontinuous on V.

On the other hand, it is clear that the convexity assumption on h cannot be dispensed with since it assures the weak lower semicontinuity of the function $u \to g(y^u) + h(u)$, a property that for infinite dimensional controllers space U is absolutely necessary to attain the infimum in problem (P).

3.2.2. *A General Approach to the Maximum Principle*

Here we shall discuss a general approach to obtain first order necessary conditions of optimality for optimal control problems and in particular for problem (P).

Most of the optimal control problems that arise in applications can be represented in the following abstract form:

($\tilde{\mathrm{P}}$) *Minimize the functional*

$$L(y,u) + h(u) \tag{2.6}$$

on all $(y,u) \in X \times U$, *subject to state equation*

$$F(y,u) = 0. \tag{2.7}$$

Here, $L: X \times U \to \mathbf{R}$ is a continuous function, $h: U \to \overline{\mathbf{R}} =]-\infty, +\infty]$ is convex and lower semicontinuous, and $F: X \times U \to Y$ is a given operator, where X, Y, U are Banach spaces.

We shall assume of course that for every $u \in U$, Eq. (2.7) has a unique solution $y^u \in X$.

Let us first assume that the map $u \to y^u$ is Gâteaux differentiable on U and that its differential in $v \in U$, $z = D_u y^u(v)$, is the solution to the equation

$$F_y(y^u,u)z + F_u(y^u,u)v = 0, \tag{2.8}$$

where F_y and F_u are the differentials of $y \to F(y,u)$ and $u \to F(y,u)$, respectively. (This happens always if the operator F is differentiable.)

Let (y^*, u^*) be an optimal pair for problem (P). We have

$$\lambda^{-1}(L(y^{u^*+\lambda v}, u^* + \lambda v) - L(y^*, u^*)) + \lambda^{-1}(h(u^* + \lambda v) - h(u^*)) \geq 0.$$

$\forall v \in U$, $\lambda > 0$, and if L is (Gâteaux) differentiable on $X \times U$ this yields

$$\big(L_y(y^*, u^*), z\big) + \langle L_u(y^*, u^*), v\rangle + h'(u^*, v) \geq 0 \qquad \forall v \in U, \quad (2.9)$$

where h' is the directional derivative of h and $(\cdot,\cdot)$, $\langle \cdot,\cdot \rangle$ are the dualities between X, X^* and U, U^*, respectively.

Denote by F_y^* the adjoint of F_y, and further assume that the linear equation

$$F_y^*(y^*, u^*)p = L_y(y^*, u^*) \quad (2.10)$$

(the dual state equation) has a solution $p \in Y^*$. Then, by (2.8) and (2.10) we see that

$$\big(L_y(y^*, u^*), z\big) = \big(p, F_y(y^*, u^*)z\big) = -\langle F_u^*(y^*, u^*)p, v\rangle,$$

and substituting in (2.9) yields

$$\langle F_u^*(y^*, u^*)p - L_u(y^*, u^*), v\rangle \leq h'(u^*, v) \qquad \forall v \in U.$$

By Proposition 2.8 in Chapter 2, this implies

$$F_u^*(y^*, u^*)p - L_u(y^*, u^*) \in \partial h(u^*). \quad (2.11)$$

Equations (2.7), (2.10), and (2.11) taken together represent the first order optimality system for problem (P), and if F is linear and L is convex these are also sufficient for optimality.

If L and F are nonsmooth, one might hope to obtain a maximum principle type result in terms of generalized derivatives of L and F by using the following method:

Assume that the control space U is reflexive and that there are a family $\{F_\varepsilon\}_{\varepsilon > 0}$ of smooth operators from $X \times U$ to Y and a set of differentiable functions $L_\varepsilon \colon X \times U \to \mathbf{R}$ such that

(i)′ For every $\varepsilon > 0$ and $u \in U$ the equation $F_\varepsilon(y, u) = 0$ has a unique solution $y = y_\varepsilon^u$ and there exists $p \geq 1$ such that

$$\|y_\varepsilon^u\|_X \leq C(1 + |u|_U^p) \qquad \forall u \in U. \quad (2.12)$$

(ii)′ If $\{u_\varepsilon\}$ is weakly convergent to u in U then $\{y_\varepsilon^{u_\varepsilon}\}$ is strongly convergent to y^u in X and

$$\liminf_{\varepsilon \to 0} L_\varepsilon(y_\varepsilon^{u_\varepsilon}, u_\varepsilon) \geq L(y^u, u). \tag{2.13}$$

Moreover, $\lim_{\varepsilon \to 0} L_\varepsilon(y_\varepsilon^u, u) = L(y^u, u)\ \forall u \in U$.

(iii)′ $L_\varepsilon(y, u) \geq -C(|y| + |u|)\ \forall (y, u) \in X \times U$, where C is independent of ε.

Let (y^*, u^*) be any optimal pair in problem $(\tilde{\mathrm{P}})$. We consider the penalized problem

$$(\mathrm{P}_\varepsilon)\quad \inf\left\{L_\varepsilon(y, u) + h(u) + \frac{1}{2p}|u - u^*|_U^{2p}\,;\, F_\varepsilon(y, u) = 0, u \in U\right\},$$

and assume that this problem has a solution $(y_\varepsilon, u_\varepsilon)$ (this happens, for instance, if the function $u \to L_\varepsilon(y_\varepsilon^u, u)$ is weakly lower semicontinuous). Then, we have

Lemma 2.1. *For $\varepsilon \to 0$, we have*

$$u_\varepsilon \to u^* \quad \textit{strongly in } U,$$
$$y_\varepsilon \to y^* \quad \textit{strongly in } X.$$

Proof. For all $\varepsilon > 0$ we have

$$L_\varepsilon(y_\varepsilon, u_\varepsilon) + h(u_\varepsilon) + \frac{1}{2p}|u_\varepsilon - u^*|_U^{2p} \leq L_\varepsilon(y_\varepsilon^{u^*}, u^*) + h(u^*),$$

and by condition (2.12) and assumption (iii)′, $\{u_\varepsilon\}$ is bounded in U and so, on a subsequence $\varepsilon_n \to 0$.

$$u_{\varepsilon_n} \to \bar{u} \quad \text{weakly in } U,$$
$$y_{\varepsilon_n} \to \bar{y} = y^{\bar{u}} \quad \text{strongly in } X.$$

Since h is weakly lower semicontinuous, we have

$$\liminf_{\varepsilon_n \to 0} h(u_{\varepsilon_n}) \geq h(\bar{u})$$

and so, by assumption (ii)′,

$$L(\bar{y}, \bar{u}) + h(\bar{u}) + \liminf_{\varepsilon_n \to o} \frac{1}{2p}|u_{\varepsilon_n} - u^*|_U^2 \leq L(y^*, u^*) = \inf P.$$

Hence, $\bar{y} = y^*$, $\bar{u} = u^*$, and $u_{\varepsilon_n} \to u^*$ strongly in U on some subsequence $\{\varepsilon_n\}$. Since the limit u^* is independent of the subsequence, we conclude that $u_\varepsilon \to u^*$ strongly in U, as claimed. ■

Now, since problem $(\tilde{\mathrm{P}}_\varepsilon)$ is a smooth optimization problem of the form $(\tilde{\mathrm{P}})$, $(y_\varepsilon, u_\varepsilon)$ satisfy an optimality system of the type (2.10), (2.11). More precisely, there is $p_\varepsilon \in Y^*$ that satisfies along with $(y_\varepsilon, u_\varepsilon)$ the system

$$\begin{gathered} F_\varepsilon(y_\varepsilon, u_\varepsilon) = 0, \\ (F_\varepsilon)_y^*(y_\varepsilon, u_\varepsilon)p_\varepsilon = (L_\varepsilon)_y(y_\varepsilon, u_\varepsilon), \\ (F_\varepsilon)_u^*(y_\varepsilon, u_\varepsilon)p_\varepsilon - (L_\varepsilon)_u(y_\varepsilon, u_\varepsilon) \in \partial h(u_\varepsilon) + J(u_\varepsilon - u^*)|u_\varepsilon - u^*|_U^{2p-2}, \end{gathered} \tag{2.14}$$

where $J: U \to U^*$ is the duality mapping of the space U.

By virtue of Lemma 2.1, we may view (2.14) as an approximating optimality system for problem (P). If one could obtain from (2.14) sufficiently sharp *a priori* estimates on p_ε, one might pass to limit in (2.14) to obtain a system of first order optimality conditions for problem $(\tilde{\mathrm{P}})$. We shall see that this is possible in most of the important cases, but let us see first how this scheme looks for problem (P) considered in the previous section.

We shall assume that g and h satisfy assumptions (i), (ii) (with the eventual exception of condition (2.5)), $B \in L(U, V')$, $f \in V'$, and the injection of V into H is compact.

Let (y^*, u^*) be an optimal pair in problem (P). Then, we associate with this pair the penalized optimal control problem (*adapted penalty*):

(P_ε) *Minimize* $g^\varepsilon(y) + h(u) + \frac{1}{2}|u - u^*|_U^2$ *on all* $(y, u) \in V \times H$, *subject to* $Ay + \nabla\varphi^\varepsilon(y) = Bu + f$.

Here, g^ε is defined by (see formula (2.79) in Chapter 2)

$$g^\varepsilon(y) = \int_{\mathbf{R}^n} g(P_n y - \varepsilon\Lambda_n\tau)\rho_n(\tau)\,d\tau \qquad \forall y \in H, \tag{2.15}$$

where $P_n u = \sum_{i=1}^n u_i e_i$, $u = \sum_{i=1}^\infty u_i e_i$, $\Lambda_n\tau = \sum_{i=1}^n u_i e_i$, $\{e_i\}$ an orthonormal basis in H, and $\varphi^\varepsilon: V \to \mathbf{R}$ is a family of convex functions of class C^2 on V such that

$$\begin{gathered} \varphi^\varepsilon(v) \ge -C(\|v\| + 1) \qquad \forall v \in V,\ \varepsilon > 0, \\ \lim_{\varepsilon \to 0} \varphi^\varepsilon(v) = \varphi(v) \qquad \forall v \in V \end{gathered}$$

whilst, for any weakly convergent sequence $v_\varepsilon \to v$ in V,

$$\liminf_{\varepsilon \to 0} \varphi^\varepsilon(v_\varepsilon) \geq \varphi(v).$$

By Proposition 2.1, we know that problem (P_ε) has at least one solution $(y_\varepsilon, u_\varepsilon)$. Since our conditions on φ^ε clearly imply assumptions (i)′, (ii)′, (iii)′ of Lemma 2.1, where $X = H$, $Y = V'$, and

$$F_\varepsilon(y, u) = Ay + \nabla\varphi^\varepsilon(y) - Bu - f,$$
$$L_\varepsilon(y, u) = g^\varepsilon(y) + h(u),$$

we conclude that, for $\varepsilon \to 0$,

$$u_\varepsilon \to u^* \qquad \text{strongly in } U,$$
$$y_\varepsilon \to y^* \qquad \text{strongly in } H, \text{weakly in } V.$$

On the other hand, the optimality system for problem (P_ε) is (see (2.14))

$$Ay_\varepsilon + \nabla\varphi^\varepsilon(y_\varepsilon) = Bu_\varepsilon + f, \tag{2.16}$$
$$-A^*p_\varepsilon - \nabla^2\varphi^\varepsilon(y_\varepsilon)p_\varepsilon = \nabla g^\varepsilon(y_\varepsilon), \tag{2.17}$$
$$B^*p_\varepsilon \in \partial h(u_\varepsilon) + u_\varepsilon - u^*. \tag{2.18}$$

Note that since $\nabla^2\varphi^\varepsilon(y_\varepsilon) \in L(V, V')$ is a positive operator, Eq. (2.17) has a unique solution $p_\varepsilon \in V$. Moreover, since $\{\nabla g^\varepsilon(y_\varepsilon)\}$ is bounded on H (because g is locally Lipschitz), we have the estimate

$$\omega\|p_\varepsilon\| \leq |\nabla g^\varepsilon(y_\varepsilon)| \leq C \qquad \forall \varepsilon > 0.$$

Hence, on a subsequence $\varepsilon_n \to 0$, we have

$$p_\varepsilon \to p \qquad \text{weakly in } V, \text{strongly in } H,$$
$$A^*p_\varepsilon \to A^*p \qquad \text{weakly in } V',$$
$$\nabla g^\varepsilon(y_\varepsilon) \to \xi \qquad \text{weakly in } H.$$

By Proposition 2.15 in Chapter 2 we know that $\xi \in \partial g(y^*)$, where ∂g is the generalized gradient of g whilst by (2.18) it follows that $B^*p \in \partial h(u^*)$ or, equivalently, $u^* \in \partial h^*(B^*p)$.

Summarizing, we have proved:

Proposition 2.2. *Let (y^*, u^*) be an optimal pair or problem* (P). *Then there is $p \in V$ such that*

$$-A^*p - \eta \in \partial g(y^*),$$
$$u^* \in \partial h^*(B^*p), \tag{2.19}$$

where $\eta = w - \lim_{\varepsilon \to 0} \nabla^2 \varphi^{\varepsilon}(y_{\varepsilon}) p_{\varepsilon}$ *in* V', $p_{\varepsilon} \to p$, *and* $y_{\varepsilon} \to y^*$ *weakly in* V. ■

We may view (2.19) as the optimality system of problem (P).

There is a large variety of possibilities in choosing the approximating family $\{\varphi^{\varepsilon}\}$. One of these is to take φ^{ε} of the form (2.15), i.e.,

$$\varphi^{\varepsilon}(y) = \int_{\mathbf{R}^n} \varphi_{\varepsilon}(P_n y - \varepsilon \Lambda_n \tau) \rho_n(\tau)\, d\tau \qquad \forall y \in V, \tag{2.20}$$

where $n = [\varepsilon^{-1}]$, $P_n : V \to X_n$, $\Lambda_n : \mathbf{R}^n \to V$ and

$$\varphi_{\varepsilon}(y) = \inf\left\{\frac{\|y - v\|^2}{2\varepsilon} + \varphi(v);\ v \in V\right\} \qquad \forall y \in V, \tag{2.21}$$

or

$$\varphi_{\varepsilon}(y) = \inf\left\{\frac{|y - v|^2}{2\varepsilon} + \varphi(v);\ v \in H\right\} \qquad \forall y \in H, \tag{2.22}$$

if φ is lower semicontinuous on H (this happens, for instance, if φ is coercive on V).

One of the interesting features of the adapted penalty problem (P_{ε}) is that every sequence u_{ε} of corresponding optimal controllers is strongly convergent to u^*. If instead of (P_{ε}) we consider the problem

$$(P^{\varepsilon}) \qquad \min\{g^{\varepsilon}(y) + h(u);\ Ay + \nabla\varphi^{\varepsilon}(y) = Bu + f \quad u \in U\},$$

then under assumptions (i)–(iii), (P^{ε}) admits at least one optimal pair $(y^{\varepsilon}, u^{\varepsilon})$ and, on a sequence $\varepsilon_n \to 0$,

$$u^{\varepsilon_n} \to u_1^* \qquad \text{weakly in } H,$$

$$y^{\varepsilon_n} \to y_1^* \qquad \text{strongly in } V,$$

where (y_1^*, u_1^*) is an optimal pair of (P). The result remains true if replace h by a smooth approximation, for instance, h_{ε} as in (2.22), and one might try to calculate u^{ε} by a gradient type algorithm taking in account that the differential $\partial\phi$ of the function $\phi: u \to \mathbf{R}$,

$$\phi(u) = g^{\varepsilon}\big((A + \nabla\varphi^{\varepsilon})^{-1}(Bu + f)\big) + h_{\varepsilon}(u), \qquad u \in U,$$

is given by

$$\partial\phi(u) = \partial h_\varepsilon(u) - B^*p,$$

where

$$-A^*p - \nabla^2\varphi^\varepsilon(y)p = \nabla g^\varepsilon(u),$$
$$Ay + \nabla\varphi^\varepsilon(y) = Bu + f.$$

Now, we shall present another approach to problem (P), apparently different from the scheme developed in the preceding, but keeping similar features.

To this end, we shall assume that, beside the assumptions (i) and (ii), the injection of V into H is compact, φ is lower semicontinuous on H, $B \in L(U, H)$, $f \in H$, and hypothesis (1.14) holds. Then, as seen earlier, the solution y to variational inequality (2.1) belongs to $D(A_H)$, i.e., $Ay \in H$.

Let $(y^*, u^*) \in D(A_H) \times U$ be an optimal pair for problem (P).

We associate with (y^*, u^*) the family of optimization problems

$$(\mathrm{Q}_\varepsilon) \qquad \inf\big\{g(y) + h(u) + \varepsilon^{-1}(\varphi(y) + \varphi^*(v) - (y, v)) + \tfrac{1}{2}|v - v^*|^2 + \tfrac{1}{2}|u - u^*|_U^2\big\},$$

where the infimum is taken on the set of all $(y, u, v) \in V \times U \times H$, subject to

$$Ay = Bu - v + f.$$

Here, $v^* = Bu^* + f - Ay^* \in \partial\varphi(y^*) \subset H$ and $\varphi^*: H \to \overline{\mathbf{R}}$ is the conjugate of φ.

Lemma 2.2. *Problem* Q_ε *has at least one solution* $(y_\varepsilon, u_\varepsilon, v_\varepsilon) \in V \times U \times H$.

Proof. Let $d = \inf \mathrm{Q}_\varepsilon$ and let (y_n, u_n, v_n) be such that

$$d \le g(y_n) + h(u_n) + \varepsilon^{-1}(\varphi(y_n) + \varphi^*(v_n) - (y_n, v_n)) + \tfrac{1}{2}\big(|v_n - v^*|^2 + |u_n - u^*|_U^2\big) \le d + \frac{1}{n}. \tag{2.23}$$

Since $\{v_n\}$ is bounded in H and $\{y_n\}$ is bounded in V, we may pass to limit in the previous inequality to get the existence of a minimum point. ■

Lemma 2.3. *For* $\varepsilon \to 0$,

$$\begin{aligned} u_\varepsilon &\to u^* && \text{strongly in } U, \\ v_\varepsilon &\to v^* && \text{strongly in } H, \\ y_\varepsilon &\to y^* && \text{strongly in } V. \end{aligned} \tag{2.24}$$

Proof. We have

$$\begin{aligned} g(y_\varepsilon) + h(u_\varepsilon) &+ \varepsilon^{-1}(\varphi(y_\varepsilon) + \varphi^*(v_\varepsilon) - (y_\varepsilon, u_\varepsilon)) \\ &+ \tfrac{1}{2}\left(|u_\varepsilon - u^*|_U^2 + |v_\varepsilon - v^*|^2\right) \le g(y^*) + h(u^*) \qquad \forall \varepsilon > 0, \end{aligned} \tag{2.25}$$

because (see Proposition 2.5 in Chapter 2)

$$\varphi(y^*) + \varphi^*(v^*) = (y^*, v^*).$$

Note also that

$$\varphi(y_\varepsilon) + \varphi^*(v_\varepsilon) - (y_\varepsilon, v_\varepsilon) \ge 0 \qquad \forall \varepsilon > 0. \tag{2.26}$$

Hence, on a subsequence $\varepsilon_n \to 0$, we have

$$\begin{aligned} u_{\varepsilon_n} &\to \bar{u} && \text{weakly in } U, \\ v_{\varepsilon_n} &\to \bar{v} && \text{weakly in } H, \\ y_{\varepsilon_n} &\to \bar{y} && \text{strongly in } H, \text{weakly in } V. \end{aligned} \tag{2.27}$$

Now, letting ε_n tend to zero in (2.25) and using (2.26) and (2.27) we get

$$\begin{aligned} g(\bar{y}) + h(\bar{u}) &+ \limsup_{\varepsilon_n \to 0} \tfrac{1}{2}\left(|u_{\varepsilon_n} - u^*|_U^2 + |v_{\varepsilon_n} - v^*|^2\right) \\ &\le g(y^*) + h(u^*) = \inf P, \end{aligned} \tag{2.28}$$

respectively,

$$\varphi(\bar{y}) + \varphi^*(\bar{v}) - (\bar{y}, \bar{v}) = 0,$$

because φ and φ^* are weakly lower semicontinuous and

$$\begin{aligned} \liminf_{\varepsilon_n \to 0}\left(-(y_{\varepsilon_n}, v_{\varepsilon_n})\right) &= -\lim_{\varepsilon_n \to 0}(y_{\varepsilon_n}, Bu_{\varepsilon_n} + f) + \liminf_{\varepsilon_n \to 0}(y_{\varepsilon_n}, Ay_{\varepsilon_n}) \\ &\ge (\bar{y}, A\bar{y}) - (\bar{y}, B\bar{u} + f) = -(\bar{y}, \bar{v}). \end{aligned}$$

Then, by Proposition 2.5, Chapter 2, we infer that $\bar{v} \in \partial\varphi(\bar{y})$, and so by (2.28) we get (2.24). ■

Lemma 2.4. *Let $(y_\varepsilon, u_\varepsilon, v_\varepsilon)$ be optimal for problem φ_ε. Then there is $p_\varepsilon \in V$ that satisfies the system*

$$-A^* p_\varepsilon \in \partial g(y_\varepsilon) + \varepsilon^{-1}(\partial\varphi(y_\varepsilon) - v_\varepsilon), \tag{2.29}$$

$$v_\varepsilon \in \partial\varphi(y_\varepsilon - \varepsilon p_\varepsilon + \varepsilon(v^* - v_\varepsilon)), \tag{2.30}$$

$$B^* p_\varepsilon \in \partial h(u_\varepsilon) + u_\varepsilon - u^*. \tag{2.31}$$

Here, $\partial\varphi: V \to V'$ is the subdifferential of $\varphi: V \to \overline{\mathbf{R}}$.

Proof. Let us denote by $L_\lambda(y, u, v)$ the function

$$\begin{aligned} L_\lambda(y,u,v) = g^\lambda(y) &+ h(u) + \varepsilon^{-1}(\varphi_\lambda(y) + \varphi^*(v) - (y,v)) \\ &+ \tfrac{1}{2}(|v - v^*|^2 + |u - u^*|_U^2) \\ &+ \tfrac{1}{2}\left(|u - u_\varepsilon|_U^2 + |v - v_\varepsilon|^2\right), \qquad \lambda > 0, \end{aligned}$$

where g^λ is defined by (2.15) and φ_λ by (2.22). Clearly, for every $\lambda > 0$ the problem

$$\inf\{L_\lambda(y,u,v);\, u \in U, v \in H,\, Ay = Bu - v + f\} \tag{2.32}$$

has a solution $(y^\lambda, u^\lambda, v^\lambda)$. We have

$$\begin{aligned} L_\lambda(y^\lambda, u^\lambda, v^\lambda) \le g^\lambda(y_\varepsilon) &+ h(u_\varepsilon) + \varepsilon^{-1}(\varphi(y_\varepsilon) + \varphi^*(v_\varepsilon) - (y_\varepsilon, v_\varepsilon)) \\ &+ \tfrac{1}{2}\left(|v_\varepsilon - v^*|^2 + |u_\varepsilon - u^*|_U^2\right) \qquad \forall \lambda > 0. \end{aligned}$$

Since $\{u^\lambda\}$, $\{v^\lambda\}$ are bounded in U and H, respectively, we may assume that

$$\begin{aligned} u^\lambda &\to \bar{u} && \text{weakly in } U \\ v^\lambda &\to \bar{v} && \text{weakly in } H \\ y^\lambda &\to \bar{y} && \text{weakly in } V, \text{ strongly in } H. \end{aligned}$$

Then we have

$$g^\lambda(y^\lambda) \to g(\bar{y}), \qquad \liminf_{\lambda \to 0} \varphi^*(v^\lambda) \ge \varphi^*(\bar{v}),$$

$$\liminf_{\lambda \to 0} \varphi_\lambda(y^\lambda) \ge \varphi(\bar{y}), \qquad \liminf_{\lambda \to 0} h(u^\lambda) \ge h(\bar{u})$$

because, by Theorem 2.2 in Chapter 2, $\varphi_\lambda(y^\lambda) \ge \varphi(J_\lambda y^\lambda)$ and

$$\|J_\lambda y^\lambda - y^\lambda\|^2 \le 2\lambda\varphi_\lambda(y^\lambda) - 2\lambda\varphi(J_\lambda y^\lambda) \le C\lambda.$$

This yields

$$\liminf_{\lambda\to 0} \tfrac{1}{2}\big(|u^\lambda - u_\varepsilon|_U^2 + |v^\lambda - v_\varepsilon|^2\big) + g(\bar{y}) + h(\bar{u})$$
$$+ \varepsilon^{-1}(\varphi(\bar{y}) + \varphi^*(\bar{v}) - (\bar{y},\bar{v})) + \tfrac{1}{2}\big(|\bar{v} - v^*|^2 + |\bar{u} - u^*|_U^2\big)$$
$$\le \inf Q_\varepsilon.$$

Hence, $\bar{y} = y_\varepsilon$, $\bar{u} = u_\varepsilon$, $\bar{v} = v_\varepsilon$ and for $\lambda \to 0$

$$u^\lambda \to u_\varepsilon \quad \text{strongly in } U,$$
$$v^\lambda \to v_\varepsilon \quad \text{strongly in } H,$$
$$y^\lambda \to y_\varepsilon \quad \text{strongly in } V. \tag{2.33}$$

Now, since problem (2.32) is smooth one can easily find the corresponding optimality conditions. Indeed, we have

$$(\nabla g^\lambda(y^\lambda), z) + h'(u^\lambda, w) + \varepsilon^{-1}\big((\nabla\varphi_\lambda(y^\lambda), z)$$
$$+(\varphi^*)'(v^\lambda, \theta) - (z, v^\lambda) - (y^\lambda, \theta)\big) + (v^\lambda - v^*, \theta)$$
$$+ \langle u^\lambda - u^*, w\rangle + \langle u^\lambda - u_\varepsilon, w\rangle \ge 0 \tag{2.34}$$

for all $w \in U$, $\theta \in H$ and z satisfying the equation

$$Az = Bw - \theta.$$

Now, let $p^\lambda \in V$ be the solution to

$$-A^*p^\lambda = \nabla g^\lambda(y^\lambda) + \varepsilon^{-1}\big(\nabla\varphi_\lambda(y^\lambda) - v^\lambda\big). \tag{2.35}$$

Substituting (2.35) in (2.34), we get

$$-\langle B^*p^\lambda, w\rangle + (\theta, p^\lambda) + h'(u^\lambda, w) + \varepsilon^{-1}((\varphi^*)'(v^\lambda, \theta) - (y^\lambda, \theta))$$
$$+(v^\lambda - v_\varepsilon, \theta) + (v^\lambda - v^*, \theta) + \langle u^\lambda - u^*, w\rangle + \langle u^\lambda - u_\varepsilon, w\rangle \ge 0$$

for all $w \in U$, $\theta \in H$. This yields

$$B^*p^\lambda \in \partial h(u^\lambda) + 2u^\lambda - u_\varepsilon - u^*, \tag{2.36}$$
$$-p^\lambda \in \varepsilon^{-1}(\partial\varphi^*(v^\lambda) - y^\lambda) + 2v^\lambda - v^* - v_\varepsilon. \tag{2.37}$$

We may equivalently write (2.37) as

$$v^\lambda \in \partial\varphi\big(y^\lambda - \varepsilon p^\lambda + \varepsilon(v^* + v_\varepsilon - 2v^\lambda)\big). \tag{2.38}$$

Then, substituting in (2.35) and multiplying the resulting equation by $(I + \lambda\,\partial\varphi)^{-1}y_\lambda - y_\lambda + \varepsilon p^\lambda - \varepsilon(v^* + v_\varepsilon - 2v_\lambda)$, we get the estimate

$$\|p^\lambda\| \le C \qquad \forall \lambda > 0,$$

because $\partial\varphi$ is monotone and $\{\nabla g^\lambda\}$ is uniformly bounded on bounded subsets.

Hence on a subsequence, $\lambda \to 0$, we have

$$p^\lambda \to p_\varepsilon \qquad \text{weakly in } V,$$
$$A^*p^\lambda \to A^*p_\varepsilon \qquad \text{weakly in } V',$$

whilst by Propositions 2.15 and 1.3, in Chapter 2,

$$\nabla g^\lambda(y^\lambda) \to \xi \in \partial g(y_\varepsilon) \qquad \text{weakly in } H,$$

and

$$\nabla\varphi_\lambda(y^\lambda) \to \eta \in \partial\varphi(y_\varepsilon) \qquad \text{weakly in } V'$$

because $\{\nabla\varphi_\lambda(y^\lambda)\}$ is bounded in V' and $y^\lambda \to y_\varepsilon$ strongly in V. Finally, by (2.36) it follows that

$$B^*p_\varepsilon \in \partial h(u_\varepsilon) + u_\varepsilon - u^*,$$

and

$$v_\varepsilon \in \partial\varphi(y_\varepsilon - \varepsilon p_\varepsilon + \varepsilon(v^* - v_\varepsilon))$$

thereby completing the proof. ■

By virtue of Lemma 2.4 we may view (2.29)–(2.31) as an approximating optimality system for problem (P).

Let $\theta_\varepsilon = v^* - v_\varepsilon$. Then, we may rewrite (2.29) as

$$-A^*p_\varepsilon \in \partial g(y_\varepsilon) + \varepsilon^{-1}(\partial\varphi(y_\varepsilon) - \partial\varphi(y_\varepsilon - \varepsilon p_\varepsilon + \varepsilon\theta_\varepsilon)). \qquad (2.39)$$

If multiply this equation by $p_\varepsilon - \theta_\varepsilon$ and use the monotonicity of $\partial\varphi$ along with the coercivity condition (2.2), we get

$$\|p_\varepsilon\| \le C \qquad \forall \varepsilon > 0,$$

because $\{\nabla g(y_\varepsilon)\}$ is bounded.

Hence, there is a sequence $\varepsilon_n \to 0$ such that

$$p_{\varepsilon_n} \to p \qquad \text{weakly in } V, \text{ strongly in } H,$$
$$A^*p_{\varepsilon_n} \to A^*p \qquad \text{weakly in } V',$$
$$u_{\varepsilon_n} \to u^* \qquad \text{strongly in } U,$$
$$v_{\varepsilon_n} \to v^* \in \partial\varphi(y^*) \qquad \text{strongly in } H.$$

Then, letting $\varepsilon = \varepsilon_n$ tend to zero in (2.29)–(2.31), we get

$$-A^*p \in \partial g(y^*) + w^*,$$
$$B^*p \in \partial h(u^*), \tag{2.40}$$

where

$$w^* = \lim_{\varepsilon_n \to 0} \frac{1}{\varepsilon_n}\left(\partial\varphi(y_{\varepsilon_n}) - \partial\varphi(y_{\varepsilon_n} - \varepsilon_n p_{\varepsilon_n} + \varepsilon_n \theta_{\varepsilon_n})\right) \tag{2.41}$$

in the weak topology of V'.

These conditions can be made more explicit in the specific problems that will be studied in what follows.

3.2.3. *Optimal Control of Semilinear Elliptic Equations*

In this section, we shall study the following particular case of problem (P):

Minimize $g(y) + h(u)$ *on all* $y \in H_0^1(\Omega) \cap H^2(\Omega)$ *and* $u \in U$, (2.42)

subject to

$$A_0 y + \beta(y) = f + Bu \qquad \text{in } \Omega,$$
$$y = 0 \qquad \text{in } \partial\Omega. \tag{2.43}$$

Here, A_0 is the elliptic differential operator (1.21), i.e.,

$$A_0 y = -\sum_{i,j=1}^{N} \left(a_{ij}(x) y_{x_i}\right)_{x_j} + a_0(x) y, \tag{2.44}$$

where $a_{ij}, a_0 \in L^\infty(\Omega)$, $a_{ij} = a_{ji}$ for all $i, j = 1, \ldots, N$, $a_0 \geq 0$ a.e. in Ω,

$$\sum_{i,j=1}^{N} a_{ij}(x)\xi_i \xi_j \geq \omega \|\xi\|_N^2 \qquad \forall \xi \in \mathbf{R}^N, \qquad \text{a.e. } x \in \Omega,$$

for some $\omega > 0$, β is a monotonically increasing continuous function on $\mathbf{R}$, $f \in L^2(\Omega)$, and $B \in L(U, L^2(\Omega))$.

We shall assume that the functions g and h satisfy assumptions (i) and (ii) of problem (P). More precisely, we shall assume:

(a) The function $g: L^2(\Omega) \to \mathbf{R}$ is Lipschitz on bounded subsets and there are $C \in \mathbf{R}$ and $f_0 \in L^2(\Omega)$ such that

$$g(y) \geq (f_0, y) + C \qquad \forall y \in L^2(\Omega). \tag{2.45}$$

(b) The function $h: U \to \overline{\mathbf{R}}$ is convex, lower semicontinuous, and satisfies condition (2.5).

Problem (2.42), (2.43) is a particular case of problem (P), where $H = L^2(\Omega)$, $V = H_0^1(\Omega)$, $A: V \to V'$ is defined by (1.19), i.e.,

$$(Ay, z) = \sum_{i,j=1}^{N} \int_\Omega a_{ij}(x) y_{x_i} z_{x_j}\, dx + \int_\Omega a_0 yz\, dx \qquad \forall y, z \in H_0^1(\Omega), \tag{2.46}$$

and

$$\varphi(y) = \int_\Omega j(y(x))\, dx, \quad y \in H_0^1(\Omega); \qquad \partial j = \beta, \quad j: \mathbf{R} \to \overline{\mathbf{R}}. \tag{2.47}$$

The function $g: L^2(\Omega) \to \mathbf{R}$ may arise as an integral functional of the form

$$g(y) = \int_\Omega g_0(x, y(x))\, dx, \qquad y \in L^2(\Omega), \tag{2.48}$$

where $g_0: \Omega \times \mathbf{R} \to \mathbf{R}$ is measurable in x and Lipschitz in y.

We shall assume that Ω is a bounded and open subset of $\mathbf{R}^N$, either with smooth boundary (of class C^2 for instance) or convex. As seen earlier, this implies that for every $u \in U$, problem (2.43) has a unique solution $y \in H_0^1(\Omega) \cap H^2(\Omega)$. Moreover, if h satisfies condition (2.5), then problem (2.42) has at least one optimal pair (y, u).

Theorem 2.1 following is a maximum principle type result for this problem.

Theorem 2.1. *Let (y^*, u^*) be any optimal pair in problem (2.42), (2.43), where β is a monotonically increasing locally Lipschitz function on $\mathbf{R}$, and g, h satisfy assumptions (a), (b). Then there exist functions $p \in H_0^1(\Omega)$, $\eta \in L^1(\Omega)$, and $\xi \in L^2(\Omega)$ such that $Ap \in (L^\infty(\Omega))^*$ and*

$$-(Ap)_a - \eta = \xi \qquad \text{a.e. in } \Omega, \tag{2.49}$$

$$\eta(x) \in p(x)\, \partial\beta(y^*(x)), \qquad \xi(x) \in \partial g(y(x)) \qquad \text{a.e. } x \in \Omega, \tag{2.50}$$

$$B^* p \in \partial h(u^*). \tag{2.51}$$

If either $1 \le N \le 3$ or β satisfies the condition

$$\beta'(r) \le C(|\beta(r)| + |r| + 1) \qquad \text{a.e. } r \in \mathbf{R}, \tag{2.52}$$

then $Ap \in L^1(\Omega)$ *and* Eq. (2.49) *becomes*

$$-Ap - \eta = \xi \qquad \text{a.e. in } \Omega. \tag{2.53}$$

Here, $\partial\beta$ and ∂g are the Clarke generalized gradients of β and g, respectively (see Section 2.3 in Chapter 2), $(L^\infty(\Omega))^*$ is the dual space of $L^\infty(\Omega)$ and $(Ap)_a$ is the absolutely continuous part of $Ap \in (L^\infty(\Omega))^*$. We can apply the Lebesgue decomposition theorem to the elements μ of the space $(L^\infty(\Omega))^*$: $\mu = \mu_a + \mu_s$, where $\mu_a \in L^1(\Omega)$ is the absolutely continuous part of μ and μ_s the singular part (see, e.g., Ioffe and Levin [1]). This means that there exists an increasing sequence of measurable sets $\Omega_k \subset \Omega$ such that $m(\Omega \setminus \Omega_k) \overset{k\to\infty}{\to} 0$ and $\mu_s(\varphi) = 0$ for all $\varphi \in L^\infty(\Omega)$ having support in Ω_k.

Thus, (2.49) should be understood in the following sense: There exists a singular measure $\nu_s \in (L^\infty(\Omega))^*$ such that $Ap = \nu_s - \eta - \xi$, where $\eta \in L^1(\Omega)$ and $\xi \in L^2(\Omega)$ satisfy Eq. (2.50).

Proof of Theorem 2.1. We shall use the approach described in Section 2.2 by approximating problem (2.41) by a family of problems of the form (Q_ε). Namely, we shall consider the approximating problem:

Minimize

$$g(y) + h(u) + \varepsilon^{-1}(\varphi(y) + \varphi^*(v) - (y,v)) + \tfrac{1}{2}|v - v^*|_2^2 + \tfrac{1}{2}|u - u^*|_U^2 \tag{2.54}$$

on all $(y,u) \in (H_0^1(\Omega) \cap H^2(\Omega)) \times U$, *subject to*

$$Ay = Bu - v + f. \tag{2.55}$$

Here, $v^* = Bu^* + f - Ay^* \in \partial\varphi(y^*) = \beta(y^*)$ a.e. in Ω and $\partial\varphi \subset L^2(\Omega) \times L^2(\Omega)$ is the subdifferential of φ given by (2.47), as a l.s.c. convex function from $L^2(\Omega)$ to $\overline{\mathbf{R}}$. Similarly,

$$\varphi^*(v) = \sup\{(v,p) - \varphi(v);\, v \in L^2(\Omega)\} = \int_\Omega j^*(v(x))\,dx, \qquad v \in L^2(\Omega).$$

(Throughout this section $|\cdot|_2$ denotes L^2 norm and $(\cdot,\cdot)$ the L^2 scalar product.) Without any loss of generality, we may assume that $\beta(0) = 0$.

As seen in Lemmas 2.2 and 2.3, problem (2.54) has at least one solution $(y_\varepsilon, u_\varepsilon, v_\varepsilon) \in (H_0^1(\Omega) \cap H^2(\Omega)) \times U \times L^2(\Omega)$ and, for $\varepsilon \to 0$,

$$\begin{aligned} u_\varepsilon &\to u^* && \text{strongly in } U, \\ v_\varepsilon &\to v^* && \text{strongly in } L^2(\Omega), \\ y_\varepsilon &\to y^* && \text{strongly in } H_0^1(\Omega) \cap H^2(\Omega). \end{aligned} \tag{2.56}$$

Now, by Lemma 2.4 it follows that there are $p_\varepsilon \in H_0^1(\Omega) \cap H^2(\Omega)$ and $\xi_\varepsilon \in L^2(\Omega)$ such that

$$-Ap_\varepsilon = \xi_\varepsilon + \varepsilon^{-1}(\beta(y_\varepsilon) - v_\varepsilon) \qquad \text{a.e. in } \Omega, \tag{2.57}$$

$$v_\varepsilon = \beta(y_\varepsilon - \varepsilon p_\varepsilon + \varepsilon(v^* - v_\varepsilon)) \qquad \text{a.e. in } \Omega, \tag{2.58}$$

$$\xi_\varepsilon(x) \in \partial g(y_\varepsilon(x)) \qquad \text{a.e. in } \Omega, \tag{2.59}$$

$$B^* p_\varepsilon \in \partial h(u_\varepsilon) + u_\varepsilon - u^*. \tag{2.60}$$

To pass to limit in system (2.57)–(2.60), we need some *a priori* estimates on p_ε. These come down to multiplying Eq. (2.57) by p_ε and integrating on Ω. We have

$$\begin{aligned} -(Ap_\varepsilon, p_\varepsilon) &= \int_\Omega \xi_\varepsilon p_\varepsilon \, dx \\ &\quad + \frac{1}{\varepsilon} \int_\Omega (\beta(y_\varepsilon) - \beta(y_\varepsilon - \varepsilon p_\varepsilon + \varepsilon(v^* - v_\varepsilon)) \\ &\quad \times (p_\varepsilon - (v^* - v_\varepsilon)) \, dx + \int_\Omega (\beta(y_\varepsilon) - v_\varepsilon)(v^* - v_\varepsilon) \, dx \\ &\geq -\|\xi_\varepsilon\|_2 \|p_\varepsilon\|_2 - (\|\beta(y_\varepsilon)\|_2 + \|v_\varepsilon\|_2)\|v^* - v_\varepsilon\|_2 \\ &\qquad \forall \varepsilon > 0. \end{aligned}$$

Since $\{\xi_\varepsilon\}$ is bounded in $L^2(\Omega)$ (because g is locally Lipschitz), we have

$$\|p_\varepsilon\|_{H_0^1(\Omega)} \leq C \qquad \forall \varepsilon > 0, \tag{2.61}$$

and so we may suppose that

$$\begin{aligned} p_\varepsilon &\to p && \text{weakly in } H_0^1(\Omega), \text{ strongly in } L^2(\Omega), \\ Ap_\varepsilon &\to Ap && \text{weakly in } H^{-1}(\Omega). \end{aligned} \tag{2.62}$$

Extracting a further subsequence $\varepsilon_n \to 0$, we can assume by (2.56) that

$$\begin{aligned} p_{\varepsilon_n}(x) &\to p(x) && \text{a.e. } x \in \Omega, \\ v_{\varepsilon_n}(x) - v^*(x) &\to 0 && \text{a.e. } x \in \Omega, \\ y_{\varepsilon_n}(x) &\to y^*(x) && \text{a.e. } x \in \Omega. \end{aligned} \tag{2.63}$$

Now, by the Egorov theorem, for every $\delta > 0$ there is a measurable subset $E_\delta \subset \Omega$ such that $m(E_\delta) \le \delta$ and the sequences p_{ε_n}, $v_{\varepsilon_n} - v^*$, y_{ε_n} are bounded in $L^\infty(\Omega \setminus E_\delta)$ and uniformly convergent on $\Omega \setminus E_\delta$.

Now multiply Eq. (2.57) by $\zeta_\lambda(p_\varepsilon)$ where ζ_λ is a smooth approximation of signum function, i.e., for $\lambda > 0$,

$$\zeta_\lambda(r) = \begin{cases} 1 & \text{for} \quad r \ge \lambda, \\ \dfrac{1}{\lambda} r & \text{for} \ -\lambda < r < \lambda, \\ -1 & \text{for} \quad r \le -\lambda. \end{cases}$$

We have

$$(Ap_\varepsilon, \zeta_\lambda(p_\varepsilon)) = \int_\Omega A_0 p_\varepsilon \zeta_\lambda(p_\varepsilon)\, dx \ge 0$$

and, therefore,

$$\begin{aligned} &\varepsilon^{-1} \int_\Omega (\beta(y_\varepsilon) - v_\varepsilon) \zeta_\lambda(p_\varepsilon - \varepsilon(v^* - v_\varepsilon))\, dx \\ &\quad \le \varepsilon^{-1} \int_\Omega (\beta(y_\varepsilon) - v_\varepsilon)(\zeta_\lambda(p_\varepsilon - \varepsilon(v^* - v_\varepsilon)) - \zeta_\lambda(p_\varepsilon))\, dx \\ &\quad\quad + \| \xi_\varepsilon \|_{L^1(\Omega)} \qquad \forall \varepsilon > 0. \end{aligned}$$

Now, let $\delta > 0$ be arbitrary and let $E_\delta \subset \Omega$ be such that $m(E_\delta) \le \delta$ and the convergences in (2.63) are uniform on $\Omega \setminus E_\delta$. Then, letting $\varepsilon = \varepsilon_n$ tend to zero in the previous inequality, we see that

$$\begin{aligned} &\limsup_{\varepsilon_n \to 0} \varepsilon_n^{-1} \int_\Omega (\beta(y_{\varepsilon_n}) - v_{\varepsilon_n}) \zeta_\lambda(p_{\varepsilon_n} - \varepsilon_n(v^* - v_{\varepsilon_n}))\, dx \\ &\quad \le C + 2 \limsup_{\varepsilon_n \to 0} \varepsilon_n^{-1} \int_{E_\delta} |\beta(y_{\varepsilon_n}) - v_{\varepsilon_n}|\, dx, \end{aligned}$$

and this yields

$$\begin{aligned} &\limsup_{\varepsilon_n \to 0} \varepsilon_n^{-1} \int_\Omega |\beta(y_{\varepsilon_n}) - v_{\varepsilon_n}|\, dx \\ &\quad \le C\Big(1 + \limsup_{\varepsilon_n \to 0} \varepsilon_n^{-1} \int_{E_\delta} |\beta(y_{\varepsilon_n}) - v_{\varepsilon_n}|\, dx, \end{aligned} \tag{2.64}$$

where C is independent of δ (for δ sufficiently small). On the other hand, since $\{\beta(y_{\varepsilon_n}) - v_{\varepsilon_n}\}$ is bounded in $L^2(\Omega)$, we have

$$\int_{E_\delta} |\beta(y_{\varepsilon_n}) - v_{\varepsilon_n}|\, dx \le C\delta^{1/2}$$

where C is independent of n and δ. Then, for $\delta = \varepsilon_n^2$, (2.64) yields

$$\limsup_{\varepsilon_n \to 0} \varepsilon_n^{-1} \int_{\Omega} |\beta(y_{\varepsilon_n}) - v_{\varepsilon_n}|\, dx \le C, \tag{2.65}$$

from which

$$\limsup_{\varepsilon_n \to 0} \|Ap_{\varepsilon_n}\|_{L^1(\Omega)} \le C. \tag{2.66}$$

Since $\{\xi_{\varepsilon_n}\}$ is bounded in $L^2(\Omega)$, we may assume that

$$\xi_{\varepsilon_n} \to \xi \qquad \text{weakly in } L^2(\Omega)$$

and, by Proposition 2.13 in Chapter 2, $\xi \in \partial g(y^*)$. Similarly, letting ε_n tend to zero in (2.60) we get (2.51).

By the estimates (2.65) and (2.66), it follows that there exists a generalized subsequence of ε_n, say ε_λ, such that

$$Ap_{\varepsilon_\lambda} \to \mu \qquad \text{weak star in } (L^\infty(\Omega))^*,$$

$$\frac{1}{\varepsilon_\lambda}\big(\beta(y_{\varepsilon_\lambda}) - v_{\varepsilon_\lambda}\big) \to \eta \qquad \text{weak star in } (L^\infty(\Omega))^*, \tag{2.67}$$

where $\mu = Ap$ on $L^\infty(\Omega) \cap H_0^1(\Omega)$ and in particular on $C_0^\infty(\Omega)$. We have

$$-\mu = \eta + \xi. \tag{2.68}$$

On the other hand, by the mean formula (Corollary 2.1, Chapter 2), we have

$$\frac{1}{\varepsilon_\lambda}\big(\beta(y_{\varepsilon_\lambda}) - v_{\varepsilon_\lambda}\big) = \theta_\lambda\big(p_{\varepsilon_\lambda} - (v^* - v_{\varepsilon_\lambda})\big) \qquad \text{a.e. in } \Omega,$$

where $\theta_\lambda \in \partial\beta(z_\lambda)$, $y_{\varepsilon_\lambda} \le z_\lambda \le y_{\varepsilon_\lambda} - \varepsilon_\lambda(p_{\varepsilon_\lambda} - v^* + v_{\varepsilon_\lambda})$). Now since θ_λ and $\varepsilon_\lambda^{-1}(\beta(y_{\varepsilon_\lambda}) - v_{\varepsilon_\lambda})$ are uniformly bounded on $\Omega \setminus E_\delta$, we may assume that

$$\theta_{\varepsilon_\lambda} \to \theta \qquad \text{weak star in } L^\infty(\Omega \setminus E_\delta) \tag{2.69}$$

and

$$\frac{1}{\varepsilon_\lambda}\big(\beta(y_{\varepsilon_\lambda}) - v_{\varepsilon_\lambda}\big) \to \theta p \qquad \text{weak star in } L^\infty(\Omega \setminus E_\delta). \tag{2.70}$$

By (2.69) we have

$$\int_{\Omega \setminus E_\delta} \theta_{\varepsilon_\lambda}(x) w(x)\, dx \le \int_{\Omega \setminus E_\delta} \beta^0\big(z_{\varepsilon_\lambda}(x), w(x)\big)\, dx \qquad \forall w \in L^1(\Omega \setminus E_\delta).$$

This yields (β^0 is the directional derivative of β)

$$\int_{\Omega \setminus E_\delta} \theta(x) w(x)\, dx \le \limsup_{\varepsilon_\lambda \to 0} \int_{\Omega \setminus E_\delta} \beta^0\big(z_{\varepsilon_\lambda}(x), w(x)\big)\, dx$$
$$\le \int_{\Omega \setminus E_\delta} \beta^0(y^*(x), w(x))\, dx.$$

Hence,

$$\theta(x) w \le \beta^0(y^*(x), w) \qquad \forall w \in \mathbf{R}, \quad \text{a.e. } x \in \Omega \setminus E_\delta,$$

and therefore $\theta(x) \in \partial\beta(y^*(x))$ a.e. $x \in \Omega \setminus E_\delta$. We have therefore proved that $\eta \in L^\infty(\Omega \setminus E_\delta)$ and $\eta(x) \in \partial\beta(y^*(x))p(x)$ a.e. $x \in \Omega \setminus E_\delta$. Then, by (2.68), we see that the restriction of μ to $\Omega \setminus E_\delta$ belongs to $L^\infty(\Omega \setminus E_\delta)$ and

$$\mu_{\mathrm{a}}(x) = \mu(x) \in -\partial\beta(y^*(x))p(x) - \xi(x) \qquad \text{a.e. } x \in \Omega \setminus E_\delta.$$

Since δ is arbitrary, we conclude that

$$-\mu_{\mathrm{a}}(x) \in \partial\beta(y^*(x))p(x) + \xi(x) \qquad \text{a.e.} x \in \Omega,$$

as claimed.

Suppose now that $1 \le N \le 3$. Then, by the Sobolev imbedding theorem, $H^2(\Omega) \subset C(\overline{\Omega})$ and so y_ε are uniformly bounded on $\overline{\Omega}$.

On the other hand, we have, by (2.57),

$$A(\varepsilon p_\varepsilon) = -\varepsilon\xi_\varepsilon - \beta(y_\varepsilon) + v_\varepsilon$$

and, since $\{\beta(y_\varepsilon) - v_\varepsilon\}$ is bounded in $L^2(\Omega)$, we infer that $\{\varepsilon p_\varepsilon\}_{\varepsilon > 0}$ is bounded in $H^2(\Omega)$ and therefore in $C(\overline{\Omega})$. Finally, we have

$$y_\varepsilon - \varepsilon p_\varepsilon + \varepsilon(v^* - v_\varepsilon) \in \beta^{-1}(v_\varepsilon),$$
$$y^* \in \beta^{-1}(v^*).$$

Subtracting and multiplying by $v_\varepsilon - v^*$, we get

$$\varepsilon|v^* - v_\varepsilon| \le |y_\varepsilon| + \varepsilon|p_\varepsilon| \qquad \text{a.e. in } \Omega$$

and, since β is locally Lipschitz, this implies that

$$|Ap_\varepsilon| \le |\xi_\varepsilon| + L(|p_\varepsilon| + |v_\varepsilon - v^*|) \qquad \text{a.e. } x \in \Omega.$$

Hence, $\{Ap_\varepsilon\}$ is bounded in $L^2(\Omega)$ and we conclude, therefore, that $p \in H_0^1(\Omega) \cap H^2(\Omega)$, $\eta \in L^2(\Omega)$, and $Ap = (Ap^*)_a$, as claimed.

Assume now that condition (2.52) holds. We shall prove that $\{q_\varepsilon = (1/\varepsilon)(\beta(y_\varepsilon) - v_\varepsilon)\}$ is weakly compact in $L^1(\Omega)$. We have

$$q_\varepsilon = \int_0^1 \beta'(y_\varepsilon - \varepsilon\lambda(p_\varepsilon - \theta_\varepsilon))(p_\varepsilon - \theta_\varepsilon)\, d\lambda \qquad \text{a.e. in } \Omega,$$

where β' is the derivative of β and $\theta_\varepsilon = v^* - v_\varepsilon$. Then, by condition (2.52), we have

$$\begin{aligned} 0 \le q_\varepsilon(x) \le C((p_\varepsilon(x) - \theta_\varepsilon(x)) \\ \times \int_0^1 |\beta(y_\varepsilon(x) - \varepsilon\lambda(p_\varepsilon(x) - \theta_\varepsilon(x)|\, d\lambda \\ + |y_\varepsilon(x)| + \varepsilon(p_\varepsilon(x) - \theta_\varepsilon(x)) \\ \text{a.e. in } \{x;\, p_\varepsilon(x) \ge \theta_\varepsilon(x)\}, \\ 0 \le -q_\varepsilon(x) \le C((\theta_\varepsilon(x) - p_\varepsilon(x)) \\ \times \int_0^1 |\beta(y_\varepsilon(x) - \varepsilon\lambda(p_\varepsilon(x) - \theta_\varepsilon(x))|\, d\lambda \\ + |y_\varepsilon(x)| + \varepsilon|p_\varepsilon(x) - \theta_\varepsilon(x)| \\ \text{a.e. in } \{x;\, p_\varepsilon(x) < \theta_\varepsilon(x)\}. \end{aligned}$$

On the other hand, we have (by the monotonicity of β)

$$\begin{aligned} \beta(y_\varepsilon - \varepsilon\lambda(p_\varepsilon - \theta_\varepsilon)) &\le \beta(y_\varepsilon) && \text{for } p_\varepsilon \ge \theta_\varepsilon, \\ \beta(y_\varepsilon - \varepsilon\lambda(p_\varepsilon - \theta_\varepsilon)) &\le \beta(y_\varepsilon - \varepsilon(p_\varepsilon - \theta_\varepsilon)) && \text{for } p_\varepsilon \le \theta_\varepsilon. \end{aligned}$$

Hence,

$$\begin{aligned} |q_\varepsilon(x)| \le C(|p_\varepsilon(x) - \theta_\varepsilon(x)|(|\beta(y_\varepsilon(x))| \\ + |\beta(y_\varepsilon(x) - \varepsilon(p_\varepsilon(x) - \theta_\varepsilon(x)))|) \\ + |y_\varepsilon(x)| + \varepsilon|p_\varepsilon(x) - \theta_\varepsilon(x)|) \qquad \text{a.e. } x \in \Omega. \quad (2.71) \end{aligned}$$

Since, as proved earlier, $\{p_\varepsilon\}$ is compact in $L^2(\Omega)$, $y_\varepsilon \to y^*$ in $L^2(\Omega)$, $\theta_\varepsilon \to 0$ in $L^2(\Omega)$, $\beta(y_\varepsilon - \varepsilon(p_\varepsilon - \theta_\varepsilon)) \to v^*$ in $L^2(\Omega)$, and $\{\beta(y_\varepsilon)\}$ is bounded in $L^2(\Omega)$, it follows by (2.71) via the Dunford–Pettis criterion that $\{q_\varepsilon\}$ is weakly compact in $L^1(\Omega)$. Hence, $\{Ap_\varepsilon\}$ is weakly compact in $L^1(\Omega)$ and so $Ap \in L^1(\Omega)$, thereby completing the proof. ■

One might obtain the same result if one uses the approach described in Proposition 2.2. Namely, we approximate problem (2.42), (2.43) by the following family of optimal control problems:

$$\textit{Minimize}\; g^{\varepsilon}(y) + h(u) + \tfrac{1}{2}|u - u^*|_U^2$$
$$\textit{on all}\; (y,u) \in (H_0^1(\Omega) \cap H^2(\Omega)) \times U, \tag{2.72}$$

subject to

$$Ay + \beta^{\varepsilon}(y) = Bu + f \qquad \text{in } \Omega, \tag{2.73}$$

where $Ay = A_0 y$ with $D(A) = H_0^1(\Omega) \cap H^2(\Omega)$, and $\beta^{\varepsilon} \in C^{\infty}(\mathbf{R})$ is defined by

$$\beta^{\varepsilon}(r) = \int_{-\infty}^{\infty} \left(\beta_{\varepsilon}(r - \varepsilon^2\theta) - \beta_{\varepsilon}(-\varepsilon^2\theta)\right)\rho(\theta)\, d\theta + \beta_{\varepsilon}(0), \qquad r \in \mathbf{R}. \tag{2.74}$$

Here, $\beta_{\varepsilon}(r) = \varepsilon^{-1}(r - (1 + \varepsilon\beta)^{-1}r)$, $r \in \mathbf{R}$, and ρ is a C_0^{∞} mollifier in $\mathbf{R}$, i.e., $\rho \in C^{\infty}(\mathbf{R})$, $\rho(r) = 0$ for $|r| > 1$, $\rho(r) = \rho(-r)\ \forall r \in \mathbf{R}$, $\int \rho(t)\, dt = 1$. Note that β^{ε} is monotonically increasing, Lipschitz, and

$$|\beta^{\varepsilon}(r) - \beta_{\varepsilon}(r)| \le 2\varepsilon \qquad \forall r \in \mathbf{R}. \tag{2.75}$$

We are in the situation described in Section 2.2, where

$$\varphi^{\varepsilon}(y) = \int_{\Omega} j^{\varepsilon}(y)\, dx \qquad \forall y \in L^2(\Omega), \quad \nabla j^{\varepsilon} = \beta^{\varepsilon}.$$

Throughout the sequel, we set $\dot{\beta}^{\varepsilon} = (\beta^{\varepsilon})'$.

Let $(y_{\varepsilon}, u_{\varepsilon})$ be optimal for problem (2.72). Then we have

$$\begin{aligned} u_{\varepsilon} &\to u^* && \text{strongly in } U,\\ y_{\varepsilon} &\to y^* && \text{strongly in } H_0^1(\Omega), \text{ weakly in } H^2(\Omega),\\ \beta^{\varepsilon}(y_{\varepsilon}) &\to \beta(y^*) && \text{weakly in } L^2(\Omega). \end{aligned} \tag{2.76}$$

Indeed, from the inequality

$$g^{\varepsilon}(y_{\varepsilon}) + h(u_{\varepsilon}) + \tfrac{1}{2}|u_{\varepsilon} - u^*|_U^2 \le g^{\varepsilon}(y^{u^*}) + h(u^*)$$

(y^u is the solution to Eq. (2.73)), we see that $\{u_\varepsilon\}$ is bounded in U. Hence, on a subsequence $\varepsilon_n \to 0$, we have

$$u_{\varepsilon_n} \to \bar{u} \qquad \text{weakly in } U.$$

On the other hand, we may write Eq. (2.73) as

$$Ay_\varepsilon + \beta_\varepsilon(y_\varepsilon) = Bu_\varepsilon + f + \beta_\varepsilon(y_\varepsilon) - \beta^\varepsilon(y_\varepsilon)$$

and so, by Proposition 1.2, $y_{\varepsilon_n} \to \bar{y}$ strongly in $H^1(\Omega)$ and weakly in $H^2(\Omega)$, where $\bar{y}$ is the solution to (2.43), where $u = \bar{u}$. This yields

$$g(\bar{y}) + h(\bar{u}) + \limsup_{\varepsilon_n \to 0} \tfrac{1}{2}|u_{\varepsilon_n} - u^*|_U^2 \le g(y^*) + h(u^*) = \inf(P).$$

Hence, $u_{\varepsilon_n} \to u^*$ strongly in U and $\bar{y} = y^*$, $\bar{u} = u^*$. The sequence ε_n being arbitrary, we have (2.76).

Moreover, there is $p_\varepsilon \in H_0^1(\Omega) \cap H^2(\Omega)$ such that

$$-Ap_\varepsilon - \dot{\beta}^\varepsilon(y_\varepsilon)p_\varepsilon = \nabla g^\varepsilon(y_\varepsilon) \qquad \text{in } \Omega, \tag{2.77}$$

$$B^*p_\varepsilon \in \partial h(u_\varepsilon) + u_\varepsilon - u^*. \tag{2.78}$$

Now, multiplying Eq. (2.77) by p_ε and sign p_ε, we get the estimate

$$\|p_\varepsilon\|^2_{H_0^1(\Omega)} + \int_\Omega |\dot{\beta}^\varepsilon(y_\varepsilon)p_\varepsilon|\,dx \le C \qquad \forall \varepsilon > 0.$$

Hence on a subsequence, again denoted ε, we have

$$p_\varepsilon \to p \qquad \text{weakly in } H_0^1(\Omega), \text{ strongly in } L^2(\Omega),$$

$$\nabla g^\varepsilon(y_\varepsilon) \to \xi \qquad \text{weakly in } L^2(\Omega),$$

and, on a generalized subsequence $\{\varepsilon_\lambda\}$,

$$\dot{\beta}^{\varepsilon_\lambda}(y_{\varepsilon_\lambda})p_{\varepsilon_\lambda} \to \eta \qquad \text{weak star in } (L^\infty(\Omega))^*.$$

Hence, $-Ap - \eta = \xi \in \partial g(y^*)$ in Ω.

Now, by the Egorov theorem, for each $\delta > 0$ there exists a measurable subset E_δ of Ω such that $m(E_\delta) \le \delta$ and $y^*, p \in L^\infty(\Omega \setminus E_\delta)$ and

$$y_\varepsilon(x) \to y^*(x), \quad p_\varepsilon(x) \to p(x) \qquad \text{uniformly on } \Omega \setminus E_\delta.$$

Since $\{(\dot{\beta}^\varepsilon(y_\varepsilon)\}$ is bounded in $L^\infty(\Omega \setminus E_\delta)$, we may assume that, on a subsequence

$$\dot{\beta}^\varepsilon(y_\varepsilon) \to f_\delta \qquad \text{weak star in } L^\infty(\Omega \setminus E_\delta).$$

Then, by Lemma 2.5 following, we infer that $f_\delta(x) \in \partial\beta(y^*(x))$ a.e. $x \in \Omega \setminus E_\delta$ and so

$$\eta_a(x) = f_\delta(x)p(x) \in \partial\beta(y^*(x))p(x) \qquad \text{a.e. } x \in \Omega \setminus E_\delta .$$

The last part of Theorem 2.1 follows as in the previous proof since $\{\dot\beta^\varepsilon(y_\varepsilon)\}$ is bounded in $L^\infty(\Omega)$ if $1 \le N \le 3$ (because $|\dot\beta^\varepsilon(y_\varepsilon)| \le |\beta'(y_\varepsilon)| \le C$ in Ω) and is weakly compact in $L^1(\Omega)$ if β satisfies condition (2.52).

The main ingredient to pass to limit in the approximating optimality system is Lemma 2.5 following, which has intrinsic interest.

Lemma 2.5. *Let X be a locally compact space and let ν be a positive measure on X such that $\nu(X) < \infty$. Let $y_\varepsilon \in L^1(X;\nu)$ be such that $y_\varepsilon \to y$ strongly in $L^1(X;\nu)$ and*

$$\dot\beta^\varepsilon(y_\varepsilon) \to f_0 \qquad \text{weakly in } L^1(X;\nu).$$

Then

$$f_0(x) \in \partial\beta(y(x)) \qquad \text{a.e. } x \in X.$$

Proof. On a subsequence, again denoted y_ε, we have

$$y_\varepsilon(x) \to y(x) \qquad \forall x \in X \setminus A, \quad \nu(A) = 0.$$

On the other hand, by Mazur's theorem there is a sequence ω_n of convex combinations of $\{\dot\beta^\varepsilon(y_\varepsilon)\}$ such that

$$\omega_n \to f_0 \qquad \text{strongly in } L^1(X;\nu),$$

where $\omega_n(x) = \Sigma_{i\in I_n} \alpha_n^i \dot\beta^{\varepsilon_i}(y_i(x))$. Here, I_n is a finite set of positive integers in $[n, +\infty[$, $y_i = y_{\varepsilon_i}$ and $\alpha_n^i \ge 0$, $\Sigma_{i\in I_n} \alpha_n^i = 1$. Hence, there is a subsequence, again denoted ω_n, such that $\omega_n(x) \to f_0(x)$ $\forall x \in \Omega \setminus B$, $\nu(B) = 0$. In formula (2.74) we make the substitution $t = r - \varepsilon^2\theta$ to get

$$\begin{aligned}
\beta^\varepsilon(r) &= -\varepsilon^{-2} \int \beta_\varepsilon(t)\rho\left(\frac{r-t}{\varepsilon^2}\right) dt - \int \beta_\varepsilon(-\varepsilon^2\theta)\rho(\theta)\,d\theta \\
&= -\varepsilon^{-2} \int_{r-\varepsilon^2}^{r+\varepsilon^2} \beta_\varepsilon(t)\rho\left(\frac{r-t}{\varepsilon^2}\right) dt - \int \beta_\varepsilon(-\varepsilon^2\theta)\rho(\theta)\,d\theta,
\end{aligned}$$

and this yields

$$\begin{aligned}
\dot\beta^\varepsilon(r) &= \varepsilon^{-4} \int_{r-\varepsilon^2}^{r+\varepsilon^2} \beta_\varepsilon(t)\rho'\left(\frac{r-t}{\varepsilon^2}\right) dt \\
&= -\varepsilon^{-2} \int \dot\beta^\varepsilon(r - \varepsilon^2\theta)\rho'(\theta)\,d\theta.
\end{aligned}$$

Hence,

$$\begin{aligned}\dot{\beta}^{\varepsilon_i}(y_i(x)) &= -\varepsilon_i^{-2}\int \beta_{\varepsilon_i}(y_i(x)-\varepsilon_i^2\theta)\rho'(\theta)\,d\theta\\&= -\int \frac{\beta_{\varepsilon_i}(y_i(x)-\varepsilon_i^2\theta)-\beta_{\varepsilon_i}(y_i(x))}{\varepsilon_i^2}\rho'(\theta)\,d\theta\\&= \sum_{k=1}^{m_i}\delta_k^i\frac{\beta_{\varepsilon_i}(y_i(x))-\beta_{\varepsilon_i}(y_i(x)-\varepsilon_i^2\theta_k^i)}{\varepsilon_i^2\theta_k^i}, \qquad |\theta_k^i|\le 1 \quad \forall k\end{aligned}$$

where $\delta_k^i \ge 0$, $\Sigma_{k=1}^{m_i}\delta_k^i = 1$. Hence,

$$\begin{aligned}&\dot{\beta}^{\varepsilon_i}(y_i(x))\\&\quad=\sum_{k=1}^{m_i}\delta_k^i\eta_k^i\frac{(1+\varepsilon_i\beta)^{-1}(y_i(x))-(1+\varepsilon_i\beta)^{-1}(y_i(x)-\varepsilon_i^2\theta_k^i)}{\varepsilon_i^2\theta_k^i},\end{aligned} \tag{2.79}$$

where

$$\eta_k^i \in \partial\beta(\tau_k^i), \qquad \tau_k^i \in \left[(1+\varepsilon_i\beta)^{-1}(y_i(x)-\epsilon_i^2\theta_k^i),(1+\varepsilon_i\beta)^{-1}(y_i(x))\right].$$

On the other hand, we have

$$\begin{aligned}&(\varepsilon_i^2\theta_k^i)^{-1}\left((1+\varepsilon_i\beta)^{-1}y_i-(1+\varepsilon_i\beta)^{-1}(y_i-\varepsilon_i^2\theta_k^i)\right)\\&\quad=\int_0^1\left((1+\varepsilon_i\beta)^{-1}\right)'(y_i-\lambda\varepsilon_i^2\theta_k^i)\,d\lambda\\&\quad=1-\varepsilon_i\int_0^1\dot{\beta}_{\varepsilon_i}(y_i-\lambda\varepsilon_i^2\theta_k^i)\,d\lambda.\end{aligned}$$

Hence,

$$\left|(\varepsilon_i^2\theta_k^i)^{-1}\left((1+\varepsilon_i\beta)^{-1}y_i-(1+\varepsilon_i\beta)^{-1}(y_i-\varepsilon_i^2\theta_k^i)\right)-1\right|\le C\varepsilon_i$$

Because $\dot{\beta}_{\varepsilon_i}$ uniformly bounded on every bounded subset (β is locally Lipschitz).

Then, by (2.79), it follows that

$$\dot{\beta}^{\varepsilon_i}(y_i(x)) = \sum_{k=1}^{m_i}\delta_k^i\eta_k^i + \gamma_i, \qquad \text{where } \gamma_i\to 0 \text{ as } i\to\infty. \tag{2.80}$$

On the other hand, $\delta_k^i - y_i \to 0$ uniformly with respect to k, and by (2.80) we have

$$\dot{\beta}^{\varepsilon_i}(y_i(x))h \leq \sum_{k=1}^{m_i} \delta_k^i \beta^0(\tau_k^i, h) + \gamma_i h \qquad \forall h \in \mathbf{R}.$$

Hence,

$$\limsup_{i \to \infty} \dot{\beta}^{\varepsilon_i}(y_i(x))h \leq \beta^0(y_i(x), h) \qquad \forall h \in \mathbf{R}.$$

Finally,

$$\begin{aligned} f_0(x)h &= \lim_{n \to \infty} \omega_n(x)h \leq \limsup_{n \to \infty} \sum_{i \in I_n} \alpha_n^i \dot{\beta}^{\varepsilon_i}(y_i(x))h \\ &\leq \beta^0(y(x), h) \qquad \forall h \in \mathbf{R}. \end{aligned}$$

Hence, $f_0(x) \in \partial\beta(y(x))$, as claimed. ■

It should be said that the methods of Section 2.2 are applicable to more general problems of the form (2.42), for instance, for the optimal control problem:

Minimize $g(y) + h(u)$ *on all* $(y, u) \in (W_0^{1,p}(\Omega) \cap W^{2,p}(\Omega)) \times U,$

$$1 \leq p < \infty, \qquad (2.81)$$

subject to

$$\begin{aligned} -\Delta y + \beta(y) &= f + Bu \qquad \text{a.e. in } \Omega, \\ y &= 0 \qquad \text{in } \partial\Omega, \end{aligned} \qquad (2.82)$$

where $f \in L^p(\Omega)$, $B \in L(U, L^p(\Omega))$, g locally Lipschitz on $L^p(\Omega)$, h is convex and l.s.c. on U, and β is monotone and locally Lipschitz.

One gets a result of the type of Theorem 2.1 by using the approximating control process (2.72), i.e.:

$$\textit{Minimize}\ g^\varepsilon(y) + h(u) + \tfrac{1}{2}|u - u^*|_U^2,$$

subject to

$$\begin{aligned} -\Delta y + \beta^\varepsilon(y) &= f + Bu \qquad \text{in } \Omega, \\ y &= 0 \qquad \text{in } \partial\Omega. \end{aligned} \qquad (2.83)$$

Then one writes the optimality system and passes to limit as in the proof of Theorem 2.1.

3.2.4. Optimal Control of the Obstacle Problem

We shall study here the following problem:

$$\text{Minimize}\ g(y) + h(u) \text{ on all } (y,u) \in (H_0^1(\Omega) \cap H^2(\Omega)) \times U,\ y \in K, \tag{2.84}$$

subject to

$$a(y, y - z) \le (f + Bu, y - z) \qquad \forall z \in K, \tag{2.85}$$

where $a: H_0^1(\Omega) \times H_0^1(\Omega) \to \mathbf{R}$ is defined by (1.19) and K is the convex set (1.24).

As seen in the preceding, (2.85) is equivalent to the obstacle problem

$$\begin{aligned} (A_0 y - f - Bu)(y - \psi) &= 0 \qquad \text{a.e. in } \Omega, \\ A_0 y - f - Bu \ge 0, \quad y &\ge \psi \qquad \text{a.e. in } \Omega, \\ y &= 0 \qquad \text{in } \partial\Omega. \end{aligned} \tag{2.86}$$

Here, $B \in L(U, L^2(\Omega))$, $f \in L^2(\Omega)$, and $g: L^2(\Omega) \to \mathbf{R}$, $h: U \to \overline{\mathbf{R}}$ satisfy assumptions (a) and (b) in Section 2.3.

Theorem 2.2. *Let (y^*, u^*) be an optimal pair for problem (2.84). Then there exist $p \in H_0^1(\Omega)$ with $Ap \in (L^\infty(\Omega))^*$ and $\xi \in L^2(\Omega)$ such that $\xi \in \partial g(y^*)$ and*

$$(Ap)_a + \xi = 0 \qquad \text{a.e. in } [x; y^*(x) > \psi(x)], \tag{2.87}$$

$$p(Ay^* - Bu^* - f) = 0 \qquad \text{a.e. in } \Omega, \tag{2.88}$$

$$a(p, \chi(y^* - \psi)) + (\xi, (y^* - \psi)\chi) = 0 \qquad \forall \chi \in C^1(\overline{\Omega}), \tag{2.89}$$

$$B^* p \in \partial h(u^*), \tag{2.90}$$

$$a(p,p) + (\xi, p) \le 0. \tag{2.91}$$

If $1 \le N \le 3$, then Eq. (2.87) reduces to

$$(Ap + \xi)(y^* - \psi) = 0 \text{ in } \Omega. \tag{2.92}$$

We have denoted by A the operator A_0 with the domain $D(A) = H_0^1(\Omega) \cap H^2(\Omega)$ and by $(Ap)_{\mathrm{a}}$ the absolutely continuous part of A^*p. If $N \le 3$, then $y^* \in H^2(\Omega) \subset C(\overline{\Omega})$ and so $Ap(y^* - \psi)$ is well-defined as an element of $(L^\infty(\Omega))^*$. In particular, (2.92) implies that $Ap^* =$ in $[x; y^*(x) > \psi(x)]$.

The system (2.87), (2.89) represents a quasivariational inequality of elliptic type.

Proof of Theorem 2.2. Consider the penalized problem

$$\text{Minimize } g^\varepsilon(y) + h(u) + \tfrac{1}{2}|u - u^*|_U^2, \tag{2.93}$$

subject to

$$Ay + \beta^\varepsilon(y - \psi) = Bu + f, \tag{2.94}$$

where

$$\begin{aligned} \beta^\varepsilon(r) &= -\varepsilon^{-1}\int_{-\infty}^{\infty}\left((r - \varepsilon^2\theta)^- - \varepsilon^2\theta^-\right)\rho(\theta)\,d\theta \\ &= \varepsilon^{-1}\int_{\varepsilon^{-2}r}^{\infty}(r - \varepsilon^2\theta)\rho(\theta)\,d\theta + \varepsilon\int_0^1 \theta\rho(\theta)\,d\theta. \end{aligned} \tag{2.95}$$

In other words, β^ε is defined by formula (2.74), where β is the graph (1.35).

By Proposition 2.1, problem (2.93) has at least one optimal pair $(u_\varepsilon, y_\varepsilon) \in U \times (H^2(\Omega) \cap H_0^1(\Omega))$.

Arguing as in the proof of Theorem 2.1 (see problem (2.72)), it follows by Proposition 1.2 that

$$\begin{aligned} u_\varepsilon &\to u^* && \text{strongly in } U, \\ y_\varepsilon &\to y^* && \text{strongly in } H_0^1(\Omega), \text{ weakly in } H^2(\Omega), \\ \beta^\varepsilon(y_\varepsilon - \psi) &\to Bu^* - Ay^* + f && \text{weakly in } L^2(\Omega). \end{aligned}$$

Moreover, we have for (2.93) the optimality system (see (2.77), (2.78))

$$-Ap_\varepsilon - \dot{\beta}^\varepsilon(y_\varepsilon - \psi)p_\varepsilon = \nabla g^\varepsilon(y_\varepsilon) \quad \text{in } \Omega, \tag{2.96}$$

$$B^*p_\varepsilon \in \partial h(u_\varepsilon) + u_\varepsilon - u^*. \tag{2.97}$$

Then multiplying Eq. (2.96) first by p_ε and then by sign p_ε, we get the estimate

$$\|p_\varepsilon\|_{H_0^1(\Omega)} + \int_\Omega |\dot{\beta}^\varepsilon(y_\varepsilon - \psi)p_\varepsilon|\,dx \\ \leq C\|\nabla g^\varepsilon(y_\varepsilon)|_{L^2(\Omega)} \leq C. \tag{2.98}$$

Hence, there is a subsequence, again denoted ε, such that

$$\begin{aligned} p_\varepsilon &\to p && \text{weakly in } H_0^1(\Omega), \text{ strongly in } L^2(\Omega),\\ \nabla g^\varepsilon(y_\varepsilon) &\to \xi \in \partial g(y^*) && \text{weakly in } L^2(\Omega). \end{aligned}$$

Letting ε tend to zero in (2.97), we get

$$B^*p \in \partial h(u^*).$$

Now, let $\xi_\varepsilon\colon \Omega \to \mathbf{R}$ and $\eta_\varepsilon\colon \Omega \to \mathbf{R}$ be the measurable functions defined by

$$\xi_\varepsilon(x) = \begin{cases} 0 & \text{if } |y_\varepsilon(x) - \psi(x)| > \varepsilon^2,\\ 1 & \text{if } |y_\varepsilon(x) - \psi(x)| \leq \varepsilon^2, \end{cases}$$

$$\eta_\varepsilon(x) = \begin{cases} 0 & \text{if } y_\varepsilon(x) - \psi(x) > -\varepsilon^2,\\ 1 & \text{if } y_\varepsilon(x) - \psi(x) \leq -\varepsilon^2. \end{cases}$$

Since

$$\dot{\beta}^\varepsilon(r) = \varepsilon^{-1}\int_{\varepsilon^{-2}r}^{\infty} \beta(\theta)\,d\theta \qquad \forall r \in \mathbf{R}, \tag{2.99}$$

it follows by (2.95) that

$$\begin{aligned} &|(y_\varepsilon(x) - \psi(x))\dot{\beta}^\varepsilon(y_\varepsilon(x) - \psi(x))p_\varepsilon(x) - p_\varepsilon(x)\beta^\varepsilon(y_\varepsilon(x) - \psi(x))|\\ &\quad = \varepsilon|p_\varepsilon(x)|\int_{\varepsilon^{-2}(y_\varepsilon(x) - \psi(x))}^{0} \theta\rho(\theta)\,d\theta \leq \varepsilon|p_\varepsilon(x)| \qquad \text{a.e. } x \in \Omega. \end{aligned} \tag{2.100}$$

On the other hand, we have

$$\begin{aligned} p_\varepsilon\beta^\varepsilon(y_\varepsilon - \psi) &= \varepsilon^{-1}p_\varepsilon\xi_\varepsilon\int_{\varepsilon^{-2}(y_\varepsilon - \psi)}^{1} (y_\varepsilon - \varepsilon^2\theta - \psi)\rho(\theta)\,d\theta\\ &\quad + \varepsilon^{-1}p_\varepsilon(y_\varepsilon - \psi)\eta_\varepsilon + \varepsilon p_\varepsilon\int \theta\rho(\theta)\,d\theta \qquad \text{a.e. in } \Omega. \end{aligned}$$

This yields

$$|p_\varepsilon \beta^\varepsilon(y_\varepsilon - \psi)| \le \varepsilon |p_\varepsilon \dot{\beta}^\varepsilon(y_\varepsilon - \psi)|(\varepsilon^{-1}|y_\varepsilon - \psi|\xi_\varepsilon + \varepsilon^{-1}|y_\varepsilon - \psi|\eta_\varepsilon) + 2\varepsilon|p_\varepsilon| \qquad \text{a.e. in } \Omega. \quad (2.101)$$

We note that $\beta^\varepsilon(y_\varepsilon - \psi)\eta_\varepsilon = \varepsilon^{-1}(y_\varepsilon - \psi)\eta_\varepsilon + C\varepsilon\eta_\varepsilon$ remain in a bounded subset of $L^2(\Omega)$, whilst by the definition of ξ_ε wee see that

$$\varepsilon^{-1}|y_\varepsilon(x) - \psi(x)|\xi_\varepsilon(x) \le \varepsilon \qquad \text{a.e. } x \in \Omega.$$

Since $\{p_\varepsilon \dot{\beta}^\varepsilon(y_\varepsilon - \psi)\}$ is bounded in $L^1(\Omega)$, it follows by (2.99) and (2.101) that, for some subsequence $\varepsilon_n \to 0$,

$$p_{\varepsilon_n}(x)\beta^{\varepsilon_n}\big(y_{\varepsilon_n}(x) - \psi(x)\big) \to 0 \qquad \text{a.e. } x \in \Omega, \qquad (2.102)$$

whilst

$$p_{\varepsilon_n}\beta^{\varepsilon_n}(y_{\varepsilon_n} - \psi) \to p(f + Bu^* - Ay^*) \qquad \text{weakly in } L^1(\Omega).$$

Together with (2.102) and the Egorov theorem, the latter yields

$$p(f + Bu^* - Ay^*) = 0 \qquad \text{a.e. in } \Omega,$$

and therefore

$$p_{\varepsilon_n}\beta^{\varepsilon_n}(y_{\varepsilon_n} - \psi) \to p(f + Bu^* - Ay^*) \qquad \text{strongly in } L^1(\Omega). \quad (2.103)$$

Then, by (2.100), we see that

$$(y_{\varepsilon_n} - \psi)\dot{\beta}^{\varepsilon_n}(y_{\varepsilon_n} - \psi)p_{\varepsilon_n} \to 0 \qquad \text{strongly in } L^1(\Omega). \quad (2.104)$$

Inasmuch as $|\beta^\varepsilon(y_\varepsilon - \psi) + \varepsilon^{-1}(y_\varepsilon - \psi)^-| \le C\varepsilon$, it follows by (2.99), (2.103), and (2.104) that

$$(y_{\varepsilon_n} - \psi)^+ \dot{\beta}^{\varepsilon_n}(y_{\varepsilon_n} - \psi)p_{\varepsilon_n} \to 0 \qquad \text{strongly in } L^1(\Omega).$$

Since $(y_{\varepsilon_n} - \psi)^+ \in H_0^1(\Omega)$, applying Green's formula in (2.96) yields

$$a\big(p_{\varepsilon_n}, (y_{\varepsilon_n} - \psi)^+\chi\big) + \big(\nabla g^{\varepsilon_n}(y_{\varepsilon_n}), (y_{\varepsilon_n} - \psi)^+\chi\big) \to 0 \qquad \forall \chi \in C^1(\overline{\Omega}).$$

Since $p_{\varepsilon_n} \to p$ weakly in $H_0^1(\Omega)$ and $(y_{\varepsilon_n} - \psi)^+ \to y^* - \psi$ strongly in $H^1(\Omega)$, we get (2.89). Regarding inequality (2.91), it is an immediate consequence of Eq. (2.96) because we have

$$(Ap_\varepsilon + \nabla g_\varepsilon(y_\varepsilon), p_\varepsilon) \le 0 \qquad \forall \varepsilon > 0.$$

Now, selecting a further subsequence, if necessary, we may assume that

$$y_{\varepsilon_n}(x) \to y^*(x) \qquad \text{a.e. } x \in \Omega.$$

On the other hand, by estimate (2.98) it follows that there is $\mu \in (L^\infty(\Omega))^*$ and a generalized subsequence ε_λ of ε_n such that

$$\dot{\beta}^{\varepsilon_\lambda}(y_{\varepsilon_\lambda} - \psi) p_{\varepsilon_\lambda} \to \mu \qquad \text{weak star in } (L^\infty(\Omega))^*.$$

This implies that Ap admits an extension as element of $(L^\infty(\Omega))^*$, and we have

$$-A^*p - \mu = \xi \in \partial g(y^*).$$

Now, by Egorov's theorem, for every $\delta > 0$ there is a measurable subset E_δ of Ω such that $m(E_\delta) \le \delta$, $y^* - \psi$ is bounded on $\Omega \setminus E_\delta = \Omega_\delta$ and

$$y_{\varepsilon_n} - \psi \to y^* - \psi \qquad \text{uniformly on } \Omega_\delta.$$

Then, by (2.104), it follows that $\mu(y^* - \psi) = 0$ in Ω_δ, i.e.,

$$\int_{\Omega_\delta} (y^* - \psi)\mu_a \varphi \, dx + \mu_s((y^* - \psi)\varphi) = 0 \qquad \forall \varphi \in L^\infty(\Omega), \text{ supp } \varphi \subset \Omega_\delta.$$

On the other hand, there is an increasing sequence $\{\Omega^k\}$ such that $m(\Omega \setminus \Omega^k) \le k^{-1}$ and $\mu_s = 0$ on $L^\infty(\Omega^k)$. Hence,

$$\int_{\Omega_\delta \cap \Omega^k} (y^* - \psi)\mu_a \varphi \, dx = 0 \qquad \forall \varphi \in L^\infty(\Omega), \text{ supp } \varphi \subset \Omega_\delta \cap \Omega^k.$$

Thus, $(y^* - \psi)\mu_a = 0$ a.e. in Ω_δ, and letting δ tend to zero we infer that $(y^* - \psi)\mu_a = 0$. Hence,

$$-(Ap)_a = \xi \in \partial g(y^*) \qquad \text{a.e. in } [y^* > \psi].$$

If $1 \le N \le 3$, then $H^2(\Omega) \subset C(\overline{\Omega})$ and so

$$y_\varepsilon(x) \to y^*(x) \qquad \text{uniformly on } \overline{\Omega}.$$

Since $\psi \in H^2(\Omega) \subset C(\overline{\Omega})$, it follows by (2.104) that $(y^* - \psi)\mu = 0$, i.e.,

$$(y^* - \psi)(Ap + \xi) = 0.$$

This completes the proof of Theorem 2.2. ∎

Remark 2.1. Theorems 2.1 and 2.2 remain valid if one assumes that (y^*, u^*) is merely local optimal in problem (2.42) (respectively, (2.84), i.e.,

$$g(y^*) + h(u^*) \leq g(y) + h(u)$$

for all (y, u) satisfying (2.43)) (respectively, (2.85)) and such that $|u - u^*|_U \leq r$.

Indeed, in problem (2.72) (respectively, (2.94)) replace the cost functional by

$$g^{\varepsilon}(y) + h(u) + \alpha|u - u^*|_U^2,$$

where α is sufficiently large that

$$\limsup_{\varepsilon \to 0} g^{\varepsilon}(y_{\varepsilon}^{u^*}) + h(u^*) \leq \alpha r^2.$$

Then $|u_{\varepsilon} - u^*|_U \leq r$ for all $\varepsilon > 0$ and this implies as before that $u_{\varepsilon} \to u^*$ strongly in U. The rest of the proof remains unchanged.

Problems of the form (2.84) arise in a large variety of situations, and we now pause briefly to present on such an example.

Consider the model, already described in Section 2.2, of an elastic plane membrane clamped along the boundary $\partial\Omega$, inflated from above by a vertical field of forces with density u and limited from below by a rigid obstacle $y = \psi(x) < 0$, $\forall x \in \Omega$ (see Fig. 1.1). We have a desired shape of the membrane, given by the distribution $y = y_0(x)$ of the deflection, and we look for a control parameter u subject to constraints.

$$|u(x)| \leq \rho \qquad \text{a.e. } x \in \Omega, \tag{2.105}$$

such that the system response y^u has a minimum deviation of y_0. For instance, we may consider the problem of minimizing the functional $\frac{1}{2}\int_{\Omega}(y(x) - y_0(x))^2\,dx$ on all $(y, u) \in (H_0^1(\Omega) \cap H^2(\Omega)) \times L^2(\Omega)$, subject to control constraint (2.105) and to state equation (2.86), where $f = u$. This is a problem of the form (2.84), where $A_0 = -\Delta$, $B = I$, $U = L^2(\Omega)$, $f \equiv 0$, and

$$g(y) = \frac{1}{2}\int_{\Omega}(y(x) - y_0(x))^2\,dx, \qquad y \in L^2(\Omega),$$

$$h(u) = \begin{cases} 0 & \text{if } |u(x)| \leq \rho \text{ a.e. } x \in \Omega, \\ +\infty & \text{otherwise.} \end{cases}$$

By Proposition 2.1, this problem has at least one solution (y^*, u^*) whilst by Theorem 2.2 such a solution must satisfy the optimality system

$$\begin{aligned} \Delta y^* + u^* &= 0 \quad \text{in } \Omega^+ = \{x \in \Omega;\ y^*(x) > \psi(x)\}, \\ y^* &= \psi, \quad \Delta\psi + u^* \leq 0 \quad \text{a.e. in } \Omega_0 = \Omega \setminus \Omega^+, \\ y^* &= 0 \quad \text{in } \partial\Omega, \end{aligned} \tag{2.106}$$

$$\Delta p = y^* - y_0 \quad \text{a.e. in } \Omega^+, \tag{2.107}$$

$$p(u^* + \Delta y^*) = 0 \quad \text{a.e. in } \Omega, \qquad p = 0 \quad \text{in } \partial\Omega, \tag{2.108}$$

$$u^* = \rho \operatorname{sign} p \quad \text{a.e. in } \Omega. \tag{2.109}$$

Assume that $|\Delta\psi| \neq \rho$ a.e. in Ω. Then, by Eq. (2.108), we see that $p = 0$ in Ω_0. Hence, p is the solution to boundary value problem

$$\begin{aligned} \Delta p &= y^* - y_0 \quad \text{in } \Omega^+, \\ p &= 0 \quad \text{in } \Omega_0, \qquad p = 0 \quad \text{in } \partial\Omega. \end{aligned} \tag{2.110}$$

This system could be solved numerically using an algorithm of the following type:

$$\begin{aligned} (\Delta y_i + u_i)(y_i - \psi) &= 0 \quad \text{a.e. in } \Omega, \\ y_i \geq \psi, \quad \Delta y_i + u_i &\leq 0 \quad \text{a.e. in } \Omega \qquad y_i = 0 \quad \text{in } \partial\Omega, \\ \Delta p_i = y_i - y_0 &\quad \text{in } \Omega_i = \{x \in \Omega;\ y_i(x) > \psi(x)\}, \\ p_i &= 0 \quad \text{in } \partial\Omega_i, \\ u_{i+1} &= \rho \operatorname{sign} p_i \quad \text{a.e. in } \Omega_i. \end{aligned}$$

Let us assume now that $y_0 \in H_0^1(\Omega) \cap H^2(\Omega)$ and $\Delta y_0(x) \geq \max(\Delta\psi(x), \rho)$ a.e. $x \in \Omega$. Then, by Eqs. (2.106), we see that $\Delta(y^* - y_0) \leq 0$ a.e. in Ω and, therefore, by the maximum principle $y^* - y_0 \geq 0$ in Ω. Then, assuming that Ω^+ is smooth enough, we deduce by (2.110) and by virtue of the maximum principle that $p < 0$ a.e. in Ω^+, and therefore, by (2.109),

$$u^* = -\rho \quad \text{in } \Omega^+.$$

Hence, $y^* \in H_0^1(\Omega) \cap H^2(\Omega)$ satisfies the variational inequality

$$\begin{aligned} \Delta y^* &= \rho \quad \text{in } \Omega^+, & \Delta y^* &\leq \rho \quad \text{in } \Omega, \\ y^* &= \psi \quad \text{in } \partial\Omega^+, & y^* &> \psi \quad \text{in } \Omega^+, \\ y^* &= \psi \quad \text{in } \Omega_0 & y^* &= 0 \quad \text{in } \partial\Omega, \end{aligned} \tag{2.111}$$

from which we may determine Ω^+. For instance, if $\Omega = (0,1)$ and $\psi = -1$, then clearly the solution to (2.111) is convex and so it is of the following form:

$$y^*(x) = -1 \qquad \text{for } 0 < a < x < b < 1,$$
$$y^*(0) = y^*(1) = 0 \qquad -1 < y^*(x) < 0 \qquad \text{for } x \in (0,a) \cup (b,1) = \Omega^+.$$

Hence

$$y^*(x) = \begin{cases} \dfrac{\rho x^2}{2} + Cx & \text{for } x \in [a,b), \\ -1 & \text{for } x \in [a,b], \\ \dfrac{\rho}{2}(x^2 - 1) + C_0(x-1) & \text{for } x \in (b,1], \end{cases}$$

and one determines constants a, b, c, and C_0 from the continuity conditions

$$y^*_-(a) = y^*_+(b) = -1 \qquad (y^*)'_-(a) = (y^*)'_+(b) = 0.$$

A problem of great interest in the study of a physical system modeled by the obstacle problem is that of controlling the incidence set $\Omega_0 = \{x \in \Omega,\ y(x) = \psi(x)\}$ or its boundary $\partial\Omega_0$. For instance, in the case of the contact problem (1.46) recalled in the preceding a problem of interest would be to find the force field f (viewed as a control parameter) such that the set of all points where the membrane is in contact with the obstacle be as close as possible (in a certain acceptable sense) to a given measurable subset D of Ω. The least squares approach to this problem leads us to consider the optimal control problem:

Minimize

$$\int_\Omega \left(\chi_D(x) - \chi_y(x)\right)^2 dx + h(u) \tag{2.112}$$

on all $(y,u) \in H_0^1(\Omega) \times U$ *subject to* (2.85).

Here, χ_D is the characteristic function of D and χ_y is that of the set $\{x \in \Omega;\ y(x) = \psi(x)\}$. However, since the function $y \to \chi_y$ is not locally Lipschitz on $L^2(\Omega)$ we shall replace it by $g_\lambda(y) = \lambda/((y-\psi)^+ + \lambda)$, $\lambda > 0$, and so we shall consider the problem:

Minimize $\int_\Omega (g_\lambda(y) - \chi_D(x))^2\,dx + h(u)$, *subject to* (2.85).

To be more specific, we shall assume that $A_0 = -\Delta$, $B = I$, $f \equiv 0$, and $h: L^2(\Omega) \to \overline{\mathbf{R}}$ is given by

$$h(u) = \begin{cases} 0 & \text{if } |u(x)| \le \rho \text{ a.e. in } \Omega, \\ +\infty & \text{otherwise.} \end{cases}$$

This problem has at least one solution (y_λ, u_λ) and since, for $\lambda \to 0$,

$$g_\lambda(y) \to \chi_y \qquad \text{strongly in } L^2(\Omega)$$

for every $y \in L^2(\Omega)$, it is readily seen that for $\lambda \to 0$, $u_\lambda \to u^*$ weak star in $L^\infty(\Omega)$, $y_\lambda \to y^*$ weakly in $H^2(\Omega)$, where (y^*, u^*) is an optimal pair for problem (2.112).

On the other hand, the optimality system of (2.87)–(2.90) is in this case

$$\Delta y_\lambda + u_\lambda = 0 \qquad \text{a.e. in } [y_\lambda > \psi] = \Omega_\lambda^+,$$

$$y_\lambda \ge \psi, \quad \Delta y_\lambda + u_\lambda \le 0 \qquad \text{a.e. in } \Omega,$$

$$\Delta p_\lambda = \frac{\lambda}{(y_\lambda - \psi + \lambda)^2}\left(\chi_D - \frac{\lambda}{y_\lambda - \psi + \lambda}\right) \qquad \text{a.e. in } \Omega_\lambda^+,$$

$$\rho |p_\lambda| + p_\lambda\,\Delta\psi = 0 \qquad \text{a.e. in } \Omega_0^\lambda = [y_\lambda = \psi],$$

$$u_\lambda = \rho \operatorname{sign} p_\lambda \qquad \text{a.e. in } \Omega.$$

This problem can be treated as in the previous examples.

Now we shall study problem (2.84) in the case where g is a continuous convex function on $C(\overline{\Omega})$, for instance,

$$g(y) = \|y - y_0\|_{C(\overline{\Omega})}, \qquad y \in C(\overline{\Omega}). \tag{2.113}$$

The subdifferential $\partial g: C(\overline{\Omega}) \to M(\overline{\Omega})$ of g is given by (see Example 4 in Chapter 1, Section 1)

$$\partial g(y) = \left\{\mu \in M(\overline{\Omega});\ \mu(\varphi) \le \max_{x_0 \in M_{y-y_0}} \{\varphi(x_0)\operatorname{sign}(y - y_0)\}\ \forall \varphi \in C(\overline{\Omega})\right\},$$

where $M_z = \{x_0 \in \overline{\Omega};\ |z(x_0)| = \|z\|_{C(\overline{\Omega})}\}$.

For simplicity, we shall assume that Ω is a bounded, open, and smooth domain of $\mathbf{R}^3$, and $A_0 = -\Delta$.

Theorem 2.3. *Let (y^*, u^*) be optimal in problem* (2.84) *where g is given by* (2.113). *Then there exists $p \in W_0^{1,q}(\Omega)$, $1 < q < 3/2$, with $\Delta p \in M(\overline{\Omega})$ and $\mu \in \partial g(y^*)$ such that*

$$\Delta p = \mu \text{ in } \{x \in \Omega;\ y^*(x) > \psi(x)\},$$

$$p(x)(\Delta y^*(x) + Bu^*(x) + f(x)) = 0 \qquad \text{a.e. } x \in \Omega,$$

$$B^*p \in \partial h(u^*).$$

Proof. Since the proof is essentially the same as that of Theorem 2.2, it will be sketched only.

We consider the approximating control problem:

Minimize $g_\varepsilon(y) + h(u) + \frac{1}{2}|u - u^*|_U^2$, subject to having all $(y, u) \in (H_0^1(\Omega) \cap H^2(\Omega)) \times U$ satisfy

$$-\Delta y + \beta^\varepsilon(y - \psi) = Bu + f \qquad \text{in } \Omega.$$

Here, g_ε is the usual convex regularization of $g(z) = \|z - y_0\|_{L^\infty(\Omega)}$, i.e.,

$$g_\varepsilon(y) = \inf\left\{\frac{1}{2\varepsilon}|y - z|_{L^2(\Omega)}^2 + g(z);\ z \in L^\infty(\Omega)\right\}.$$

Let $(y_\varepsilon, u_\varepsilon)$ be optimal in the preceding problem. Then we have

$$u_\varepsilon \to u^* \qquad \text{strongly in } U,$$

$$y_\varepsilon \to y^* \qquad \text{strongly in } H_0^1(\Omega) \cap H^2(\Omega),$$

$$\beta^\varepsilon(y_\varepsilon - \psi) \to Bu^* + \Delta y^* + f \qquad \text{weakly in } L^2(\Omega),$$

and there are $p_\varepsilon \in H_0^1(\Omega) \cap H^2(\Omega)$ such that

$$\Delta p_\varepsilon - \dot{\beta}^\varepsilon(y_\varepsilon - \psi)p_\varepsilon = \nabla g_\varepsilon(y_\varepsilon) \qquad \text{in } \Omega,\ B^*p_\varepsilon \in \partial h(u_\varepsilon) + u_\varepsilon - u^*.$$

Since $\sup\{\|\xi\|_{L^1(\Omega)};\ \xi \in \partial g(y)\} \le 1$, we have

$$\|\nabla g_\varepsilon(y_\varepsilon)\|_{L^1(\Omega)} \le C \qquad \forall \varepsilon > 0.$$

Then, multiplying the latter equation by sign p_ε and integrating on Ω, we get

$$\int_\Omega |\dot{\beta}^\varepsilon(y_\varepsilon - \psi)p_\varepsilon|\, dx \le C \qquad \forall \varepsilon > 0,$$

or equivalently,

$$\|\Delta p_\varepsilon\|_{L^1(\Omega)} \le C \qquad \forall \varepsilon > 0.$$

Now, let $h_i \in L^\alpha(\Omega)$, $i = 0, 1, 2$ $\alpha > 3$. Then, as mentioned in Chapter 2 (see Section 3.2), the problem

$$-\Delta\theta = h_0 + \sum_{i=1}^{2} \frac{\partial h_i}{\partial x_i} \quad \text{in } \Omega, \qquad \theta = 0 \quad \text{in } \partial\Omega,$$

has a unique solution $\theta \in H_0^1(\Omega) \cap L^\alpha(\Omega)$ and

$$\|\theta\|_{L^\infty(\Omega)} \le C \sum_{i=0}^{2} \|h_i\|_{L^\alpha(\Omega)} \qquad \forall h_i \in L^\alpha(\Omega).$$

This yields

$$-\int_\Omega \theta\,\Delta p_\varepsilon\,dx = \int_\Omega \left(h_0 p_\varepsilon - \sum_{i=1}^{2} h_i (p_\varepsilon)_{x_i} \right) dx$$

and, therefore,

$$\|p_\varepsilon\|_{W_0^{1,q}(\Omega)} \le C \qquad \forall \varepsilon > 0,$$

where $1/q + 1/\alpha = 1$, i.e., $1 < q < 3/2$. Hence, on a generalized subsequence,

$$\begin{aligned} p_\varepsilon &\to p && \text{weakly in } W_0^{1,q}(\Omega), \\ \nabla g_\varepsilon(y_\varepsilon) &\to \mu && \text{vaguely in } M(\overline{\Omega}), \\ \dot\beta^\varepsilon(y_\varepsilon - \psi) p_\varepsilon &\to \nu && \text{vaguely in } M(\overline{\Omega}). \end{aligned}$$

We have, therefore,

$$\begin{aligned} \Delta p &= \nu + \mu \qquad \text{in } \mathscr{D}'(\Omega), \\ B^*p &\in \partial h(u^*). \end{aligned}$$

Since $y_\varepsilon \to y^*$ uniformly on Ω (because $H^2(\Omega) \subset C(\overline{\Omega})$ compactly), we have

$$\nu = 0 \qquad \text{in } \{x \in \Omega;\ y^*(x) > \psi(x)\}.$$

Then, arguing as in the proof of Theorem 2.2, we get

$$p(x)(Bu^*(x) + \Delta y^*(x) + f(x)) = 0 \qquad \text{a.e. } x \in \Omega.$$

Finally, letting ε tend to zero in the obvious inequality

$$\int_\Omega \nabla g_\varepsilon(y_\varepsilon)(y_\varepsilon - \varphi)\,dx \ge g_\varepsilon(y_\varepsilon) - g_\varepsilon(\varphi) \qquad \forall \varphi \in C(\overline{\Omega}),$$

we infer that

$$\mu(y^* - \varphi) \geq g(y^*) - g(\varphi) \qquad \forall \varphi \in C(\overline{\Omega}).$$

Hence $\mu \in \partial g(y^*)$, as claimed. ■

Remark 2.2. In applications, the function $g: L^2(\Omega) \to \mathbf{R}$ that occurs in the payoff of problem (P) and subsequent optimal control problems considered here is usually an integral functional of the form

$$g(y) = \int_\Omega g_0(x, y(x))\, dx, \qquad y \in L^2(\Omega),$$

where $g_0: \Omega \times \mathbf{R} \to \mathbf{R}$ is measurable in x and locally Lipschitz in y, whilst condition (i) or (a) requires that g_0 be global Lipschitz in y. However, it turns out that most of the optimality results established here remain valid if instead of (i), one merely assumes that

(i)′ $g_0 \geq 0$ on $\Omega \times \mathbf{R}$, $g_0(\cdot, 0) \in L^1(\Omega)$, and for each $r > 0$ there is $h_r \in L^1(\Omega)$ such that

$$|g_0(x,y) - g_0(x,z)| \leq h_r(x)|y - z| \qquad \text{a.e. } x \in \Omega,$$

for all $y, z \in \mathbf{R}$ such that $|y|, |z| \leq r$.

(ii)′ There exists some positive constants α, C_1, C_2 and $\beta \in L^1(\Omega)$ such that

$$\left|\frac{\partial g_0(x,y)}{\partial y}\right| \leq C_1 g_0(x,y)|y|^{-\alpha} + C_2|y|^2 + \beta(x)$$

$$\text{a.e. } x \in \Omega, y \in \mathbf{R}.$$

For the form and the proof of the optimality conditions under these general assumptions on g we refer to author's work [3].

3.2.5. *Elliptic Control Problems with Nonlinear Boundary Conditions*

We will study here the following problem:

Minimize

$$g(y) + h(u) \tag{2.114}$$

on all $(y,u) \in H^2(\Omega) \times U$, *subject to*

$$y - \Delta y = f + Bu \qquad \text{a.e. in } \Omega,$$
$$\frac{\partial y}{\partial \nu} + \beta(y) \ni 0 \qquad \text{a.e. in } \partial\Omega, \tag{2.115}$$

where $B \in L(U, L^2(\Omega))$, $f \in L^2(\Omega)$, and $\beta \subset \mathbf{R} \times \mathbf{R}$ is a maximal monotone graph. The functions $g: L^2(\Omega) \to \mathbf{R}$ and $h: U \to \overline{\mathbf{R}}$ satisfy assumptions (a), (b) of Section 2.3.

We know by Proposition 2.1 that if h satisfies the coercivity condition (2.5) then this problem admits at least one solution.

Let (y^*, u^*) be any optimal pair for problem (2.114). Then, using the standard approach, we associate with (2.114) the adapted penalized problem:

Minimize

$$g^\varepsilon(y) + h(u) + \tfrac{1}{2}|u - u^*|_U^2, \tag{2.116}$$

on all $(y,u) \in H^2(\Omega) \times U$, *subject to*

$$y - \Delta y = f + Bu \qquad \text{a.e. in } \Omega,$$
$$\frac{\partial y}{\partial \nu} + \beta^\varepsilon(y) \ni 0 \qquad \text{a.e. in } \partial\Omega, \tag{2.117}$$

where β^ε is defined by (2.74).

Let $(y_\varepsilon^*, u_\varepsilon^*)$ be an optimal pair for problem (2.116). Then, by the inequality

$$g^\varepsilon(y_\varepsilon) + h(u_\varepsilon) + \tfrac{1}{2}|u_\varepsilon - u^*|^2 \le g^\varepsilon(y_\epsilon^{u^*}) + h(u^*)$$

(y_ε^u is the solution to (2.117)), we deduce as in the previous cases that

$$u_\varepsilon \to u^* \qquad \text{strongly in } U$$
$$y_\varepsilon \to y^* \qquad \text{weakly in } H^2(\Omega), \text{ strongly in } H^1(\Omega). \tag{2.118}$$

Indeed, by Proposition 1.3 if $u_\varepsilon \to \overline{u}$ weakly in U then $y_\varepsilon \to y^{\overline{u}}$ weakly in $H^2(\Omega)$, where $y^{\overline{u}}$ is the solution to (2.115). This yields

$$\limsup_{\varepsilon \to 0} |u_\varepsilon - u^*|^2 = 0$$

because $\liminf_{\varepsilon \to 0} h(u_\varepsilon) \ge h(\bar{u})$ and $g^\varepsilon(y_\varepsilon) \to g(y^{\bar{u}})$. Now the optimal pair $(y_\varepsilon, u_\varepsilon)$ satisfies, along with some $p_\varepsilon \in H^2(\Omega)$, the first order optimality system

$$-p_\varepsilon + \Delta p_\varepsilon = \nabla g^\varepsilon(y_\varepsilon) \qquad \text{a.e. in } \Omega,$$
$$\frac{\partial p_\varepsilon}{\partial \nu} + \dot{\beta}^\varepsilon(y_\varepsilon)p_\varepsilon = 0 \qquad \text{a.e. in } \partial\Omega,$$
$$B^* p_\varepsilon \in \partial h(u_\varepsilon) + u_\varepsilon - u^*. \tag{2.119}$$

Now, multiplying Eq. (2.119) first by p_ε and then by sign p_ε, we get the estimate

$$\|p_\varepsilon\|_{H^1(\Omega)} + \int_{\partial\Omega} |\dot{\beta}^\varepsilon(y_\varepsilon)p_\varepsilon|\, dx \le C. \tag{2.120}$$

Hence, on a subsequence, again denoted ε, we have

$$\begin{aligned} p_\varepsilon &\to p && \text{weakly in } H^2(\Omega), \\ \nabla g^\varepsilon(y_\varepsilon) &\to \xi \in \partial g(y^*) && \text{weakly in } L^2(\Omega), \\ p_\varepsilon &\to p && \text{strongly in } L^2(\Omega), \text{weakly in } H^1(\Omega), \end{aligned} \tag{2.121}$$

and, by (2.118),

$$\beta^\varepsilon(y_\varepsilon) \to \eta_0 \qquad \text{weakly in } L^2(\partial\Omega),$$

where $\eta_0(x) \in \beta(y^*(x))$ a.e. $x \in \partial\Omega$.

Note also that, by (2.120),

$$\dot{\beta}^{\varepsilon_\lambda}(y_{\varepsilon_\lambda})p_{\varepsilon_\lambda} \to \mu \qquad \text{weak star in } (L^\infty(\partial\Omega))^*, \tag{2.122}$$

where $\{\varepsilon_\lambda\}$ is a generalized subsequence (directed subset) of $\{\varepsilon\}$.

Now, letting ε tend to zero in Eq. (2.119), we see that p satisfies the system

$$-p + \Delta p \in \partial g(y^*) \qquad \text{a.e. in } \Omega, \tag{2.123}$$
$$\frac{\partial p}{\partial \nu} + \mu = 0 \qquad \text{in } \partial\Omega,$$
$$B^* p \in \partial h(u^*). \tag{2.124}$$

We note that since $p \in H^1(\Omega)$ and $\Delta p \in L^2(\Omega)$, $\partial p/\partial \nu \in H^{-1/2}(\partial\Omega)$ is well-defined by the formula

$$\frac{\partial p}{\partial \nu}(\gamma_0 \varphi) = \int_\Omega \varphi \,\Delta p\, dx + \int_\Omega \nabla p \cdot \nabla \varphi\, dx \qquad \forall \varphi \in H^1(\Omega),$$

where $\gamma_0 \varphi$ is the trace of φ to $\partial\Omega$. The boundary condition in (2.123) means of course that

$$\frac{\partial p}{\partial \nu}(\varphi) + \mu(\varphi) = 0 \qquad \forall \varphi \in L^\infty(\partial\Omega) \cap H^{1/2}(\partial\Omega),$$

and in particular it makes sense in $\mathscr{D}'(\partial\Omega)$.

These equations can be made more explicit in some specific situations.

Theorem 2.4. *Let* (y^*, u^*) *be optimal for problem* (2.114), *where* β *is monotonically increasing and locally Lipschitz. Then there are* $p \in H^1(\Omega)$, $\eta \in L^1(\partial\Omega)$, *and* $\xi \in L^2(\Omega)$ *such that* $p - \Delta p \in L^2(\Omega)$, $\partial p/\partial \nu \in (L^\infty(\partial\Omega))^*$ *and*

$$\begin{aligned} -p + \Delta p &= \xi && \text{in } \Omega, \\ \xi(x) &\in \partial g(y^*(x)) && \text{a.e. } x \in \Omega, \end{aligned} \tag{2.125}$$

$$\left(\frac{\partial p}{\partial \nu}\right)_{\mathrm{a}} + \eta = 0 \qquad \text{a.e. in } \partial\Omega,$$

$$\eta(x) \in \beta(y^*(x))p(x) \qquad \text{a.e. } x \in \partial\Omega, \tag{2.126}$$

$$B^*p \in \partial h(u^*). \tag{2.127}$$

If either $1 \le N \le 3$ *or* β *satisfies condition* (2.52) *then* $\partial p/\partial \nu \in L^1(\partial\Omega)$ *and* Eq. (2.126) *becomes*

$$\frac{\partial p}{\partial \nu} + \partial\beta(y^*)\, p \ni 0 \qquad \text{a.e. in } \partial\Omega. \tag{2.128}$$

Here, $(\partial p/\partial \nu)_{\mathrm{a}}$ is the absolutely continuous part of $\partial p/\partial \nu$.

Proof. By the Egorov theorem, for every $\delta > 0$ there is $E_\delta \subset \partial\Omega$ such that $m(E_\delta) \le \delta$, p_ε, y_ε are uniformly bounded on $\partial\Omega \setminus E_\delta$, and

$$\begin{aligned} y_\varepsilon &\to y^* && \text{uniformly on } \partial\Omega \setminus E_\delta, \\ p_\varepsilon &\to p && \text{uniformly on } \partial\Omega \setminus E_\delta \end{aligned}$$

and, on a subsequence $\{\varepsilon_n\} \subset \{\varepsilon\}$,

$$\dot\beta^{\varepsilon_n}(y_{\varepsilon_n}) \to f_0 \qquad \text{weak star in } L^\infty(\partial\Omega \setminus E_0).$$

Then, by Lemma 2.5, we infer that $f_0(x) \in \partial\beta(y^*(x))$ a.e. $x \in \partial\Omega \setminus E_\delta$ and so, by (2.122) (see the proof of Theorem 2.1),

$$\mu_{\mathrm{a}}(x) \in \partial\beta(y^*(x))\, p(x) \qquad \text{a.e. } x \in \partial\Omega \setminus E_\delta.$$

Since δ is arbitrary, Eq. (2.126) follows. Now, if $1 \leq N \leq 3$ then $H^2(\Omega) \subset C(\overline{\Omega})$, and so $\{y_\varepsilon\}$ is bounded in $C(\overline{\Omega})$. We may conclude, therefore, that

$$|\dot{\beta}^\varepsilon(y_\varepsilon)| \leq C \qquad \forall x \in \partial\Omega.$$

This implies that $\{\dot{\beta}^\varepsilon(y_\varepsilon)\}$ is weak star compact in $L^\infty(\Omega)$ and so $\eta = \eta_a = -\partial p/\partial\nu \in L^\infty(\partial\Omega)$. If condition (2.52) holds then we derive as in the proof of Theorem 2.1, via the Dunford–Pettis criterion, that $\{\dot{\beta}^\varepsilon(y_\varepsilon)\}$ is weakly compact in $L^1(\Omega)$. Hence $\mu \in L^1(\partial\Omega)$, and the proof of Theorem 2.4 is complete. ∎

Now we shall consider the case where β is defined by (1.71). Then, Eq. (2.13) reduces to the Signorini problem

$$\begin{aligned} &y - \Delta y = Bu + f \qquad \text{a.e. in } \Omega,\ y \geq 0, \\ &\frac{\partial y}{\partial\nu} \geq 0, \quad y\frac{\partial y}{\partial\nu} = 0 \qquad \text{a.e. in } \partial\Omega, \end{aligned} \tag{2.129}$$

which models the equilibrium of an elastic body in contact with a rigid supporting body. The control of displacement y is achieved through a distributed field of forces with density Bu.

Theorem 2.5. *Let $(y^*, u^*) \in H^2(\Omega) \times U$ be an optimal pair for problem (2.114) governed by Signorini system (2.129). Then there exist functions $p \in H^1(\Omega)$ and $\xi \in L^2(\Omega)$ such that $\partial p/\partial\nu \in (L^\infty(\partial\Omega))^*$, $\xi \in \partial g(y^*)$, and*

$$-p + \Delta p = \xi \qquad \text{a.e. in } \Omega, \tag{2.130}$$

$$y^*\left(\frac{\partial p}{\partial\nu}\right)_a = 0, \quad p\frac{\partial y^*}{\partial\nu} = 0 \qquad \text{a.e. in } \partial\Omega, \tag{2.131}$$

$$B^*p \in h(u^*). \tag{2.132}$$

If $1 \leq N \leq 3$, then

$$y^*\frac{\partial p}{\partial\nu} = 0 \qquad \text{a.e. in } \partial\Omega. \tag{2.133}$$

Proof. The proof is almost identical with that of Theorem 2.2. However, we sketch it for reader's convenience.

Note that in this case β^ε is given by formula (2.95). Let $\xi_\varepsilon: \partial\Omega \to \mathbf{R}$ and $\lambda_\varepsilon: \partial\Omega \to \mathbf{R}$ be the measurable functions

$$\xi_\varepsilon(x) = \begin{cases} 0 & \text{if } |y_\varepsilon(x)| > \varepsilon^2, \\ 1 & \text{if } |y_\varepsilon(x)| \leq \varepsilon^2, \end{cases}$$

$$\lambda_\varepsilon(x) = \begin{cases} 0 & \text{if } y_\varepsilon(x) > -\varepsilon^2, \\ 1 & \text{if } y_\varepsilon(x) \leq -\varepsilon^2. \end{cases}$$

We have

$$|y_\varepsilon(x)\dot{\beta}^\varepsilon(y_\varepsilon(x))p_\varepsilon(x) - p_\varepsilon(x)\beta^\varepsilon(y_\varepsilon(x))| \leq \varepsilon|p_\varepsilon(x)| \qquad \text{a.e. } x \in \partial\Omega, \tag{2.134}$$

and

$$|p_\varepsilon(x)\beta^\varepsilon(y_\varepsilon(x))| \leq \varepsilon|p_\varepsilon(x)\dot{\beta}^\varepsilon(y_\varepsilon(x))|(\varepsilon^{-1}|y_\varepsilon(x)|\xi_\varepsilon(x) + \varepsilon^{-1}|y_\varepsilon(x)\lambda_\varepsilon(x)|) + 2\varepsilon|p_\varepsilon(x)| \qquad \text{a.e. } x \in \partial\Omega.$$

Since $\{\beta^\varepsilon(y_\varepsilon)\}$ is bounded in $L^2(\partial\Omega)$ and

$$\beta^\varepsilon(y_\varepsilon)\lambda_\varepsilon(x) = \varepsilon^{-1}y_\varepsilon(x)\lambda_\varepsilon(x) + \varepsilon\lambda_\varepsilon(x)\int_0^1 \theta\rho(\theta)\,d\theta,$$

we infer that $\{\varepsilon^{-1}y_\varepsilon\lambda_\varepsilon\}$ is bounded in $L^2(\partial\Omega)$. Note also that $\varepsilon^{-1}|y_\varepsilon|\xi_\varepsilon \leq \varepsilon$ a.e. in $\partial\Omega$ and, since $\{\dot{\beta}^\varepsilon(y_\varepsilon)p_\varepsilon\}$ is bounded in $L^1(\partial\Omega)$, there is a sequence $\varepsilon_n \to 0$ such that

$$p_{\varepsilon_n}(x)\beta^{\varepsilon_n}(y_{\varepsilon_n}(x)) \to 0 \qquad \text{a.e. } x \in \partial\Omega.$$

Then, by (2.121), we conclude that

$$p_{\varepsilon_n}\beta^{\varepsilon_n}(y_{\varepsilon_n}) \to -p\frac{\partial y^*}{\partial\nu} = 0 \qquad \text{strongly in } L^1(\partial\Omega).$$

Finally, by (2.134) we see that

$$y_{\varepsilon_n}\dot{\beta}^{\varepsilon_n}(y_{\varepsilon_n})p_{\varepsilon_n} \to 0 \qquad \text{strongly in } L^1(\partial\Omega). \tag{2.135}$$

Now, by the Egorov theorem, for every $\delta > 0$ there exists $E_\delta \subset \partial\Omega$ such that $m(E_\delta) \leq \delta$, y_{ε_n} are uniformly bounded on $\partial\Omega \setminus E_\delta$ and

$$y_{\varepsilon_n} \to y^* \qquad \text{uniformly on } \partial\Omega \setminus E_\delta.$$

By virtue of (2.122) this implies that

$$y^*\mu = 0 \qquad \text{in } \partial\Omega \setminus E_\delta$$

and, arguing as in the proof of Theorem 2.2, we see that $y^*\mu_a = 0$ a.e. in $\partial\Omega$, which along with (2.123) yields (2.131).

If $1 \le N \le 3$, then $H^2(\Omega) \subset C(\overline{\Omega})$ and

$$y_\varepsilon \to y^* \qquad \text{uniformly on } \overline{\Omega}.$$

Then, by (2.122) and (2.135), we deduce that

$$y^*\mu = 0,$$

where $y^*\mu$ is the product of the measure $\mu \in (L^\infty(\partial\Omega))^*$ with the function $y^* \in C(\partial\Omega)$. ■

3.2.6. *Control and Observation on the Boundary*

We consider here the following problem:

Minimize

$$g_1(y) + \int_{\Gamma_1} g_0(x, y(x))\, dx + h(u) \tag{2.136}$$

on all $(y, u) \in H^1(\Omega) \times U$, *subject to*

$$\begin{aligned} y - \Delta y &= f \qquad \text{in } \Omega, \\ \frac{\partial y}{\partial \nu} + \beta(y) &\ni B_0 u \qquad \text{in } \Gamma_1, \qquad y = 0 \qquad \text{in } \Gamma_2, \end{aligned} \tag{2.137}$$

where $\beta \subset \mathbf{R} \times \mathbf{R}$ is maximal monotone, $B_0 \in L(U, L^2(\Gamma_1))$, $f \in L^2(\Omega)$, and $\partial\Omega = \Gamma_1 \cup \Gamma_2$, where Γ_1 and Γ_2 are smooth disjoint parts of $\partial\Omega$. The functions $g_1 : L^2(\Omega) \to \mathbf{R}$ and $h : U \to \overline{\mathbf{R}}$ satisfy conditions (a), (b) of Section 2.3, whilst $g_0 : \Gamma_1 \times \mathbf{R} \to \mathbf{R}$ is measurable in x, differentiable in y, and

$$g_0(x, 0) = 0, \qquad |\nabla_y g_0(x, y)| \le C(1 + |y|) \qquad \text{a.e. } x \in \Gamma_1,\ y \in \mathbf{R}.$$

As seen in Section 1.4, for each $u \in U$, Eq. (2.137) has a unique solution $y \in H^1(\Omega)$ (if $\bar{\Gamma}_1 \cap \bar{\Gamma}_2 = \emptyset$, then $y \in H^2(\Omega)$). As a matter of fact (2.136) is a problem of the form (P) considered in Section 2.1, where

$$g(y) = g_1(x) + \int_{\Gamma_1} g_0(x, y(x))\, dx \qquad \forall y \in V,$$

and V, A are defined as in Section 1.4 (see (1.62)).

Let (y^*, u^*) be an arbitrary optimal pair for problem (2.136). Then, following the general approach developed in Section 2.3, consider the penalized problem:

$$\text{Minimize} \quad g_1^\varepsilon(y) + \int_{\Gamma_1} g_0(x, y(x))\, dx + h(u) + \tfrac{1}{2}|u - u^*|_U^2, \quad (2.138)$$

subject to

$$y - \Delta y = f \qquad \text{in } \Omega$$

$$\frac{\partial y}{\partial \nu} + \beta^\varepsilon(y) = B_0 u \qquad \text{in } \Gamma_1, \qquad y = 0 \qquad \text{in } \Gamma_2.$$

Here, g_1^ε is defined by (2.15) and β^ε by (2.74).

By a standard argument, it follows that, for $\varepsilon \to 0$,

$$\begin{aligned}
u_\varepsilon &\to u^* && \text{strongly in } U,\\
y_\varepsilon &\to y^* && \text{weakly in } H^1(\Omega), \text{ strongly in } L^2(\Omega),\\
y_\varepsilon &\to y^* && \text{strongly in } L^2(\partial\Omega),\\
\beta^\varepsilon(y_\varepsilon) &\to f_0 && \text{weakly in } L^2(\partial\Omega), \qquad (2.140)
\end{aligned}$$

where $f_0(x) \in \beta(y^*(x))$ a.e. $x \in \partial\Omega$.

On the other hand, the optimality principle for problem (2.138) has the form

$$\begin{aligned}
p_\varepsilon - \Delta p_\varepsilon &= \nabla g_1^\varepsilon(y_\varepsilon) && \text{in } \Omega,\\
\frac{\partial p_\varepsilon}{\partial \nu} + \dot{\beta}^\varepsilon(y_\varepsilon) p_\varepsilon &= -\nabla g_0(x, y_\varepsilon) && \text{in } \Gamma_1,\\
p_\varepsilon &= 0 && \text{in } \Gamma_2, \qquad (2.141a)\\
B^* p_\varepsilon &\in \partial h(u_\varepsilon) + u_\varepsilon - u^*. && \qquad (2.141b)
\end{aligned}$$

Now, multiplying Eq. (2.141a) by p_ε and sign p_ε, (more precisely, with $\zeta(p_\varepsilon)$, where ζ is a smooth approximation of sign), we find the estimates

$$\|p_\varepsilon\|_{H^1(\Omega)} + \int_{\Gamma_1} |\dot{\beta}^\varepsilon(y_\varepsilon) p_\varepsilon| \, dx \leq C \qquad \forall \varepsilon > 0,$$

and so on a subsequence, again denoted ε, we have

$$\begin{aligned}
p_\varepsilon &\to p && \text{weakly in } H^1(\Omega), \text{ strongly in } L^2(\Omega),\\
p_\varepsilon &\to p && \text{strongly in } L^2(\partial\Omega),\\
\nabla g_1(y_\varepsilon) &\to \zeta_1 \in \partial g(y^*) && \text{weakly in } L^2(\Omega),\\
\nabla g_0(x, y_\varepsilon) &\to \nabla g_0(x, y^*) && \text{in } L^2(\Gamma_1).
\end{aligned}$$

Letting ε tend to zero in Eqs. (2.141a), we get

$$p - \Delta p = \zeta_1 \in \partial g(y_1),$$

$$\frac{\partial p}{\partial \nu} + \mu = -\nabla g_0(x, y^*) \qquad \text{in } \Gamma_1, \qquad p = 0 \qquad \text{in } \Gamma_2, \tag{2.142}$$

$$B^* p \in \partial h(u^*), \tag{2.143}$$

where $\mu \in (L^\infty(\Gamma_1))^*$.

One can give explicit forms for these equations if β is locally Lipschitz or if β is the graph of the form (1.71). Since the proofs are identical with that of Theorem 2.2 and 2.4 respectively, we only mention the results.

Theorem 2.6. *Let (y^*, u^*) be optimal in problem* (2.136), (2.137), *where β is monotonically increasing and satisfies condition* (2.52). *Then there exists $p \in H^1(\Omega)$ such that $\Delta p \in L^2(\Omega)$, $\partial p/\partial \nu \in L^2(\Gamma_1)$,*

$$-p + \Delta p \in \partial g_1(y^*) \qquad \text{a.e. in } \Omega,$$

$$\frac{\partial p}{\partial \nu} + p\, \partial\beta(y^*) \ni 0 \qquad \text{a.e. in } \Gamma_1, \qquad p = 0 \qquad \text{in } \Gamma_2,$$

$$B_0^* p \in \partial h(u^*). \tag{2.144}$$

■

Consider now the case where β is given by (1.71). In this case, the state equation (2.137) reduces to the unilateral problem

$$\begin{aligned} &y - \Delta y = f \quad \text{in } \Omega, \\ &\quad y \geq 0, \quad \frac{\partial y}{\partial \nu} - B_0 u \geq 0, \quad \left(\frac{\partial y}{\partial \nu} - B_0 u\right) y = 0 \quad \text{in } \Gamma_1, \\ &\quad y = 0 \quad \text{in } \Gamma_2. \end{aligned} \tag{2.145}$$

Theorem 2.7. *Let* (y^*, u^*) *be optimal for problem* (2.136) *governed by state equations* (2.145). *Then there is* $p \in H^1(\Omega)$ *such that* $\Delta p \in L^2(\Omega)$, $\partial p / \partial \nu \in (L^\infty(\Gamma_1))^*$ *and*

$$\begin{aligned} -p + \Delta p \in \partial g_1(y^*) \quad &\text{a.e. in } \Omega, \\ \left(\frac{\partial p}{\partial \nu}\right)_{\mathrm{a}} + \nabla g_0(x, y^*) = 0 \quad &\text{a.e. in } \{x \in \Gamma_1 ; y^*(x) > 0\}, \\ p\left(B_0 u^* - \frac{\partial y^*}{\partial \nu}\right) = 0 \quad &\text{a.e. in } \Gamma_1, \\ B_0^* p \in \partial h(u^*). \quad &\blacksquare \end{aligned}$$

Similar results can be obtained if one considers problems of the form (2.135) governed by variational inequalities on Ω with Dirichlet boundary control (see V. Barbu [7], p. 107).

Bibliographical Notes and Remarks

Section 1. The results of this section are classical and can be found in standard texts and monographs devoted to variational inequalities (see, for instance, J. L. Lions [1], Kinderlehrer and G. Stampacchia [1], A. Friedman [1], and C. M. Elliott and J. R. Ockendon [1]). For other recent results on analysis and shape of free boundary in elliptic variational inequalities and nonlinear elliptic boundary value problems, we refer to J. Diaz [1].

Section 2. Most of the results presented in this section rely on author's work [3, 7]. For other related results, we refer the reader to the works of A. Friedman [2], V. Barbu and D. Tiba [1], V. Barbu and Ph. Korman [1], D. Tiba [1], G. Moroşanu and Zheng-Xu He [1], and L. Nicolaescu [1]. An attractive feature of the approach used here is that it allows the treatment of more general problems such as optimal control problems governed by not well-posed systems (J. L. Lions [2]) or by hemivariational inequalities (Haslinger and Panagiotopoulos [1], Panagiotopoulos [2]). A different approach to first order necessary conditions for optimal control problem governed by elliptic variational inequalities is due to F. Mignot [1] (see also Mignot and Puel [1]), and relies on the concept of conical derivative of the map $u \to y^u$. Let us briefly describe such a result, for the optimal control problem (2.84), where $\psi \equiv 0$, $U = L^2(\Omega)$, $B = I$, and

$$g(y) = \frac{1}{2}\int_\Omega (y(x) - y^0(x))^2\, dx, \qquad h(u) = \frac{1}{2}\int_\Omega u^2(x)\, dx.$$

If $Z_u = \{x \in \Omega,\ y^u(x) = 0\}$ and $S_u = \{\varphi \in H_0^1(\Omega),\ \varphi \geq 0 \text{ in } Z_u;\ (\varphi, A_0 y^u - uf) = 0\}$, where y^u is the solution to (2.85), then (y^*, u^*) is optimal if and only if there is $p \in -S_{u^*}$ such that $u^* = p$ and

$$a(p, \varphi) + \int_\Omega (y^* - y^0)\varphi\, dx \leq 0, \qquad \forall \varphi \in S_{u^*}$$

(Mignot and Puel [1]). A different approach related to the method developed in Section 2.2 (see problem Q_ε) has been used by Bermudez and Saguez ([1–4]). In a few words, the idea is to transform the original problem in a linear optimal control problem with nonconvex state constraints and to apply to this problem the abstract Lagrange multiplier rule in infinite dimensional spaces. A different approach involving Eckeland variational principle was used by J. Yong [4]. Optimal controllers for the dam problem were studied by A. Friedman *et al.* [1]. Optimality conditions for problems governed by general variational inequalities in infinite dimensional spaces were obtained in the work of Shuzong Shi [1] and Barbu and Tiba [1]. For some earlier results on optimal control problems governed by variational inequalities and nonlinear partial differential equations, we refer to J. L. Lions [4] (see also [2]).

There is an extensive literature on optimal control of free boundary in elliptic variational inequalities containing control parameters on variable domains (shape optimization). We mention in this direction the works of Ch. Saguez [1], J. Haslinger and P. Neittaanmäki [1], P. Neittaanmäki, *et al.* [1], J. P. Zolesio [1], V. Barbu and A. Friedman [1], W. B. Lui and J. E. Rubio [1], Hlavacek, *et al.* [1], and Hoffman and Haslinger [1]. A standard shape optimization problem involving free boundaries is the following: Let Ω_u be a domain in $\mathbf{R}^N$ that depends upon a control variable $u \in U$ and let its boundary $\partial\Omega_u = \Gamma^0 \cup \Gamma_u$, $\overline{\Gamma}^0 \cap \overline{\Gamma}_u = \emptyset$, where Γ^0 is prescribed independently of u. The problem is to find $u \in U$ such that Γ^0 becomes the free boundary and Ω_u the noncoincidence set of a given obstacle problem in a domain $\Omega \supset \Omega_u$, for instance: $\Delta y = f$ in $[y > 0]$, $\Delta y \le f$ in Ω_u, $y > 0$ in Ω_u, $y = 1$ in Γ_u. Such a problem is studied by the methods of this chapter in the work of Barbu and Friedman [1] and an explicit form of the solution is found in a particular case. A different approach has been developed in Barbu and Tiba [2] and Barbu and Stojanovic [1]. The idea is to reduce the problem to a linear control problem of the following type:

Find $u \in U$ such that $\partial y/\partial \nu = 0$ in Γ^0 and $y > 0$ in Ω_u, where y is the solution to Dirichlet problem $\Delta z = f$ in Γ_u, $z = 0$ in Γ^0, $z = 1$ in Γ_u.

First order necessary conditions for state constraint optimal control problems governed by semilinear elliptic problems have been obtained by Bonnans and Casas [1] using methods of convex analysis (see also Bonnans and Tiba [1]).

Chapter 4 | Nonlinear Accretive Differential Equations

This chapter is devoted to the Cauchy problem associated with nonlinear accretive operators in Banach space. The main result is related to the Crandall–Liggett exponential formula for autonomous equations, from which practically all existence results for the nonlinear accretive Cauchy problem follow in a more or less straightforward way. A large part deals with applications to nonlinear partial differential equations, which include nonlinear parabolic equations and variational inequalities of parabolic type, first order quasilinear equations, the nonlinear diffusion equations, and nonlinear hyperbolic equations.

4.1. The Basic Existence Results

4.1.1. Mild Solutions

Let X be a real Banach space with the norm $\|\cdot\|$ and dual X^* and let $A \subset X \times X$ be an ω-accretive set of $X \times X$; i.e., $A + \omega I$ is accretive for some $\omega \in \mathbf{R}$ (Section 3 in Chapter 2).

Consider the Cauchy problem

$$\frac{dy(t)}{dt} + Ay(t) \ni f(t), \qquad t \in [0, T],$$

$$y(0) = y_0, \tag{1.1}$$

where $f \in L^1(0, T; X)$ and $y_0 \in \overline{D(A)}$. Frequently, we shall write Eq. (1.1) in the form $y' + Ay \ni f$, $y(0) = y_0$.

Definition 1.1. A strong solution of (1.1) is a function $y \in W^{1,1}((0,T];X) \cap C([0,T];X)$ such that

$$f(t) - \frac{dy}{dt}(t) \in Ay(t) \qquad \text{a.e. } t \in (0,T), \qquad y(0) = y_0.$$

Here, $W^{1,1}((0,T];X) = \{y \in L^1(0,T;X);\ y' \in L^1(\delta,T;X)\ \forall \delta \in (0,T)\}$

It is readily seen that any strong solution to (1.1) is unique and is a continuous function of f and y_0. More precisely, we have:

Proposition 1.1. *Let A be ω-accretive, $f_i \in L^1(0,T;X)$, $y_0^i \in \overline{D(A)}$, $i = 1,2$ and let $y_i \in W^{1,1}((0,T];X)$, $i = 1,2$, be corresponding strong solutions to problem* (1.1). *Then,*

$$\begin{aligned} &\|y_1(t) - y_2(t)\| \\ &\quad \le e^{\omega t}\|y_0^1 - y_0^2\| \\ &\qquad + \int_0^t e^{\omega(t-s)}[y_1(s) - y_2(s), f_1(s) - f_2(s)]_s\,ds \\ &\quad \le e^{\omega t}\|y_0^1 - y_0^2\| + \int_0^t e^{\omega(t-s)}\|f_1(s) - f_2(s)\|\,ds. \end{aligned} \tag{1.2}$$

The main ingredient of the proof is Lemma 1.1 following.

Lemma 1.1. *Let $y = y(t)$ be an X-valued function on $[0,T]$. Assume that $y(t)$ and $\|y(t)\|$ are differentiable at $t = s$. Then*

$$\|y(s)\|\frac{d}{ds}\|y(s)\| = \left(\frac{dy}{ds}(s), w\right) \qquad \forall w \in J(y(s)). \tag{1.3}$$

Here, $J: X \to X^*$ is the duality mapping of X.

Proof. Let $\varepsilon > 0$. We have

$$(y(s+\varepsilon) - y(s), w) \le (\|y(s+\varepsilon)\| - \|y(s)\|)\|w\| \qquad \forall w \in J(y(s)),$$

and this yields

$$\left(\frac{dy}{ds}(s), w\right) \le \frac{d}{ds}\|y(s)\|\,\|y(s)\|.$$

Similarly, from the inequality

$$(y(s-\varepsilon) - y(s), w) \le (\|y(s-\varepsilon)\| - \|y(s)\|)\|w\|,$$

we get

$$\left(\frac{d}{ds}y(s), w\right) \ge \frac{d}{ds}\|y(s)\|\,\|y(s)\|,$$

as claimed. ■

In particular, it follows by (1.3) that

$$\frac{d}{ds}\|y(s)\| = \left[y(s), \frac{dy}{ds}(s)\right]_s, \tag{1.4}$$

where (see Proposition 3.7 in Chapter 2)

$$[x, y]_s = \inf_{\lambda>0} \lambda^{-1}(\|x + \lambda y\| - \|x\|) = \max\{(y, x^*); x^* \in \Phi(x)\} \tag{1.5}$$

and $\Phi(x) = \partial(\|x\|)$.

Proof of Proposition 1.1. We have

$$\frac{d}{ds}(y_1(s) - y_2(s)) + Ay_1(s) - Ay_2(s) \ni f_1(s) - f_2(s)$$

$$\text{a. e. } s \in (0, T).$$

On the other hand, since A is ω-accretive,

$$[y_1(s) - y_2(s), Ay_1(s) - Ay_2(s)]_s \ge -\omega\|y_1(s) - y_2(s)\|$$

and so, by (1.4), we see that

$$\frac{d}{ds}\|y_1(s) - y_2(s)\|$$

$$\le [y_1(s) - y_2(s), f_1(s) - f_2(s)]_s + \omega\|y_1(s) - y_2(s)\|$$

$$\text{a. e. } s \in (0, T).$$

Then integrating on $[0, t]$ we get (1.2), as claimed. ■

Proposition 1.1 shows that, as far as concerns uniqueness and continuous dependence of solution of data, the class of ω-accretive operators A offers a suitable framework for the Cauchy problem. However, for existence we must extend the notion of solution for Cauchy problem (1.1).

We now define this extended notion.

Let $f \in L^1(0, T; X)$ and $\varepsilon > 0$ be given. An *ε-discretization* on $[0, T]$ of the equation $y' + Ay \ni f$ consists of a partition $0 = t_0 \le t_1 \le t_2 \le \cdots \le t_N$ of the interval $[0, t_N]$ and a finite sequence $\{f_i\}_{i=1}^N \subset x$ such that

$$t_i - t_{i-1} < \varepsilon \qquad \text{for } i = 1, \dots, N, T - \varepsilon < t_N \le T, \tag{1.6}$$

$$\sum_{i=1}^{N} \int_{t_{i-1}}^{t_i} \|f(s) - f_i\| \, ds < \varepsilon. \tag{1.7}$$

We shall denote by $D_A^\varepsilon(0 = t_0, t_1, \dots, t_N; f_1, \dots, f_N)$ this ε-discretization.

A *solution* to the ε-discretization $D_A^\varepsilon(0 = t_0, t_1, \dots, t_N; f_1, \dots, f_N)$ is a piecewise constant function $z: [0, t_N] \to X$ whose values z_i on $(t_{i-1}, t_i]$ satisfy the equation

$$\frac{z_i - z_{i-1}}{t_i - t_{i-1}} + Az_i \ni f_i, \qquad i = 1, \dots, N. \tag{1.8}$$

Such a function $z = \{z_i\}_{i=1}^N$ is called *ε-approximate solution* to the Cauchy problem (1.1) if it further satisfies

$$\|z(0) - y_0\| \le \varepsilon. \tag{1.9}$$

Definition 1.2. A *mild solution* of the Cauchy problem (1.1) is a function $y \in C([0, T]; X)$ with the property that for each $\varepsilon > 0$ there is an ε-approximate solution z of $y' + Ay \ni f$ on $[0, T]$ such that $\|y(t) - z(t)\| \le \varepsilon$ for all $t \in [0, T]$ and $y(0) = x$.

Let us notice that every strong solution $y \in C([0, T]; X) \cap W^{1,1}((0, T]; X)$ to (1.1) is a mild solution. Indeed, let $0 = t_0 \le t_1 \le \cdots \le t_N$ be an ε-discretization of $[0, T]$ such that

$$\left\| \frac{d}{dt} y(t) - \frac{y(t_i) - y(t_{i-1})}{t_i - t_{i-1}} \right\| \le \varepsilon, \qquad t_i - t_{i-1} \le \delta, i = 1, 2, \dots, N,$$

and $\int_{t_{i-1}}^{t_i} \|f(t) - f(t_i)\| \, dt \le \varepsilon(t_i - t_{i-1})$. Then $z = y(t_i)$ on $(t_{i-1}, t_i]$ is a solution to the ε-discretization $D_A^\varepsilon(0 = t_0, t_1, \dots, t_N; f_1, \dots, f_N)$, and if we choose the discretization $\{t_j\}$ so that $\|y(t) - y(s)\| \le \varepsilon$ for $t, s \in (t_{i-1}, t_i)$ we have $\|y(t) - z(t)\| \le \varepsilon$ for all $t \in [0, T]$, as claimed.

Theorem 1.1 following is the main result of this section.

Theorem 1.1. *Let A be ω-accretive, $y_0 \in \overline{D(A)}$, and $f \in L^1(0, T; X)$. For each $\varepsilon > 0$, let problem* (1.1) *have an ε-approximate solution. Then the Cauchy problem* (1.1) *has a unique mild solution y. Moreover, there is a* *continuous function* $\delta = \delta(\varepsilon)$ *such that* $\delta(0) = 0$ *and if* z *is an* ε*-approximate solution of* (1.1) *then*

$$\|y(t) - z(t)\| \le \delta(\varepsilon) \qquad \text{for } t \in [0, T - \varepsilon]. \tag{1.10}$$

Let $f, g \in L^1(0, T; X)$ *and* $y, \bar{y}$ *be mild solutions to* (1.1) *corresponding to* f *and* g, *respectively. Then*

$$\|y(t) - \bar{y}(t)\| \le e^{\omega(t-s)}\|y(s) - \bar{y}(s)\| + \int_s^t e^{\omega(t-\tau)}[y(\tau) - \bar{y}(\tau), f(\tau) - g(\tau)]_s d\tau \quad \text{for } 0 \le s < t \le T. \tag{1.11}$$

This important result, which represents the core of existence theory of evolution processes governed by accretive operators will be proved in several steps. Let us first present some of its immediate consequences.

Theorem 1.2. *Let* C *be a closed convex cone of* X *and let* A *be* ω*-accretive in* $X \times X$ *such that*

$$D(A) \subset C \subset \bigcap_{0<\lambda<\lambda_0} R(I + \lambda A) \qquad \text{for some } \lambda_0 > 0. \tag{1.12}$$

Let $y_0 \in \overline{D(A)}$ *and* $f \in L^1(0, T; X)$ *be such that* $f(t) \in C$ a.e. $t \in (0, T)$. *Then problem* (1.1) *has a unique mild solution* y. *If* y *and* $\bar{y}$ *are two mild solutions to* (1.1) *corresponding to* f *and* g, *respectively, then*

$$\|y(t) - \bar{y}(t)\| \le e^{\omega(t-s)}\|y(s) - \bar{y}(s)\| + \int_s^t e^{\omega(t-\tau)}[y(\tau) - \bar{y}(\tau), f(\tau) - g(\tau)]_s d\tau \quad \text{for } 0 \le s < t \le T. \tag{1.13}$$

Proof of Theorem 1.2. Let $f \in L^1(0, T; X)$ and

$$f_i = \frac{1}{t_i - t_{i-1}} \int_{t_{i-1}}^{t_i} f(s)\, ds, \qquad i = 1, 2, \dots, N,$$

where $\{t_i\}_{i=0}^N$, $t_0 = 0$, is a partition of the interval $[0, t_N]$ such that $t_i - t_{i-1} < \varepsilon$, $T - \varepsilon < t_N < T$. By assumption (1.12), it follows that for ε small enough the function $z = z_i$ on $(t_{i-1}, t_i]$, $z_0 = y_0$, defined by (1.8) is an ε-approximate solution to (1.1). (It is readily seen by assumption (1.2) and the ω-accretivity of A that Eq. (1.8) has a unique solution $\{z_i\}_{i=0}^N$.) Thus,

Theorem 1.1 is applicable and so problem (1.1) has a unique mild solution satisfying (1.13). ■

In the sequel, we shall frequently refer to the map $(y_0, f) \to y$ from $\overline{D(A)} \times L^1(0, T; X)$ to $C([0, T]; X)$ as *the evolution associated to* A. It should be noted that in particular the range condition (1.12) holds if $C = X$ and A is ω-m-accretive in $X \times X$.

In the particular case when $f \equiv 0$, if A is ω-accretive and

$$R(I + \lambda A) \supset \overline{D(A)} \qquad \text{for all small } \lambda > 0, \tag{1.14}$$

then Eq. (1.8) has a unique solution $\{z_i\}_{i=0}^N$, $z_0 \in D(A)$, if ε is sufficiently small. Hence, for all $\varepsilon > 0$ sufficiently small, problem (1.1) with $f \equiv 0$ has an ε-approximate solution, and so by Theorem 1.1 we have:

Theorem 1.3 (Crandall and Liggett). *Let A be ω-accretive, satisfying the range condition* (1.14) *and $y_0 \in \overline{D(A)}$. Then the Cauchy problem*

$$\frac{dy}{dt} + Ay \ni 0, \qquad t > 0,$$

$$y(0) = y_0, \tag{1.15}$$

has a unique mild solution y. Moreover

$$y(t) = \lim_{n \to \infty} \left(I + \frac{t}{n} A\right)^{-n} y_0 \tag{1.16}$$

uniformly in t on compact intervals. ■

Indeed, in this case if $t_0 = 0$, $t_i = i\varepsilon$, $i = 1, \ldots, N$, then the solution z_ε to the ε-discretization $D_A(0 = t_0, t_1, \ldots, t_N)$ is given by

$$z_\varepsilon = (I + \varepsilon A)^{-i} y_0 \qquad \text{on } ((i-1)\varepsilon, i\varepsilon J].$$

Hence, by (1.10), we have

$$\|y(t) - (I + \varepsilon A)^{-i} y_0\| \le \delta(\varepsilon) \qquad \text{for } (i-1)\varepsilon < t \le i\varepsilon,$$

which implies the exponential formula (1.16). We note that in particular the range conditions (1.12), (1.14) are automatically satisfied if A is ω-m-accretive, i.e., if $\omega I + A$ is m-accretive for some real ω.

We shall apply now Theorem 1.2 to the mild solutions $y = y(t)$ and $y = x$ to the equations

$$y' + Ay \ni f \qquad \text{in } (0, T),$$

and

$$y' + Ay \ni v \qquad \text{in } (0, T), \qquad v \in Ax,$$

respectively. We have, by (1.13),

$$\|y(t) - x\| \le e^{\omega(t-s)}\|y(s) - x\| + \int_s^t [y(\tau) - y, f(\tau) - v]_s e^{\omega(t-\tau)}\, d\tau$$
$$\forall 0 \le s < t \le T,\ [x, v] \in A. \tag{1.17}$$

Such a function $y \in C([0, T]; X)$ is called (Bénilan [1]) an *integral solution* to Eq. (1.1).

We may conclude, therefore, that under assumptions of Theorem 1.2 the Cauchy problem (1.1) has an integral solution, which coincides with the mild solution of this problem. On the other hand, it turns out that the integral solution is unique (Bénilan [1]) and under the assumptions of Theorem 1.2 (in particular, if A is ω-m-accretive) these two notions coincide.

Let us now come back to the proof of Theorem 1.1.

Let z be a solution to an ε-discretization $D_A^\varepsilon(0 = t_0, t_1, \ldots, t_N; f_1, \ldots, f_N)$ and let w be a solution to $D_A^\varepsilon(0 = s_0, s_1, \ldots, s_M; g_1, \ldots, g_M)$ with the nodal values z_i and w_j, respectively. We set $a_{ij} = \|z_i - w_j\|$, $\delta_i = (t_i - t_{i-1})$, $\gamma_j = (s_j - s_{j-1})$.

Lemma 1.2. *For all* $1 \le i \le N$, $1 \le j \le M$, *we have*

$$a_{ij} \le \left(1 - \omega\frac{\delta_i\gamma_j}{\delta_i + \gamma_j}\right)^{-1}\left(\frac{\gamma_j}{\delta_i + \gamma_j}a_{i-1,j} + \frac{\delta_i}{\delta_i + \gamma_j}a_{i,j-1}\right.$$
$$\left. + \frac{\delta_i\gamma_j}{\delta_i + \gamma_j}[z_i - w_j, f_i - g_j]_s\right). \tag{1.18}$$

Moreover, for all $[x, v] \in A$ *we have*

$$a_{i,0} \le \alpha_{i,1}\|z_0 - x\| + \|w_0 - x\| + \sum_{k=1}^{i} \alpha_{i,k}\delta_k(\|f_k\| + \|v\|),$$
$$0 \le i \le N, \tag{1.19}$$

and

$$a_{0,j} \le \beta_{j,1}\|w_0 - x\| + \|z_0 - x\| + \sum_{k=1}^{j} \beta_{j,k}\gamma_k(\|g_k\| + \|v\|),$$
$$0 \le j \le M, \quad (1.20)$$

where

$$\alpha_{i,k} = \prod_{m=k}^{i} (1 - \omega\delta_m)^{-1}, \qquad \beta_{j,k} = \prod_{m=k}^{j} (1 - \omega\gamma_m)^{-1}. \quad (1.21)$$

Proof. We have

$$f_i + \delta_i^{-1}(z_{i-1} - z_i) \in Az_i, \qquad g_j + \gamma_j^{-1}(w_{j-1} - w_j) \in Aw_j, \quad (1.22)$$

and since A is ω-accretive this yields

$$\left[z_i - w_j, f_i + \delta_i^{-1}(z_{i-1} - z_i) - g_j - \gamma_j^{-1}(w_{j-1} - w_j)\right]_s$$
$$\ge -\omega\|z_i - w_j\|.$$

Hence,

$$\begin{aligned} -\omega\|z_i - w_j\| &\le [z_i - w_j, f_i - g_j]_s + \delta_i^{-1}[z_i - w_j, z_{i-1} - z_i]_s \\ &\quad + \gamma_j^{-1}[z_i - w_j, w_j - w_{j-1}]_s \\ &\le [z_i - w_j, f_i - g_j]_s - \delta_i^{-1}\big(\|z_i - w_j\| - \|z_{i-1} - w_j\|\big) \\ &\quad - \gamma_j^{-1}\big(\|z_i - w_j\| - \|z_i - w_{j-1}\|\big), \end{aligned}$$

and rearranging gives (1.18).

To get estimates (1.19), (1.20), we note that since A is ω-accretive we have

$$\|z_i - x\| \le (1 - \delta_i\omega)^{-1}\|z_i - x + \delta_i(f_i + \delta_i^{-1}(z_{i-1} - z_i) - v)\|,$$

respectively

$$\|w_j - x\| \le (1 - \gamma_j\omega)^{-1}\|w_j - x + \gamma_j\big(g_j + \gamma_j^{-1}(w_{j-1} - w_j) - v\big)\|,$$

for all $[x, v] \in A$. Hence,

$$\|z_i - x\| \le (1 - \delta_i\omega)^{-1}\|z_{i-1} - x\| + (1 - \delta_i\omega)^{-1}\delta_i(\|f_i\| + \|v\|),$$
$$\|w_j - x\| \le (1 - \gamma_j\omega)^{-1}\|w_{j-1} - x\| + (1 - \gamma_j\omega)^{-1}\gamma_j\big(\|g_j\| + \|v\|\big),$$

and (1.19), (1.20) follow by a simple calculation. ■

Now, consider the functions ψ and φ on $[0, T]$ that satisfy the linear equation

$$\frac{\partial \psi}{\partial t}(t, s) + \frac{\partial \psi}{\partial s}(t, s) - \omega \psi(t, s) = \varphi(t, s)$$
$$\text{for } 0 \leq t \leq T, 0 \leq s \leq T, \quad (1.23)$$

and the boundary conditions

$$\psi(t, s) = b(t - s) \qquad \text{for } t = 0 \text{ or } s = 0, \quad (1.24)$$

where $b \in C([-T, T])$.

There is a close relationship between Eq. (1.23) and inequality (1.18). Indeed, if define the grid

$$\Delta = \{(t_i, s_j); 0 = t_0 \leq t_1 \leq \cdots \leq t_N < T,$$
$$0 = s_0 \leq s_1 \leq \cdots \leq s_M < T\}$$

and approximate (1.23) by the difference equations

$$\frac{\psi_{i,j} - \psi_{i-1,j}}{\delta_i} + \frac{\psi_{i,j} - \psi_{i,j-1}}{\gamma_j} - \omega \psi_{ij} = \phi_{i,j}$$
$$\text{for } i = 1, \ldots, N, j = 1, \ldots, M. \quad (1.25)$$

where $\delta_i = t_i - t_{i-1}$, $\gamma_j = s_j - s_{j-1}$, and $\phi_{i,j} = \varphi(t_i, s_j)$, we get after some rearrangement

$$\psi_{i,j} = \left(1 - \omega \frac{\delta_i \gamma_j}{\delta_i + \gamma_j}\right)^{-1} \left(\frac{\gamma_j}{\delta_i + \gamma_j} \psi_{i-1,j} + \frac{\psi_i}{\delta_i + \gamma_j} \psi_{i,j-1} + \frac{\delta_i \gamma_j}{\delta_i + \gamma_j} \phi_{i,j}\right),$$
$$i = 1, \ldots, N, j = 1, \ldots, M. \quad (1.26)$$

Define the functions

$$\varphi(t, s) = \|f(t) - g(s)\|, \quad \phi_{i,j} = \|f_i - g_j\|,$$
$$i = 1, \ldots, N, j = 1, \ldots, M,$$

where f_i and g_j are the nodal approximations of f and $g \in L^1(0, T; X)$, respectively.

Integrating Eqs. (1.23) and (1.24), we get

$$\psi(t,s) = G(b,\varphi)(t,s)$$
$$= \begin{cases} e^{\omega s} b(t-s) + \int_0^s e^{\omega(s-\tau)} \varphi(t-s+\tau,\tau)\, d\tau \\ \qquad \text{if } 0 \le s < t \le T, \\ e^{\omega t} b(t-s) + \int_0^t e^{\omega(t-\tau)} \varphi(\tau, s-t+\tau)\, d\tau \\ \qquad \text{if } 0 \le t < s \le T. \end{cases} \tag{1.27}$$

We set $\Omega = (0,T) \times (0,T)$, and for every measurable function $\varphi : [0,T] \times [0,T] \to \mathbf{R}$ we set

$$\|\varphi\|_\Omega = \inf\{\|f\|_{L^1(0,T)} + \|g\|_{L^1(0,T)};$$
$$|\varphi(t,s)| \le |f(t)| + |g(s)| \text{ a.e. } (t,s) \in \Omega\}. \tag{1.28}$$

Let $\Omega(\Delta) = [0, t_N] \times [0, s_M]$ and $B : [-s_M, t_N] \to \mathbf{R}$, $\phi : \Omega(\Delta) \to \mathbf{R}$ be piecewise constant functions; i.e., there are $B_{i,j}$, $\phi_{i,j} \in \mathbf{R}$ such that $B(0) = B_{0,0}$ and

$$B(r) = B_{ij} \qquad \text{for } t_{i-1} < r \le r_i,\ -s_j \le s < -s_{j-1},$$
$$\phi(t,s) = \phi_{i,j} \qquad \text{for } (t,s) \in (t_{i-1}, t_i] \times (s_{j-1}, s_j].$$

Observe by (1.27) that if the mesh $m(\Delta) = \max\{(\delta_i, \gamma_j);\ i,j\}$ of Δ is sufficiently small, then the system (1.26) with the boundary value conditions

$$\psi_{i,j} = B_{i,j} \qquad \text{for } i = 0 \text{ or } j = 0, \tag{1.29}$$

has a unique solution $\{\psi_{ij}\} i = 1,\dots,N,\ j = 1,\dots,M$.
Denote by $\Psi = H_\Delta(B,\phi)$ the piecewise constant function on Ω defined by

$$\Psi = \psi_{ij} \qquad \text{on } (t_{i-1}, t_i] \times (s_{j-1}, s_j], \tag{1.30}$$

i.e., the solution to (1.26), (1.29).

Lemma 1.3 following gives the convergence of the finite difference scheme (1.25), (1.29) as $m(\Delta) \to 0$.

Lemma 1.3. *Let $b \in C([-T,T])$ and $\varphi \in L^1(\Omega)$ be given. Then*

$$\|G(b,\varphi) - H_\Delta(B,\phi)\|_{L^\infty(\Omega(\Delta))} \to 0 \tag{1.31}$$

as

$$m(\Delta) + \|b - B\|_{L^\infty(-s_M, t_N)} + \|\varphi - \phi\|_{\Omega(\Delta)} \to 0.$$

Proof. In order to avoid a tedious calculus, we shall prove (1.31) in the accretive case (i.e., $\omega = 0$). Let us prove first the estimate

$$\|H_\Delta(B, \phi)\|_{L^\infty(\Omega(\Delta))} \leq \|B\|_{L^\infty(-s_M, t_N)} + \|\phi\|_{\Omega(\Delta)}. \tag{1.32}$$

Indeed, we have $H_\Delta(B, \phi) = H_\Delta(B, 0) + H_\Delta(0, \phi)$, and by (1.28), (1.30) we see that the values of $H_\Delta(B, 0)$ are convex combinations of the values of B. Hence,

$$\|H_\Delta(B, 0)\|_{L^\infty(\Omega(\Delta))} \leq \|B\|_{L^\infty(-s_M, t_N)}.$$

It remains to show that

$$\|H_\Delta(0, \phi)\|_{L^\infty(\Omega(\Delta))} \leq \|\phi\|_{\Omega(\Delta)}.$$

By the definition of the $\|\cdot\|_{\Omega(\Delta)}$-norm, we have

$$\|\phi\|_{\Omega(\Delta)} = \inf\left\{\sum_{i=1}^{N} \delta_i \alpha_i + \sum_{j=1}^{m} \gamma_j \beta_j ;\ \alpha_i + \beta_j \geq |\phi_{i,j}|,\ \alpha_i, \beta_j \geq 0\right\}.$$

Now, let $g_{i,j} = \alpha_i + \beta_j \geq |\phi_{i,j}|$ and

$$d_{i,j} = \sum_{k=1}^{i} \alpha_k \delta_k + \sum_{k=1}^{j} \beta_k \gamma_k.$$

It is readily seen that $\psi_{i,j} = d_{i,j}$ satisfy the system (1.26) where $\phi_{i,j} = g_{i,j}$. Hence, $d = H_\Delta(\tilde{B}, g)$ provided $d_{i,j} = \tilde{b}_{i,j}$ for $i = 0$ or $j = 0$, where $d = \{d_{i,j}\}$, $\tilde{B} = \{\tilde{b}_{i,j}\}$ and $g = \{g_{i,j}\}$. Inasmuch as $g_{i,j} \geq |\phi_{i,j}|$, we have

$$d = H_\Delta(\tilde{B}, g) \geq H_\Delta(0, \phi) \geq |H_\Delta(0, \phi)|$$

if $\tilde{b}_{i,j} \geq 0$. Hence

$$\|H_\Delta(0, \phi)\|_{L^\infty(\Omega(\Delta))} \leq \|d\|_{L^\infty(\Omega(\Delta))} \leq \|\phi\|_{\Omega(\Delta)},$$

as claimed.

Now, let $\tilde{\psi} = G(\tilde{\mathrm{b}}, \tilde{\varphi})$ and assume first that $\tilde{\psi}_{tt}, \tilde{\psi}_{ss} \in L^\infty(\Omega)$. Then, by (1.23), we see that $\tilde{\psi}_{i,j} = \tilde{\psi}(t_i, s_j)$ satisfy the system

$$\frac{\tilde{\psi}_{i,j} - \tilde{\psi}_{i-1,j}}{\delta_i} + \frac{\tilde{\psi}_{i,j} - \tilde{\psi}_{i,j-1}}{\gamma_j} = \tilde{\varphi}(t_i, s_j) + e_{i,j},$$

$$\tilde{\psi}_{i,0} = \tilde{b}(t_i), \qquad \tilde{\psi}_{0,j} = \tilde{b}(-s_j), \qquad i = 0, 1, \ldots, N,\ j = 0, 1, \ldots, M,$$

where

$$|e_{ij}| \le \gamma_j \|\tilde{\psi}_{ss}\|_{L^\infty(\Omega)} + \delta_i \|\tilde{\psi}_{tt}\|_{L^\infty(\Omega)} \qquad \forall i, j.$$

Then, by (1.32), we see that

$$\begin{aligned}
&\|G(\tilde{b}, \tilde{\varphi}) - H_\Delta(B, \phi)\|_{L^\infty(\Omega(\Delta))} \\
&\quad \le \|B - \tilde{b}\|_{L^\infty(-s_M, t_N)} + \|\tilde{\varphi} - \phi\|_{\Omega(\Delta)} + \|e\|_{\Omega(\Delta)} \\
&\quad \le \|B - \tilde{b}\|_{L^\infty(-s_M, t_N)} + \|\tilde{\varphi} - \phi\|_{\Omega(\Delta)} \\
&\qquad + Cm(\Omega)\left(\|\tilde{\psi}_{tt}\|_{L^\infty(\Omega)} + \|\tilde{\psi}_{ss}\|_{L^\infty(\Omega)}\right).
\end{aligned} \tag{1.33}$$

Now, let $\varphi \in L^1(\Omega)$, $b \in C([-T, T])$, and $\tilde{b} \in C^2([-T, T])$, $\tilde{\varphi} \in C^2(\bar{\Omega})$. Then $\tilde{\psi} = G(\tilde{b}, \tilde{\varphi})$ is smooth, and by (1.33) and (1.29) we have

$$\begin{aligned}
&\|G(b, \varphi) - H_\Delta(B, \phi)\|_{L^\infty(\Omega(\Delta))} \\
&\quad \le \|G(b, \varphi) - G(\tilde{b}, \tilde{\varphi})\|_{L^\infty(\Omega(\Delta))} + \|G(\tilde{b}, \tilde{\varphi}) - H_\Delta(B, \phi)\|_{L^\infty(\Omega(\Delta))} \\
&\quad \le 2\|b - \tilde{b}\|_{L^\infty(-s_M, t_N)} + C\|\varphi - \tilde{\varphi}\|_{\Omega(\Delta)} + \|B - b\|_{L^\infty(-s_M, t_N)} \\
&\qquad + \|\tilde{\varphi} - \phi\|_{\Omega(\Delta)} + Cm(\Delta)\left(\|\tilde{\psi}_{tt}\|_{L^\infty(\Omega)} + \|\tilde{\psi}_{ss}\|_{L^\infty(\Omega)}\right).
\end{aligned} \tag{1.34}$$

Given $\eta > 0$, we may choose $\tilde{b}$ and $\tilde{\varphi}$ such that $\|b - \tilde{b}\|_{L^\infty(-s_M, t_N)}$, $\|\varphi - \tilde{\varphi}\|_{\Omega(\Delta)} \le \eta$. Then (1.34) implies (1.31), as desired. ∎

Proof of Theorem 1.1. (continued). We shall apply Lemma 1.3, where $\varphi(t,s) = \|f(t) - g(s)\|$, $\phi = \{\phi_{i,j}\}$, $\phi_{i,j} = \|f_i - g_j\|$, $1 \le j \le M$, $1 \le i \le N$, where f_i and g_j are the nodal values of f and g, respectively, and

$$\begin{aligned}
B(t) &= B_{i,0} \qquad \text{for } t_{i-1} < t \le t_i, \qquad i = 1, \dots, N, \\
B(s) &= B_{0,j} \qquad \text{for } -s_j < s \le -s_{j-1}, \quad j = 1, \dots, M.
\end{aligned}$$

Here, $B_{i,0}$ is the right hand side of (1.19) and $B_{0,j}$ is the right hand side of (1.20). It is easily seen that, for $\varepsilon \to 0$,

$$\begin{aligned}
B(t) \to b(t) &= e^{\omega t}\|z_0 - x\| + \|w_0 - x\| \\
&\quad + \int_0^t e^{\omega(t-\tau)}(\|f(\tau)\| + \|v\|)\, d\tau \qquad \forall t \in [0, T],
\end{aligned}$$

and

$$\begin{aligned}
B(s) \to b(-s) &= e^{\omega s}\|w_0 - x\| + \|z_0 - x\| \\
&\quad + \int_0^s e^{\omega(s-\tau)}(\|g(\tau)\| + v\|)\, d\tau, \qquad s \in [-T, 0].
\end{aligned}$$

By (1.7) and (1.31), we have

$$\|\varphi - \phi\|_{\Omega(\Delta)} \leq 2\varepsilon$$

whilst, by Lemma 1.2,

$$a_{i,j} = \|z_i - w_j\| \leq H_\Delta(B, \phi)_{i,j} \qquad \forall i, j.$$

Then, by Lemma 1.3, we see that for every $\eta > 0$, we have

$$\|z(t) - w(s)\| \leq G(b, \varphi)(t, s) + \eta \qquad \forall s, t \in [0, T], \tag{1.35}$$

as soon as $0 < \varepsilon < \nu(\eta)$.

If $f \equiv g$ and $z_0 = w_0$, then $G(b, \varphi)(t, t) = e^{\omega t} b(0) = 2e^{\omega t}\|z_0 - x\|$ and so, by (1.35),

$$\|z(t) - w(t)\| \leq \eta + 2e^{\omega t}\|z_0 - x\| \qquad \forall x \in D(A), t \in [0, T],$$

for all $0 < \varepsilon \leq \nu(\eta)$. Since $\|z_0 - s_0\| \leq \varepsilon$, $y_0 \in \overline{D(A)}$, and x is arbitrary in $D(A)$, it follows that the sequence z_ε of ε-approximate solutions satisfies the Cauchy criterion and so $y(t) = \lim_{\varepsilon \to 0} z_\varepsilon(t)$ exists uniformly on $[0, T]$. Now we take the limit as $\varepsilon \to 0$ in (1.35) with $s = t + h$, $g \equiv f$, and $z_0 = w_0 = y_0$. We get

$$\begin{aligned} \|y(t + h) - y(t)\| \leq G(b, \varphi)(t + h, t) &= e^{\omega t}(e^{\omega h} + 1)\|y_0 - x\| \\ &\quad + \int_0^h e^{\omega(h-\tau)}(\|f(\tau)\| + \|v\|)\, d\tau \\ &\quad + \int_0^t e^{\omega(t-\tau)}\|f(\tau + h) - f(\tau)\|\, d\tau \qquad \forall [x, v] \in A, \end{aligned}$$

and therefore y is continuous on $[0, T]$.

Now by (1.35) we have, for $f \equiv g$, $t = s$,

$$\|z(t) - y(t)\| \leq \delta(\varepsilon) \qquad \forall t \in [0, T],$$

where z is an ε-approximate solution and $\delta(\varepsilon) \to 0$ as $\varepsilon \to 0$. Finally, we take $t = s$ in (1.35) and let ε tend to zero. We get

$$\|y(t) - \bar{y}(t)\| \leq e^{\omega t}\|y(0) - \bar{y}(0)\| + \int_0^t e^{\omega(t-\tau)}\|f(\tau) - g(\tau)\|\, d\tau.$$

To obtain (1.11) we apply now inequality (1.35), where

$$\varphi(t, s) = [y(t) - \bar{y}(s), f(t) - g(s)]_s \qquad \text{and } t = s.$$

Then

$$G(h,\varphi)(t,t) = e^{\omega t}\|y(0) - \bar{y}(0)\| + \int_0^t e^{\omega(t-s)}[y(s) - \bar{y}(s), f(s) - g(s)]_s,$$

and so (1.11) follows for $s = 0$ and consequently for all $s \in (0,t)$.

Thus, the proof of Theorem 1.1 is complete. ■

The convergence theorem can be made more precise in the autonomous case of (1.15) i.e., for $f \equiv 0$.

Corollary 1.1. *Let A be ω-accretive and satisfy condition* (1.14), *and let $y_0 \in \overline{D(A)}$. Let y be the mild solution to problem* (1.15) *and let y_ε be an ε-approximate solution to* (1.15) *with $y_\varepsilon(0) = y_0$. Then*

$$\|y_\varepsilon(t) - y(t)\| \le C_T\big(\|y_0 - x\| + |Ax|(\varepsilon + t^{1/2}\varepsilon^{1/2})\big) \qquad \forall t \in [0,T], \tag{1.36}$$

for all $x \in D(A)$. In particular, we have

$$\|y(t) - \left(I + \frac{t}{n}A\right)^{-n} y_0\| \le C_T(\|y_0 - x\| + tn^{-1/2}|Ax|) \qquad (1.36)'$$

for all $t \in [0,T]$ and $x \in D(A)$. Here, C_T is a positive constant independent of x and y_0.

Proof. Since the mappings $y_0 \to y$ and $y_0 \to y_\varepsilon$ are Lipschitz with Lipschitz constant $e^{\omega T}$, it suffices to prove estimate (1.36) for $y_0 \in D(A)$.

By the estimate (1.34) we have, for all $T > 0$,

$$\|G(b,0) - H_\Delta(B,0)\|_{L^\infty(\Omega(\Delta))} \le \|b - \tilde{b}\|_{L^\infty(-T,T)} + \|B - \tilde{b}\|_{L^\infty(-T,T)} + C\varepsilon\left(\|\tilde{\psi}_{tt}\|_{L^\infty(\Omega)} + \|\tilde{\psi}_{ss}\|_{L^\infty(\Omega)}\right),$$

where $\bar{\psi} = G(\tilde{b},0)$, $\tilde{b}$ is a sufficiently smooth function on $[-T,T]$, $\Omega = (0,T) \times (0,T)$, and C is independent of ε, b, and B. We shall apply this inequality for B and b as in the proof of Theorem 1.1, i.e.,

$$b(t) = \omega^{-1}(e^{\omega|t|} - 1)|Ax| \qquad \forall t \in [-T,T].$$

Then we have

$$b'(t) = e^{\omega|t|}|Ax|\,\mathrm{sign}\,t.$$

We shall approximate the signum function by

$$\theta(t) = \begin{cases} \dfrac{t}{\lambda} & \text{for } |t| \leq \lambda, \\ \dfrac{t}{|t|} & \text{for } |t| > \lambda, \end{cases}$$

and so we construct a smooth function $\tilde{b}$ such that $\tilde{b}(0) = 0$ and

$$\tilde{b}''(t) = \omega \operatorname{sign} t |Ax| e^{\omega|t|} + \theta'(t)|Ax|e^{\omega|t|}.$$

Hence,

$$\sup\{|\tilde{b}''(s)|; 0 \leq s \leq t\} \leq e^{\omega|t|}|Ax|(\omega + \lambda^{-1})$$

and, therefore,

$$\begin{aligned} &\|b - \tilde{b}\|_{L^\infty(-t,t)} + C\varepsilon t\Big(\|\tilde{\psi}_{tt}\|_{L^\infty((0,t)\times(0,t))} \\ &\quad + \|\tilde{\psi}_{ss}\|_{L^\infty((0,t)\times(0,t))}\Big) \\ &\quad \leq Ct\varepsilon|Ax|(1 + \lambda^{-1}) + C\lambda|Ax| \qquad \forall t \in [0,T], \end{aligned}$$

where C depends on T only.

Similarly, we have

$$\|B - \tilde{b}\|_{L^\infty(-t,t)} \leq C(\varepsilon + \lambda)|Ax|.$$

Finally,

$$\|G(b,0) - H_\Delta(B,0)\|_{L^\infty(\Omega_t(\Delta))} \leq C(\varepsilon + \lambda + t\varepsilon\lambda^{-1})|Ax|,$$

where $\Omega_t = (0,t) \times (0,t)$. This implies that (see the proof of Theorem 1.1)

$$\|y_\varepsilon(t) - y(t)\| \leq G(b,0)(t,t) + C|Ax|(\varepsilon + \lambda + t\varepsilon\lambda^{-1})$$

for all $t \in [0,T]$ and all $\lambda > 0$. For $\lambda = (t\varepsilon)^{1/2}$ this yields

$$\|y_\varepsilon(t) - y(t)\| \leq C|Ax|(\varepsilon + t^{1/2}\varepsilon^{1/2}) \qquad \forall t \in [0,T],$$

which completes the proof. ■

4.1.2. *Regularity of Mild Solutions*

A question of great interest is that of circumstances under which the mild solutions are strong solutions. One may construct simple examples that show that in a general Banach space this might be false. However, if the space is reflexive then under natural assumptions on A, f, and y_0 the answer is positive.

Theorem 1.4. *Let X be reflexive and let A be closed, ω-accretive and let it satisfy assumption* (1.12). *Let $y_0 \in D(A)$ and let $f \in W^{1,1}([0,T];X)$ be such that $f(t) \in C$ $\forall t \in [0,T]$. Then problem* (1.1) *has a unique strong solution $y \in W^{1,\infty}([0,T];X)$. Moreover, y satisfies the estimate*

$$\left\| \frac{dy}{dt}(t) \right\| \le e^{\omega t}|f(0) - Ay_0| + \int_0^t e^{\omega(t-s)} \left\| \frac{df}{ds}(s) \right\| ds \qquad \text{a.e. } t \in (0,T), \tag{1.37}$$

where $|f(0) - Ay_0| = \inf\{\|w\|;\ w \in f(0) - Ay_0\}$.

Proof. Let y be the mild solution to problem (1.1) provided by Theorem 1.2. We shall apply estimate (1.13) where $y(t) := y(t+h)$ and $g(t) := f(t+h)$. We get

$$\begin{aligned}\|y(t+h) - y(t)\| &\le \|y(h) - y(0)\|e^{\omega t} \\ &\quad + \int_0^t \|f(s+h) - f(s)\|e^{\omega(t-s)}\, ds \\ &\le Ch + \|y(h) - y(0)\|e^{\omega t},\end{aligned}$$

because $f \in W^{1,1}([0,T];X)$ (see Theorem 3.3 and Remark 3.1 in Chapter 1). Now, applying the same estimate (1.13) to y and y_0, we get

$$\|y(h) - y_0\| \le \int_0^h \|f(s) - \xi\|e^{\omega(h-s)}\, ds \le \int_0^h |Ay_0 - f(s)|\, ds \qquad \forall \xi \in Ay_0,\ h \in [0,T].$$

We may conclude, therefore, that the mild solution y is Lipschitz on $[0,T]$. Then, by Theorem 3.1 of Chapter 1, it is a.e. differentiable and belongs to $W^{1,\infty}([0,T];X)$. Moreover, we have

$$\begin{aligned}\left\| \frac{dy}{dt}(t) \right\| &= \lim_{h \to 0} \frac{\|y(t+h) - y(t)\|}{h} \\ &\le e^{\omega t}|Ay_0 - f(0)| + \int_0^t \left\| \frac{df}{ds} \right\| e^{\omega(t-s)}\, ds \qquad \text{a.e. } t \in (0,T).\end{aligned}$$

Now, let $t \in [0, T]$ be such that

$$\frac{dy}{dt}(t) = \lim_{h \to 0} \frac{1}{h}(y(t+h) - y(t))$$

exists. By inequality (1.17), we have

$$\|y(t+h) - x\| \le e^{\omega h}\|y(t) - x\| + \int_t^{t+h} e^{\omega(t+h-s)}[y(\tau) - x, f(\tau) - w]_s \, d\tau \qquad \forall [x, w] \in A.$$

Noticing that

$$[v - x, u - v]_s \le \|u - x\| - \|v - x\| \qquad \forall u, v, x \in X,$$

we get

$$[y(t) - x, y(t+h) - y(t)]_s \le (e^{\omega h} - 1)\|y(t) - x\| + \int_t^{t+h} e^{\omega(t+h-\tau)}[y(\tau) - x, f(\tau) - w]_s \, d\tau.$$

Since the bracket $[u, v]_s$ is upper semicontinuous in (u, v), and positively homogeneous and continuous in v, this yields

$$\left[y(t) - x, \frac{dy}{dt}(t)\right]_s - \omega\|y(t) - x\| \le [y(t) - x, f(t) - w]_s \qquad \forall [x, w] \in A.$$

This implies that there is $\xi \in J(y(t) - x)$ such that (J is the duality mapping)

$$\left(\frac{dy}{dt}(t) - \omega(y(t) - x) - f(t) - w, \xi\right) \le 0. \tag{1.38}$$

Now we have

$$y(t-h) = y(t) - h\frac{d}{dt}y(t) + hg(h), \tag{1.39}$$

where $g(h) \to 0$ as $h \to 0$. On the other hand, by condition (1.12), for every h sufficiently small and positive, there are $[x_h, w_h] \in A$ such that

$$y(t-h) + hf(t) = x_h + hw_h .$$

Substituting successively in (1.38) and in (1.39), we get

$$(1 - \omega h)\|y(t) - x_h\| \leq h\|g(h)\| \qquad \forall h \in (0, \lambda_0).$$

Hence, $x_h \to y(t)$ and $w_h \to f(t) - (dy(t))/dt$ as $h \to 0$. Since A is closed we conclude that

$$\frac{dy}{dt}(t) + Ay(t) \ni f(t),$$

as claimed. ■

Remark 1.1. If A is ω-m-accretive, then as seen earlier it is closed in $X \times X$ and so assumptions (1.12) hold automatically.

In particular, we have:

Theorem 1.5. *Let A be ω-accretive, closed and satisfy the range condition* (1.14). *Let X be reflexive and $y_0 \in D(A)$. Then problem* (1.15) *has a unique strong solution $y \in W^{1,\infty}([0,\infty); X)$. Moreover,*

$$\left\|\frac{dy}{dt}(t)\right\| \leq e^{\omega t}|Ay_0| \qquad \text{a.e. } t > 0,$$

where $|Ay_0| = \inf\{\|z\|;\ z \in Ay_0\}$.

More can be said about the regularity of strong solution to problem (1.1) if the space X is uniformly convex.

Theorem 1.6. *Let A be ω-m-accretive, $f \in W^{1,1}([0,T]; X)$, $y_0 \in D(A)$, and X be uniformly convex along with the dual X^*. Then the strong solution y to problem* (1.1) *is everywhere differentiable from the right, $(d^+/dt)y$ is right* *continuous and*

$$\frac{d^+}{dt}y(t) + (Ay(t) - f(t))^0 = 0 \qquad \forall t \in [0,T), \tag{1.41}$$

$$\left\|\frac{d^+}{dt}y(t)\right\| \le e^{\omega t}\|(Ay_0 - f(0))^0\| + \int_0^t e^{\omega(t-s)}\left\|\frac{df}{ds}(s)\right\| ds \qquad \forall t \in [0,T). \tag{1.42a}$$

Here, $(Ay - f)^0$ is the element of minimum norm in the set $Ay - f$.

Proof. Since X and X^* are uniformly convex, Ay is a closed convex subset of X for every $x \in D(A)$ (see Section 3 in Chapter 2) and so $(Ay(t) - f(t))^0$ is well-defined.

Let $y \in W^{1,\infty}([0,T]; X)$ be the strong solution to (1.1). We have

$$\frac{d}{dh}(y(t+h) - y(t)) + Ay(t+h) \ni f(t+h) \qquad \text{a.e. } h > 0,\ t \in (0,T),$$

and since A is ω-accretive, this yields

$$\left(\frac{d}{dh}(y(t+h) - y(t)), \xi\right) \le \omega\|y(t+h) - y(t)\|^2 + (f(t+h) - \eta(t), \xi) \qquad \forall \eta(t) \in Ay(t),$$

where $\xi \in J(y(t+h) - y(t))$.

Then, by Lemma 1.1, we get

$$\|y(t+h) - y(t)\| \le \int_0^h e^{\omega(h-s)}\|\eta(t) - f(t+s)\|\, ds, \tag{1.42b}$$

which yields

$$\left\|\frac{dy}{dt}(t)\right\| \le \|f(t) - \eta(t)\| \qquad \forall \eta(t) \in Ay(t), \text{ a.e. } t \in (0,T).$$

In other words,

$$\left\|\frac{dy}{dt}(t)\right\| \le \|(Ay(t) - f(t))^0\| \qquad \text{a.e. } t \in (0,T),$$

and since $dy(t)/dt + Ay(t) \ni f(t)$ a.e. $t \in (0,T)$, we conclude that

$$\frac{dy}{dt} + (Ay(t) - f(t))^0 = 0 \qquad \text{a.e. } t \in (0,T). \tag{1.43}$$

Observe also that y satisfies the equation

$$\frac{d}{dt}(y(t+h) - y(t)) + Ay(t+h) - Ay(t) \ni f(t+h) - f(t)$$
$$\text{a.e. in } (0,T).$$

Multiplying this by $J(y(t+h) - y(t))$ and using the ω-accretivity of A, we see by Lemma 1.1 that

$$\frac{d}{dt}\|y(t+h) - y(t)\| \leq \omega\|y(t+h) - y(t)\| + \|f(t+h) - f(t)\|$$
$$\text{a.e. } t, t+h \in (0,T),$$

and therefore

$$\|y(t+h) - y(t)\| \leq e^{\omega(t-s)}\|y(s+h) - y(s)\| + \int_s^t e^{\omega(t-\tau)}\|f(\tau+h) - f(\tau)\|d\tau. \tag{1.44}$$

Finally,

$$\left\|\frac{dy}{dt}(t)\right\| \leq e^{\omega(t-s)}\left\|\frac{dy}{ds}(s)\right\| + \int_s^t e^{\omega(t-\tau)}\left\|\frac{df}{d\tau}(\tau)\right\|d\tau \qquad \text{a.e. } 0 < s < t < T. \tag{1.45}$$

Similarly, multiplying the equation

$$\frac{d}{dt}(y(t) - y_0) + Ay(t) \ni f(t) \qquad \text{a.e. } t \in (0,T),$$

by $J(y(t) - y_0)$ and integrating on $(0,t)$, we get the estimate

$$\|y(t) - y_0\| \leq \int_0^t e^{\omega(t-s)}\|(Ay_0 - f(s))^0\|\, ds \qquad \forall t \in [0,T], \tag{1.46}$$

and substituting in (1.44) with $s = 0$, we get

$$\left\|\frac{d}{dt}y(t)\right\| \leq e^{\omega t}\|(Ay_0 - f(0))^0\| + \int_0^t e^{\omega(t-s)}\left\|\frac{df}{ds}(s)\right\| ds \qquad \text{a.e. } t \in (0,T). \tag{1.47}$$

Since A is demiclosed (Proposition 3.4 Chapter 2) and X is reflexive, it follows by (1.43) and (1.47) that $y(t) \in D(A)\ \forall t \in [0,T]$ and

$$\|(Ay(t) - f(t))^0\| \le C \qquad \forall t \in [0,T]. \tag{1.48}$$

For t arbitrary but fixed, consider $h_n \to 0$ such that $h_n > 0$ for all n and

$$\frac{y(t+h_n) - y(t)}{h_n} \rightharpoonup \xi \ \text{ in } X \qquad \text{as } n \to \infty.$$

By estimate (1.42), we see that

$$\|\xi\| \le \|(Ay(t) - f(t))^0\|, \tag{1.49}$$

whilst $\xi \in f(t) - Ay(t)$ because A is demiclosed. Indeed,

$$f(t) - \xi = \text{w} - \lim \frac{1}{h_n} \int_t^{t+h_n} \eta(s)\, ds,$$

where $\eta \in L^\infty(0,T)$ and $\eta(t) \in Ay(t)\ \forall t \in [0,T]$. We set $\eta_n(s) = \eta(t + sh_n)$ and $y_n(s) = y(t + sh_n)$. If we denote again by A the realization of A in $L^2(0,1;X) \times L^2(0,1;X)$, we have $y_n \to y(t)$ in $L^2(0,1;X)$, $\eta_n \to f(t) - \xi$ weakly in $L^2(0,1;X)$. Since A is demiclosed in $L^2(0,1;X) \times L^2(0,1;X)$ we have that $f(t) - \xi \in Ay(t)$, as claimed. Then, by (1.49), we conclude that $\xi = (Ay(t) - f(t))^0$ and, therefore,

$$\frac{d^+}{dt} y(t) = \lim_{h \downarrow 0} \frac{y(t+h) - y(t)}{h} = -(Ay(t) - f(t))^0 \qquad \forall t \in [0,T).$$

Next, we see by (1.44) that

$$\left\|\frac{d^+}{dt} y(t)\right\| \le e^{\omega(t-s)} \left\|\frac{d^+}{dt} y(s)\right\| + \int_s^t e^{\omega(t-\tau)} \left\|\frac{df}{d\tau}(\tau)\right\| d\tau \qquad 0 \le s \le t \le T. \tag{1.50}$$

Let $t_n \to t$ be such that $t_n > t$ for all n. Then, on a subsequence, again denoted by t_n,

$$\frac{d^+ y(t_n)}{dt} = -(Ay(t_n) - f(t_n))^0 \rightharpoonup \xi,$$

where $-\xi \in Ay(t) - f(t)$ (because A is demiclosed). On the other hand, it follows by (1.50) that

$$\|\xi\| \le \limsup_{n\to\infty} \|(Ay(t_n) - f(t_n))^0\| \le \|(Ay(t) - f(t))^0\|.$$

Hence, $\xi = -(Ay(t) - f(t))^0$ and $(d^+/dt)\, y(t_n) \to \xi$ strongly in X (because X is uniformly convex). We have therefore proved that $(d^+/dt)\, y(t)$ is right continuous on $[0, T)$, thereby completing the proof. ■

In particular, it follows by Theorem 1.6 that if A is ω-m-accretive, $y_0 \in D(A)$, and X, X^* are uniformly convex, then the solution y to the autonomous problem (1.15) is everywhere differentiable from the right and

$$\frac{d^+}{dt} y(t) + A^0 y(t) = 0 \qquad \forall t \geq 0, \tag{1.51}$$

where A^0 is the minimal section of A. Moreover, the function $t \to A^0 y(t)$ is continuous from the right on $\mathbf{R}^+$.

It turns out that this result remains true under weaker conditions on A. Namely, one has:

Theorem 1.7. *Let A be ω-accretive, closed, and satisfy the condition*

$$\overline{\operatorname{conv} D(A)} \subset \bigcap_{0<\lambda<\lambda_0} R(I + \lambda A) \qquad \text{for some } \lambda_0 > 0. \tag{1.52}$$

Let X and X^* be uniformly convex. Then for every $x \in D(A)$ the set Ax has a unique element of minimum norm $A^0 x$, and for every $y_0 \in D(A)$ the Cauchy problem (1.1) *has a unique strong solution $y \in W^{1,\infty}([0,\infty); X)$, which is everywhere differentiable from the right and*

$$\frac{d^+}{dt} y(t) + A^0 y(t) = 0 \qquad \forall t \geq 0. \tag{1.53}$$

Moreover, the function $t \to A^0 y(t)$ is continuous from the right and

$$\left\| \frac{d^+}{dt} y(t) \right\| \leq e^{\omega t} \|A^0 y_0\| \qquad \forall t \geq 0. \tag{1.54}$$

Proof. We shall assume first that A is demiclosed in $X \times X$. Define the set $B \subset X \times X$ by

$$Bx = \overline{\operatorname{conv} Ax}, \qquad x \in D(B) = D(A).$$

It is readily seen that B is ω-accretive. Moreover, by (1.52) it follows that

$$D(A) \subset \bigcap_{0<\lambda<\lambda_0} R(I + \lambda B).$$

Let $x \in D(A)$. Then, $x_\lambda = (I + \lambda A)^{-1} x$ and $y_\lambda = A_\lambda x$ are well-defined for $0 < \lambda < \lambda_0$. Moreover $\|A_\lambda x\| \leq |Ax| = \inf\{\|w\|;\ w \in Ax\}$ and $x_\lambda \to x$

and $\lambda \to 0$ (see Proposition 3.2 in Chapter 2). Let $\lambda_n \to 0$ be such that $A_{\lambda_n} x \rightharpoonup y$. Since $A_{\lambda_n} x \in Ax_{\lambda_n}$ and A is demiclosed, it follows that $y \in Ax$. On the other hand, we have

$$\|A_\lambda x\| = \|B_\lambda x\| \leq |Bx| = \|B^0 x\|.$$

($B^0 x$ exists and is unique because Bx is convex.) This implies that $y = B^0 x \in Ax$. Hence, Ax has a unique element of minimum norm $A^0 x$. Then we may apply Theorem 1.6 to deduce that the strong solution y to problem (1.15) (which exists and is unique by Theorem 1.5) satisfies (1.53) and (1.54). (In the proof of Theorem 1.6, the ω-m-accretivity has been used only to assure the existence of a strong solution, the demiclosedness of A, and the existence of A^0.)

To complete the proof, we turn now to the case where A is only closed. Let $\tilde{A}$ be the closure of A in $X \times X_w$, i.e., the smallest demiclosed extension of A. Clearly, $D(A) \subset D(\tilde{A}) \subset \overline{D(A)}$ and $\tilde{A}$ satisfies condition (1.52). Moreover, since the duality mapping J is continuous we easily see that $\tilde{A}$ is ω-accretive. Then, applying the first part of the proof, we conclude that problem

$$\frac{d^+ u}{dt} + \tilde{A}^0 u = 0 \qquad \text{in } [0, \infty),$$

$$u(0) = y_0,$$

has a unique solution u satisfying all conditions of the theorem. To conclude the proof it suffices to show that $D(\tilde{A}) = D(A)$ and $\tilde{A}^0 = A^0$.

Let $x \in D(\tilde{A})$. Then, there is $[x_\lambda, y_\lambda] \in A \subset \tilde{A}$ such that $x = x_\lambda - \lambda y_\lambda$ for $0 < \lambda < \lambda_0$.

We have $x_\lambda = (I + \lambda A)^{-1} x$ and $y_\lambda = A_\lambda x = \tilde{A}_\lambda x$. Since $x \in D(\tilde{A})$, $x_\lambda \xrightarrow{\lambda \to 0} x$ and $\|y_\lambda\| \leq |\tilde{A}x| = \|\tilde{A}^0 x\|$. As $\tilde{A}$ is demiclosed and X is uniformly convex this implies by a standard device that $y_\lambda \to \tilde{A}^0 x$ as $\lambda \to 0$. Since A is closed, this yields $\tilde{A}^0 x \in Ax$ and $x \in D(A)$. Hence, $D(\tilde{A}) = D(A)$ and $\tilde{A}^0 x = A^0 x \quad \forall x \in D(A)$. The proof of Theorem 1.7 is complete. ■

Remark 1.2. If the space X^* is uniformly convex, A is ω-m-accretive, $f \in W^{1,1}([0, T]; X)$, and $y_0 \in D(A)$, then the strong solution $y \in W^{1,\infty}([0, T]; X)$ to problem (1.1) (Theorem 1.4) can be obtained as

$$y(t) = \lim_{\lambda \to 0} y_\lambda(t) \qquad \text{in } X, \text{ uniformly on } [0, T], \tag{1.55}$$

where $y_\lambda \in C^1([0,T];X)$ are the solutions to the Yosida approximating system

$$\frac{dy_\lambda}{dt} + A_\lambda y_\lambda = f, \qquad t \in [0,T],$$
$$y_\lambda(0) = y_0. \tag{1.56}$$

Here, $A_\lambda = \lambda^{-1}(I - (I + \lambda A)^{-1})$ for $0 < \lambda < \lambda_0$. Indeed, by Lemma 1.2, we have

$$\frac{1}{2}\frac{d}{dt}\|y_\lambda(t) - y_\mu(t)\|^2 + \big(A_\lambda y_\lambda(t) - A_\mu y_\mu(t), J\big(y_\lambda(t) - y_\mu(t)\big)\big) = 0$$
$$\text{a.e. } t \in (0,T),$$

for all $\lambda, \mu \in (0, \lambda_0)$. Since A is ω-accretive and $A_\lambda y \in A(I + \lambda A)^{-1}y$, we get

$$\begin{aligned}\frac{1}{2}\frac{d}{dt}\|y_\lambda(t) - y_\mu(t)\|^2 &+ \big(A_\lambda y_\lambda(t) - A_\mu y_\mu(t), J\big(y_\lambda(t) - y_\mu(t)\big)\\ &- J\Big((I + \lambda A)^{-1}y_\lambda(t) - (1 + \mu A)^{-1}y_\mu(t)\Big)\\ &\le \omega\|(1 + \lambda A)^{-1}y_\lambda(t) - (1 + \mu A)^{-1}y_\mu(t)\|^2 \qquad \text{a.e. } t \in (0,T).\end{aligned} \tag{1.57}$$

On the other hand, multiplying the equation

$$\frac{d^2y_\lambda}{dt^2} + \frac{d}{dt}A_\lambda y_\lambda(t) = \frac{df}{dt} \qquad \text{a.e. } t \in (0,T),$$

by $J(dy_\lambda/dt)$ yields

$$\frac{1}{2}\frac{d}{dt}\left\|\frac{dy_\lambda}{dt}(t)\right\|^2 \le \left\|\frac{df}{dt}(t)\right\|\left\|\frac{dy_\lambda}{dt}(t)\right\| + \omega\left\|\frac{dy_\lambda}{dt}(t)\right\| \qquad \text{a.e. } t \in (0,T),$$

because A_λ is ω-accretive. This implies

$$\begin{aligned}\left\|\frac{dy_\lambda}{dt}(t)\right\| &\le e^{\omega t}\left\|\frac{dy_\lambda}{dt}(0)\right\| + \int_0^t e^{\omega(t-s)}\left\|\frac{df}{ds}(s)\right\| ds\\ &\le e^{\omega t}|Ay_0 - f(0)| + \int_0^t e^{\omega(t-s)}\left\|\frac{df}{ds}\right\| ds.\end{aligned} \tag{1.58a}$$

Hence, $\|A_\lambda y_\lambda(t)\| \le C \ \forall \lambda \in (0, \lambda_0)$, and $\|y_\lambda(t) - (1 + \lambda A)^{-1}y_\lambda(t)\| \le C\lambda$. Since J is uniformly continuous on bounded sets, it follows by (1.57) that

$\{y_\lambda\}$ is a Cauchy sequence in the space $C([0,T];X)$ and $y(t) = \lim_{\lambda \to 0} y_\lambda(t)$ exists in X uniformly on $[0,T]$. Let $[x,w]$ be arbitrary in A and let $x_\lambda = x + \lambda w$. Multiplying Eq. (1.56) by $J(y_\lambda(t) - x_\lambda)$ and integrating on $[s,t]$, we get

$$\tfrac{1}{2}\|y_\lambda(t) - x_\lambda\|^2 \le \tfrac{1}{2}\|y_\lambda(s) - x_\lambda\|^2 e^{\omega(t-s)}$$
$$+ \int_s^t e^{\omega(t-\tau)}(f(\tau) - w, J(y_\lambda(\tau) - x))\, d\tau,$$

and, letting $\lambda \to 0$,

$$\tfrac{1}{2}\|y(t) - x\|^2 \le \tfrac{1}{2}\|y(s) - x\|^2 e^{\omega(t-s)}$$
$$+ \int_s^t e^{\omega(t-\tau)}(f(\tau) - w, J(y(\tau) - x))\, d\tau,$$

because J is continuous. This yields

$$\left(\frac{y(t) - y(s)}{t - s}, J(y(s) - x)\right)$$
$$\le \frac{1}{2}\|y(s) - x\|^2 (e^{\omega(t-s)} - 1)(t - s)^{-1}$$
$$+ \frac{1}{t - s}\int_s^t e^{\omega(t-\tau)}(f(\tau) - w, J(y(\tau) - x))\, d\tau. \quad (1.58\text{b})$$

By estimate (1.57), we see that y is absolutely continuous on $[0,T]$ and $dy/dt \in L^\infty(0,T;X)$. Hence, y is a.e. differentiable on $(0,T)$. If $s = t_0$ is a point where y is differentiable, by (1.58b) we see that

$$\left(f(t_0) + \frac{dy}{dt}(t_0) - w + \omega(y(t_0) - x), J(y(t_0) - x)\right) \ge 0$$
$$\forall [x,w] \in A.$$

Since $A + \omega I$ is m-accretive, this implies that

$$f(t_0) - \frac{dy}{dt}(t_0) \in Ay(t_0).$$

Hence, y is the strong solution in problem (1.1). ■

4.1.3. The Cauchy Problem Associated with Demicontinuous Monotone Operators

We are given a Hilbert space H and a reflexive Banach space V such that $V \subset H$ continuously and densely. Denote by V' the dual space. Then, identifying H with its own dual, we may write

$$V \subset H \subset V'$$

algebraically and topologically.

The norms of V and H will be denoted $\|\cdot\|$ and $|\cdot|$, respectively. We shall denote by (v_1, v_2) the pairing between $v_1 \in V'$ and $v_2 \in V$; if $v_1, v_2 \in H$, this is the ordinary inner product in H. Finally, we shall denote by $\|\cdot\|_*$ the norm of V' (which is the dual norm).

Besides these spaces we are given a single valued, monotone operator $A: V \to V'$. We shall assume that A is demicontinuous and coercive from V to V'.

We begin with the following simple application of Theorem 1.6.

Theorem 1.8. *Let $f \in W^{1,1}([0,T]; H)$ and $y_0 \in V$ be such that $Ay_0 \in H$. Then there exists one and only one function $y: [0,T] \to V$ that satisfies*

$$y \in W^{1,\infty}([0,T]; H), \qquad Ay \in L^\infty(0,T; H), \tag{1.59}$$

$$\frac{dy}{dt}(t) + Ay(t) = f(t) \qquad \text{a.e. } t \in (0,T),$$

$$y(0) = y_0. \tag{1.60}$$

Moreover, y is everywhere differentiable from the right (in H) and

$$\frac{d^+}{dt}y(t) + Ay(t) = f(t) \qquad \forall t \in [0,T).$$

Proof. Define the operator $A_H: H \to H$,

$$A_H u = Au \qquad \forall u \in D(A_H) = \{u \in V;\ Au \in H\}. \tag{1.61}$$

By hypothesis, the operator $u \to u + Au$ is monotone, demicontinuous, and coercive from V to V'. Hence, it is surjective (see, e.g., Corollary 1.3 in Chapter 2) and so A_H is m-accretive (maximal monotone) in $H \times H$. Then we may apply Theorem 1.6 to conclude the proof. ∎

Now, we shall use Theorem 1.8 to deduce a classical result due to J. L. Lions [1].

Theorem 1.9. *Let $A: V \to V'$ be a demicontinuous monotone operator that satisfies the conditions*

$$(Au, u) \geq \omega \|u\|^p + C_1 \qquad \forall u \in V, \tag{1.62}$$

$$\|Au\|_* \leq C_2(1 + \|u\|^{p-1}) \qquad \forall u \in V, \tag{1.63}$$

where $\omega > 0$ and $p > 2$. Given $y_0 \in H$ and $f \in L^q(0, T; V')$, $1/p + 1/q = 1$, there exists a unique absolutely continuous function $y:[0, T] \to V'$ that satisfies

$$y \in C([0,T]; H) \cap L^p(0,T;V) \cap W^{1,q}([0,T];V'), \tag{1.64}$$

$$\frac{dy}{dt}(t) + Ay(t) = f(t) \qquad \text{a.e. } t \in (0,T), \qquad y(0) = y_0, \tag{1.65}$$

where d/dt is considered in the strong topology of V'.

Proof. Assume that $y_0 \in D(A_H)$ and $f \in W^{1,1}([0, T]; H)$. Then, by Theorem 1.8, there is $y \in W^{1,\infty}([0, T]; H)$ with $Ay \in L^\infty(0, T; H)$ satisfying (1.65). Then, by assumption (1.62), we have

$$\frac{1}{2}\frac{d}{dt}|y(t)|^2 + \omega\|y(t)\|^p \leq \|f(t)\|_*\|y(t)\| \qquad \text{a.e. } t \in (0,T),$$

and therefore

$$|y(t)|^2 + \int_0^t \|y(s)\|^p\, ds \leq C\left(|y_0|^2 + \int_0^t \|f(s)\|_*^q\, ds\right)$$
$$\forall t \in [0,T]. \tag{1.66}$$

Then, by (1.63), we get

$$\int_0^T \left\|\frac{dy}{dt}(t)\right\|_*^q dt \leq C\left(\|y_0\|^2 + \int_0^T \|f(t)\|_*^q\, dt\right). \tag{1.67}$$

(We shall denote by C several positive constants independent of y_0 and f.)

Let us show now that $D(A_H)$ is a dense subset of H. Indeed, if x is any element of H we set $x_\varepsilon = (I + \varepsilon A_H)^{-1}x$ (I is the unity operator in H). Multiplying the equation $x_\varepsilon + \varepsilon A x_\varepsilon = x$ by x_ε, it follows by (1.62) and

(1.63) that

$$|x_\varepsilon|^2 + \omega\varepsilon\|x_\varepsilon\|^p \le |x_\varepsilon|\,|x| + C\varepsilon \qquad \forall \varepsilon > 0,$$

and

$$\|x_\varepsilon - x\|_* \le \varepsilon\|Ax_\varepsilon\|_* \le C\varepsilon\left(\|x_\varepsilon\|^{p-1} + 1\right) \qquad \forall \varepsilon > 0.$$

Hence, $\{x_\varepsilon\}$ is bounded in H and $x_\varepsilon \to x$ in V' as $\varepsilon \to 0$. Hence, $x_\varepsilon \rightharpoonup x$ in H as $\varepsilon \to 0$, which implies that $D(A_H)$ is dense in H.

Now, let $y_0 \in H$ and $f \in L^q(0,T;V')$. Then there are the sequences $\{y_0^n\} \subset D(A_H)$, $\{f_n\} \subset W^{1,1}([0,T];H)$ such that

$$y_0^n \to y_0 \quad \text{in } H, \qquad f_n \to f \quad \text{in } L^q(0,T;V'),$$

as $n \to \infty$. Let $y_n \in W^{1,\infty}([0,T];H)$ be the solution to problem (1.65), where $y_0 = y_0^n$ and $f = f_n$. Since A is monotone, we have

$$\frac{1}{2}\frac{d}{dt}|y_n(t) - y_m(t)|^2 \le (f_n(t) - f_m(t), y_n(t) - y_m(t)) \qquad \text{a.e. } t \in (0,T).$$

Integrating from 0 to t, we get

$$|y_n(t) - y_m(t)|^2 \le |y_n^0 - y_m^0|^2 + 2\left(\int_0^t \|f_n(s) - f_m(s)\|_*^q\,ds\right)^{1/q} \times \left(\int_0^t \|y_n(s) - y_m(s)\|^p\,ds\right)^{1/p}. \tag{1.68}$$

On the other hand, it follows by estimates (1.66) and (1.67) that $\{y_n\}$ is bounded in $L^p(0,T;V)$ and $\{dy_n/dt\}$ is bounded in $L^q(0,T;V')$. Then, it follows by (1.68) that $y(t) = \lim_{n\to\infty} y_n(t)$ exists in H uniformly in t on $[0,T]$. Moreover, extracting a further subsequence if necessary, we have

$$y_n \to y \qquad \text{weakly in } L^p(0,T;V),$$
$$\frac{dy_n}{dt} \to \frac{dy}{dt} \qquad \text{weakly in } L^q(0,T;V'),$$

where dy/dt is considered in the sense of V'-valued distributions on $(0,T)$. In particular, we have proved that $y \in C([0,T];H) \cap L^p(0,T;V) \cap W^{1,q}([0,T];V')$. It remains to prove that y satisfies a.e. on $(0,T)$ Eq. (1.65).

Let $x \in V$ be arbitrary but fixed. Multiplying the equation

$$\frac{dy_n}{dt} + Ay_n = f_n \qquad \text{a.e. } t \in (0,T)$$

by $y_n - x$ and integrating on (s, t), we get

$$\tfrac{1}{2}\left(|y_n(t) - x|^2 - |y_n(s) - x|^2\right) \le \int_s^t (f_n(\tau) - Ax, y_n(\tau) - x)\, d\tau.$$

Letting $n \to \infty$ yields

$$\tfrac{1}{2}\left(|y(t) - x|^2 - |y(s) - x|^2\right) \le \int_s^t (f(\tau) - Ax, y(\tau) - x)\, d\tau.$$

Hence,

$$\left(\frac{(y(t) - y(s)}{t - s}, y(s) - x\right) \le \frac{1}{t - s} \int_s^t (f(\tau) - Ax, y(\tau) - x)\, d\tau. \tag{1.69}$$

We know that y is a.e. differentiable from $(0, T)$ into V' and

$$f(t_0) = \lim_{h \downarrow 0} \frac{1}{h} \int_{t_0}^{t_0 + h} f(s)\, ds \qquad \text{a.e. } t_0 \in (0, T).$$

Let t_0 be such a point where y is differentiable. By (1.69), it follows that

$$\left(\frac{dy}{dt}(t_0) - f(t_0) + Ax, y(t_0) - x\right) \le 0,$$

and since x is arbitrary in V and A is maximal monotone in $V \times V'$ this implies that

$$\frac{dy}{dt}(t_0) + Ay(t_0) = f(t_0),$$

thereby completing the proof. ■

Remark 1.3. Theorm 1.9 applies neatly to the parabolic boundary value problem

$$\frac{dy}{dt} + \sum_{|\alpha| \le m} (-1)^{\alpha} D^{\alpha} A_{\alpha}(x, y, \ldots, D_y^m) = f \qquad \text{in } \Omega \times (0, T),$$

$$D^{\beta} y = 0 \qquad \text{in } \partial\Omega \times (0, T), \qquad |\beta| \le m - 1,$$

$$y(x, 0) = y_0(x), \qquad x \in \Omega,$$

where $f \in L^q(0, T; W^{-m,q}(\Omega))$, $y_0 \in L^2(\Omega)$ and the A_α satisfy conditions (i)–(iii) of Proposition 1.2 in Chapter 2.

Theorem 1.9 remains true for time-dependent operator $A(t): V \to V'$ satisfying assumptions (1.62) and (1.63). More precisely, we have:

Theorem 1.9′. *Let $\{A(t);\ t \in [0,T]\}$ be a family of nonlinear, monotone, and demicontinuosu operators from V to V' satisfying the assumptions:*

(i) *The function $t \to A(t)u(t)$ is measurable from $[0,T]$ to V' for every measurable function $u: [0,T] \to V$;*
(ii) $(A(t)u, u) \geq \omega\|u\|^p + C_1 \ \forall u \in V,\ t \in [0,T]$;
(iii) $\|A(t)u\|_* \leq C_2(1 + \|u\|^{p-1}) \ \ \forall u \in V,\ \ t \in [0,T]$, *where* $\omega > 0$, $p \geq 2$.

Then for every $y_0 \in H$ and $f \in L^q(0,T;V')$ $1/p + 1/q = 1$, there is a unique absolutely continuous function $y \in W^{1,q}([0,T];V')$ that satisfies

$$y \in C([0,T];H) \cap L^p(0,T;V),$$
$$\frac{dy}{dt}(t) + A(t)y(t) = f(t) \qquad \text{a.e. } t \in (0,T),$$
$$y(0) = y_0. \tag{1.70}$$

Proof. Consider the spaces

$$\mathscr{V} = L^p(0,T;V), \qquad \mathscr{H} = L^2(0,T;H), \qquad \mathscr{V}' = L^q(0,T;V').$$

Clearly, $\mathscr{V}$ and $\mathscr{V}'$ are dual pairs and

$$\mathscr{V} \subset \mathscr{H} \subset \mathscr{V}'$$

algebraically and topologically.

Let $y_0 \in H$ be arbitrary and fixed and let $B: \mathscr{V} \to \mathscr{V}'$ be the operator

$$Bu = \frac{du}{dt}, \qquad u \in D(B) = \left\{u \in \mathscr{V};\ \frac{du}{dt} \in \mathscr{V}',\ u(0) = y_0\right\},$$

where d/dt is considered in the sense of vectorial distributions on $(0,T)$. We note that $D(B) \subset W^{1,q}(0,T;V') \cap L^p(0,T;V) \subset C([0,T];H)$, so that $y(0) = y_0$ makes sense.

Let us check that B is maximal monotone in $\mathscr{V} \times \mathscr{V}'$. Since B is clearly monotone, it suffices to show that $R(B + \phi) = \mathscr{V}'$, where

$$\phi(u(t)) = F(u(t))\|u(t)\|^{p-2}, \qquad u \in \mathscr{V},$$

and $F: V \to V'$ is the duality mapping of V. Indeed, for every $f \in \mathscr{V}'$ the equation

$$Bu + \phi(u) = f,$$

or equivalently

$$\frac{du}{dt} + F(u)\|u\|^{p-2} = f \qquad \text{in } [0,T], \qquad u(0) = y_0,$$

has by virtue of Theorem 1.9 a unique solution

$$u \in C([0,T];H) \cap L^p(0,T;V), \qquad \frac{du}{dt} \in L^q(0,T;V').$$

(Renorming the spaces V and V', we may assume that F is demicontinuous and that so is the operator $u \to F(u)\|u\|^{p-2}$.) Hence, B is maximal monotone in $\mathscr{V} \times \mathscr{V}'$.

Define the operator $A_0: \mathscr{V} \to \mathscr{V}'$ by

$$(A_0 u)(t) = A(t)u(t) \qquad \text{a.e. } t \in (0,T).$$

Clearly, A_0 is monotone, demicontinuous, and coercive from $\mathscr{V}$ to $\mathscr{V}'$. Then, by Corollaries 1.2 and 1.5 of Chapter 2, $A_0 + B$ is maximal monotone and surjective. Hence, $R(A_0 + B) = \mathscr{V}'$, which completes the proof. ■

4.1.4. *Continuous Semigroups of Contractions*

Definition 1.3. Let C be a closed subset of a Banach space X. A *continuous semigroup of contractions on* C is a family of mappings $\{S(t); t \geq 0\}$ that maps C into itself with the properties:

(i) $S(t+s)x = S(t)S(s)x \ \forall x \in C,\ t, s \geq 0$;

(ii) $S(0)x = x \ \forall x \in C$;

(iii) For every $x \in C$, the function $t \to S(t)x$ is continuous on $[0,\infty)$.

(iv) $\|S(t)x - s(t)y\| \leq \|x - y\| \ \forall t \geq 0;\ x, y \in C$.

More generally, if instead of (iv) we have

(iv) $\|S(t)x - S(t)y\| \leq e^{\omega t}\|x - y\| \ \forall t \geq 0;\ x, y \in C$,
we say that $S(t)$ is a continuous ω-quasicontractive semigroup on C. The operator $A_0: D(A) \subset C \to X$ defined by

$$A_0 x = \lim_{t \downarrow 0} \frac{S(t)x - x}{t}, \qquad x \in D(A_0), \tag{1.71}$$

where $D(A_0)$ is the set of all $x \in C$ for which the limit (1.71) exists, is called the *infinitesimal generator* of the semigroup $S(t)$.

There is a close relationship between the continuous semigroups of contractions and accretive operators. Indeed, it is easily seen that $-A_0$ is accretive in $X \times X$. More generally, if $S(t)$ is quasicontractive, then $-A_0$ is ω-accretive. Keeping in mind the theory of C_0-semigroups of contractions, one might suspect that there is an one-to-one correspondence between the class of continuous semigroups of contractions and that of m-accretive subsets.

As seen in Theorem 1.3, if X is a Banach space and A is an ω-accretive mapping satisfying the range condition (1.14), then for every $y_0 \in \overline{D(A)}$ the Cauchy problem (1.15) has a unique mild solution $y(t) = S_A(t)y_0$ given by the exponential formula (1.16), i.e.,

$$S_A(t)y_0 = \lim_{n \to \infty} \left(I + \frac{t}{n}A\right)^{-n} y_0. \tag{1.72}$$

For this reason, we shall also denote $S_A(t)$ by e^{-At}. We have:

Proposition 1.2. *$S_A(t)$ is a continuous ω-quasicontractive semigroup on $C = \overline{D(A)}$.*

Proof. It is obvious that conditions (ii)–(iv) are satisfied as a consequence of the existence theorem, Theorem 1.3. To prove (i), we note that, for a fixed $s > 0$, $y_1(t) = S_A(t+s)x$ and $y_2(t) = S_A(t)S_A(s)x$ are both mild solutions to problem

$$\frac{dy}{dt} + Ay = 0, \qquad t \geq 0,$$
$$y(0) = S_A(s)x,$$

and so by uniqueness $y_1 \equiv y_2$.

Let us assume now that X, X^* are uniformly convex and that A is an ω-accretive set that is closed and satisfies the condition (1.52), i.e.,

$$\overline{\operatorname{conv} D(A)} \subset \bigcap_{0<\lambda<\lambda_0} R(I + \lambda A) \qquad \text{for some } \lambda_0 > 0. \tag{1.73}$$

Then, by Theorem 1.7, for every $x \in D(A)$, $S_A(t)x$ is differentiable from the right on $[0, +\infty)$ and

$$-A^0x = \lim_{t \downarrow 0} \frac{S_A(t)x - x}{t} \qquad \forall x \in D(A).$$

Hence $-A^0 \subset A_0$, where A_0 is the infinitesimal generator of $S_A(t)$. ∎

As a matter of fact we may prove in this case the following partial extension of Hille–Yosida–Philips theorem to continuous semigroups of contractions.

Proposition 1.3. *Let X and X^* be uniformly convex and let A be an ω-accretive and closed set of $X \times X$ satisfying condition* (1.73). *Then there is a continuous ω-quasicontractive semigroup $S(t)$ on $\overline{D(A)}$ whose generator A_0 coincides with $-A^0$.*

Proof. For simplicity, we shall assume that $\omega = 0$. We have already seen that A^0 (the minimal section of A) is single valued, everywhere defined on $D(A)$, and $-A_0 x = A^0 x \ \forall x \in D(A)$. Here, A_0 is the infinitesimal generator of the semigroup $S_A(t)$ defined on $\overline{D(A)}$ by the exponential formula (1.72). We shall prove that $D(A_0) = D(A)$. Let $x \in D(A_0)$. Then

$$\limsup_{h \downarrow 0} \frac{\|S_A(t+h)x - S_A(t)x\|}{h} < \infty \qquad \forall t \geq 0,$$

and by the semigroup property (i) it follows that $t \to S_A(t)x$ is Lipschitz on every compact interval $[0, T]$. Hence, $t \to S_A(t)x$ is a.e. differentiable on $(0, \infty)$ and

$$\frac{d}{dt} S_A(t)x = A_0 S_A(t)x \qquad \text{a.e. } t > 0.$$

Now, since $y(t) = S_A(t)x$ is a mild solution to (1.15) that is a.e. differentiable and $(d/dt)\, y(0) = A_0 x$ it follows by Theorem 1.5 that $S_A(t)x$ is a strong solution to (1.15), i.e.,

$$\frac{d}{dt} S_A(t)x + A^0 S_A(t)x \ni 0 \qquad \text{a.e. } t > 0.$$

Now,

$$-A_0 x = \lim_{h \downarrow 0} \frac{1}{h} \int_0^h A^0 S_A(t)x \, dt,$$

and this implies as in the proof of Theorem 1.6 that $x \in D(A)$ and $-A_0 x \in Ax$ (as seen in the proof of Theorem 1.7, we may assume that A is demiclosed). This completes the proof. ■

If X is a Hilbert space it has been proved by Y. Komura [2] that every continuous semigroup of contractions $S(t)$ on a closed convex set $C \subset X$ is generated by an m-accretive set A, i.e., there is an m-accretive set $A \subset X \times X$ such that $-A^0$ is the infinitesimal generator of $S(t)$. This result has been extended by S. Reich [1] to uniformly convex Banach spaces X with uniformly convex dual X^*.

Remark 1.4. There is a simple way due to Dafermos and Slemrod [1] to transform the nonhomogeneous Cauchy problem (1.1) into a homogeneous problem. Let us assume that $f \in L^1(0, \infty; X)$ and denote by Y the product space $Y = X \times L^1(0, \infty; X)$ endowed with the norm

$$\|(x, f)\|_Y = \|x\| + \int_0^\infty \|f(t)\| dt, \qquad (x, f) \in Y.$$

Let $\mathscr{A}: Y \to Y$ be the (multivalued) operator

$$\mathscr{A}(x, f) = \{Ax - f(0), -f'\}, \qquad (x, f) \in D(\mathscr{A}),$$
$$D(\mathscr{A}) = D(A) \times W^{1,1}([0, \infty); X),$$

where $f' = df/dt$. It is readily seen that if y is a solution to problem (1.1) then $Y(t) = (y(t), f_t)$, $f_t(s) = f(t + s)$, is the solution to the homogeneous Cauchy problem

$$\frac{d}{dt} Y(t) + \mathscr{A}Y(t) \ni 0, \qquad t \geq 0,$$
$$Y(0) = (y_0, f).$$

On the other hand, if A is ω-m-accretive in $X \times X$, so is $\mathscr{A}$ in $Y \times Y$.

4.1.5. *Nonlinear Evolution Associated with Subgradient Operators*

We shall study here problem (1.1) in the case in which A is the subdifferential $\partial\varphi$ of a lower semicontinuous convex function φ from a Hilbert space H to $\overline{\mathbf{R}} =]-\infty, +\infty]$. In other words, consider the problem

$$\frac{dy}{dt} + \partial\varphi(y) \ni f(t) \qquad \text{in } (0, T),$$
$$y(0) = y_0, \tag{1.74}$$

in a real Hilbert space H with the scalar product $(\cdot, \cdot)$ and norm $|\cdot|$. It turns out that the nonlinear semigroup generated by $A = \partial\varphi$ on $\overline{D(A)}$ has regularity properties that in the linear case are characteristic to analytic semigroups.

If $\varphi: H \to \overline{\mathbf{R}}$ is a lower semicontinuous, convex function, then its subdifferential $A = \partial\varphi$ is maximal monotone (equivalently, m-accretive) in $H \times H$ and $\overline{D(A)} = \overline{D(\varphi)}$ (see Section 2.1 in Chapter 2). Then, by Theorem 1.2, for every $y_0 \in \overline{D(A)}$ and $f \in L^1(0,T;H)$ the Cauchy problem (1.74) has a unique mild solution $y \in C([0,T];H)$, which is a strong solution if $y_0 \in D(A)$ and $f \in W^{1,1}([0,T];H)$ (Theorem 1.4).

Theorem 1.10 following amounts to saying that y remains a strong solution to (1.74) on every interval $[\delta, T]$ even if $y_0 \notin D(A)$ and f is not absolutely continuous. In other words, the evolution generated by $\partial\varphi$ has a smoothing effect on initial data and on the right hand side f of (1.74).

Theorem 1.10. *Let $f \in L^2(0,T;H)$ and $y_0 \in \overline{D(A)}$. Then the mild solution y to problem* (1.1) *belongs to $W^{1,2}([\delta,T];H)$ for every $0 < \delta < T$, and*

$$y(t) \in D(A) \qquad \text{a.e. } t \in (0,T), \tag{1.75}$$

$$t^{1/2}\frac{dy}{dt} \in L^2(0,T;H) \qquad \varphi(u) \in L^1(0,T), \tag{1.76}$$

$$\frac{dy}{dt}(t) + \partial\varphi(y(t)) \ni f(t) \qquad \text{a.e. } t \in (0,T). \tag{1.77}$$

Moreover, if $y_0 \in D(\varphi)$ then

$$\frac{dy}{dt} \in L^2(0,T;H), \qquad \varphi(y) \in W^{1,1}([0,T]). \tag{1.78}$$

The main ingredient of the proof is the following chain rule differentiation lemma.

Lemma 1.4. *Let $u \in W^{1,2}([0,T];H)$ and $g \in L^2(0,T;H)$ be such that $g(t) \in \partial\varphi(u(t))$ a.e. $t \in (0,T)$. Then the function $t \to \varphi(u(t))$ is absolutely continuous on $[0,T]$ and*

$$\frac{d}{dt}\varphi(u(t)) = \left(g(t), \frac{du}{dt}(t)\right) \qquad \text{a.e. } t \in (0,T). \tag{1.79}$$

Proof. Let φ_λ be the regularization of φ, i.e.,

$$\varphi_\lambda(u) = \inf\left\{\frac{|u-v|^2}{2\lambda} + \varphi(v); v \in H\right\}, \qquad u \in H, \quad \lambda > 0.$$

We recall (see Theorem 2.2 in Chapter 2) that φ_λ is Fréchet differentiable on H and

$$\nabla\varphi_\lambda = (\partial\varphi)_\lambda = \lambda^{-1}\big(I - (I + \lambda\partial\varphi)^{-1}\big), \qquad \lambda > 0.$$

Obviously, the function $t \to \varphi_\lambda(u(t))$ is absolutely continuous (in fact, it belongs to $W^{1,2}([0,T]; H)$) and

$$\frac{d}{dt}\varphi_\lambda(u(t)) = \left((\partial\varphi)_\lambda(u(t)), \frac{du}{dt}(t)\right) \qquad \text{a.e. } t \in (0,T).$$

Hence

$$\varphi_\lambda(u(t)) - \varphi_\lambda(u(s)) = \int_s^t \left((\partial\varphi)_\lambda(u(\tau)), \frac{du}{d\tau}(\tau)\right) d\tau \qquad \forall s < t,$$

and, letting λ tend to zero,

$$\varphi(u(t)) - \varphi(u(s)) = \int_s^t \left((\partial\varphi)^0(u(\tau)), \frac{du}{d\tau}(\tau)\right) d\tau, \qquad 0 \le s < t.$$

By the Lebesgue dominated convergence theorem, the function $t \to (\partial\varphi)^0(u(t))$ is in $L^2(0,T; H)$ and so $t \to \varphi(u(t))$ is absolutely continuous on $[0,T]$. Let t_0 be such that $\varphi(u(t))$ is differentiable at $t = t_0$. We have

$$\varphi(u(t_0)) \le \varphi(v) + (g(t_0), u(t_0) - v) \qquad \forall v \in H.$$

This yields, for $v = u(t_0 - \varepsilon)$,

$$\frac{d}{dt}\varphi(u(t_0)) \le \left(g(t_0), \frac{du}{dt}(t_0)\right).$$

Now, by taking $v = u(t_0 + \varepsilon)$ we get the opposite inequality, and so (1.79) follows. ■

Proof of Theorem 1.10. Let x_0 be an element of $D(\partial\varphi)$ and $y_0 \in \partial\varphi(x_0)$. If we replace the function φ by $\tilde{\varphi}(y) = \varphi(y) - \varphi(x_0) - (y_0, u - x_0)$, Eq. (1.74) reads

$$\frac{dy}{dt} + \partial\tilde{\varphi}(y) \ni f(t) - y_0.$$

Hence, without any loss of generality, we may assume that

$$\min\{\varphi(u); u \in H\} = \varphi(x_0) = 0.$$

Let us assume first that $y_0 \in D(\partial\varphi)$ and $f \in W^{1,2}([0,T];H)$, i.e., $df/dt \in L^2(0,T;H)$. Then, by Theorem 1.2, the Cauchy problem (1.74) has a unique strong solution $y \in W^{1,\infty}([0,T];H)$. The idea of the proof is to obtain *a priori* estimates in $W^{1,2}([\delta,T];H)$ for y, and after this to pass to the limit together with the initial values and forcing term f.

To this end, we multiply Eq. (1.74) by $t\,dy/dt$. By Lemma 1.4, we have

$$t\left|\frac{dy}{dt}(t)\right|^2 + t\frac{d}{dt}\varphi(y(t)) = t\left(f(t), \frac{dy}{dt}(t)\right) \qquad \text{a.e. } t \in (0,T).$$

Hence

$$\int_0^T t\left|\frac{dy}{dt}(t)\right|^2 dt + T\varphi(y(T)) = \int_0^T t\left(f(t), \frac{dy}{dt}(t)\right) dt + \int_0^T \varphi(y(t))\,dt$$

and, therefore,

$$\int_0^T t\left|\frac{dy}{dt}(t)\right|^2 dt \leq \int_0^T t|f(t)|^2\,dt + 2\int_0^T \varphi(y(t))\,dt \tag{1.80}$$

because $\varphi \geq 0$ in H.

Next, we use the obvious inequality

$$\varphi(y(t)) \leq (w(t), y(t) - x_0) \qquad \forall w(t) \in \partial\varphi(y(t))$$

to get

$$\varphi(y(t)) \leq \left(f(t) - \frac{dy}{dt}(t), y(t) - x_0\right) \qquad \text{a.e. } t \in (0,T),$$

which yields

$$\int_0^T \varphi(y(t))\,dt \leq \tfrac{1}{2}|y(0) - x_0|^2 + \int_0^T |f(t)|\,|y(t) - x_0|\,dt.$$

Now, multiplying Eq. (1.74) by $y(t) - x_0$ and integrating on $[0,t]$ yields

$$|y(t) - x_0| \leq |y(0) - x_0| + \int_0^t |f(s)|\,ds \qquad \forall t \in [0,T].$$

Hence,

$$2\int_0^T \varphi(y(t))\,dt \leq \left(|y(0) - x_0| + \int_0^T |f(t)|\,dt\right)^2. \tag{1.81}$$

Now, combining estimates (1.80) and (1.81), we get

$$\int_0^T t\left|\frac{dy}{dt}(t)\right|^2 dt \le \int_0^T t|f(t)|^2\, dt + 2\left(|y_0 - x_0| + \int_0^T |f(t)|\, dt\right)^2. \quad (1.82)$$

Multiplying Eq. (1.74) by dy/dt and integrating on $(0, t)$, we get

$$\left|\frac{dy}{dt}(t)\right|^2 + \frac{d}{dt}\varphi(y(t)) = \left(f(t), \frac{dy}{dt}(t)\right) \qquad \text{a.e. } t \in (0,T).$$

Hence,

$$\frac{1}{2}\int_0^t \left|\frac{dy}{dt}(s)\right|^2 ds + \varphi(y(t)) \le \frac{1}{2}\int_0^t |f(s)|^2\, ds + \varphi(y_0). \quad (1.83)$$

Now, let us assume that $y_0 \in \overline{D(\partial\varphi)}$ and $f \in L^2(0,T;H)$. Then there exist sequences $\{y_0^n\} \subset D(\partial\varphi)$ and $\{f_n\} \subset W^{1,2}([0,T];H)$ such that $y_0^n \to y_0$ in H and $f_n \to f$ in $L^2(0,T;H)$ as $n \to \infty$. Denote by $y_n \in W^{1,\infty}([0,T];H)$ the corresponding solutions to (1.74). Since $\partial\varphi$ is monotone, we have (see Proposition 1.1)

$$|y_n(t) - y_m(t)| \le |y_0^n - y_0^m| + \int_0^t |f_n(s) - f_m(s)|\, ds.$$

Hence, $y_n \to y$ in $C([0,T];H)$. On the other hand, this clearly implies that

$$\frac{dy_n}{dt} \to \frac{dy}{dt} \qquad \text{in } \mathscr{D}'(0,T;H),$$

and by estimate (1.82) it follows that $t^{1/2}\, dy/dt \in L^2(0,T;H)$. Hence, y is absolutely continuous on every interval $[\delta, T]$ and $y \in W^{1,2}([\delta,T];H)$ for all $0 < \delta < T$.

Moreover, by estimate (1.81) we deduce by virtue of Fatou's lemma that $\varphi(y) \in L^1(0,T)$ and

$$\int_0^T \varphi(y(t))\, dt \le \liminf_{n\to\infty} \int_0^T \varphi(y_n(t))\, dt$$
$$\le \left(|y_0 - x| + \int_0^T |f(t)|\, dt\right)^2.$$

We may infer, therefore, that y satisfies estimates (1.81) and (1.82). Moreover, y satisfies Eq. (1.77). Indeed, we have

$$\tfrac{1}{2}|y_n(t) - x|^2 \le \tfrac{1}{2}|y_n(s) - x|^2 + \int_s^t (f_n(\tau) - w, y_n(\tau) - x)\, d\tau$$

for all $0 \le s < t \le T$ and all $[x, w] \in \partial\varphi$. This yields

$$\tfrac{1}{2}\left(|y(t) - x|^2 - |y(s) - x|^2\right) \le \int_s^t (f(\tau) - w, y(\tau) - x)\, d\tau$$

and, therefore,

$$\left(\frac{y(t) - y(s)}{t - s}, y(s) - x\right) \le \frac{1}{t - s} \int_s^t (f(\tau) - w, y(\tau) - x)\, d\tau.$$

Letting $s \to t$ we get, a.e. $t \in (0, T)$,

$$\left(\frac{dy}{dt}(t), y(t) - x\right) \le (f(t) - w, y(t) - x)$$

for all $[x, w] \in A$, and since $A = \partial\varphi$ is maximal monotone this implies that $y(t) \in D(A)$ and $(d/dt)\, y(t) \in f(t) - Ay(t)$ a.e. $t \in (0, T)$, as desired.

Assume now that $y_0 \in D(\varphi)$. We choose in this case $y_0^n = (I + n^{-1}\, \partial\varphi)^{-1} y_0$ and note that $y_0^n \to y_0$ as $n \to \infty$, and

$$\varphi(y_0^n) \le \varphi(y_0) + \left(\partial\varphi_n(y_0), (I + n^{-1}\, \partial\varphi)^{-1} y_0 - y_0\right) \le \varphi(y_0) \qquad \forall n \in \mathbf{N}^*.$$

Then, by estimate (1.83), we have

$$\frac{1}{2} \int_0^t \left|\frac{dy_n}{ds}\right|^2 ds + \varphi(y_n(t)) \le \frac{1}{2} \int_0^t \left|\frac{df_n}{ds}(s)\right|^2 ds + \varphi(y_0)$$

and, letting $n \to \infty$, we find the estimate

$$\frac{1}{2} \int_0^t \left|\frac{dy}{ds}\right|^2 ds + \varphi(y(t)) \le \frac{1}{2} \int_0^t \left|\frac{df}{ds}(s)\right|^2 ds + \varphi(y_0), \qquad t \in [0, T], \tag{1.84}$$

because φ is lower semicontinuous.

This completes the proof of Theorem 1.10. ∎

In the sequel, we shall denote by $W^{1,p}((0, T]; H)$, $1 \le p \le \infty$, the space of all $y \in L^p(0, T; H)$ such that $dy/dt \in L^p(\delta, T; H)$ for every $\delta \in (0, T)$.

Theorem 1.11. *Assume that* $y_0 \in \overline{D(A)}$ *and* $f \in W^{1,1}([0,T]; H)$. *Then the solution* y *to problem* (1.74) *satisfies*

$$t\frac{dy}{dt} \in L^\infty(0,\infty; H), \qquad y(t) \in D(A), \qquad \forall t \in (0,T], \tag{1.85}$$

$$\frac{d^+}{dt}y(t) + (Ay(t) - f(t))^0 = 0 \qquad \forall t \in (0,T]. \tag{1.86}$$

Proof. By Eq. (1.74), we have

$$\frac{d}{dt}|y(t+h) - y(t)| \le |f(t+h) - f(t)| \qquad \text{a.e. } t, t+h \in (0,T).$$

Hence,

$$\left|\frac{dy}{dt}(t)\right| \le \left|\frac{dy}{dt}(s)\right| + \int_s^t \left|\frac{df}{d\tau}(\tau)\right| d\tau \qquad \text{a.e. } 0 < s < t < T. \tag{1.87}$$

This yields

$$\frac{1}{2}s\left|\frac{dy}{dt}(t)\right|^2 \le s\left|\frac{dy}{ds}(s)\right|^2 + s\left(\int_s^t \left|\frac{df}{d\tau}\right| d\tau\right)^2 \qquad \text{a.e. } 0 < s < t < T.$$

Then integrating from 0 to t and using estimate (1.82), we get

$$t\left|\frac{dy}{dt}(t)\right| \le \left(\int_0^t s|f(s)|^2\, ds + 2\left(|y(0) - x_0| + \int_0^t |f(s)|\, ds\right)^2 + \frac{t^2}{2}\left(\int_0^t \left|\frac{df}{d\tau}\right| d\tau\right)^2\right)^{1/2} \qquad \text{a.e. } t \in (0,T). \tag{1.88}$$

In particular, it follows by (1.88) that

$$\limsup_{\substack{h \to 0 \\ h > 0}} \left|\frac{y(t+h) - y(t)}{h}\right| < \infty \qquad \forall t \in [0,T].$$

Hence, the weak closure E of $\{(y(t+h) - y(t))/h\}$ for $h \to 0$ is nonempty for every $t \in [0,T)$. Let η be an element of E. We have proved earlier the inequality

$$\left(\frac{y(t+h) - y(t)}{h}, y(t) - x\right) \le \frac{1}{h}\int_t^{t+h} (f(\tau) - w, u(\tau) - x)\, d\tau$$

for all $[x, w] \in \partial\varphi$ and $t, t+h \in (0, T)$. This yields

$$(\eta, y(t) - x) \le (f(t) - w, u(t) - x) \qquad \forall t \in (0, T),$$

and since $[x, w]$ is arbitrary in $\partial\varphi$ we conclude that $y(t) \in D(A)$ and $f(t) - \eta \in Ay(t)$. Hence, $y(t) \in D(A)$ for every $t \in (0, T)$. Then, by Theorem 1.6, it follows that

$$\frac{d^+}{dt} y(t) + (Ay(t) - f(t))^0 = 0 \qquad \forall t \in (0, T), \tag{1.89}$$

because for every $\varepsilon > 0$ sufficiently small $y(\varepsilon) \in D(A)$ and so (1.89) holds for all $t > \varepsilon$.

This completes the proof of Theorem 1.11. ∎

In particular, it follows by Theorem 1.11 that the semigroup $S(t)$ generated by $A = \partial\varphi$ on $\overline{D(A)}$ maps $\overline{D(A)}$ into $D(A)$ for all $t > 0$ and

$$t\left|\frac{d^+}{dt} S(t) y_0\right| \le C \qquad \forall t > 0.$$

More precisely, we have:

Corollary 1.2. *Let $S(t)$ be the continuous semigroup of contractions generated by $A = \partial\varphi$. Then $S(t)\overline{D(A)} \subset D(A)$ for all $t > 0$, and*

$$\left|\frac{d^+}{dt} S(t) y_0\right| = |A^0 S(t) y_0| \le |A^0 x| + \frac{1}{t}|x - y_0| \qquad \forall t > 0, \tag{1.90}$$

for all $y_0 \in \overline{D(A)}$ and $x \in D(A)$.

Proof. Multiplying Eq. (1.74) (where $f \equiv 0$) by dy/dt and integrating on $(0, t)$, we get

$$\int_0^t s\left|\frac{dy}{ds}(s)\right|^2 ds + t\varphi(y(t)) \le \int_0^T \varphi(y(s))\, ds \qquad \forall t > 0.$$

Next, we multiply the same equation by $y(t) - x$ and integrate on $(0, t)$. We get

$$\tfrac{1}{2}|y(t) - x|^2 + \int_0^t \varphi(y(s))\, ds \le \tfrac{1}{2}|y(0) - x|^2 + t\varphi(x).$$

Combining these two inequalities, we obtain

$$\begin{aligned}\int_0^t s\left|\frac{dy}{ds}(s)\right|^2 ds &\le \frac{1}{2}\left(|y(0)-x|^2 - |y(t)-x|^2\right)\\ &\quad + t(\varphi(x) - \varphi(y(t)))\\ &\le \frac{1}{2}\left(|y(0)-x|^2 - |y(t)-x|^2\right)\\ &\quad + t(A^0x, x - y(t))\\ &\le \frac{1}{2}|y(0)-x|^2 + \frac{t^2|A^0x|^2}{2} \qquad \forall t > 0.\end{aligned}$$

Since by (1.87) the function $t \to |(dy/dt)(t)|$ (and consequently $t \to |(d^+/dt)\, y(t)|$) is monotonically decreasing, this implies (1.90), thereby completing the proof. ∎

Remark 1.5. Theorems 1.10 and 1.11 clearly remain true for equations of the form

$$\frac{dy}{dt} + \partial\varphi(y) - \omega y \ni f \qquad \text{a.e. in } (0,T),$$
$$y(0) = y_0,$$

where $\omega \in \mathbf{R}$. The proof is exactly the same.

4.2. Approximation and Convergence of Nonlinear Evolutions and Semigroups

4.2.1. The Trotter–Kato Theorem for Nonlinear Evolutions

Consider in a general Banach space X a sequence A_n of subsets of $X \times X$. The set $\liminf A_n$ is defined as $[x, y] \in \liminf A_n$ if there are sequences x_n, y_n such that $y_n \in A_n x_n$, $x_n \to x$ and $y_n \to y$ as $n \to \infty$.

If the A_n are ω-m-accretive there is a simple resolvent characterization of $\liminf A_n$.

Proposition 2.1. *Let $A_n + \omega I$ be m-accretive for $n = 1, 2, \ldots$. Then $A \subset \liminf A_n$ if and only if*

$$\lim_{n\to\infty} (I + \lambda A_n)^{-1} x = (I + \lambda A)^{-1} x \qquad \forall x \in X, \tag{2.1}$$

for $0 < \lambda < \omega^{-1}$.

Proof. Assume that (2.1) holds and let $[x, y] \in A$ be arbitrary but fixed. Then we have

$$(I + \lambda A)^{-1}(x + \lambda y) = x \qquad \forall \lambda \in (0, \omega^{-1})$$

and, by (2.1),

$$(I + \lambda A_n)^{-1}(x + \lambda y) \to (I + \lambda A)^{-1}(x + \lambda y) = x.$$

In other words, $x_n = (I + \lambda A_n)^{-1}(x + \lambda y) \to x$ as $n \to \infty$ and $x_n + \lambda y_n = x + \lambda y$, $y_n \in Ax_n$. Hence, $y_n \to y$ as $n \to \infty$, and so $[x, y] \in \liminf A_n$.

Conversely, let us assume now that $A \subset \liminf A_n$. Let x be arbitrary in X and let $x_0 = (I + \lambda A)^{-1}x$, i.e.,

$$x_0 + \lambda y_0 = x, \qquad \text{where } y_0 \in Ax_0.$$

Then, there are $[x_n, y_n] \in A_n$ such that $x_n \to x_0$ and $y_n \to y_0$ as $n \to \infty$. We have

$$x_n + \lambda y_n = z_n \to x_0 + \lambda y_0 = x \qquad \text{as } n \to \infty.$$

Hence

$$(I + \lambda A_n)^{-1}x \to x_0 = (I + \lambda A)^{-1}y_0 \qquad \text{for } 0 < \lambda < \omega^{-1},$$

as claimed. ■

Theorem 2.1 following is the nonlinear version of the Trotter–Kato theorem from the theory of C_0-semigroups (see T. Kato [1]).

Theorem 2.1. *Let A_n be ω-accretive in $X \times X$, $f^n \in L^1(0; T; X)$ for $n = 1, 2, \ldots$. Let y_n be the mild solution to*

$$\frac{dy_n}{dt} + A_n y_n \ni f^n \qquad \text{in } [0, T], \qquad y_n(0) = y_0^n. \tag{2.2}$$

Let $A \subset \liminf A_n$ and

$$\lim_{n \to \infty} \left(\int_0^T \|f^n(t) - f(t)\| \, dt + \|y_0^n - y_0\| \right) = 0. \tag{2.3}$$

Then $y_n(t) \to y(t)$ uniformly on $[0, T]$, where y is the mild solution to problem (1.1).

Proof. Let $D^{\varepsilon}_{A^n}(0 = t_0, t_1, \ldots, t_N; f_n^n, \ldots, f_N^n)$ be an ε-discretization of problem (2.2) and let $D^{\varepsilon}_{A}(0 = t_0, t_1, \ldots, t_N; f_1, \ldots, f_N)$ be the corresponding ε-discretization for (1.1). We shall take $t_i = i\varepsilon$ for all i. Let $y_{\varepsilon,n}$ and y_ε be the corresponding ε-approximate solutions, i.e.,

$$y_{\varepsilon,n}(t) = y^i_{\varepsilon,n}, \qquad y_\varepsilon(t) = y^i_\varepsilon, \qquad \text{for } t \in (t_{i-1}, t_i],$$

where $y^0_{\varepsilon,n} = y^n_0$, $y^0_\varepsilon = y_0$, and

$$y^i_{\varepsilon,n} + \varepsilon A_n y^i_{\varepsilon,n} \ni y^{i-1}_{\varepsilon,n} + \varepsilon f^n_i, \qquad i = 1,\dots,N, \tag{2.4}$$

$$y^i_\varepsilon + \varepsilon A y^i_\varepsilon \ni y^{i-1}_\varepsilon + \varepsilon f_i, \qquad i = 1,\dots,N. \tag{2.5}$$

By the definition of lim inf A_n, for every $\eta > 0$ there is $[\bar{y}^i_{\varepsilon,n}, w^i_{\varepsilon,n}] \in A_n$ such that

$$\|\bar{y}^i_{\varepsilon,n} - y^i_\varepsilon\| + \|w^i_{\varepsilon,n} - w^i_\varepsilon\| \le \eta \qquad \text{for } n \ge \delta(\eta, \varepsilon). \tag{2.6}$$

Here, $w^i_\varepsilon = (1/\varepsilon)(y^{i-1}_\varepsilon + \varepsilon f_i - y^i_\varepsilon) \in Ay^i_\varepsilon$. Then, using the ω-accretivity of A_n, by (2.4)–(2.6) it follows that

$$\begin{aligned}\|\bar{y}^i_{\varepsilon,n} - y^i_{\varepsilon,n}\| \le{}& (1-\varepsilon\omega)^{-1}\|\bar{y}^{i-1}_{\varepsilon,n} - y^{i-1}_{\varepsilon,n}\| \\ &+ \varepsilon(1-\varepsilon\omega)^{-1}\|f^n_i - f_i\| + C\varepsilon\eta \qquad \forall i,\end{aligned}$$

for $n \ge \delta(\eta, \varepsilon)$. This yields

$$\|\bar{y}^i_{\varepsilon,n} - y^i_{\varepsilon,n}\| \le C\eta + C\varepsilon \sum_{k=1}^{i} (1-\varepsilon\omega)^{-k}\|f^n_k - f_k\|, \qquad i = 1,\dots,N.$$

Hence,

$$\|y^i_{\varepsilon,n} - y^i_\varepsilon\| \le C\eta + C\varepsilon \sum_{k=1}^{i} (1-\varepsilon\omega)^{-k}\|f^n_k - f_k\|, \qquad i = 1,\dots,N,$$

for $n \ge \delta(\varepsilon, \eta)$.

We have shown, therefore, that, for $n \ge \delta(\varepsilon, \eta)$,

$$\|y_{\varepsilon,n}(t) - y_\varepsilon(t)\| \le C\left(\eta + \int_0^T \|f^n(t) - f(t)\|\,dt\right) \qquad \forall t \in [0,T], \tag{2.7}$$

where C is independent of n and ε.

Now, we have

$$\begin{aligned}\|y_n(t) - y(t)\| \le{}& \|y_n(t) - y_{\varepsilon,n}(t)\| + \|y_{\varepsilon,n}(t) - y_\varepsilon(t)\| \\ &+ \|y_\varepsilon(t) - y(t)\| \qquad \forall t \in [0,T).\end{aligned} \tag{2.8}$$

Let η be arbitrary but fixed. Then, by Theorem 1.1,

$$\|y_\varepsilon(t) - y(t)\| \leq \eta \quad \forall t \in [0, T], \qquad \text{if } 0 < \varepsilon < \varepsilon_0(\eta).$$

Also, by estimate (1.35) in the proof of Theorem 1.1, we have

$$\|y_{\varepsilon,n}(t) - y_n(t)\| \leq \eta \qquad \forall t \in [0, T],$$

for all $0 < \varepsilon < \varepsilon_1(\eta)$ where $\varepsilon_1(\eta)$ does not depend of n. Thus, by (2.7) and (2.8), we have

$$\|y_n(t) - y(t)\| \leq C\left(\eta + \int_0^T \|f^n(t) - f(t)\|\,dt\right) \qquad \forall t \in [0, T]$$

for all n sufficiently large. This completes the proof of Theorem 2.1. ■

Corollary 2.1. *Let A be ω-m-accretive, $f \in L^1(0, T; X)$, and $y_0 \in \overline{D(A)}$. Let $y_\lambda \in C^1([0, T]; X)$ be the solution to the approximating Cauchy problem*

$$\frac{dy}{dt} + A_\lambda y = f \qquad \text{in } [0, T], \qquad y(0) = y_0, \qquad 0 < \lambda < \frac{1}{\omega}, \quad (2.9)$$

where $A_\lambda = \lambda^{-1}(I - (I + \lambda A)^{-1})$. Then $\lim_{\lambda \to 0} y_\lambda(t) = y(t)$ uniformly in t on $[0, T]$, where y is the mild solution to problem (1.1).

Proof. It is easily seen that $A \subset \liminf_{\lambda \to 0} A_\lambda$. Indeed, for $\alpha \in (0, 1/\omega)$ we set

$$x_\lambda = (I + \alpha A_\lambda)^{-1} x, \qquad u = (I + \alpha A)^{-1} x.$$

After some calculation, we see that

$$x_\lambda + \alpha A\left(\left(1 + \frac{\lambda}{\alpha}\right)x - \frac{\lambda}{\alpha}x\right) \ni x.$$

Subtracting this equation from $u + \alpha Au \ni x$ and using the ω-accretivity of A, we get

$$\|x_\lambda - u\|^2 \leq \alpha\omega\left\|\left(1 + \frac{\lambda}{\alpha}\right)x - \frac{\lambda}{\alpha}x - u\right\|^2 + \frac{\lambda}{\alpha}(x_\lambda - u, x - x_\lambda).$$

Hence, $\lim_{\lambda \to 0} x_\lambda = u = (I + \alpha A)^{-1}x$ for $0 < \alpha < 1/\lambda$, and so we may apply Theorem 2.1. ■

Remark 2.1. If X is a Hilbert space and S_n is the semigroup generated by A_n, then condition (2.1) is equivalent to the following one (H. Brézis

[6]): For every $x \in \overline{D(A)}, \exists \{x_n\} \subset D(A_n)$ such that $x_n \to x$ and $S_n(t)x_n \to S(t)x \ \forall t > 0$, where $S(t)$ is the semigroup generated by A.

Theorem 2.1 is useful in proving the convergence of many approximation schemes for problem (1.1). If A is a nonlinear partial differential operator on a certain space of functions defined on a domain $\Omega \subset \mathbf{R}^m$, then very often the A_n arise as finite element approximations of A on a subspace X_n of X (see e.g., H. T. Banks and K. Kunisch [1], and F. Kappel and W. Schappacher [1].

4.2.2. The Nonlinear Chernoff Theorem and Lie–Trotter Products

We shall prove here the nonlinear version of the famous Chernoff theorem (Chernoff [1]), along with some implications for the convergence of the Lie–Trotter product formula for nonlinear semigroups of contractions.

Theorem 2.2. *Let X be a real Banach space, A be an accretive operator satisfying the range condition* (1.14), *and let $C = \overline{D(A)}$ be convex. For each $t > 0$, let $F(t): C \to C$ satisfy*:

(i) $\|F(t)x - F(t)u\| \le \|x - u\| \qquad \forall x, u \in C$ and $t \in [0, T]$;

(ii) $\displaystyle \lim_{t \downarrow 0} \left(I + \lambda \frac{I - F(t)}{t} \right)^{-1} x = (I + \lambda A)^{-1} x \qquad \forall x \in C, \lambda > 0.$

Then, for each $x \in C$ and $t > 0$,

$$\lim_{n \to \infty} \left(F\left(\frac{t}{n}\right) \right)^n x = S_A(t)x, \tag{2.10}$$

uniformly in t on compact intervals.

Here, $S_A(t)$ is the semigroup generated by A on $C = \overline{D(A)}$.

The main ingredient of the proof is the following convergence result.

Proposition 2.2. *Let $C \subset X$ be nonempty, closed, and convex, let $F: C \to C$ nonexpansive, and let $h > 0$. Then the Cauchy problem*

$$\frac{du}{dt} + h^{-1}(I - F)u = 0 \qquad u(0) = x \in C, \tag{2.11}$$

has a unique solution $u \in C^1([0, \infty); X)$, $u(t) \in C$, for all $t \ge 0$.

Moreover, the following estimate holds:

$$\|F^n x - u(t)\| \le \left(\left(n - \frac{t}{h}\right)^2 + n\right)^{1/2} \|x - Fx\| \qquad \forall t \ge 0, \quad (2.12)$$

for all $n \ge 0$. In particular, for $t = nh$ we have

$$\|F^n x - u(nh)\| \le n^{1/2}\|x - Fx\|, \qquad n = 1,2,\dots,\ t \ge 0. \quad (2.13)$$

Proof. The initial value problem (2.11) can be written equivalently as

$$u(t) = e^{-t/h}x + \int_0^t e^{-(t-s)/h}Fu(s)\,ds \qquad \forall t \ge 0,$$

and it has a unique solution $u(t) \in C$ $\forall t \ge 0$, by the Banach fixed point theorem. Making the substitution $t \to t/h$, we can reduce the problem to the case $h = 1$.

Multiplying Eq. (2.11) by $w \in J(u(t) - x)$, we get

$$\frac{d}{dt}\|u(t) - x\| \le \|Fx - x\| \qquad \text{a.e. } t > 0,$$

because $I - F$ is accretive. Hence,

$$\|u(t) - x\| \le t\|Fx - x\| \qquad \forall t \ge 0. \quad (2.14)$$

On the other hand, we have

$$u(t) - F^n x = e^{-t}(x - F^n x) + \int_0^t e^{s-t}(Fu(s) - F^n x)\,ds$$

and

$$\|x - F^n x\| \le \sum_{k=1}^{n} \|F^{k-1}x - F^k x\| \le n\|x - Fx\| \qquad \forall n.$$

Hence,

$$\|u(t) - F^n x\| \le ne^{-t}\|x - Fx\| + \int_0^t e^{s-t}\|u(s) - F^{n-1}x\|\,ds.$$

We set $\varphi_n(t) = \|u(t) - F^n x\|\,\|x - Fx\|^{-1}e^t$. Then, we have

$$\varphi_n(t) \le n + \int_0^t \varphi_{n-1}(x)\,ds \qquad \forall t \ge 0, \quad n = 1,2,\dots, \quad (2.15)$$

and, by (2.14),

$$\varphi_0(t) \le te^t \qquad \forall t \ge 0. \quad (2.16)$$

Solving (2.15), (2.16), we get

$$\begin{aligned}
\varphi_n(t) &\le \sum_{k=1}^{n} \frac{kt^{n-k}}{(n-k)!} + \frac{1}{(n-1)!}\int_0^t (t-s)^{n-1}\varphi_0(s)\,ds \\
&= \sum_{k=1}^{n} \frac{kt^{n-k}}{(n-k)!} + \frac{1}{(n-1)!}\int_0^t (t-s)^{n-1}\sum_{j=0}^{\infty}\frac{s^{j+1}}{j!}\,ds \\
&= \sum_{k=1}^{n} \frac{kt^{n-k}}{(n-k)!} + \sum_{j=0}^{\infty}\frac{1}{(n-1)!j!}\int_0^t (t-s)^{n-1}s^{j+1}\,ds.
\end{aligned}$$

Since

$$\int_0^t (t-s)^{n-1}s^{j+1}\,ds = \frac{t^{n+j+1}(j+1)!(n-1)!}{(n+j+1)!},$$

we get

$$\begin{aligned}
\varphi_n(t) &\le \sum_{k=0}^{n} \frac{(n-k)t^k}{k!} + \sum_{j=0}^{\infty}\frac{(j+1)t^{n+j+1}}{(n+j+1)!} = \sum_{k=0}^{\infty}\frac{(n-k)t^k}{k!} \\
&= \sum_{k=0}^{\infty}\frac{t^k}{k!}|n-k| \le \left(\sum_{k=0}^{\infty}\frac{(n-k)^2t^k}{k!}\right)^{1/2} e^{t/2}.
\end{aligned}$$

Hence

$$\varphi_n(t) \le e^t\big((n-t)^2+t\big)^{1/2} \qquad \forall t \ge 0,$$

as claimed. ■

Proof of Theorem 2.2. We set $A_h = h^{-1}(I - F(h))$ and denote by $S_h(t)$ the semigroup generated by A_h on $C = \overline{D(A)}$. We shall also use the notation

$$J_\lambda = (I+\lambda A)^{-1}, \qquad J_\lambda^h = (I+\lambda A_h)^{-1}.$$

Since $J_\lambda^h x \to J_\lambda x$ $\forall x \in C$, as $h \to 0$, it follows by Theorem 2.1 that for every $x \in C$,

$$S_h(t)x \to S_A(t)x \qquad \text{uniformly in } t \text{ on compact intervals.} \quad (2.17)$$

Next, by Proposition 2.2 we have

$$S_h(nh)x - F^n(h)x\| \le \|S_h(nh)J_\lambda^h x - F^n(h)J_\lambda^h x\| + 2\|x - J_\lambda^h x\|$$
$$\le \|x - J_\lambda^h x\|(2 + \lambda^{-1}hn^{1/2}).$$

Now we fix $x \in D(A)$ and $h = n^{-1}t$. Then the previous inequality yields

$$\|S_{t/n}(t)x - F^n\left(\frac{t}{n}\right)x\|$$
$$\le (2 + \lambda^{-1}tn^{-1/2})(\|x - J_\lambda x\| + \|J_\lambda^{t/n}x - J_\lambda x\|)$$
$$\le (2 + \lambda^{-1}tn^{-1/2})(\lambda|Ax| + \|J_\lambda^{t/n}x - J_\lambda x\|) \qquad \forall t > 0,\ \lambda > 0.$$

Finally,

$$\|S_{t/n}(t)x - F^n\left(\frac{t}{n}\right)x\| \le 2\lambda|Ax| + tn^{-1/2}|Ax|$$
$$+(2 + \lambda^{-1}tn^{-1/2})\|J_\lambda^{t/n}x - J_\lambda x\|$$
$$\forall t > 0,\ \lambda > 0. \tag{2.18}$$

Now fix $\lambda > 0$ such that $2\lambda|\mathrm{Ax}| \le \varepsilon/3$. Then, by (ii), we have

$$(2 + \lambda^{-1}tn^{-1/2})\|J_\lambda^{t/n}x - J_\lambda x\| \le \varepsilon/3 \text{ for } n > N(\varepsilon),$$

and so by (2.17) and (2.18) we conclude that, for $n \to \infty$,

$$F^n\left(\frac{t}{n}\right)x \to S_A(t)x \qquad \text{uniformly in } t \text{ on every } [0,T]. \tag{2.19}$$

Now, since

$$\|S_A(t)x - S_A(t)y\| \le |x - y| \qquad \forall t \ge 0,\ x, y \in C,$$

and

$$\left\|F^n\left(\frac{1}{n}\right)x - F^n\left(\frac{t}{n}\right)y\right\| \le \|x - y\| \qquad \forall t \ge 0,\ x, y \in C,$$

(2.19) extends to all $x \in \overline{D(A)} = C$.

The proof of Theorem 2.2 is complete. ∎

Remark 2.2. The conclusion of Theorem 2.2 remains unchanged if A is ω-accretive, satisfies the range condition (1.14), and $F(t): C \to C$ are Lipschitz with Lipschitz constant $L(t) = 1 + \omega t + 0(t)$ as $t \to 0$. The proof is essentially the same and relies on an appropriate estimate of the form (2.13) for Lipschitz mappings on C.

Given two m-accretive operators $A, B \subset X \times X$ such that $A + B$ is m-accretive, one might expect that

$$S_{A+B}(t)x = \lim_{n \to \infty} \left(S_A\left(\frac{t}{n}\right) S_B\left(\frac{t}{n}\right) \right)^n x \qquad \forall t \geq 0, \tag{2.20}$$

for all $x \in \overline{D(A) \cap D(B)}$. This is the *Lie–Trotter product* formula and one knows that it is true for C_0-semigroups of contractions and in other situations (see A. Pazy [1], p. 92). It is readily seen that (2.20) is equivalent to the convergence of the fractional step method scheme for the Cauchy problem

$$\begin{aligned} &\frac{dy}{dt} + Ay + By \ni 0 \qquad \text{in } [0, T], \\ &\qquad y(0) = y_0, \end{aligned} \tag{2.21}$$

i.e.,

$$\begin{aligned} &\frac{dy}{dt} + Ay \ni 0 \qquad \text{in } [i\varepsilon, (i+1)\varepsilon], i = 0, 1, \ldots, N-1, T = N\varepsilon, \\ &y^+(i\varepsilon) = z(\varepsilon), \qquad i = 0, 1, \ldots, N-1, \\ &y^+(0) = y_0, \end{aligned} \tag{2.22}$$

$$\begin{aligned} &\frac{dz}{dt} + Bz \ni 0 \qquad \text{in } [0, \varepsilon], \\ &z(0) = y^-(i\varepsilon). \end{aligned} \tag{2.23}$$

In a general Banach space, the Lie–Trotter formula (2.20) is not convergent even for regular operators B unless $S_A(t)$ admits a graph infinitesimal generator A, i.e., for all $[x, y] \in A$ there is $x_h \to x$ as $h \to 0$ such that $h^{-1}(x_h - S_A(h)x) \to y$ (Bėnilan and Ismail [1]). However, there are known several situations in which formula (2.20) is true and one is described in Theorem 2.3 following.

Theorem 2.3. *Let X and X^* be uniformly convex and let A, B be m-accretive single valued operators on X such that $A + B$ is m-accretive and $S_A(t), S_B(t)$ map $\overline{D(A) \cap D(B)}$ into itself. Then*

$$S_{A+B}(t)x = \lim_{n \to \infty} \left(S_A\left(\frac{t}{n}\right) S_B\left(\frac{t}{n}\right) \right)^n x \qquad \forall x \in \overline{D(A) \cap D(B)}, \tag{2.24}$$

and the limit is uniform in t on compact intervals.

Proof. We shall verify the hypotheses of Theorem 2.2, where $F(t) = S_A(t)S_B(t)$ and $C = \overline{D(A) \cap D(B)}$. To prove (ii), it suffices to show that

$$\lim_{t \downarrow 0} \frac{x - F(t)x}{x} = Ax + Bx \qquad \forall x \in D(A) \cap D(B). \tag{2.25}$$

Indeed, if

$$x_t = \left(I + \lambda \frac{I - F(t)}{t}\right)^{-1} x$$

and

$$x_0 = (I + \lambda(A + B))^{-1} x$$

then we have

$$x_t + \frac{\lambda}{t}(x_t - F(t)x_t) = x \tag{2.26}$$

and, respectively,

$$x_0 + \lambda A x_0 + \lambda B x_0 = x. \tag{2.27}$$

Subtracting (2.26) from (2.27), we may write

$$x_t - x_0 + \frac{\lambda}{t}((I - F(t)x_t - (I - F(t)x_0))$$
$$+ \lambda\left(Ax_0 + Bx_0 - \frac{x_0 - F(t)x_0}{t}\right) = 0.$$

Multiplying this by $J(x_t - x_0)$, where J is the duality mapping of X, and using (2.25) and the accretiveness of $I - F(t)$, it follows that

$$\lim_{t \downarrow 0} \|x_t - x_0\| \le \lambda \lim_{t \downarrow 0} \left\| Ax_0 + Bx_0 - \frac{x_0 - F(t)x_0}{t} \right\| = 0.$$

Hence, $\lim_{t \downarrow 0} x_t = x_0$, which implies (ii).

To prove (2.25), we write $t^{-1}(x - F(t)x)$ as

$$t^{-1}(x - F(t)x) = t^{-1}(x - S_A(t)x) + t^{-1}(S_A(t)x - S_A(t)S_B(t)x).$$

Since $t^{-1}(x - S_A(t)x) \to Ax$ as $t \to 0$ (Theorem 1.7), it remains to prove that

$$z_t = t^{-1}(S_A(t)x - S_A(t)S_B(t)x) \to Bx \qquad \text{as } t \to 0. \tag{2.28}$$

Since $S_A(t)$ is nonexpansive, we have

$$\|z_t\| \le t^{-1}\|S_B(t)x - x\| \le \|Bx\| \qquad \forall t > 0. \tag{2.29}$$

On the other hand, inasmuch as $I - S_A(t)$ is accretive, we have

$$\left(\frac{u - S_A(t)u}{t} + (S_A(t) - I)S_B(t)x - z_t, J(u - S_A(t)x)\right) > 0$$
$$\forall u \in C, t > 0. \tag{2.30}$$

Let $t_n \to 0$ be such that $z_{t_n} \rightharpoonup z$. Then, by (2.30), we have

$$(Au + Bx - Ax - z, J(u - x)) \ge 0 \qquad \forall u \in D(A),$$

and since A is m-accretive, this implies that $Ax + z - Bx = Ax$, i.e., $z = Bx$. By (2.29), recalling that X is uniformly convex, it follows that $z_{t_n} \to Bx$ (strongly). Then (2.28) follows, and the proof of Theorem 2.3 is complete. ■

Remark 2.3. Theorem 2.3, which is essentially due to H. Brézis and A. Pazy [1] was extended by Y. Kobayashi [2] to multivalued operators A and B in a Hilbert space H. More precisely, if A, B and $A + B$ are maximal monotone and if there is a nonempty closed convex set $C \subset \overline{D(A) \cap D(B)}$ such that $(I + \lambda A)^{-1}C \subset C$ and $(I + \lambda B)^{-1}C \subset C$ $\forall \lambda > 0$, then

$$S_{A+B}(t)x = \lim_{n\to\infty}\left(S_A\left(\frac{t}{n}\right)S_B\left(\frac{t}{n}\right)\right)^n x \qquad \forall x \in C,$$

uniformly in t on compact intervals.

4.2.3. Null Controllability of Nonlinear Accretive Equations

Let A be an m-accretive subset of $X \times X$, where X is a Banach space. Consider the controlled system

$$\frac{dy}{dt} + Ay \ni u \qquad \text{in } \mathbf{R}^+ = [0, +\infty)$$
$$y(0) = y_0, \tag{2.31}$$

where $y_0 \in \overline{D(A)}$ and $u \in L^1_{\text{loc}}(\mathbf{R}^+; X)$. Consider the set

$$\mathscr{U}_\rho = \{u \in L^\infty(\mathbf{R}^+; X); \|u(t)\| \le \rho \text{ a.e. } t > 0\}. \tag{2.32}$$

The parameter function u is called the *control* and the corresponding mild solution $y = y^u(t)$ to problem (2.31) is called the *state* of the system. A problem of great interest is to find $u \in \mathcal{U}_\rho$ that steers y_0 into the origin in a finite time T. If this happens, we say that the system (2.31) is $\mathcal{U}_\rho$ *null controllable*.

Let us define the mapping sgn: $X \to X$,

$$\operatorname{sgn} y = \begin{cases} y/\|y\| & \text{if } y \neq 0, \\ \{z; \|z\| \leq 1\} & \text{if } y = 0. \end{cases} \tag{2.33}$$

It is readily seen that this mapping is m-accretive in $X \times X$. Indeed, for every $f \in X$ the equation

$$y + \lambda \operatorname{sgn} y \ni f$$

has the unique solution

$$y_f = \begin{cases} \dfrac{f(\|f\| - \lambda)}{\|f\|} & \text{if } \|f\| > \lambda, \\ 0 & \text{if } \|f\| \leq \lambda, \end{cases} \tag{2.34}$$

and after some computation we see that

$$\|y_f - y_g\| \leq \|f - g\| \qquad \forall f, g \in X, \quad \lambda > 0.$$

Proposition 2.3. *Assume that* $0 \in A0$ *and that the operator* $A + \rho \operatorname{sgn}$ *is m-accretive in* $X \times X$. *Then the feedback law* $u = -\rho \operatorname{sgn} y$ *steers* $y_0 \in \overline{D(A)}$ *into the origin in a finite time* $T \leq \rho^{-1}\|y_0\|$, *i.e., the solution* y *to the system*

$$\frac{dy}{dt} + Ay + \rho \operatorname{sgn} y \ni 0 \qquad \text{in } \mathbf{R}^+,$$
$$y(0) = y_0, \tag{2.35}$$

has the support in $[0, T]$, *i.e.*, $y(t) = 0$ *for* $t \geq T$.

Proof. If $A + \rho \operatorname{sgn}$ is m-accretive then the Cauchy problem (2.35) has a unique mild solution $y \in C(\mathbf{R}^+; X)$, given by

$$y(t) = \lim_{\varepsilon \to 0} y_\varepsilon(t) \qquad \forall t \geq 0, \tag{2.36}$$

uniformly on compact intervals, where y_ε is the solution to the difference equation

$$y_\varepsilon(t) + \varepsilon A y_\varepsilon(t) + \varepsilon\rho \operatorname{sgn} y_\varepsilon(t) \ni y_\varepsilon(t - \varepsilon) \qquad \text{for } t \geq \varepsilon,$$
$$y_\varepsilon(t) = y_0 \qquad \forall t \leq 0. \tag{2.37}$$

Multiplying this by $\theta_\varepsilon(t) \in \operatorname{sgn} y_\varepsilon(t)$, we get

$$\|y_\varepsilon(t)\| + \varepsilon\rho \leq \|y_\varepsilon(t-s)\| \qquad \forall t \geq \varepsilon,$$

because A is accretive. This yields

$$y_\varepsilon(k\varepsilon) + k\varepsilon\rho \leq \|y_0\|, \qquad k = 0, 1, \dots,$$

and so $y_\varepsilon(k\varepsilon) = 0$ for $k\varepsilon \geq \rho^{-1}\|y_0\|$.

Now, letting ε tend to zero, it follows by (2.36) that $y(t) = 0$ for $t \geq \rho^{-1}\|y_0\|$, as claimed. ∎

Proposition 2.3, which can be deduced from a more general result regarding the finite extinction time property for the solution to nonlinear m-accretive equations $y' + By \ni 0$ with $\operatorname{int} B0 \neq \emptyset$, provides a simple feedback law that steers all initial states $y_0 \in \overline{D(A)}$ into the origin in finite time. We do not know whether $A + \rho \operatorname{sgn}$ is m-accretive in a general Banach space X. However, this happens if X^* is uniformly convex (see Proposition 2.4 following) and in several other significant situations that will be discussed later.

Proposition 2.4. *Let X^* be uniformly convex and A be m-accretive. Then $A + \rho \operatorname{sgn}$ is m-accretive.*

Proof. Without loss of generality, we may assume that $0 \in A0$. Put $B = \rho \operatorname{sgn}$. By (2.34), we see that the Yosida approximation B_λ of B is given by

$$B_\lambda f = \lambda^{-1}\left(f - (I + \lambda B)^{-1} f\right) = \begin{cases} \dfrac{\rho f}{\|f\|} & \text{if } \|f\| \geq \lambda, \\[2ex] \dfrac{\rho}{\lambda} f & \text{if } \|f\| < \lambda. \end{cases}$$

Hence,

$$(Af, J(B_\lambda f)) \geq 0 \qquad \forall \lambda > 0, f \in X,$$

and so by Proposition 3.8 in Chapter 2 we infer that $A + B$ is m-accretive, as claimed. ∎

By Proposition 2.3 we may conclude, therefore, that if X^* is uniformly convex then the system (2.31) is $\mathscr{U}_\rho$ null controllable.

4.2.4. Compact Evolutions

Let A be an ω-m-accretive mapping in a Banach space X and let $S(t)\colon \overline{D(A)} \to \overline{D(A)}$, $t \geq 0$, be the semigroup generated by on $\overline{D(A)}$. The semigroup $S(t)$ is said to be compact if $S(t)$ is a compact mapping in X for each $t > 0$.

Theorem 2.4. *The semigroup $S(t)$ is compact if and only if two following two conditions are satisfied*:

(i) $J_\lambda = (I + \lambda A)^{-1}$ *is compact for all* $0 < \lambda < \omega^{-1}$;
(ii) *For each bounded subset* $M \subset \overline{D(A)}$ *and* $t_0 > 0$, $\lim_{t \to t_0} S(t)x = S(t_0)x$ *uniformly in* $x \in M$ *(i.e., $S(t)$ is equicontinuous).*

Proof. Let us assume first that (i) and (ii) are satisfied. By inequality (1.11) in Theorem 1.1, we have

$$\|S(t)y_0 - x\| \leq e^{\omega t}\|y_0 - x\| + \int_0^t e^{\omega(t-\tau)}[S(\tau)y_0 - x, -y]_s \, d\tau$$

for all $[x, y] \in A$ and $t > 0$. Since

$$[S(\tau)y_0 - x, -y]_s \leq \frac{\|S(\tau)y_0 - x - \lambda y\| - \|S(\tau)y_0 - x\|}{\lambda},$$

we get

$$\|S(t)y_0 - x\| \leq e^{\omega t}\|y_0 - x\| + \frac{1}{\lambda}\int_0^t e^{\omega(t-\tau)}(\|S(\tau)y_0 - x - \lambda y\| - \|S(\tau)y_0 - x\|)\, d\tau.$$

If we take $x = J_\lambda y_0$ and $y = A_\lambda y_0$, we get

$$\|S(t)y_0 - J_\lambda y_0\| \leq e^{\omega t}\|J_\lambda y_0 - y_0\| + \frac{1}{\lambda}\int_0^t e^{\omega(t-\tau)}(\|S(\tau)y_0 - y_0\| - \|S(\tau)y_0 - J_\lambda y_0\|)\, ds.$$

Using the obvious inequality

$$\|J_\lambda y_0 - y_0\| \le \|S(\tau)y_0 - J_\lambda y_0\| + \|S(\tau)y_0 - y_0\|,$$

we get

$$\|J_\lambda y_0 - y_0\| \le \frac{(e^{\omega t} - 1)^{-1}}{1 - \lambda\omega}\Bigg(\|S(t)y_0 - y_0\| + \frac{2}{\lambda}\int_0^t e^{\omega(t-\tau)}\|S(\tau)y_0 - y_0\|\,d\tau\Bigg) \qquad \forall t > 0,$$

and this yields

$$\|S(s)y_0 - J_\lambda S(s)y_0\| \le \frac{(e^{\omega t} - 1)^{-1}\lambda\omega}{1 - \lambda\omega}\Bigg(\|S(t+s)y_0 - S(s)y_0\| + \frac{2}{\lambda}\int_0^t e^{\omega(t-\tau)}\|S(\tau+s)y_0 - S(s)y_o\|\,d\tau\Bigg)$$

for all $\lambda, t, s > 0$. If in the latter inequality that $t = \lambda$ and let λ tend to zero, it follows by assumption (ii) that

$$\lim_{\lambda\to 0} J_\lambda S(s)y_0 = S(s)y_0 \qquad \forall s > 0,$$

uniformly on every bounded subset of $\overline{D(A)}$. Since J_λ is compact we conclude that $S(s)$ is compact for every $s > 0$, as claimed.

Assume now that $S(t)$ is compact. Again by inequality (1.11), we have

$$\begin{aligned}\|S(t)J_\lambda x - J_\lambda x\| &\le \int_0^t e^{\omega(t-\tau)}[S(\tau)J_\lambda x - J_\lambda x, -A_\lambda x]_s\,d\tau \\ &\le \frac{\lambda}{\omega}\|x - J_\lambda x\|(e^{\omega t} - 1) \qquad \forall \lambda > 0, t > 0.\end{aligned}$$

This yields

$$\lim_{t\to 0} S(t)J_\lambda x = J_\lambda x \qquad \forall t > 0,$$

uniformly in x on bounded subsets. Hence, J_λ is compact for every $\lambda > 0$. Regarding (ii), it follows by a standard argument we do not reproduce here. ■

An example of compact semigroup is that generated by $A = \partial\varphi$ on $\overline{D(A)}$, where $\varphi: H \to \overline{\mathbf{R}}$ is a lower semicontinuous convex function on a Hilbert space H such that for all λ, $R > 0$ the set $\{x \in H;\ \|x\| \le R,\ \varphi(x) \le \lambda\}$ is compact. Indeed, it is readily seen that in this case $J_\lambda = (I + \lambda A)^{-1}$ is compact because

$$\varphi(J_\lambda x) \le \varphi(y_0) + (A_\lambda x, J_\lambda x - y_0) \le M_R \qquad \text{for } \|x\| \le R,$$

where $y_0 \in D(\varphi)$.

On the other hand, by Corollary 1.2 we have

$$|S(t)y_0 - S(t+\varepsilon)y_0| \le \varepsilon|A^0 x| + \frac{\varepsilon}{t}\|x - y_0\| \qquad \forall \varepsilon > 0, t > 0,$$

for all $y_0 \in \overline{D(A)}$ and $x \in D(A)$. This clearly implies that $S(t)$ is equicontinuous.

Regarding the compactness of evolutions generated by ω-m-accretive operators, we mention the following result due to Baras [1].

Theorem 2.5. *Let A be ω-m-accretive in a Banach space X and let $Q: \overline{D(A)} \times L^1(0,T;X) \to L^1(0,T;X)$ be the evolution generated by A, i.e., $Q(y_0, f) = y$ where y is the mild solution to* (1.1). *If the semigroup $S(t)$ generated by A on $\overline{D(A)}$ is compact then Q is compact, too, from $\overline{D(A)} \times L^1(0,T;X)$ to $L^1(0,T;X)$.* ■

4.3. Applications to Partial Differential Equations

In this section we present a variety of applications to nonlinear partial differential equations illustrating the ideas and general existence theory developed in the previous sections.

4.3.1 Semilinear Parabolic Equations

Let $\beta \subset \mathbf{R} \times \mathbf{R}$ be a maximal monotone graph such that $0 \in D(\beta)$, and let Ω be an open and bounded subset of $\mathbf{R}^N$ with a sufficiently smooth boundary $\partial\Omega$ (for instance, of class C^2). Consider the boundary value problem

$$\begin{aligned}
\frac{\partial y}{\partial t} - \Delta y + \beta(y) \ni f \qquad & \text{in } \Omega \times (0,T) = Q, \\
y(x,0) = y_0(x) \qquad & \forall x \in \Omega, \\
y = 0 \qquad & \text{in } \partial\Omega \times (0,T) = \Sigma
\end{aligned} \tag{3.1}$$

where $y_0 \in L^2(\Omega)$ and $f \in L^2(Q)$.

We may represent (3.1) as a nonlinear differential equation in the space $H = L^2(\Omega)$:

$$\begin{aligned}
\frac{dy}{dt} + Ay \ni f \qquad & \text{in } [0,T], \\
y(0) = y_0, &
\end{aligned} \tag{3.2}$$

where $A: L^2(\Omega) \to L^2(\Omega)$ is the operator defined by

$$\begin{aligned}
Ay &= \{z \in L^2(\Omega);\ z = -\Delta y + w, w(x) \in \beta(y(x)) \text{ a.e. } x \in \Omega\}, \\
D(A) &= \{y \in H_0^1(\Omega) \cap H^2(\Omega);\ \exists w \in L^2(\Omega), w(x) \in \beta(y(x)) \\
&\qquad \text{a.e. } x \in \Omega\}.
\end{aligned} \tag{3.3}$$

Recall (see Proposition 2.10 in Chapter 2) that A is maximal monotone (i.e., m-accretive) in $L^2(\Omega) \times L^2(\Omega)$ and $A = \partial\varphi$, where

$$\varphi(y) = \tfrac{1}{2} \int_\Omega |\nabla y|^2\, dx + \int_\Omega g(y)\, dx, \qquad y \in L^2(\Omega),$$

and $\partial g = \beta$. Moreover, we have

$$\|y\|_{H^2(\Omega)} + \|y\|_{H_0^1(\Omega)} \le C\big(\|A^0 y\|_{L^2(\Omega)} + 1\big) \qquad \forall y \in D(A). \tag{3.4}$$

Writing Eq. (3.1) in the form (3.2), we view its solution y as a function of t from $[0,T]$ to $L^2(\Omega)$. The boundary conditions that appear in (3.1) are implicitly incorporated into problem (3.2) through the condition $y(t) \in D(A)\ \forall t \in [0,T]$.

The function $y: \Omega \times [0,T] \to \mathbf{R}$ is called a *strong solution* to problem (3.1) if $y: [0,T] \to L^2(\Omega)$ is continuous on $[0,T]$, absolutely continuous on $(0,T)$, and satisfies

$$\begin{aligned}
&\frac{d}{dt} y(x,t) - \Delta y(x,t) + \beta(y(x,t)) \ni f(x,t) && \text{a.e. } t \in (0,T), x \in \Omega, \\
&y(x,0) = y_0(x) && \text{a.e. } x \in \Omega, \\
&y(x,t) = 0 && \text{a.e. } x \in \partial\Omega, t \in (0,T).
\end{aligned} \tag{3.5}$$

Here, $(d/dt)\,y$ is the strong derivative of $y:[0,T] \to L^2(\Omega)$ and Δy is considered in the sense of distributions on Ω.

As a matter of fact, it is readily seen that if y is absolutely continuous from $[a,b]$ to $L^1(\Omega)$, then $dy/dt = \partial y/\partial t$ in $\mathcal{D}'((a,b); L^1(\Omega))$, and so a strong solution to Eq. (3.1) satisfies this equation in the sense of distributions in $(0,T) \times \Omega$. For this reason, whenever there will not be any danger of confusion we shall write $\partial y/\partial t$ instead of dy/dt.

Proposition 3.1. *Let $y_0 \in L^2(\Omega)$ and $f \in L^2(0,T; L^2(\Omega)) = L^2(Q)$ be such that*

$$y_0(x) \in \overline{D(\beta)} \qquad \text{a.e. } x \in \Omega.$$

Then problem (3.1) *has a unique strong solution $y \in C([0,T]; L^2(\Omega)) \cap W^{1,1}((0,T]; L^2(\Omega))$ that satisfies*

$$t^{1/2}y \in L^2(0,T; H_0^1(\Omega) \cap H^2(\Omega)) \qquad t^{1/2}\frac{dy}{dt} \in L^2(0,T; L^2(\Omega)). \tag{3.6}$$

If, in addition, $f \in W^{1,1}([0,T]; L^2(\Omega))$ then $y(t) \in H_0^1(\Omega) \cap H^2(\Omega)$ for every $t \in (0,T]$ and

$$t\frac{dy}{dt} \in L^\infty(0,T; L^2(\Omega)). \tag{3.7}$$

If $y_0 \in H_0^1(\Omega)$, $g(y_0) \in L^1(\Omega)$, and $f \in L^2(0,T; L^2(\Omega))$, then

$$\frac{dy}{dt} \in L^2(0,T; L^2(\Omega)), \qquad y \in L^\infty(0,T; H_0^1(\Omega)) \cap L^2(0,T; H^2(\Omega)). \tag{3.8}$$

Finally, if $y_0 \in D(A)$ and $f \in W^{1,1}([0,T]; L^2(\Omega))$, then

$$\frac{dy}{dt} \in L^\infty(0,T; L^2(\Omega)), \qquad y \in L^\infty(0,T; H^2(\Omega) \cap H_0^1(\Omega)) \tag{3.9}$$

and

$$\frac{d^+}{dt}y(t) + (-\Delta y(t) + \beta(y(t)) - f(t))^0 = 0 \qquad \forall t \in [0,T). \tag{3.10}$$

Proof. This is direct consequence of Theorems 1.10, 1.11, and 1.6. Here, we have used the fact that

$$\overline{D(A)} = \{u \in L^2(\Omega);\ u(x) \in \overline{D(\beta)} \text{ a.e. } x \in \Omega\},$$

which follows by Theorem 2.4 in Chapter 2 because

$$\int_\Omega g\big((1+\lambda A_2)^{-1}y(x)\big)\,dx \le \int_\Omega g(y(x))\,dx \qquad \forall y\in L^2(\Omega),\ \forall\lambda>0,$$

where $A_2 = -\Delta$, $D(A_2) = H_0^1(\Omega)\cap H^2(\Omega)$, and $\beta = \partial g$ (we may assume that $0\in\beta(0)$). ∎

In particular, it follows that for $y_0 \in H_0^1(\Omega)$, $g(y_0)\in L^1(\Omega)$, and $f\in L^2(\Omega\times(0,T))$, the solution y to problem (3.1) belongs to the space $H^{2,1}(Q) = \{y\in L^2(0,T;H^2(\Omega)),\ \partial y/\partial t\in L^2(Q)\}$, $Q=\Omega\times(0,T)$.

Problem (3.1) can be studied in the L^p setting, $1\le p<\infty$, if one defines $A: L^p(\Omega)\to L^p(\Omega)$ as

$$Ay = \{z\in L^p(\Omega);\ z = -\Delta y + w,\ w(x)\in\beta(y(x))\ \text{a.e. } x\in\Omega\}, \tag{3.11}$$

$$D(A) = \{y\in W_0^{1,p}(\Omega)\cap W^{2,p}(\Omega);\ w\in L^p(\Omega)\ \text{such that}\ w(x)\in\beta(y(x))\ \text{a.e. } x\in\Omega\} \quad \text{if } p>1, \tag{3.12}$$

$$D(A) = \{y\in W_0^{1,1}(\Omega);\ \Delta y\in L^1(\Omega),\ \exists w\in L^1(\Omega)\ \text{such that}\ w(x)\in\beta(y(x))\ \text{a.e. } x\in\Omega\} \quad \text{if } p=1 \tag{3.13}$$

As seen earlier (Proposition 3.9 in Chapter 2), the operator A is m-accretive in $L^p(\Omega)\times L^p(\Omega)$ and so the general existence theory is applicable.

Proposition 3.2. *Let $y_0\in D(A)$ and $f\in W^{1,1}([0,T];L^p(\Omega))$, $1<p<\infty$. Then problem* (3.1) *has a unique strong solution $y\in C([0,T];L^p(\Omega))$, which satisfies*

$$\frac{d}{dt}y\in L^\infty(0,T;L^p(\Omega)), \qquad y\in L^\infty(0,T;W_0^{1,p}(\Omega)\cap W^{2,p}(\Omega)), \tag{3.14}$$

$$\frac{d^+}{dt}y(t) + (-\Delta y(t) + \beta(y(t)) - f(t))^0 = 0 \qquad \forall t\in[0,T). \tag{3.15}$$

Proof. Proposition 3.2 follows by Theorem 1.6 (recall that $X = L^p(\Omega)$ is uniformly convex for $1<p<\infty$). ∎

If $y_0\in\overline{D(A)}$ and $f\in L^1(0,T;L^p(\Omega))$, then according to the general theorem, 1.1, the Cauchy problem (3.2), and implicitly problem (3.1), has a unique mild solution $y\in C([0,T];L^p(\Omega))$.

Since the space $X = L^1(\Omega)$ is not reflexive, by the general theory, the mild solution to the Cauchy problem (3.2) in $L^1(\Omega)$ is only continuous, even if y_0 and f are regular. However, also in this case we have a differentiability property of mild solutions comparable with the situation encountered in the linear case (see Proposition 4.4 in Chapter 1).

Proposition 3.3. *Let $\beta\colon \mathbf{R} \to \mathbf{R}$ be a maximal monotone graph,* $0 \in D(\beta)$, *and* $\beta = \partial g$. *Let $f \in L^2(0,T; L^\infty(\Omega))$ and $y_0 \in L^1(\Omega)$ be such that $y_0(x) \in \overline{D(\beta)}$ a.e. $x \in \Omega$. Then the mild solution $y \in C([0,T]; L^1(\Omega))$ to problem* (3.1) *satisfies*

$$\|y(t)\|_{L^\infty(\Omega)} \le C\big(t^{-N/2}\|y_0\|_{L^1(\Omega)} + \int_0^t \|f(s)\|_{L^\infty(\Omega)}\,ds, \tag{3.16}$$

$$\begin{aligned}
&\int_0^T \int_\Omega \left(t^{(N+4)/2}y_t^2 + t^{(N+2)/2}|\nabla y|^2\right) dx\,dt \\
&\qquad + T^{(N+4)/2}\int_\Omega |\nabla y(x,T)|^2\,dx \\
&\quad \le C\left(\left(\|y_0\|_{L^1(\Omega)}^{4/(N+2)} + \int_0^T\int_\Omega |f|\,dx\,dt\right)^{(N+2)/2}\right. \\
&\qquad \left. + T^{(N+4)/2}\int_0^T\int_\Omega f^2\,dx\,dt\right).
\end{aligned} \tag{3.17}$$

Proof. Without loss of generality, we may assume that $0 \in \beta(0)$. Also, let us assume first that $y_0 \in H_0^1(\Omega) \cap H^2(\Omega)$. Then, as seen in Proposition 3.1, the problem (3.1) has a unique strong solution such that $t^{1/2}y_t \in L^2(Q)$, $t^{1/2}y \in L^2(0,T; H_0^1(\Omega) \cap H^2(\Omega))$, i.e.,

$$\begin{aligned}
&\frac{\partial y}{\partial t}(x,t) - \Delta y(x,t) + \beta(y(x,t)) \ni f(x,t) && \text{a.e. } (x,t) \in Q, \\
&y(x,0) = y_0(x), && x \in \Omega, \\
&y = 0 && \text{in } \partial\Omega \times (0,T).
\end{aligned} \tag{3.18}$$

Consider the linear problem

$$\begin{aligned}
\frac{\partial z}{\partial t} - \Delta z &= \|f(t)\|_{L^\infty(\Omega)} && \text{in } Q, \\
z(x,0) &= |y_0(x)|, && x \in \Omega, \\
z &= 0 && \text{in } \partial\Omega \times (0,T).
\end{aligned} \tag{3.19}$$

Subtracting these two equations and multiplying the resulting equation by $(y-z)^+$, after integration on Ω we get

$$\frac{1}{2}\frac{d}{dt}\|(y-z)^+\|^2_{L^2(\Omega)} + \int_\Omega |\nabla(y-z)^+|^2\,dx \le 0 \text{ a.e. } t \in (0,T),$$

$$(y-z)^+(0) \le 0 \qquad \text{in } \Omega,$$

because $z \ge 0$ and β is monotonically increasing. Hence, $y(x,t) \le z(x,t)$ a.e. in Q and so $|y(x,t)| \le z(x,t)$ a.e. $(x,t) \in Q$. On the other hand, the solution z to problem (3.19) can be represented as

$$z(x,t) = S(t)(|y_0|)(x) + \int_0^t S(t-s)(\|f(s)\|_{L^\infty(\Omega)})\,ds$$

$$\text{a.e. } (x,t) \in Q,$$

where $S(t)$ is the semigroup generated on $L^1(\Omega)$ by $-\Delta$ with Dirichlet homogeneous conditions on $\partial\Omega$. We know by Proposition 4.4 in Chapter 1 that

$$\|S(t)u_0\|_{L^\infty(\Omega)} \le Ct^{-N/2}\|u_0\|_{L^1(\Omega)} \qquad \forall u_0 \in L^1(\Omega),\ t>0.$$

Hence,

$$|y(x,t)| \le Ct^{-N/2}\|y_0\|_{L^1(\Omega)} + \int_0^t \|f(s)\|_{L^\infty(\Omega)}\,ds, \qquad (t,x) \in Q. \quad (3.20)$$

Now, for an arbitrary $y_0 \in L^1(\Omega)$ such that $y_0 \in \overline{D(\beta)}$ a.e. in Ω we choose a sequence $\{y_0^n\} \subset H_0^1(\Omega) \cap H^2(\Omega)$, $y_0^n \in \overline{D(\beta)}$ a.e. in Q, such that $y_0^n \to y_0$ in $L^1(\Omega)$ as $n \to \infty$. (We may take, for instance, $y_0^n = S(n^{-1})((1+n^{-1}\beta)^{-1}y_0)$.) If y_n is the corresponding solution to the problem (3.1), then we know that $y_n \to y$ strongly in $C([0,T]; L^1(\Omega))$, where y is the solution with the initial value y_0. By (3.20), it follows that y satisfies the estimate (3.16).

Since $y(t) \in L^\infty(\Omega) \subset L^2(\Omega)$ for all $t>0$, it follows by Proposition 3.1 that $y \in W^{1,2}([\delta,T]; L^2(\Omega)) \cap L^2(\delta,T; H_0^1(\Omega) \cap H^2(\Omega))$ for all $0<\delta<T$ and it satisfies Eq. (3.18) a. e. in $Q = \Omega \times (0,T)$. (Arguing as before, we may assume that $y_0 \in H_0^1(\Omega) \cap H^2(\Omega)$ and so y_t, $y \in L^2(0,T; L^2(\Omega))$.) To get the desired estimate (3.17), we multiply Eq. (3.18) by $y_t t^{k+2}$ and integrate on Q to get

$$\int_0^T\int_\Omega t^{k+2}y_t^2\,dx\,dt + \frac{1}{2}\int_0^T\int_\Omega t^{k+2}|\nabla y|_t^2\,dx\,dt + \int_0^T\int_\Omega t^{k+2}\frac{\partial}{\partial t}g(y)\,dx\,dt$$

$$= \int_0^T\int_\Omega t^{k+2}y_t f\,dx\,dt,$$

where $y_t = \partial y/\partial t$ and $\partial g = \beta$.
This yields

$$\int_Q t^{k+2} y_t^2 \,dx\,dt + \frac{T^{k+2}}{2} \int_\Omega |\nabla y(x,T)|^2 \,dx$$

$$+T^{k+2} \int_\Omega g(y(x,T)) \,dx$$

$$\leq \frac{k+2}{2} \int_Q t^{k+1} |\nabla y|^2 \,dx\,dt + (k+2) \int_Q t^{k+1} g(y) \,dx\,dt$$

$$+\frac{1}{2} \int_0^T t^{k+2} y_t^2 \,dx\,dt + \frac{1}{2} \int_Q t^{k+2} f^2 \,dx\,dt.$$

Hence,

$$\int_Q t^{k+2} y_t^2 \,dx\,dt + T^{k+2} \int_\Omega |\nabla y(x,T)|^2 \,dx$$

$$\leq (k+2) \int_Q t^{k+1} |\nabla y|^2 \,dx\,dt + 2(k+2) \int_Q t^{k+1} \beta(y) y \,dx$$

$$+ T^{k+2} \int_Q f^2 \,dx\,dt.$$

Finally, writing $\beta(y)y$ as $(f + \Delta y - y_t)y$ and using Green's formula, we get

$$\int_Q t^{k+2} y_t^2 \,dx\,dt + T^{k+2} \int_\Omega |\nabla y(x,T)|^2 \,dx$$

$$+ \int_Q t^{k+1} |\nabla y|^2 \,dx\,dt$$

$$\leq (k+2)(k+1) \int_Q y^2 t^k \,dx\,dt + T^{k+2} \int_Q f^2 \,dx\,dt$$

$$+2(k+2) \int_Q t^{k+1} |f|\,|y| \,dx\,dt$$

$$\leq C\left(\int_Q t^k y^2 \,dx\,dt + T^{k+2} \int_Q f^2 \,dx\,dt\right). \tag{3.21}$$

Next, we have, by the Hölder inequality

$$\int_\Omega y^2\,dx \le \|y\|_{L^p(\Omega)}^{(N-2)/(N+2)}\|y\|_{L^1(\Omega)}^{4/(N+2)}$$

for $p = 2N(N-2)^{-1}$. Then, by the Sobolev imbedding theorem,

$$\int_\Omega |y(x,t)|^2\,dx$$
$$\le \left(\int_\Omega |\nabla y(x,t)|^2\,dx\right)^{N/(N+2)}\left(\int_\Omega |y(x,t)|\,dx\right)^{4/(N+2)}. \quad (3.22)$$

On the other hand, multiplying Eq. (3.18) by sign y and integrating on $\Omega \times (0,t)$, we get

$$\|y(t)\|_{L^1(\Omega)} \le \|y_0\|_{L^1(\Omega)} + \int_Q |f(x,s)|\,dx\,ds, \qquad t \ge 0,$$

because, as seen earlier

$$\int_\Omega \Delta y \text{ sign } y\,dx \le 0.$$

Then, by estimates (3.21) and (3.22), we get

$$\int_Q t^{k+2}y_t^2\,dx\,dt + T^{k+2}\int_\Omega |\nabla y(x,T)|^2\,dx$$
$$+\int_Q t^{k+1}|\nabla y(x,t)|^2\,dx\,dt$$
$$\le C\bigg(\bigg(\|y_0\|_{L^1(\Omega)}^{4/(N+2)} + \int_0^T dt\int_\Omega |f(x,t)|\,dx\,dt$$
$$\times \int_0^T t^k\|\nabla y(t)\|_{L^2(\Omega)}^{2N/(N+2)}\,dt + T^{k+2}\int_Q f^2 dx\,dt.$$

On the other hand we have, for $k = N/2$,

$$\int_0^T t^k|\nabla y(t)|^{2N/(N+2)}\,dt \le \left(\int_0^T t^{k+1}|\nabla y(t)|^2\,dt\right)^{N/(N+2)} T).$$

Substituting in the latter inequality, we get after some calculation

$$\int_Q t^{(N+4)/2} y_t^2 \, dx\, dt + \int_Q t^{(N+2)/2} |\nabla y(x,t)|^2 \, dx\, dt$$

$$+ T^{(N+4)/2} \int_\Omega |\nabla y(x,T)|^2 \, dx$$

$$\leq C_1 \left(\|y_0\|_{L^1(\Omega)}^{4/(N+2)} + \int_Q |f(x,t)| \, dx\, dt \right)^{(N+2)/2}$$

$$+ C_2 T^{(N+4)/2} \int_Q f^2(x,t) \, dx\, dt, \tag{3.23}$$

as claimed. ■

In particular, it follows by Proposition 3.3 that the semigroup $S(t)$ generated by A (defined by (3.11), (3.13)) on $L^1(\Omega)$ has a smoothing effect on initial data, i.e., for all $t > 0$ it maps $L^1(\Omega)$ into $D(A)$ and is differentiable on $(0, \infty)$.

In the special case where

$$\beta(r) = \begin{cases} 0 & \text{if } r > 0, \\ \mathbf{R}^- & \text{if } r = 0, \end{cases}$$

problem (3.1) reduces to the parabolic variational inequality (the obstacle problem)

$$\frac{\partial y}{\partial t} - \Delta y = f \qquad \text{in } \{(x,t);\ y(x,t) > 0\},$$

$$y \geq 0, \quad \frac{\partial y}{\partial t} - \Delta y \geq f \qquad \text{in } Q,$$

$$y(x,0) = y_0(x) \quad \text{in } \Omega, \qquad y = 0 \quad \text{in } \partial\Omega \times (0,T) = \Sigma. \tag{3.24}$$

More will be said about this in the next section.

We also point out that Proposition 3.1 and partially Proposition 3.2 remain true for equations of the form

$$\frac{\partial y}{\partial t} - \Delta y + \beta(x,y) \ni f \qquad \text{in } Q,$$

$$y(x,0) = y_0(x) \qquad \text{in } \Omega,$$

$$y = 0 \qquad \text{in } \Sigma,$$

where $\beta: \Omega \times \mathbf{R} \to 2^{\mathbf{R}}$ is of the form $\beta(x, y) = \partial_y g(x, y)$ and $g: \Omega \times \mathbf{R} \to \mathbf{R}$ is a normal convex integrand on $\Omega \times \mathbf{R}$ (see Section 2.2 in Chapter 2). The details are left to the reader.

Now, we consider the equation

$$\frac{\partial y}{\partial t} - \Delta y = f \qquad \text{in } \Omega \times (0, T) = Q,$$

$$\frac{\partial}{\partial \nu} y + \beta(y) \ni 0 \qquad \text{in } \Sigma,$$

$$y(x, 0) = y_0(x) \qquad \text{in } \Omega, \tag{3.25}$$

where $\beta \subset \mathbf{R} \times \mathbf{R}$ is a maximal monotone graph, $0 \in D(\beta)$, $y_0 \in L^2(\Omega)$, and $f \in L^2(Q)$. As seen earlier (Proposition 2.11 in Chapter 2) we may write (3.25) as

$$\frac{dy}{dt} + Ay = f \qquad \text{in } (0, T),$$

$$y(0) = y_0,$$

where $Ay = -\Delta y$, $\forall y \in D(A) = \{y \in H^2(\Omega);\ 0 \in \partial y/\partial \nu + \beta(y) \text{ a.e. in } \partial\Omega\}$.

More precisely, $A = \partial\varphi$, where $\varphi: L^2(\Omega) \to \overline{\mathbf{R}}$ is defined by

$$\varphi(y) = \tfrac{1}{2} \int_\Omega |\nabla y|^2\, dx + \int_{\partial\Omega} j(y)\, d\sigma, \qquad \forall y \in L^2(\Omega),$$

and $\partial j = \beta$.

Then, applying Theorems 1.10 and 1.11, we get:

Proposition 3.4. *Let $y_0 \in \overline{D(A)}$ and $f \in L^2(Q)$. Then problem (3.25) has a unique strong solution $y \in C([0, T]; L^2(\Omega))$ such that*

$$t^{1/2} \frac{\partial y}{\partial t} \in L^2(0, T; L^2(\Omega)), \qquad t^{1/2} y \in L^2(0, T; H^2(\Omega)).$$

If $y_0 \in H^1(\Omega)$ and $j(y_0) \in L^1(\Omega)$, then

$$\frac{\partial y}{\partial t} \in L^2(0, T; L^2(\Omega)), \qquad y \in L^2(0, T; H^2(\Omega)) \cap L^\infty(0, T; H^1(\Omega)).$$

Finally, if $y_0 \in D(A)$ *and* $f, \partial f/\partial t \in L^1(0,T; L^2(\Omega))$, *then*

$$\frac{\partial y}{\partial t} \in L^\infty(0,T; L^2(\Omega)), \qquad y \in L^\infty(0,T; H^2(\Omega))$$

and

$$\frac{d^+}{dt} y(t) - \Delta y(t) = f(t) \qquad \forall t \in [0,T). \qquad \blacksquare$$

It should be mentioned that one also uses here the estimate

$$\|u\|_{H^2(\Omega)} \le C(\|u - \Delta u\|_{L^2(\Omega)} + 1) \qquad \forall u \in D(A).$$

An important special case is

$$\beta(y) = \begin{cases} 0 & \text{if } y > 0, \\ (-\infty, 0] & \text{if } y = 0. \end{cases}$$

Then, problem (3.25) reads

$$\frac{\partial y}{\partial t} - \Delta y = f \qquad \text{in } Q,$$

$$y \frac{\partial y}{\partial \nu} = 0, \quad y \ge 0, \quad \frac{\partial y}{\partial \nu} \ge 0 \qquad \text{in } \Sigma,$$

$$y(x,0) = y_0(x) \qquad \text{in } \Omega. \tag{3.26}$$

A problem of this type arises in the control of a heat field. More generally, the thermostat control process is modeled by Eq. (3.25), where

$$\beta(r) = \begin{cases} a_1(r - \theta_1) & \text{if } -\infty < r < \theta_1, \\ 0 & \text{if } \theta_1 \le r \le \theta_2, \\ a_2(r - \theta_2) & \text{if } \theta_2 < r < \infty, \end{cases}$$

$a_i \ge 0$, $\theta_1 \in \mathbf{R}$, $i = 1, 2$. In the limit case, we obtain (3.26).

The black body radiation heat emission on $\partial\Omega$ is described by Eq. (3.25) where β is given by (the Stefan–Boltzman law)

$$\beta(y) = \begin{cases} \alpha(y^4 - y_1^4) & \text{for } y \ge 0, \\ -y_1^4 & \text{for } y < 0, \end{cases}$$

whilst in the case of natural convection

$$\beta(y) = \begin{cases} ay^{5/4} & \text{for } y \geq 0, \\ 0 & \text{for } y < 0. \end{cases}$$

Note also that the Michaelis–Menten dynamic model of enzyme diffusion reaction is described by Eq. (3.1), where

$$\beta(y) = \begin{cases} \dfrac{y}{\lambda(y+k)} & \text{for } y > 0, \\ (-\infty, 0] & \text{for } y = 0, \\ \emptyset & \text{for } y < 0, \end{cases}$$

where λ, k are positive constants.

We note that more general boundary values problems of the form

$$\begin{aligned} \frac{\partial y}{\partial t} - \Delta y + \gamma(y) &\ni f && \text{in } Q, \\ y(x,0) &= y_0(x) && \text{in } \Omega, \\ \frac{\partial y}{\partial \nu} + \beta(y) &\ni 0 && \text{in } \Sigma, \end{aligned}$$

where β and γ are maximal monotone graphs in $\mathbf{R} \times \mathbf{R}$ such that $0 \in D(\beta)$, $0 \in D(\gamma)$, can be written in the form (3.2) where $A = \partial\varphi$ and $\varphi: L^2(\Omega) \to \overline{\mathbf{R}}$ is defined by

$$\varphi(y) = \begin{cases} \dfrac{1}{2}\displaystyle\int_\Omega |\nabla y|^2\,dx + \int_\Omega g(y)\,dx + \int_{\partial\Omega} j(y)\,d\sigma & \text{if } y \in H^1(\Omega), \\ +\infty & \text{otherwise}, \end{cases}$$

and $\partial g = \gamma$, $\partial j = \beta$.

We may conclude, therefore, that for $f \in L^2(\Omega)$ and $y_0 \in H^1(\Omega)$ such that $g(y_0) \in L^1(\Omega)$, $j(y_0) \in L^1(\partial\Omega)$ the preceding problem has a unique solution $y \in W^{1,2}([0,T]; L^2(\Omega)) \cap L^2(0,T; H^2(\Omega))$.

On the other hand, semilinear parabolic problems of the form (3.1) or (3.25) arise very often as feedback systems associated with the linear heat equation. For instance, the feedback control

$$u = -\rho \operatorname{sign} y, \tag{3.27}$$

where

$$\operatorname{sign} r = \begin{cases} \dfrac{r}{|r|} & \text{if } r \neq 0, \\ [-1,1] & \text{if } r = 0, \end{cases}$$

applied to the controlled heat equation

$$\begin{aligned} \frac{\partial y}{\partial t} - \Delta y &= u && \text{in } \Omega \times \mathbf{R}^+, \\ y &= 0 && \text{in } \partial\Omega \times \mathbf{R}^+, \\ y(x,0) &= y_0(x) && \text{in } \Omega, \end{aligned} \tag{3.28}$$

transforms it into a nonlinear equation of the form (3.1), i.e.,

$$\begin{aligned} \frac{\partial y}{\partial t} - \Delta y + \rho \operatorname{sign} y &\ni 0 && \text{in } \Omega \times \mathbf{R}^+, \\ y &= 0 && \text{in } \partial\Omega \times \mathbf{R}^+, \\ y(x,0) &= y_0(x) && \text{in } \Omega. \end{aligned} \tag{3.29}$$

This is the closed loop system associated with the feedback law (3.27) and, according to Proposition 3.2, for every $y_0 \in L^1(\Omega)$ it has a unique strong solution $y \in C(\mathbf{R}^+; L^2(\Omega))$ satisfying

$$\begin{gathered} y(t) \in L^\infty(\Omega) \qquad \forall t > 0, \\ t^{(N+4)/4} y_t, t^{(N+4)/2} y \in L^2_{\text{loc}}(\mathbf{R}^+; L^2(\Omega)). \end{gathered}$$

Let us observe that the feedback control (3.27) belongs to constraint set $\{u \in L^\infty(\Omega \times \mathbf{R}^+); \|u\|_{L^\infty(\Omega \times \mathbf{R}^+)} \leq \rho\}$ and *steers the initial state y_0 into the origin in a finite time T*. Here is the argument. We shall assume first that $y_0 \in L^\infty(\Omega)$ and consider the function $w(x,t) = \|y_0\|_{L^\infty(\Omega)} - \rho t$. On the domain $\Omega \times (0, \rho^{-1}\|y_0\|_{L^\infty(\Omega)}) = Q_0$, we have

$$\begin{aligned} \frac{\partial w}{\partial t} - \Delta w + \rho \operatorname{sign} w \ni 0 \quad & \text{in } Q_0, \\ w(0) = \|y_0\|_{L^\infty(\Omega)} \quad & \text{in } \Omega, \\ w \geq 0 \quad & \text{in } \partial\Omega \times \left(0, \rho^{-1}\|y_0\|_{L^\infty(\Omega)}\right). \end{aligned} \tag{3.30}$$

Then subtracting equations (3.29) and (3.30) and multiplying by $(y - w)^+$ (or simply applying the maximum principle), we get

$$(y - w)^+ \leq 0 \qquad \text{in } Q_0.$$

Hence, $y \le w$ in Q_0. Similarly, it follows that $y \ge -w$ in Q_0 and, therefore,

$$|y(x,t)| \le \|y_0\|_{L^\infty(\Omega)} - \rho t \qquad \forall (x,t) \in Q_0.$$

Hence, $y(t) \equiv 0$ for all $t \ge T = \rho^{-1}\|y_0\|_{L^\infty(\Omega)}$. Now if $y_0 \in L^1(\Omega)$, then applying in the system (3.28) the control

$$u(t) = \begin{cases} 0 & \text{for } 0 \le t \le \varepsilon, \\ -\rho \operatorname{sign} y(t) & \text{for } t > \varepsilon, \end{cases}$$

we get a trajectory $y(t)$ that steers y_0 into the origin in the time

$$T(y_0) < \varepsilon + \rho^{-1}\|y(\varepsilon)\|_{L^\infty(\Omega)} \le \varepsilon + C(\rho\varepsilon^{N/2})^{-1}\|y_0\|_{L^1(\Omega)},$$

where C is independent of ε and y_0 (see estimate (3.16)). If we choose $\varepsilon > 0$ that minimizes the right hand side of the latter inequality, we get

$$T(y_0) = \left(\frac{CN}{2\rho}\|y_0\|_{L^1(\Omega)}\right)^{2/(N+2)} + \left(\frac{N}{2}\right)^{-N/(N+2)}\left(\frac{C}{\rho}\|y_0\|_{L^1(\Omega)}\right)^{2/(N+2)}.$$

We have therefore proved the following null controllability result for the system (3.28):

Proposition 3.5. *For any $y_0 \in L^1(\Omega)$ and $\rho > 0$ there is $u \in L^\infty(\Omega \times \mathbf{R}^+)$, $\|u\|_{L^\infty(\Omega\times\mathbf{R}^+)} < \rho$, that steers y_0 into the origin in a finite time $T(y_0)$.* ■

This result extends to controlled semilinear equations of the form (3.1) (see Barbu [10]).

Remark 3.1. Consider the nonlinear parabolic equation

$$\begin{aligned} \frac{\partial y}{\partial t} - \Delta y + |y|^{p-1}y = 0 \qquad & \text{in } \Omega \times \mathbf{R}^+, \\ y(x,0) = y_0(x), \qquad & x \in \Omega, \\ y = 0 \qquad & \text{in } \partial\Omega \times \mathbf{R}^+, \end{aligned} \tag{3.31}$$

where $0 < p < (N+2)/N$ and $y_0 \in L^1(\Omega)$. By Proposition 3.3, we know that the solution y satisfies the estimates

$$\|y(t)\|_{L^\infty(\Omega)} \le Ct^{-N/2}\|y_0\|_{L^1(\Omega)}, \qquad \|y(t)\|_{L^1(\Omega)} \le C\|y_0\|_{L^1(\Omega)},$$

for all $t > 0$. Now, if y_0 is a bounded Radon measure on Ω, i.e., $y_0 \in M(\Omega) = (C_0(\overline{\Omega}))^*$ ($C_0(\overline{\Omega})$ is the space of continuous functions on $\overline{\Omega}$ that vanish on $\partial\Omega$), there is a sequence $\{y_0^j\} \subset C_0(\Omega)$ such that $\|y_0^j\|_{L^1(\Omega)} \le C$ and $y_0^j \to y_0$ weak star in $M(\Omega)$. Then if y^j is the corresponding solution to Eq. (3.31) it follows from the previous estimates that (see Brézis and Friedman [1])

$$y^j \to y \qquad \text{in } L^q(Q), 1 < q < (N+2)/N,$$
$$|y^j|^{p-1}y^j \to |y|^{p-1}y \qquad \text{in } L^1(Q).$$

This implies that y is a generalized (mild) solution to Eq. (3.31). If $p > (N+2)/N$, then there is no solution to (3.31).

Remark 3.2. Consider the semilinear parabolic equation (3.1), where β is a continuous monotonically increasing function, $f \in L^p(Q)$, $p > 1$, and $y_0 \in W_0^{p,2-2/p}(\Omega)$, $j(y_0) \in L^1(\Omega)$ where $j(r) = \int_0^r |\beta(s)|^{p-2}\beta(s)\,ds$. Then the solution y to problem (3.1) belongs to $W_p^{2,1}(Q)$ and

$$\|y\|_{W_p^{2,1}(Q)}^p \le C\left(\|f\|_{L^p(\Omega)}^p + \|y_0\|_{W_0^{p,2-2/p}(\Omega)}^p + \int_\Omega j(y_0)\,dx\right). \tag{3.32}$$

Here, $W_p^{2,1}(Q)$ is the space $\{y \in L^p(Q); \partial^{r+s}/\partial t^r\,\partial x^s)y \in L^p(Q), 2r + s \le 2\}$. For $p = 2$, $W_2^{2,1}(Q) = H^{2,1}(Q)$.

Indeed, if we multiply Eq. (3.1) by $|\beta(y)|^{p-2}\beta(y)$ we get the estimate (for f and y_0 smooth enough this problem has a unique solution $y \in W^{1,\infty}([0,T]; L^p(\Omega))$, $y \in L^\infty(0,T; W_2^p(\Omega))$

$$\int_\Omega j(y(x,t))\,dx + \int_0^t\int_\Omega |\beta(y(x,t))|^p\,dx\,dt$$
$$\le \int_0^t\int_\Omega |\beta(y(x,s))|^{p-1}|f(x,s)|\,dx\,ds + \int_\Omega j(y_0(x))\,dx$$
$$\le \left(\int_0^t\int_\Omega |\beta(y(x,s))|^p\,dx\,ds\right)^{1/q}\left(\int_0^t\int_\Omega |f(x,s)|^p\,dx\,ds\right)^{1/p},$$

where $1/p + 1/q = 1$. In particular, this implies that

$$\|\beta(y)\|_{L^p(Q)} \le C(\|f\|_{L^p(Q)} + \|j(y_0)\|_{L^1(\Omega)})$$

and by the L^p estimates for linear parabolic equations (see, e.g., O. A. Ladyzenskaya *et al.* [1] and A. Friedman [1]) we find the estimate (3.32), which clearly extends to all $f \in L^p(Q)$ and $y_0 \in W_0^{p,2-2/p}(\Omega)$, $j(y_0) \in L^1(\Omega)$.

4.3.2. Parabolic Variational Inequalities

Here and throughout the sequel, V and H will be real Hilbert spaces such that V is dense in H and $V \subset H \subset V'$ algebraically and topologically. We shall denote by $|\cdot|$ and $\|\cdot\|$ the norms of H and V, respectively, and by $(\cdot,\cdot)$ the scalar product in H and the pairing between V and its dual V'. The norm of V' will be denoted $\|\cdot\|_*$.

We are given a linear continuous and symmetric operator A from V to V' satisfying the coercivity condition

$$(Ay, y) + \alpha|y|^2 \geq \omega\|y\|^2 \qquad \forall y \in V, \tag{3.33}$$

for some $\omega > 0$ and $\alpha \in \mathbf{R}$. We are also given a lower semicontinuous convex function $\varphi: V \to \overline{\mathbf{R}} =]-\infty, +\infty]$, $\varphi \not\equiv +\infty$.

For $y_0 \in V$ and $f \in L^2(0,T;V')$, consider the following problem:

Find $y \in L^2(0,T;V) \cap C([0,T];H) \cap W^{1,2}([0,T];V')$ *such that*

$$(y'(t) + Ay(t), y(t) - z) + \varphi(y(t)) - \varphi(z) \leq (f(t), y(t) - z)$$
$$\text{a.e.} \quad t \in (0,T), \forall z \in V,$$
$$y(0) = y_0. \tag{3.34}$$

Here, $y' = dy/dt$ is the strong derivative of the function $y:[0,T] \to V'$. In terms of the subgradient mapping $\partial\varphi: V \to V'$ problem (3.34) can be written as

$$y'(t) + Ay(t) + \partial\varphi(y(t)) \ni f(t) \qquad \text{a.e. } t \in (0,T),$$
$$y(0) = y_0. \tag{3.35}$$

This is an abstract variational inequality of parabolic type. In applications to partial differential equations, V is a Sobolev subspace of $H = L^2(\Omega)$ (Ω is an open subset of $\mathbf{R}^N$), A is an elliptic operator on Ω, and the unknown function $y: \Omega \times [0,T] \to \mathbf{R}$ is viewed as a function of t from $[0,T]$ to $L^2(\Omega)$. Finally, the derivative y' is the partial derivative of $y = y(x,t)$ in t in the sense of distributions.

In the special case where $\varphi = I_K$ is the indicator function of a closed convex subset K of V, i.e.

$$\varphi(y) = 0 \quad \text{if } y \in K, \qquad \varphi(y) = +\infty \quad \text{if } y \notin K, \tag{3.36}$$

the variational inequality (3.34) reduces to

$$
\begin{aligned}
&y(t) \in K \qquad \forall t \in [0,T],\\
&(y'(t) + Ay(t), y(t) - z) \le (f(t), y(t) - z)\\
&\qquad\qquad\qquad\qquad \text{a.e. } t \in (0,T),\ \forall z \in K,\\
&y(0) = y_0.
\end{aligned} \tag{3.37}
$$

Regarding existence for problem (3.34), we have:

Theorem 3.1. *Let $f \in W^{1,2}([0,T]; V')$ and $y_0 \in V$ be such that*

$$\{Ay_0 + \partial\varphi(y_0) - f(0)\} \cap H \neq \emptyset. \tag{3.38}$$

Then problem (3.34) has a unique solution $y \in W^{1,2}([0,T]; V) \cap W^{1,\infty}([0,T]; H)$ and the map $(y_0, f) \to y$ is Lipschitz from $H \times L^2(0,T; V')$ to $C([0,T]; H) \cap L^2(0,T; V)$. If $f \in W^{1,2}([0,T]; V')$ and $\varphi(y_0) < \infty$, then the problem (3.34) has a unique solution $y \in W^{1,2}([0,T]; H) \cap C_w([0,T]; V)$. If $f \in L^2(0,T; H)$ and $\varphi(y_0) < \infty$, then the problem (3.34) has a unique solution $y \in W^{1,2}([0,T]; H) \cap C_w([0,T]; V)$, which satisfies

$$y'(t) = (f(t) - Ay(t) - \partial\varphi(y(t)))^0 \qquad \text{a.e. } t \in (0,T).$$

Proof. Consider the operator $L: D(L) \subset H \to H$,

$$
\begin{aligned}
Ly &= \{Ay + \partial\varphi(y)\} \cap H \qquad \forall y \in D(L),\\
D(L) &= \{y \in V; \{Ay + \partial\varphi(y)\} \cap H \neq \emptyset\}.
\end{aligned}
$$

Note that $\alpha I + L$ is maximal monotone in $H \times H$ (I is the identity operator in H). Indeed, by hypothesis (3.33) the operator $\alpha I + A$ is continuous and positive definite from V to V'. Since $\partial\varphi: V \to V'$ is maximal monotone we infer by Theorem 1.5 (or Corollary 1.5) in Chapter 2 that $\alpha I + L$ is maximal monotone from V to V' and, consequently, in $H \times H$.

Then, by Theorem 1.4, for every $y_0 \in D(L)$ and $g \in W^{1,1}([0,T]; H)$ the Cauchy problem

$$
\begin{aligned}
&\frac{dy}{dt} + Ly \ni g \qquad \text{a.e. in } (0,T),\\
&y(0) = y_0,
\end{aligned}
$$

has a unique strong solution $y \in W^{1,\infty}([0,T];H)$. Let us observe that $\partial\varphi_\alpha = \alpha I + L$, where $\varphi_\alpha: H \to \overline{\mathbf{R}}$ is given by

$$\varphi_\alpha(y) = \tfrac{1}{2}(Ay + \alpha y, y) + \varphi(y) \qquad \forall y \in H. \tag{3.39}$$

Indeed, φ_α is convex and lower semicontinuous in H because

$$\lim_{\|y\|\to\infty} \varphi_\alpha(y)/\|y\| = \infty$$

and φ_α is lower semicontinuous on V.

On the other hand, it is readily seen that $\alpha I + L \subset \partial\varphi_\alpha$, and since $\alpha I + L$ is maximal monotone, we infer that $\alpha I + L = \partial\varphi_\alpha$, as claimed. In particular, this implies that $\overline{D(L)} = \overline{D(\varphi_\alpha)} = \overline{D(\varphi)}$ (in the topology of H).

Now, let $y_0 \in V$ and $f \in W^{1,2}([0,T];V')$, satisfying condition (3.38). Let $\{y_0^n\} \subset D(L)$ and $\{f_n\} \subset W^{1,2}([0,T];H)$ be such that

$$\begin{aligned} y_0^n &\to y_0 && \text{strongly in } H, \text{ weakly in } V, \\ f_n &\to f && \text{strongly in } L^2(0,T;V'), \\ f_n' &\to f' && \text{strongly in } L^2(0,T;V') \end{aligned}$$

Let $y_n \in W^{1,\infty}([0,T];H)$ be the corresponding solution to the Cauchy problem

$$\begin{aligned} \frac{dy_n}{dt} + Ly_n &\ni f_n \qquad \text{a.e. in } (0,T), \\ y_n(0) &= y_0^n . \end{aligned} \tag{3.40}$$

If we multiply Eq. (3.40) by $y_n - y_0$ and use condition (3.33), we get

$$\begin{aligned} &\frac{1}{2}\frac{d}{dt}|y_n(t) - y_0|^2 + \omega\|y_n(t) - y_0\|^2 \\ &\quad \le \alpha|y_n(t) - y_0|^2 + (f_n(t) - \xi, y_n(t) - y_0) \qquad \text{a.e. } t \in (0,T), \end{aligned} \tag{3.41}$$

where $\xi \in Ay_0 + \partial\varphi(y_0) \subset V'$. After some calculation involving Gronwall's lemma, this yields

$$|y_n(t) - y_0|^2 + \int_0^t \|y_n(s) - y_0\|^2\, ds \le C \qquad \forall n, t \in [0,T]. \tag{3.42}$$

Now we use the monotonicity of $\partial\varphi$ along with condition (3.33) to get by (3.40) that

$$\frac{1}{2}\frac{d}{dt}|y_n(t) - y_m(t)|^2 + \omega\|y_n(t) - y_m(t)\|^2$$
$$\leq \alpha|y_n(t) - y_m(t)|^2 + \|f_n(t) - f_m(t)\|_*\|y_n(t) - y_m(t)$$
$$\text{a.e. } t \in (0,T).$$

Integrating on $(0,t)$ and using Gronwall's lemma, we obtain the inequality

$$|y_n(t) - y_m(t)|^2 + \int_0^T \|y_n(t) - y_m(t)\|^2\,dt$$
$$\leq C\left(|y_0^n - y_0^m|^2 + \int_0^T \|f_n(t) - f_m(t)\|^2\,dt\right).$$

Thus, there is $y \in C([0,T]; H) \cap L^2(0,T;V)$ such that

$$y_n \to y \qquad \text{in } C([0,T];H) \cap L^2(0,T;V). \tag{3.43}$$

Now, again using Eq. (3.40), we get

$$\frac{1}{2}\frac{d}{dt}|y_n(t+h) - y_n(t)|^2 + \omega\|y_n(t+h) - y_n(t)\|^2$$
$$\leq \alpha|y_n(t+h) - y_n(t)|^2$$
$$+ \|f_n(t+h) - f_n(t)\|_*\|y_n(t+h) - y_n(t)\|$$
$$\text{a.e. } t,h \in (0,T).$$

This yields

$$|y_n(t+h) - y_n(t)|^2 + \int_0^{T-h} \|y_n(t+h) - y_n(t)\|^2\,dt$$
$$\leq C\left(|y_n(h) - y_0^n|^2 + \int_0^{T-h} \|f_n(t+h) - f_n(t)\|_*^2\,dt\right)$$

and, letting n tend to $+\infty$,

$$|y(t+h) - y(t)|^2 + \int_0^{T-h} \|y(t+h) - y(t)\|^2\,dt$$
$$\leq C\left(|y(h) - y_0|^2 + \int_0^{T-h} \|f(t+h) - f(t)\|_*^2\,dt\right)$$
$$\forall t \in [0, T-h]. \tag{3.44}$$

Next, by (3.41) we see that if $\xi \in Ay_0 + \partial\varphi(y_0)$ is such that $f(0) - \xi \in H$, then we have

$$\frac{1}{2}\frac{d}{dt}|y_n(t) - y_0|^2 + \omega\|y_n(t) - y_0\|^2$$
$$\leq \alpha|y_n(t) - y_0|^2 + \|f_n(t) - f_n(0)\|_* \|y_n(t) - y_0^n\|$$
$$+ |f_n(0) - \xi|\,|y_n(t) - y_0^n|.$$

Integrating and letting $n \to \infty$ we get, by the Gronwall inequality,

$$|y(t) - y_0| \leq C\left(\int_0^t \|f(s) - f(0)\|_* \, ds + |f(0) - \xi|t\right) \qquad \forall t \in [0,T].$$

This yields

$$|y(t) - y_0| \leq Ct \qquad \forall t \in [0,T].$$

Along with (3.44) the latter inequality implies that y is H-valued, absolutely continuous on $[0,T]$, and

$$|y'(t)|^2 + \int_0^t \|y'(t)\|^2 \, dt \leq C\left(|y_0|^2 + \int_0^T \|f'(t)\|_*^2 \, dt + 1\right) \qquad \text{a.e. } t \in (0,T).$$

Hence, $y \in W^{1,\infty}([0,T]; H) \cap W^{1,2}([0,T]; V)$.

Let us show now that y satisfies equation (3.34) (equivalently, (3.35)). By (3.40) we have

$$\frac{1}{2}\frac{d}{dt}|y_n(t) - z|^2 \leq (f_n(t) + \alpha y_n(t) - \eta, y_n(t) - z) \qquad \text{a.e. } t \in (0,T),$$

where $z \in D(L)$ and $\eta \in Lz$. This yields

$$\tfrac{1}{2}\left(|y_n(t+\varepsilon) - z|^2 - |y_n(t) - z|^2\right)$$
$$\leq \int_t^{t+\varepsilon} (f_n(s) + \alpha y_n(s) - \eta, y_n(s) - z) \, ds$$

and, letting $n \to \infty$,

$$\tfrac{1}{2}\left(|y(t+\varepsilon)-z|^2-|y(t)-z|^2\right) \le \int_t^{t+\varepsilon}(f(s)+\alpha y(s)-\eta, y(s)-z\,ds.$$

Finally, this yields

$$(y(t+\varepsilon)-y(t), y(t)-z) \le \int_t^{t+\varepsilon}(f(s)+\alpha y(s)-\eta, y(s)-z)\,ds.$$

Since y is a.e. H-differentiable on $(0,T)$, we get

$$(y'(t)-\alpha y(t)+\eta-f(t), y(t)-z) \le 0 \qquad \text{a.e. } t \in (0,T),$$

for all $[z,\eta] \in L$. Now, since L is maximal monotone in $H \times H$, we conclude that

$$f(t) \in y'(t)+Ly(t) \qquad \text{a.e. } t \in (0,T),$$

as desired.

Now, if (y_0^i, f_i), $i=1,2$, satisfy condition (3.38) and the y_i are the corresponding solution to Eq. (3.35), by assumption (3.33) it follows that

$$|y_1(t)-y_2(t)|^2+\int_0^T\|y_1(t)-y_2(t)\|^2\,dt \le C\left(|y_0^1-y_0^2|^2+\int_0^T\|f_1(t)-f_2(t)\|_*^2\,dt\right) \qquad \forall t \in [0,T].$$

Now assume that $f \in W^{1,2}([0,T];V')$ and $y_0 \in D(\varphi)$. Then, as seen earlier, we may rewrite Eq. (3.35) as

$$y'+\partial\varphi_\alpha(y)-\alpha y \ni f \qquad \text{a.e. } t \in (0,T),$$
$$y(0)=y_0, \tag{3.45}$$

where $\varphi_\alpha: H \to \overline{\mathbf{R}}$ is defined by (3.39). For $f=f_n$ and $y_0=y_0^n$, $y=y_n$, we have the estimate

$$|y_n'(t)|^2+\frac{d}{dt}\varphi_\alpha(y_n(t))-\frac{\alpha}{2}\frac{d}{dt}|y_n(t)|^2 \le (f_n(t), y_n'(t)) \qquad \text{a.e. } t \in (0,T).$$

This yields

$$\int_0^T|y_n'(t)|^2\,dt+\varphi_\alpha(y_n(t)|) \le (f_n(0), y_n^0)+\int_0^T\|f_n'(t)\|_*\|y_n(t)\|\,dt-\frac{\alpha}{2}|y_n^0|^2.$$

Finally,

$$\int_0^T |y_n'(t)|^2 \, dt + \|y_n(t)\|^2 \le C\left(\|f_n\|_{W^{1,2}([0,T];V')} + |y_n^0|^2\right) \le C.$$

Then, arguing as before, we see that the function y given by (3.43) belongs to $W^{1,2}([0,T]; H) \cap L^\infty(0,T;V)$ and is a solution to Eq. (3.34). Since $y \in C([0,T]; H) \cap L^\infty(0,T;V)$, it is readily seen that y is weakly continuous from $[0,T]$ to V.

If $f \in L^\infty(0,T; H)$ and $y_0 \in D(\varphi_\alpha)$ we may apply Theorem 1.10 to Eq. (3.45) to arrive at the same result. ■

Theorem 3.2. *Let $y_0 \in K$ and $f \in W^{1,2}([0,T]; V')$ be given such that*

$$(f(0) - Ay_0 - \xi_0, y_0 - v) \ge 0 \qquad \forall v \in K, \tag{3.46}$$

for some $\xi_0 \in H$. Then problem (3.37) has a unique solution $y \in W^{1,\infty}([0,T]; H) \cap W^{1,2}([0,T]; V)$.

If $y_0 \in K$ and $f \in W^{1,2}([0,T]; V')$, then problem (3.37) has a unique solution $y \in W^{1,2}([0,T]; H) \cap C_w([0,T]; V)$. If $f \in L^2(0,T; H)$ and $y_0 \in K$, then problem (3.37) has a unique solution $y \in W^{1,2}([0,T]; H) \cap C_w([0,T]; V)$. Assume in addition that

$$(Ay, y) \ge \omega \|y\|^2 \qquad \forall y \in V, \tag{3.47}$$

for some $\omega > 0$, and that there is $h \in H$ such that

$$(I + \varepsilon A_H)^{-1}(y + \varepsilon h) \in K \qquad \forall \varepsilon > 0, \forall y \in K. \tag{3.48}$$

Then $Ay \in L^2(0,T; H)$.

Proof. The first part of the theorem is an immediate consequence of Theorem 3.1. Now assume that $f \in L^2(0,T; H)$, $y_0 \in K$, and conditions (3.47), (3.48) hold. Let $y \in W^{1,2}([0,T]; H) \cap C_w([0,T]; V)$ be the solution to (3.37). If in (3.37) we take $z = (I + \varepsilon A_H)^{-1}(y + \varepsilon h)$ (we recall that $A_H y = Ay \cap H$), we get

$$\begin{aligned}
&\left(y'(t) + A(t), A_\varepsilon y(t) - (I + \varepsilon A_H)^{-1} h\right) \\
&\qquad \le \left(f(t), A_\varepsilon y(t) - (I + \varepsilon A_H)^{-1} h\right) \qquad \text{a.e. } t \in (0,T),
\end{aligned}$$

where $A_\varepsilon = A(I + \varepsilon A_H)^{-1} = \varepsilon^{-1}(I - (I + \varepsilon A_H)^{-1})$. Since

$$(Ay, A_\varepsilon y) \ge |A_\varepsilon y|^2 \qquad \forall y \in D(A_H) = \{y;\ Ay \in H\}$$

and

$$\frac{1}{2}\frac{d}{dt}(A_\varepsilon y(t), y(t)) = (y'(t), A_\varepsilon y(t)) \qquad \text{a.e. } t \in (0,T)$$

we get

$$\begin{aligned}(A_\varepsilon y(t), y(t)) &+ \int_0^t |A_\varepsilon y(s)|^2\, ds \\ &\leq (A_\varepsilon y_0, y_0) + 2\int_0^t \big(A_\varepsilon y(s) - (I + \varepsilon A_H)^{-1} f(s), h\big)\, ds \\ &\quad + \int_0^t |f(s)|^2\, ds + 2\big(y(t) - y_0, (I + \varepsilon A_H)^{-1} h\big) \\ &\qquad\qquad \text{a.e. } t \in (0,T).\end{aligned}$$

Hence,

$$\int_0^T |A_\varepsilon y(t)|^2\, dt + (A_\varepsilon y(t), y(t)) \leq C \qquad \forall \varepsilon > 0,\ t \in [0,T],$$

and by Proposition 1.3 in Chapter 2 we conclude that $Ay \in L^2(0,T;H)$, as claimed. ■

Now, we shall prove a variant of Theorem 3.1 in the case where $\varphi: V \to \overline{\mathbf{R}}$ is lower semicontinuous on H. (It is easily seen that this happens, for instance, if $\varphi(u)/\|u\| \to +\infty$ as $\|u\| \to \infty$.

Proposition 3.6. *Let $A: V \to V'$ be a linear, continuous, symmetric operator satisfying condition* (3.33) *and let $\varphi: H \to \overline{\mathbf{R}}$ be a lower semicontinuous convex function. Further, assume that there is C independent of ε such that either*

$$(Ay, \nabla\varphi_\varepsilon(y)) \geq -C(1 + |\nabla\varphi_\varepsilon(y)|)(1 + |y|) \qquad \forall y \in D(A_H), \tag{3.49}$$

or

$$\varphi\big((I + \varepsilon A_\alpha)^{-1}(y + \varepsilon h)\big) \leq \varphi(y) + C \qquad \forall \varepsilon > 0\ \forall y \in H, \tag{3.50}$$

for some $h \in H$, where $A_\alpha = \alpha I + A_H$.

Then for every $y_0 \in \overline{D(\varphi) \cap V}$ and every $f \in L^2(0,T;H)$, problem (3.37) *has a unique solution $y \in W^{1,2}((0,T];H) \cap C([0,T];H)$ such that $t^{1/2}y' \in L^2(0,T;H)$, $t^{1/2}Ay \in L^2(0,T;H)$. If $y_0 \in D(\varphi) \cap V$, then $y \in W^{1/2}([0,T];H) \cap C([0,T];V)$. Finally, if $y_0 \in D(A_H) \cap D(\partial\varphi)$ and $f \in W^{1,1}([0,T];H)$, then $y \in W^{1,\infty}([0,T];H)$.*

Here, φ_ε is the regularization of φ.

Proof. As seen previously, the operator

$$A_\alpha y = \alpha y + Ay \qquad \forall y \in D(A_\alpha) = D(A_H),$$

is maximal monotone in $H \times H$. Then, by Theorems 1.6 (if condition (3.49) holds) and, respectively, 2.4 in Chapter 2 (under assumption (3.50)), $A_\alpha + \partial\varphi$ is maximal monotone in $H \times H$ and

$$|A_\alpha y| \leq C\big(|(A_\alpha + \partial\varphi)^0(y)| + |y| + 1\big) \qquad \forall y \in D(A_H) \cap D(\partial\varphi).$$

Moreover, $A_\alpha + \partial\varphi = \partial\varphi^\alpha$, where (see (3.39))

$$\varphi^\alpha(y) = \frac{1}{2}(Ay, y) + \varphi(y) + \frac{\alpha}{2}|y|^2 \qquad \forall y \in V,$$

and writing Eq. (3.35) as

$$y' + \partial\varphi^\alpha(y) - \alpha y \ni f \qquad \text{a.e. in } (0,T),$$
$$y(0) = y_0,$$

it follows by Theorem 1.10 that there is a strong solution y to Eq. (3.39) satisfying the conditions of the theorem. Note, for instance, that if $y_0 \in D(\varphi) \cap V$ then $y \in W^{1,2}([0,T]; H)$ and $\varphi^\alpha(y) \in W^{1,1}([0,T])$. Since y is continuous from $[0,T]$ to H this implies that y is weakly continuous from $[0,T]$ to V. Now, since $t \to \varphi^\alpha(y(t))$ is continuous and $\varphi: H \to \overline{\mathbf{R}}$ is lower semicontinuous, we have

$$\lim_{t_n \to t} (Ay(t_n), y(t_n)) \leq (Ay(t), y(t)) \qquad \forall t \in [0,T],$$

and this implies that $y \in C([0,T]; V)$, as claimed. ■

Corollary 3.1. *Let $A: V \to V'$ be a linear, continuous, and symmetric operator satisfying condition* (3.33) *and let K be a closed convex subset of H with*

$$(I + \varepsilon A_\alpha)^{-1}(y + \varepsilon h) \in K \qquad \forall \varepsilon > 0, \forall y \in K, \tag{3.51}$$

for some $h \in H$.

Then, for every $y_0 \in K$ and $f \in L^2(0,T; H)$, the variational inequality (3.37) *has a unique solution $y \in W^{1,2}([0,T]; H) \cap C([0,T]; V) \cap$*

$L^2(0,T;D(A_H))$. Moreover, one has

$$\frac{dy}{dt}(t) + (A_H y(t) - f(t) - N_K(y(t)))^0 = 0 \qquad \text{a.e. } t \in (0,T),$$

where $N_K(y) \subset L^2(\Omega)$ is the normal cone at K in y. ■

As an example, consider the *obstacle parabolic problem*

$$\begin{aligned}
\frac{\partial y}{\partial t} - \Delta y &= f && \text{in } \{(x,t) \in Q;\ y(x,t) > \psi(x)\},\\
\frac{\partial y}{\partial t} - \Delta y &\geq f && \text{in } Q = \Omega \times (0,T),\\
y(x,t) &\geq \psi(x) && \forall (x,t) \in Q,\\
\alpha_1 y + \alpha_2 \frac{\partial y}{\partial \nu} &= 0 && \text{in } \Sigma = \partial\Omega \times (0,T),\\
y(x,0) &= y_0(x), && x \in \Omega,
\end{aligned} \tag{3.52}$$

where Ω is an open bounded subset of $\mathbf{R}^N$ with smooth boundary (of class $C^{1,1}$, for instance), $\psi \in H^2(\Omega)$, and $\alpha_1, \alpha_2 \geq 0$, $\alpha_1 + \alpha_2 > 0$.

This is a problem of the form (3.37), where $H = L^2(\Omega)$, $V = H^1(\Omega)$, and $A \in L(V,V')$ is defined by

$$(Ay,z) = \int_\Omega \nabla y \cdot \nabla z \, dx + \frac{\alpha_1}{\alpha_2} \int_{\partial\Omega} yz \, d\sigma \qquad \forall y,z \in H^1(\Omega), \tag{3.53}$$

if $\alpha_2 \neq 0$, or

$$(Ay,z) = \int_\Omega \nabla y \cdot \nabla z \, dx \qquad \forall y,z \in H_0^1(\Omega), \tag{3.54}$$

if $\alpha_2 = 0$. (In this case, $V = H_0^1(\Omega)$, $V' = H^{-1}(\Omega)$.)

The set $K \subset V$ is given by

$$K = \{y \in H^1(\Omega);\ y(x) \geq \psi(x) \text{ a.e. } x \in \Omega\}, \tag{3.55}$$

and condition (3.51) is satisfied if (see Proposition 1.1 in Chapter 3)

$$\alpha_1 \psi + \alpha_2 \frac{\partial \psi}{\partial \nu} \leq 0 \qquad \text{a.e. in } \partial\Omega. \tag{3.56}$$

Note also that $A_H: D(A_H) \subset L^2(\Omega) \to L^2(\Omega)$ is defined by

$$A_H y = -\Delta y \qquad \text{a.e. in } \Omega, \forall y \in D(A_H),$$

$$D(A_H) = \left\{ y \in H^2(\Omega);\ \alpha_1 y + \alpha_2 \frac{\partial y}{\partial \nu} = 0 \text{ a.e. in } \partial\Omega \right\},$$

and

$$\|y\|_{H^2(\Omega)} \le C(\|A_H y\|_{L^2(\Omega)} + \|y\|_{L^2(\Omega)}) \qquad \forall y \in D(A_H).$$

Then we may apply Corollary 3.1 to get:

Corollary 3.2. *Let $f \in L^2(Q)$, $y_0 \in H^1(\Omega)$ ($y_0 \in H_0^1(\Omega)$ if $\alpha_2 = 0$) be such that $y_0 \ge \psi$ a.e. in Ω. Assume also that $\psi \in H^2(\Omega)$ satisfies condition* (3.56). *Then problem* (3.52) *has a unique solution $y \in W^{1,2}([0,T]; L^2(\Omega)) \cap L^2(0,T; H^2(\Omega)) \cap C([0,T]; H_0^1(\Omega))$.* ■

Noticing that (see Section 1.1 in Chapter 3), $N_K(y) = \{v \in L^2(\Omega);\ v(x) \in \beta(y(x) - \psi(x)) \text{ a.e. } x \in \Omega\}$, where $\beta: \mathbf{R} \to 2^{\mathbf{R}}$ is given by

$$\beta(r) = \begin{cases} 0 & r > 0, \\ \mathbf{R}^- & r = 0, \\ \varnothing & r < 0, \end{cases}$$

it follows by Corollary 3.1 that the solution y satisfies the equation

$$\frac{d}{dt} y(t) + (-\Delta y(t) + \beta(y(t) - \psi) - f(t))^0 = 0 \qquad \text{a.e. } t \in (0,T).$$

Hence, the solution y to problem (3.52) given by Corollary 3.2 satisfies the system

$$\frac{\partial}{\partial t} y(x,t) - \Delta y(x,t) = f(x,t)$$
$$\text{a.e. in } \{(x,t) \in Q;\ y(x,t) > \psi(x)\},$$
$$\frac{\partial}{\partial t} y(x,t) = \max\{f(x,t) + \Delta\psi(x), 0\}$$
$$\text{a.e. in } \{(x,t);\ y(x,t) = \psi(x)\}, \quad (3.57)$$

because $y(\cdot,t) \in H^2(\Omega)$ and so $\Delta y(x,t) = \Delta\psi(x)$ a.e. in $\{y(x,t) = \psi(x)\}$.

Since the previous problem can be equivalently written as

$$\frac{\partial y}{\partial t} - \Delta y + \beta(y - \psi) \ni f \qquad \text{in } Q,$$
$$y(x,0) = y_0(x) \qquad \text{in } \Omega,$$
$$\alpha_1 y + \alpha_2 \frac{\partial y}{\partial \nu} = 0 \qquad \text{in } \Sigma,$$

we may recover Corollary 3.2 as a particular case of Theorem 1.10 (see Proposition 3.1). It follows, therefore, that the solution y to the obstacle problem (3.52) is given by

$$y(t) = \lim_{\varepsilon \to 0} y_\varepsilon(t) \qquad \text{in } C([0,T]; L^2(\Omega)),$$

where y_ε is the solution to penalized problem

$$\begin{aligned} \frac{\partial y}{\partial t} - \Delta y - \frac{1}{\varepsilon}(y - \psi)^- = f \qquad & \text{in } Q, \\ y(x,0) = y_0(x) \qquad & \text{in } \Omega, \\ \alpha_1 y + \alpha_2 \frac{\partial y}{\partial \nu} = 0 \qquad & \text{in } \Sigma. \end{aligned} \tag{3.58}$$

Now let us consider the obstacle problem (3.52) with nonhomogeneous boundary conditions, i.e.,

$$\begin{aligned} &\frac{\partial y}{\partial t} - \Delta y = f && \text{in } \{(x,t) \in Q;\ y(x,t) > \psi(x)\}, \\ &\frac{\partial y}{\partial t} - \Delta y \geq f, \quad y \geq 0 && \text{in } Q, \\ &\alpha y + \frac{\partial y}{\partial \nu} = g && \text{in } \Sigma_1 = \Gamma_1 \times (0,T) \\ &y = 0 && \text{in } \Sigma_2 = \Gamma_2 \times (0,T), \\ &y(x,0) = y_0(x) && \text{in } \Omega, \end{aligned} \tag{3.59}$$

where $\partial\Omega = \Gamma_1 \cup \Gamma_2$, $\Gamma_1 \cap \Gamma_2 = \emptyset$, and $g \in L^2(\Sigma_1)$. If we take $V = \{y \in H^1(\Omega);\ y = 0 \text{ in } \Gamma_2\}$, define $A: V \to V'$ by

$$(Ay, z) = \int_\Omega \nabla y \cdot \nabla z \, dx + \alpha \int_{\Gamma_1} yz \, d\sigma \qquad \forall y, z \in V,$$

and $f_0: [0,T] \to V'$ by

$$(f_0(t), z) = \int_{\Gamma_1} g(x,t) z(x) \, dx \qquad \forall z \in V,$$

we may write problem (3.59) as

$$\left(\frac{dy}{dt} + Ay(t), y(t) - z\right) \le (F(t), y(t) - z) \qquad \forall z \in K, \text{ a.e. } t \in (0,T),$$

$$y(0) = y_0, \tag{3.60}$$

where $F = f + f_0 \in L^2(0,T;V')$ and K is defined by (3.55).

Equivalently,

$$\begin{aligned}
&\int_\Omega \frac{\partial y}{\partial t}(x,t)\,(y(x,t) - z(x))\,dx + \int_\Omega \nabla y(x,t)\,\nabla(y(x,t) - z(x))\,dx \\
&\quad + \alpha \int_{\Gamma_1} y(x,t)\,(y(x,t) - z(x))\,dx \\
&\le \int_\Omega f(x,t)\,(y(x,t) - z(x))\,dx \\
&\quad + \int_{\Gamma_1} g(x,t)\,(y(x,t) - z(x))\,dx \\
&\quad \forall z \in K,\ t \in [0,T].
\end{aligned} \tag{3.61}$$

Applying Theorem 3.2, we get:

Corollary 3.3. *Let $f \in W^{1,2}([0,T]; L^2(\Omega))$, $g \in W^{1,2}([0,T]; L^2(\Gamma_1))$, and $y_0 \in K$. Then the problem (3.61) has a unique solution $y \in W^{1,2}([0,T]; V) \cap C_w([0,T]; V)$. If, in addition,*

$$\frac{\partial y_0}{\partial \nu} + \alpha y_0 = g(x,0) \qquad \text{a.e. in } \{x \in \Gamma_1;\ y_0(x) > \psi(x)\},$$

$$\frac{\partial \psi}{\partial \nu} + \alpha\psi \le g(x,0) \qquad \text{a.e. in } \{x \in \Gamma_1;\ y_0(x) = \psi(x)\}, \tag{3.62}$$

then $y \in W^{1,2}([0,T]; V) \cap W^{1,\infty}([0,T]; L^2(\Omega))$. ■

(We note that condition (3.62) implies (3.46).)

It is readily seen that the solution y to (3.61) satisfies (3.59) in a certain generalized sense. Indeed, assuming that the set $E = \{(x,t);\ y(x,t) > \psi(x)\}$ is open and taking $z = y(x,t) \pm \rho\varphi$ in (3.61), where $\varphi \in C_0^\infty(E)$ and

ρ is sufficiently small, we see that

$$\frac{\partial y}{\partial t} - \Delta y = f \qquad \text{in } \mathscr{D}'(E). \tag{3.63}$$

It is also obvious that

$$\frac{\partial y}{\partial t} - \Delta y \geq f \qquad \text{in } \mathscr{D}'(Q). \tag{3.64}$$

Regarding the boundary conditions, by (3.61), (3.63), and (3.64) it follows that

$$\frac{\partial y}{\partial \nu} + \alpha y = g \qquad \text{in } \mathscr{D}'(E \cap \Sigma_1),$$

respectively

$$\frac{\partial y}{\partial \nu} + \alpha y \geq g \qquad \text{in } \mathscr{D}'(\Sigma_1).$$

In other words,

$$\frac{\partial y}{\partial \nu} + \alpha y = g \qquad \text{in } \{(x,t) \in \Sigma_1;\, y(x,t) > \psi(x)\},$$

$$\frac{\partial \psi}{\partial \nu} + \alpha \psi \geq g \qquad \text{in } \{(x,t) \in \Sigma_1;\, y(x,t) = \psi(x)\}.$$

Hence, if g satisfies the compatibility condition

$$\frac{\partial \psi}{\partial \nu} + \alpha \psi \leq \qquad g \text{ in } \Sigma_1,$$

then the solution y to problem (3.61) satisfies the required boundary conditions on Σ_1.

Also in this case, the solution y given by Corollary 3.3 can be obtained as limit as $\varepsilon \to 0$ of the solution y_ε to the equation

$$\frac{\partial y_\varepsilon}{\partial t} - \Delta y_\varepsilon + \beta_\varepsilon(y_\varepsilon - \psi) = f \qquad \text{in } \Omega \times (0,T),$$

$$y_\varepsilon(x,0) = y_0(x) \quad \text{in } \Omega,$$

$$\frac{\partial y_\varepsilon}{\partial \nu} + \alpha y_\varepsilon = g \qquad \text{in } \Sigma_1, \qquad y_\varepsilon = 0 \qquad \text{in } \Sigma_2, \tag{3.65}$$

where $\beta_\varepsilon(r) = -(1/\varepsilon)r^- \ \forall r \in \mathbf{R}$.

If $Q^+ = \{(x,t) \in Q;\ y(x,t) > \psi(x)\}$, we may view y as the solution to free boundary problem

$$\frac{\partial y}{\partial t} - \Delta y = f \quad \text{in } Q^+,$$
$$y(x,0) = y_0(x) \quad \text{in } \Omega,$$
$$\alpha_1 y + \alpha_2 \frac{\partial y}{\partial \nu} = 0 \quad \text{in } \Sigma, \qquad y = \psi, \quad \frac{\partial y}{\partial \nu} = \frac{\partial \psi}{\partial \nu} \quad \text{in } \partial Q^+(t), \tag{3.66}$$

where $\partial Q^+(t)$ is the boundary of the set $Q^+(t) = \{x \in \Omega;\ y(x,t) > \psi(x)\}$. We call $\partial Q^+(t)$ the *moving boundary* and ∂Q^+ *the free boundary* of the problem (3.66).

In problem (3.66), the noncoincidence set Q^+ as well as the free boundary ∂Q^+ are not known *a priori* and represent unknowns of the problem. In problems (3.37) or (3.61) the free boundary does not appear explicitly, but in this formulation the problem is nonlinear and multivalued.

Perhaps the best-known example of a parabolic free boundary problem is the classical Stefan problem, which we will briefly describe in what follows and which has provided one of the principle motivations of the theory of parabolic variational inequalities.

The Stefan Problem. This problem describes the conduction of heat in a medium involving a phase charge. To be more specific, consider a unit volume of ice Ω at temperature $\theta < 0$. If a uniform heat source of intensity F is applied, then the temperature increases at rate E/C_1 until it reaches the melting point $\theta = 0$. Then the temperature remains at zero until ρ units of heat have been supplied to transform the ice into water (ρ is the latent heat). After all the ice has melted the temperature begins to increase at the rate h/C_2 (C_1 and C_2 are specific heats of ice and water, respectively). During this process the variation of the internal energy $e(t)$ is therefore given by

$$e(t) = C(\theta(t)) + \rho H(\theta(t)),$$

where

$$C(\theta) = \begin{cases} C_1\theta & \text{for } \theta \le 0, \\ C_2\theta & \text{for} \theta > 0, \end{cases}$$

and H is the Heaviside graph,

$$H(\theta) = \begin{cases} 1 & \theta > 0, \\ [0,1] & \theta = 0, \\ 0 & \theta < 0. \end{cases}$$

In other words,

$$e = \gamma(\theta) = \begin{cases} C_1\theta & \text{if } \theta < 0, \\ [0,\rho] & \text{if } \theta = 0, \\ C_2\theta & \text{if } \theta > 0. \end{cases} \tag{3.67}$$

The function γ is called the enthalpy of the system.

Now, let $Q = \Omega \times (0,\infty)$ and denote by Q_-, Q_+, Q_0 the regions of Q where $\theta < 0$, $\theta > 0$, and $\theta = 0$, respectively. We set $S_+ = \partial Q_+$, $S_- = \partial Q_-$, and $S = S_+ \cup S_-$.

If $\theta = \theta(x,t)$ is the temperature distribution in Q and $q = q(x,t)$ the heat flux, then, according to the Fourier law,

$$q(x,t) = -k\,\nabla\theta(x,t), \tag{3.68}$$

where k is the thermal conductivity. Consider the function

$$K(\theta) = \begin{cases} k_1\theta & \text{if } \theta \le 0, \\ k_2\theta & \text{if } \theta \le 0, \end{cases}$$

where k_1, k_2 are the thermal conductivity of the ice and water, respectively.

If f is the external heat source, then the conservation law yields

$$\frac{d}{dt}\int_{\Omega^*} e(x,t)\,dx = -\int_{\partial\Omega^*} (q(x,t),\nu)\,d\sigma + + \int_{\tilde{\Omega}} F(x,t)\,dx$$

for any subdomain $\Omega^* \times (t_1,t_2) \subset Q$ (ν is the normal to $\partial\Omega^*$) if e and q are smooth. Equivalently,

$$\begin{aligned} &\int_{\Omega^*} e_t(x,t)\,dx + \int_{S\cap\Omega^*} [\![e(t)]\!]V(t)\,dt \\ &\quad = -\int_{\Omega^*} \operatorname{div} q(x,t)\,dx \\ &\quad\quad + \int_{\partial\Omega^*\cap S} [\![(q(t),\nu)]\!]\,d\sigma + \int_{\Omega^*} F(x,t)\,dx, \end{aligned}$$

where $V(t) = -N_t/\|N_x\|$ is the true velocity of the interface S ($N = (N_t, N_x)$ is the unit normal to S) and $[\![\cdot]\!]$ is the jump along S.

The previous inequality yields

$$\frac{\partial}{\partial t}e(x,t) + \operatorname{div} q(x,t) = F(x,t) \qquad \text{in } Q \setminus S,$$

$$[\![e(t)]\!]N_t + [\![(q(t), N_x)]\!] = 0 \qquad \text{on } S. \tag{3.69}$$

Taking into account Eqs. (3.67)–(3.69), we get the system

$$C_1 \frac{\partial \theta}{\partial t} - k_1 \,\Delta\theta = f \qquad \text{in } Q_-,$$

$$C_2 \frac{\partial \theta}{\partial t} - k_2 \,\Delta\theta = f \qquad \text{in } Q_+, \tag{3.70}$$

$$(k_2 \nabla\theta^+ - k_1 \nabla\theta^-, N_x) = \rho N_t \qquad \text{in } S,$$

$$\theta^+ = \theta^- = 0 \qquad \text{in } S. \tag{3.71}$$

If we represent the interface S by the equation $t = \sigma(x)$, then Eqs. (3.71) read

$$(k_1 \nabla\theta^+ - k_2 \nabla\theta^-, \nabla\sigma) = -\rho \qquad \text{in } S,$$

$$\theta^+ = \theta^- = 0. \tag{3.72}$$

To Eqs. (3.70), (3.72), usual boundary and initial value conditions can be associated, for instance,

$$\theta = 0 \qquad \text{in } \partial\Omega \times (0,T), \tag{3.73}$$

$$\theta(x,0) = \theta_0(x) \qquad \text{in } \Omega, \tag{3.74}$$

or Neumann boundary conditions on $\partial\Omega$.

This is the classical two phase Stefan problem. Here we shall study with the methods of variational inequalities a simplified model described by the one phase Stefan problem

$$\frac{\partial \theta}{\partial t} - \Delta\theta = 0 \qquad \text{in } Q_+ = \{(x,t) \in Q;\ \sigma(x) < t < T\},$$

$$\theta = 0 \qquad \text{in } Q_- = \{(x,t) \in Q;\ 0 < t < \sigma(x)\},$$

$$\nabla_x \theta(x,t) \cdot \nabla\sigma(x) = -\rho \qquad \text{in } S = \{(x,t);\ t = \sigma(x)\},$$

$$\theta = 0 \qquad \text{in } S \cup Q_-,$$

$$\theta \geq 0 \qquad \text{in } Q_+. \tag{3.75}$$

These equations model the melting of a body of ice $\Omega \subset \mathbf{R}^3$ maintained at 0°C in contact with a region of water along a portion Γ_2 of the boundary $\partial\Omega$. So assume that $\partial\Omega = \Gamma_1 \cup \Gamma_2$, where Γ_1 and Γ_2 have no common boundary and Γ_1 is in contact with a heating medium with temperature θ_1; $t = \sigma(x)$ is the equation of the interface (moving boundary) S_t, which separates the liquid phase (water) and solid phase (ice) (see Fig. 4.1).Thus, to Eqs. (3.75) we must add the boundary conditions

$$\frac{\partial\theta}{\partial\nu} + \alpha(\theta - \theta_1) = 0 \qquad \text{in } \Sigma_1 = \Gamma_1 \times (0,T),$$

$$\theta = 0 \qquad \text{in } \Sigma_2 = \Gamma_2 \times (0,T) \tag{3.76}$$

and the initial value conditions

$$\theta(x,0) = \theta_0(x) > 0 \qquad \forall x \in \Omega_0, \qquad \theta(x,0) = 0 \qquad \forall x \in \Omega \setminus \Omega_0. \tag{3.77}$$

Instead of (3.76), we may impose the Dirichlet boundary conditions

$$\theta = v \quad \text{in } \Sigma_1, \qquad \theta = 0 \quad \text{in } \Sigma_2. \tag{3.76$'$}$$

We set $\Omega^t = \{x \in \Omega;\, 0 < t < \sigma(x)\}$. Then, $\partial\Omega^t = S_t$, and we assume that $\Omega^0 = \Omega_0$ (see Fig. 4.2).

There is a simple device due to G. Duvaut [1] that permits us to reduce problem (3.75)–(3.77) to a parabolic variational inequality via a Baiocchi

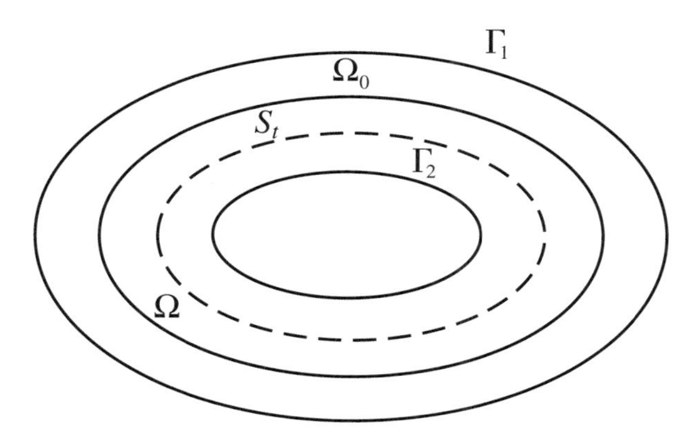

Figure 4.1.

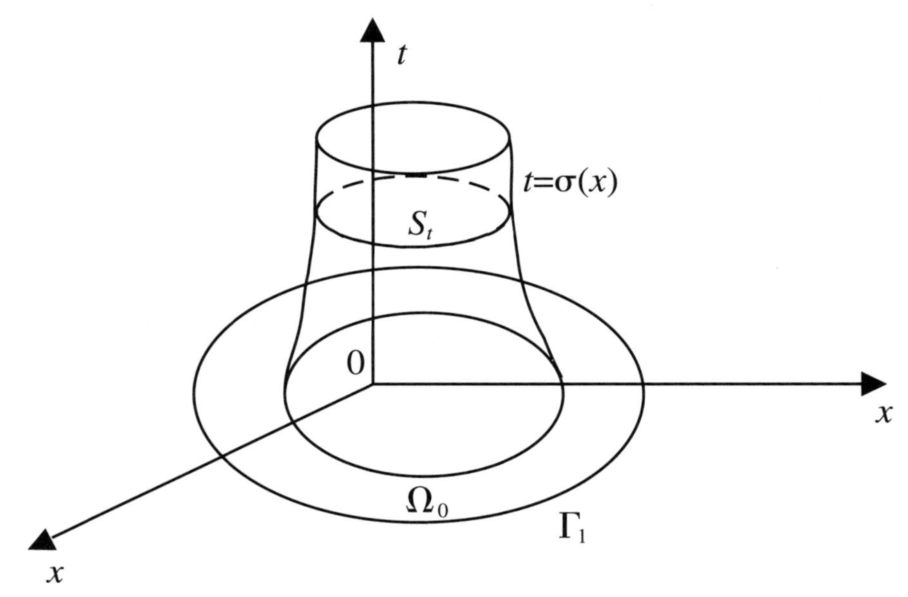

Figure 4.2.

transformation (see Section 1.2, Chapter 3). To this end, consider the function

$$y(x,t) = \begin{cases} \int_{\sigma(x)}^{t} \theta(x,s)\, ds & \text{if } x \in \Omega \setminus \Omega_0, t > \sigma(x), \\ \int_0^t \theta(x,s)\, ds & \text{if } x \in \Omega_0, t \in [0,T], \\ 0 & \text{if } (x,t) \in Q_-, \end{cases} \tag{3.78}$$

and let

$$f_0(x,t) = \begin{cases} -\rho & \text{if } x \in \Omega \setminus \Omega_0, 0 < t < T, \\ \theta_0(x) & \text{if } x \in \Omega_0, 0 < t < T. \end{cases} \tag{3.79}$$

Lemma 3.1. *Let $\theta \in H^1(Q)$ and $\sigma \in H^1(\Omega)$. Then*

$$\frac{\partial y}{\partial t} - \Delta y = f_0 \chi \quad \text{in } \mathscr{D}'(Q), \tag{3.80}$$

where χ is the characteristic function of Q_+.

Proof. By (3.78), we have

$$\frac{\partial y}{\partial t}(\varphi) = \int_{Q_+} \theta(x,t)\varphi(x,t)\,dx\,dt \qquad \forall \varphi \in C_0^\infty(Q).$$

On the other hand, we have

$$\begin{aligned}
y_x(\varphi) = -y(\varphi_x) &= -\int_{\Omega\setminus\Omega_0} dx \int_{\sigma(x)}^{T} \varphi_x(x,t)\,dt \int_{\sigma(x)}^{t} \theta(x,s)\,ds \\
&\quad - \int_{\Omega_0} dx \int_0^T \varphi_x(x,t)\,dt \int_0^t \theta(x,s)\,ds \\
&= -\int_{\Omega\setminus\Omega_0} dx \operatorname{div}\left(\int_{\sigma(x)}^{T} \varphi(x,t)\,dt \int_{\sigma(x)}^{t} \theta(x,s)\,ds\right) \\
&= \int_{\Omega\setminus\Omega_0} dx \left(\int_{\sigma(x)}^{T} \varphi(x,t) \int_{\sigma(x)}^{t} \theta_x(x,s)\,ds\right) \\
&\quad - \int_{\Omega_0} dx \operatorname{div}\left(\int_0^T \varphi(x,t)dt \int_0^t \theta(x,s)\,ds\right) \\
&= \int_{\Omega\setminus\Omega_0} dx \int_{\sigma(x)}^{T} \varphi(x,t)\,dt \int_{\sigma(x)}^{t} \theta_x(x,s)\,ds \\
&\quad + \int_{\Omega_0} dx \int_0^T \varphi(x,t)\,dt \int_0^t \theta_x(x,s)\,ds.
\end{aligned}$$

This yields

$$\begin{aligned}
\Delta y(\varphi) = -y_x(\varphi_x) &= -\int_{\Omega\setminus\Omega_0} dx \int_{\sigma(x)}^{T} \varphi_x(x,t)\,dt \int_{\sigma(x)}^{T} \theta_x(x,s)\,ds \\
&\quad - \int_{\Omega_0} dx \int_0^T \varphi_x(x,t)\,dt \int_0^t \theta_x(x,s)\,ds
\end{aligned}$$

and, by the divergence formula, we get

$$\begin{aligned}
\Delta y(\varphi) &= \int_{\Omega\setminus\Omega_0} dx \int_{\sigma(x)}^{T} dt \left(\int_{\sigma(x)}^{t} \Delta\theta(x,s)\,ds\,\varphi(x,t)\right) \\
&\quad + \int_{\Omega_0} dx \int_0^T dt \left(\int_0^t \Delta\theta(x,s)\,ds\,\varphi(x,t)\right) \qquad \forall \varphi \in C_0^\infty(Q),
\end{aligned}$$

because $\nabla_x \theta(x, \sigma(x)) \cdot \nabla\sigma(x) = -\rho \ \forall x \in \Omega \setminus \Omega_0$. Then, by Eqs. (3.75), we see that

$$\begin{aligned}\left(\frac{\partial y}{\partial t} - \Delta y\right)(\varphi) = &-\int_{\Omega\setminus\Omega_0} dx \int_{\sigma(x)}^{T} dt \left(\int_{\sigma(x)}^{t} \theta_t(x,s)\,ds - \theta(x,t)\right)\varphi(x,t)\\ &-\int_{\Omega_0} dx \int_0^T dt \left(\int_0^t \theta_t(x,s)\,ds - \theta(x,t)\right)\varphi(x,t))\\ &-\rho\int_{\Omega\setminus\Omega_0} dx \int_{\sigma(x)}^{T} \varphi(x,t)\,dt\\ = &\int_{Q_+} f_0(x,t)\varphi(x,t)\,dx\,dt,\end{aligned}$$

as claimed. ■

By Lemma 3.1 we see that the function y satisfies the obstacle problem

$$\begin{aligned} y \geq 0, \quad \frac{\partial y}{\partial t} - \Delta y \geq f_0 \quad &\text{in } Q,\\ \frac{\partial y}{\partial t} - \Delta y = f_0 \quad &\text{in } \{(x,t) \in Q;\ y(x,t) > 0\}.\\ y = 0 \quad &\text{in } \{(x,t) \in Q;\ \sigma(x) > t\},\end{aligned} \tag{3.81}$$

and the boundary value conditions

$$\frac{\partial y}{\partial \nu} = -\alpha\left(y - \tilde{\theta}_1\right) \quad \text{in } \Sigma_1, \qquad y = 0 \quad \text{in } \Sigma_2, \tag{3.82}$$

where $\tilde{\theta}_1(x,t) = \int_0^t \theta_1(x,s)\,ds \ \ \forall (x,t) \in \Sigma_1$. Then, by Corollary 3.2, we have:

Corollary 3.4. *Let $\theta_1 \in L^2(\Sigma_1)$ be given. Then the problem* (3.81), (3.82) *has a unique (generalized) solution $y \in W^{1,\infty}([0,T]; L^2(\Omega)) \cap W^{1,2}([0,T]; H^1(\Omega))$.* ■

Keeping in mind that $S_t = \partial\{(x,t);\ y(x,t) = 0\}$, we can derive from Corollary 3.4 an existence result for the one phase Stefan problem (3.75)–(3.77).

Other mathematical models for physical problems involving free boundary such as the oxygen diffusion in an absorbing tissue (E. Magenes [1], C. M. Elliott and J. R. Ockendon [1]) or electrochemical machining processes lead by similar devices to parabolic variational inequalities of the same type.

4.3.3. *The Nonlinear Diffusion Equation*

We shall study here the nonlinear Cauchy problem

$$\frac{\partial y}{\partial t} - \Delta \beta(y) \ni f \qquad \text{in } \Omega \times (0,T) = Q,$$

$$y = 0 \qquad \text{in } \partial\Omega \times (0,T) = \Sigma,$$

$$y(x,0) = y_0(x) \qquad \text{in } \Omega, \tag{3.83}$$

where Ω is a bounded and open subset of $\mathbf{R}^N$ with smooth boundary and $\beta: \mathbf{R} \to 2^{\mathbf{R}}$ is a maximal monotone graph in $\mathbf{R} \times \mathbf{R}$ such that $0 \in D(\beta)$.

The function $y \in C([0,T]; L^1(\Omega))$ is called a generalized solution to Eq. (3.83) if

$$\int_Q (y\varphi_t + \beta(y)\,\Delta\varphi)\,dx\,dt + \int_Q f\varphi\,dx\,dt + \int_\Omega y_0\,\varphi(x,0)\,dx = 0 \tag{3.84}$$

for all $\varphi \in C^{2,1}(\overline{Q})$ such that $\varphi(x,T) = 0$ in Ω and $\varphi = 0$ in Σ.

Let us first briefly describe some physical problems that lead to equation of this type.

1. *The flow of gases in porous media*. Let y be the density of a gas that flows through a porous medium that occupies a domain $\Omega \subset \mathbf{R}^3$. If p denotes the pressure, we have $p = p_0 y^\alpha$ for some $\alpha \geq 1$. Then, the conservation law equation

$$k_1 \frac{\partial y}{\partial t} + \operatorname{div} y\,\bar{v} = 0$$

combined with Darcy's law

$$\nu\bar{v} = -k_2\,\nabla p$$

(k_1 is the porosity of the medium, k_2 the permeability, and ν viscosity) yields the *porous medium equation*

$$\frac{\partial y}{\partial t} - \delta\,\Delta y^{\alpha+1} = 0 \qquad \text{in } Q, \tag{3.85}$$

where $\delta = k_2(k_1(\alpha + 1)\gamma p_0)^{-1}$.

2. *Two phase Stefan problem*. Consider the problem (3.70), (3.71), (3.73), (3.74), i.e.,

$$
\begin{aligned}
C_1\theta_t - k_1\,\Delta\theta = f \qquad & \text{in } Q_- = \{(x,t);\, \theta(x,t) < 0\},\\
C_2\theta_t - k_2\,\Delta\theta = f \qquad & \text{in } Q_+ = \{(x,t);\, \theta(x,t) > 0\},\\
(k_2\,\nabla\theta^+ - k_1\,\nabla\theta^-, \nabla\sigma(x)) = -\rho, &
\end{aligned}
\tag{3.86}
$$

where $t = \sigma(x)$ is the equation of the interface S.

We may write system (3.86) as

$$
\frac{\partial}{\partial t}\gamma(\theta) - \Delta K(\theta) \ni f \qquad \text{in } Q, \tag{3.87}
$$

where $\gamma: \mathbf{R} \to 2^{\mathbf{R}}$ is

$$
\gamma(r) = \begin{cases} C_1 r & \text{for } r < 0,\\ [0, \rho] & \text{for } r = 0,\\ C_2 r + \rho & \text{for } r > 0, \end{cases} \tag{3.88a}
$$

and

$$
K(\theta) = \begin{cases} k_1\theta & \text{for } \theta < 0,\\ k_2\theta & \text{for } \theta \geq 0. \end{cases} \tag{3.88b}
$$

Indeed, for every test function $\varphi \in C_0^\infty(Q)$ we have

$$
\begin{aligned}
\left(\frac{\partial}{\partial t}\gamma(\theta) - \Delta K(\theta)\right)(\varphi)
&= -\int_Q (\gamma(\theta)\varphi_t + K(\theta)\,\Delta\varphi)\,dx\,dt\\
&= C_1\int_{Q_-}\theta_t\varphi\,dx\,dt + C_2\int_{Q_+}\theta_t\varphi\,dx\,dt - k_1\int_{Q_-}\varphi\,\Delta\theta\,dx\,dt\\
&\quad - k_2\int_{Q_+}\varphi\,\Delta\theta\,dx\,ds + \int_S\left(k_2\frac{\partial\theta^+}{\partial\nu} - k_1\frac{\partial\theta^-}{\partial\nu}\right)\varphi\,ds - \rho\int_{Q_+}\varphi_t\,dx\,dt\\
&= \int_{Q_-}(C_1\theta_t - k_1\,\Delta\theta)\varphi\,dx\,dt + \int_{Q_+}(C_2\theta_t - k_2\,\Delta\theta)\varphi\,dx\,dt\\
&\quad + \int_S((k_2\,\nabla\theta^+ - k_1\,\nabla\theta^-, \nabla\sigma) + \rho)\,ds = 0.
\end{aligned}
$$

If we denote by β the function $\gamma^{-1}K$, i.e.,

$$\beta(r) = \begin{cases} C_1^{-1}r & \text{for } r < 0, \\ 0 & \text{for } 0 \leq r < k_2^{-1}\rho, \\ C_2^{-1}(k_2 r - \rho) & \text{for } r \geq k_2^{-1}\rho, \end{cases} \tag{3.89}$$

we may write (3.87) in the form (3.83).

Equation (3.83) is also relevent in the study of other mathematical models, such as population dynamics (Gurtin and MacCamy [1]). Problem (3.83) can be treated as a nonlinear accretive Cauchy problem in two functional spaces: $H^{-1}(\Omega)$ and $L^1(\Omega)$.

3. *The Hilbert Space Approach.* In the space $H^{-1}(\Omega)$, consider the operator

$$A = \{[y, w] \in (H^{-1}(\Omega) \cap L^1(\Omega)) \times H^{-1}(\Omega); w = -\Delta v, \\ v \in H_0^1(\Omega), v(x) \in \beta(y(x)) \text{ a.e. } x \in \Omega\}.$$

We shall assume that

β^{-1} *is everywhere defined and bounded on the bounded subsets of* $\mathbf{R}$. (3.90)

Then, by Proposition 2.12 in Chapter 2, A is maximal monotone in $H^{-1}(\Omega) \times H^{-1}(\Omega)$. More precisely, $A = \partial\varphi$ where $\varphi: H^{-1}(\Omega) \to \overline{\mathbf{R}}$ is defined by

$$\varphi(y) = \begin{cases} \int_\Omega j(y(x))\, dx & \text{if } y \in L^1(\Omega) \cap H^{-1}(\Omega), j(y) \in L^1(\Omega), \\ +\infty & \text{otherwise}, \end{cases}$$

where $\partial j = \beta$.

We may write problem (3.83) as

$$\frac{dy}{dt} + Ay \ni f \quad \text{in } (0, T),$$
$$y(0) = y_0, \tag{3.91}$$

and so by Theorem 1.10, we obtain:

Theorem 3.3. *Let β be a maximal monotone graph in $\mathbf{R} \times \mathbf{R}$ satisfying condition (3.90). Let $f \in L^1(0,T; H^{-1}(\Omega))$ and let $y_0 \in H^{-1}(\Omega) \cap L^1(\Omega)$ be such that $y_0(x) \in \overline{D(\beta)}$ a.e. $x \in \Omega$. Then there is a unique pair of functions $y \in C([0,T]; H^{-1}(\Omega)) \cap W^{1,2}((0,T]; H^{-1}(\Omega))$ and $v: Q \to \mathbf{R}$ satisfying*

$$\frac{\partial y}{\partial t} - \Delta v = f \qquad \text{a.e. in } Q = \Omega \times (0,T),$$

$$v(x,t) \in \beta(y(x,t)) \qquad \text{a.e. } (x,t) \in Q, \qquad v = 0 \quad \text{in } \Sigma,$$

$$y(x,0) = y_0(x) \qquad \text{a.e.in } \Omega. \tag{3.92}$$

$$t^{1/2} \frac{\partial y}{\partial t} \in L^2(0,T; H^{-1}(\Omega)), \qquad t^{1/2} v \in L^2(0,T; H_0^1(\Omega)). \tag{3.93}$$

Moreover, if $j(y_0) \in L^1(\Omega)$ then

$$\frac{\partial y}{\partial t} \in L^2(0,T; H^{-1}(\Omega)), \qquad v \in L^2(0,T; H_0^1(\Omega)). \tag{3.94}$$

If $y_0 \in D(A)$ and $f \in W^{1,1}([0,T]; H^{-1}(\Omega))$, then

$$\frac{\partial y}{\partial t} \in L^\infty(0,T; H^{-1}(\Omega)), \qquad v \in L^\infty(0,T; H_0^1(\Omega)). \qquad \blacksquare \tag{3.95}$$

We note that the derivative $\partial y/\partial t$ in (3.92) is the strong derivative of the function $t \to y(\cdot,t)$ from $[0,T]$ into $H^{-1}(\Omega)$, and it coincides with the derivative $\partial y/\partial t$ in the sense of distributions on Q. It is readily seen that the solution y provided by Theorem 3.3 is a generalized solution to (3.83) in the sense of definition (3.84).

The L^1-Approach. In the space $X = L^1(\Omega)$, consider the operator

$$A = \{[y,w] \in L^1(\Omega) \times L^1(\Omega); w = -\Delta v,$$
$$v \in W_0^{1,1}(\Omega), v(x) \in \beta(y(x)) \text{ a.e. } x \in \Omega\}. \tag{3.96}$$

We have seen earlier (Proposition 3.10 in Chapter 2) that A is m-accretive in $L^1(\Omega) \times L^1(\Omega)$. Then, applying the general existence Theorem 1.2, we obtain:

Proposition 3.7. *Let β be a maximal monotone graph in $\mathbf{R} \times \mathbf{R}$ such that $0 \in \beta(0)$. Then for every $f \in L^1(0,T; L^1(\Omega))$ and every $y_0 \in L^1(\Omega)$, $y_0(x)$*

$\in \overline{D(\beta)}$ *a.e.* $x \in \Omega$, *the Cauchy problem*

$$\frac{dy}{dt} + Ay \ni f \qquad \text{in } (0, T),$$

$$y(0) = y_0, \tag{3.97}$$

has a unique mild solution $y \in C([0, T]; L^1(\Omega))$. ■

We note that $\overline{D(A)} = \{y_0 \in L^1(\Omega);\ y_0(x) \in \overline{D(\beta)} \text{ a.e. } x \in \Omega\}$. Indeed, $(1 + \varepsilon\beta)^{-1} y_0 \to y_0$ in $L^1(\Omega)$ as $\varepsilon \to 0$, if $y_0(x) \in \overline{D(\beta)}$ a.e. $x \in \Omega$, and $(I + \varepsilon A)^{-1} y_0 \to y_0$ if $j(y_0) \in L^1(\Omega)$.

Proposition 3.7 amounts to saying that

$$y(t) = \lim_{\varepsilon \to 0} y_\varepsilon(t) \qquad \text{in } L^1(\Omega), \text{ uniformly on } [0, T],$$

where y_ε is the solution to difference equations

$$\frac{1}{\varepsilon}(y_\varepsilon(t) - y_\varepsilon(t - \varepsilon)) - \Delta v_\varepsilon(t) = f_\varepsilon(t), \qquad t \geq \varepsilon, x \in \Omega,$$

$$v_\varepsilon = 0 \qquad \text{in } \partial\Omega \times (0, T),$$

$$v_\varepsilon(x, t) \in \beta(y_\varepsilon(x, t)) \qquad \text{a.e. in } \Omega \times (0, T),$$

$$y_\varepsilon(t) = y_0 \qquad \text{for } t \leq 0, x \in \Omega. \tag{3.98}$$

The function $t \to v_\varepsilon(t) \in W_0^{1,1}(\Omega)$ is piecewise constant on $[0, T]$ and $f_\varepsilon(t) = f_i \ \ \forall t \in [i\varepsilon, (i + 1)\varepsilon]$ is a piecewise constant approximation of $f: [0, T] \to L^1(\Omega)$.

By (3.98) it is readily seen that y is a generalized solutiion to problem (3.83). In particular, it follows by proposition 3.7 that the operator A defined by (3.96) generates a semigroup of nonlinear contractions $S(t)$: $\overline{D(A)} \to \overline{D(A)}$. This semigroup is not differentiable in $L^1(\Omega)$, but in some special situations it has regularity properties comparable with those of the semigroup generated by the Laplace operator on $L^1(\Omega)$ under Dirichlet boundary conditions (see Proposition 4.4 in Chapter 1).

Theorem 3.4. *Let* $\beta \in C^1(\mathbf{R} \setminus \{0\} \cap C(\mathbf{R})$ *be a monotone function satisfying the conditions*

$$\beta(0) = 0, \qquad \beta'(r) \geq C|r|^{\alpha - 1} \qquad \forall r \neq 0, \tag{3.99}$$

where $\alpha > 0$ if $N \leq 2$ and $\alpha > (N-2)/N$ if $N \geq 3$. Then $S(t)(L^1(\Omega)) \subset L^\infty(\Omega)$ for every $t > 0$,

$$\|S(t)y_0\|_{L^\infty(\Omega)} \leq Ct^{-N/(N\alpha+2-N)}\|y_0\|_{L^1(\Omega)}^{2/(2+N(\alpha-1))} \qquad \forall t > 0, \quad (3.100)$$

and $S(t)(L^p(\Omega)) \subset L^p(\Omega)$ for all $t > 0$ and $1 \leq p < \infty$.

Proof. We shall establish first the estimates

$$\|(I+\lambda A)^{-1}f\|_p + C\lambda\left(\int_\Omega |(I+\lambda A)^{-1}f|^{(p+\alpha-1)N/(N-2)}\,dx\right)^{(N-2)/N}$$
$$\leq \|f\|_p^p \qquad \forall f \in L^p(\Omega),\ \lambda > 0, \qquad (3.101)$$

for $N > 2$, and

$$\|(I+\lambda A)^{-1}f\|_p + C\lambda\left(\int_\Omega |(I+\lambda A)^{-1}f|^{(p+1-\alpha)q}\right)^{1/q} dx$$
$$\leq \int_\Omega |f|^p\,dx \qquad \forall q > 1, \qquad (3.102)$$

if $N = 2$. Here, $\|\cdot\|_p$ is the L^p norm in Ω, C is independent of $p \geq 1$, and A is the operator defined by (3.96).

We set $u = (I+\lambda A)^{-1}f$, i.e.,

$$u - \lambda\Delta\beta(u) = f \quad \text{in } \Omega,$$
$$\beta(u) = 0 \quad \text{in } \partial\Omega. \qquad (3.103)$$

We recall that $\beta(u) \in W_0^{1,q}(\Omega)$, where $1 < q < N/(N-2)$ (see Corollary 3.1 in Chapter 2).

Multiplying Eq. (3.103) by $|u|^{p-1}\operatorname{sign} u$ and integrating on Ω, we get

$$\int_\Omega |u|^p\,dx + \lambda p(p-1)\int_\Omega \beta'(u)|u|^{p-2}|\nabla u|^2\,dx \leq \int_\Omega |f|^p\,dx.$$

Now using the identity

$$|y|^{p+\alpha-3}|\nabla y|^2 = \frac{2}{(p+\alpha-1)^2}|\nabla|y|^{(p+\alpha-1)/2}|^2 \qquad \text{a.e. in } \Omega$$

and condition (3.99), we get

$$\int_\Omega |u|^p\,dx + \frac{4\lambda p(p-1)}{(p+\alpha-1)^2}C\int_\Omega |\nabla|u|^{(p+\alpha-1)/2}|^2\,dx$$
$$\leq \int_\Omega |f|^p\,dx. \qquad (3.104)$$

On the other hand, by Sobolev imbedding theorem

$$\int_\Omega |\nabla |y|^{(p+1-\alpha)/2}|^2\,dx \le C\left(\int_\Omega |y|^{(p+1-\alpha)N/(N-2)}\,dx\right)^{(N-2)/N}$$

if $N > 2$,

and

$$\int_\Omega |\nabla |y|^{(p+1-\alpha)/2}|^2\,dx \le C\left(\int_\Omega |y|^{(p+1-\alpha)q}\,dx\right)^{1/q} \qquad \forall q > 1,$$

for $N = 2$. Then, substituting these inequalities into (3.104), we get (3.101) and (3.102), respectively.

We set $J_\lambda = (I + \lambda A)^{-1}$ and denote $\varphi(u) = \|u\|_p^p$, $\psi(u) = C\|u\|_{(p+\alpha-1)N/(N-2)}^{p+\alpha-1}$.

Then inequality (3.101) can be written as

$$\varphi(J_\lambda f) + \lambda\psi(J_\lambda f) \le \varphi(f) \qquad \forall f \in L^p(\Omega).$$

This yields

$$\varphi(J_\lambda^k f) + \lambda\psi(J_\lambda^k f) = \varphi(J_\lambda^{k-1} f) \qquad \forall k.$$

Summing these equations from $k = 1$ to $k = n$ and taking $\lambda = t/n$ yields

$$\varphi(J_{t/n}^n f) + \sum_{k=1}^n \frac{t}{n}\psi\left(J_{t/n}^k f\right) = \varphi(f).$$

Recalling that $J_{t/n}^n f \to S(t)f$ for $n \to \infty$, the last equation implies that

$$\varphi(S(t)f) + \int_0^t \psi(S(\tau)f)\,d\tau = \varphi(f) \qquad \forall t \ge 0. \qquad (3.105)$$

In particular, it follows that the function $t \to \varphi(S(t)f)$ is decreasing and so is $t \to \psi(S(t)f)$. Then, by (3.105), we see that

$$\varphi(S(t)f) + t\psi(S(t)f) \le \varphi(f) \qquad \forall t > 0,$$

i.e.,

$$\|S(t)f\|_p^p + Ct\|S(t)f\|_{(p+\alpha-1)N/(N-2)}^{p+\alpha-1} \le \|f\|_p^p \qquad \forall t > 0, \quad (3.106)$$

where C is independent of p and f.

Let $p_{n+1} = (p_n + \alpha - 1)N/(N-2)$. Then, by (3.106), we see that

$$\|S(t_{n+1})f\|_{p_{n+1}}^{(N/N-2)p_{n+1}} \le \frac{\|S(t_n)f\|_{p_n}^{p_n}}{C(t_{n+1} - t_n)},$$

where $t_0 = 0$ and $t_{n+1} > t_n$. Choosing $t_{n+1} - t_n = t/2^{n+1}$, we get after some calculation

$$\limsup_{n\to\infty} \|S(t)f\|_{p_{n+1}}^{(N-2/N)n_{p_{n+1}}} \leq C\|f\|_{p_0}\left(\frac{2}{t}\right)^{\mu} \qquad \forall t > 0,$$

where $\mu = N/2$. Since

$$p_n = \left(\frac{N}{N-2}\right)^n p_0 + \frac{N\alpha}{2(N-2)}\left(\left(\frac{N}{N-2}\right)^n - 1\right)$$

(here, we have used the fact that $\alpha > (N-2)/N$, we get

$$\|S(t)f\|_\infty \leq C\|f\|_{p_0}^{2p_0/(2p_0+N(\alpha-1))} t^{-N/(2p_0+N(\alpha-1))} \qquad \forall p_0 \geq 1,$$

as claimed.

The case $N = 2$ follows similarly.

Moreover, by by inequality (3.101) and the exponential formula defining $S(t)$, it follows that

$$\|S(t)f\|_p \leq \|f\|_p \qquad \forall f \in L^p(\Omega),\, t \geq 0.$$

This complete the proof of Theorem 3.4. ∎

4.3.4. *First Order Quasilinear Partial Differential Equations*

We shall consider here the Cauchy problem

$$\frac{\partial y}{\partial t} + \sum_{i=1}^{N} \frac{\partial}{\partial x_i} a_i(y) = 0 \qquad \text{in } \mathbf{R}^N \times \mathbf{R}^+,$$

$$y(x,0) = y_0(x), \qquad x \in \mathbf{R}^N. \tag{3.107}$$

in the space $X = L^1(\mathbf{R}^N)$, where $a = (a_1, \ldots, a_N)$ is a continuous map from $\mathbf{R}$ to $\mathbf{R}^N$ satisfying the condition

$$\limsup_{|r|\to 0} \|a(r)\|/|r| < \infty,$$

and $y_0 \in L^1(\mathbf{R}^N)$.

We have seen earlier (see Proposition 3.11 in Chapter 2) that the first order differential operator $y \to \Sigma_{i=1}^N (\partial/\partial x_i)\, a_i(y)$ admits an m-accretive extension $A \subset L^1(\mathbf{R}^N) \times L^1(\mathbf{R}^N)$ defined as the closure in $L^1(\mathbf{R}^N) \times L^1(\mathbf{R}^N)$ of the operator A_0 given by Definition 3.2 (Chapter 2).

Then, by the generation theorem, 1.3, the Cauchy problem

$$\frac{dy}{dt} + Ay \ni 0 \qquad \text{in } [0, +\infty),$$
$$y(0) = y_0,$$

has for every $y_0 \in \overline{D(A)}$ a unique mild solution $y(t) = S(t)y_0$ given by the exponential formula (1.16) or, equivalently,

$$y(t) = \lim_{\varepsilon \to 0} y_\varepsilon(t) \qquad \text{uniformly on compact intervals,}$$

where y_ε is the solution to difference equation

$$\begin{aligned} \varepsilon^{-1}(y_\varepsilon(t) - y_\varepsilon(t-\varepsilon)) + Ay_\varepsilon(t) &= 0 \qquad \text{for } t > \varepsilon, \\ y_\varepsilon(t) &= y_0 \qquad \text{for } t < 0. \end{aligned} \tag{3.108}$$

We call such a function $y(t) = S(t)y_0$ a *semigroup solution* to the Cauchy problem (3.107).

Theorem 3.5. *Let $S(t)$ be the semigroup of contractions generated by A on $\overline{D(A)}$ and let $y = S(t)y_0$ be the mild solution to problem* (3.107). *Then*

(i) *$S(t)L^p(\mathbf{R}^N) \subset L^p(\mathbf{R}^N)$ for all $1 \le p \le \infty$ and*

$$\|S(t)y_0\|_{L^p(\mathbf{R}^N)} \le \|y_0\|_{L^p(\mathbf{R}^N)} \qquad \forall y_0 \in \overline{D(A)} \cap L^p(\mathbf{R}^N). \tag{3.109}$$

(ii) *If $y_0 \in \overline{D(A)} \cap L^\infty(\mathbf{R}^N)$, then*

$$\begin{aligned} \int_0^T \int_{\mathbf{R}^N} (|y(x,t) - k|\varphi_t(x,t) + \operatorname{sign}_0(y(x,t) - k)\,(a(y(x,t)) \\ -a(k), \varphi_x(x,t)))\,dx\,dt \ge 0 \end{aligned} \tag{3.110}$$

for every $\varphi \in C_0^\infty(\mathbf{R}^N \times (0,T))$ such that $\varphi \ge 0$, and all $k \in \mathbf{R}^N$ and $T > 0$.

Inequality (3.110) is Kružkov's [1] definition of generalized solution to the Cauchy problem (3.107).

Proof of Theorem 3.5. Since, as seen in the proof of Proposition 3.11 (Chapter 2), $(I + \lambda A)^{-1}$ maps $L^p(\mathbf{R}^N)$ into itself and

$$\|(I + \lambda A)^{-1}u\|_{L^p(\mathbf{R}^N)} \le \|u\|_{L^p(\mathbf{R}^N)}$$
$$\forall \lambda > 0,\ u \in L^p(\mathbf{R}^N) \text{ for } 1 \le p \le \infty,$$

we deduce (i) by the exponential formula (1.16).

To prove inequality (3.109), consider the solution y to Eq. (3.108), where $y_0 \in L^1(\mathbf{R}^N) \cap L^\infty(\mathbf{R}^N)$ and $A_0 = A$. (Recall that $L^1(\mathbf{R}^N) \cap L^\infty(\mathbf{R}^N) \subset R(I + \lambda A)^{-1}$ for all $\lambda > 0$.) Then, $\|y_\varepsilon(t)\|_{L^p(\mathbf{R}^N)} \leq \|y_0\|_{L^p(\mathbf{R}^N)}$ for $p = 1, \infty$ and, by Definition 3.2 in Chapter 2 we have

$$\begin{aligned}\int_{\mathbf{R}^N} &(\text{sign}_0(y_\varepsilon(x,t) - k)\,(a(y_\varepsilon(x,t)) - a(k)), \varphi_x(x,t))\\ &+ \varepsilon^{-1}(y_\varepsilon(x,t-\varepsilon) - y_\varepsilon(x,t))\,\text{sign}_0(y_\varepsilon(x,t) - k)\\ &\times \varphi(x,t))\,dx \geq 0 \qquad \forall k \in \mathbf{R},\ \varphi \in C_0^\infty(\mathbf{R}^N \times (0,T)),\ \varphi \geq 0.\end{aligned} \tag{3.111}$$

On the other hand, we have

$$\begin{aligned}(y_\varepsilon(x,t-\varepsilon) &- y_\varepsilon(x,t))\,\text{sign}_0(y_\varepsilon(x,t) - k)\\ &= (y_\varepsilon(x,t-\varepsilon) - k)\,\text{sign}_0(y_\varepsilon(x,t) - k)\\ &\quad - (y_\varepsilon(x,t) - k)\,\text{sign}_0(y_\varepsilon(x,t) - k)\\ &\leq z_\varepsilon(x,t-\varepsilon) - z_\varepsilon(x,t)\end{aligned}$$

where $z_\varepsilon(x,t) = |y_\varepsilon(x,t) - k|$.

Substituting this into (3.111) and integrating on $\mathbf{R}^N \times [0,T]$, we get

$$\begin{aligned}\int_0^T \int_{\mathbf{R}^N} &\text{sign}_0(y_\varepsilon(x,t) - k)\,(a(y_\varepsilon(x,t) - a(k), \varphi_x(x,t))\\ &+ \varepsilon^{-1}(z_\varepsilon(x,t-\varepsilon) - z_\varepsilon(x,t))\varphi(x,t))\,dx\,dt \geq 0.\end{aligned}$$

Now letting ε tend to zero, we get (3.110) because $y_\varepsilon(t) \to y(t)$ uniformly on $[0,T]$ in $L^1(\mathbf{R}^N)$ and

$$\begin{aligned}\varepsilon^{-1} &\int_0^T \int_{\mathbf{R}^N} (z_\varepsilon(x,t-\varepsilon) - z_\varepsilon(x,t))\varphi(x,t)\,dx\,dt\\ &= \varepsilon^{-1} \int_0^\varepsilon \int_{\mathbf{R}^N} |y(x,t) - k|\varphi(x,t)dx\,dt - \varepsilon^{-1} \int_{T-\varepsilon}^T \int_{\mathbf{R}^N} z_\varepsilon(x,t)\varphi(x,t)dx\,dt\\ &\quad + \varepsilon^{-1} \int_\varepsilon^{T-\varepsilon} \int_{\mathbf{R}^N} z_\varepsilon(x,t)(\varphi(x,t+\varepsilon) - \varphi(x,t))\,dx\,dt\\ &\overset{\varepsilon \to 0}{\to} \int_0^T \int_{\mathbf{R}^N} |y(x,t) - k|\varphi_t(x,t)\,dx\,dt.\end{aligned}$$

This complete the proof of Theorem 3.5. ■

Equation (3.107) is known in literature as the *law conservation equation* and has a large spectrum of applications in fluid mechanics and applied sciences. Controlled systems of the form

$$\frac{\partial y}{\partial t} + \sum_{i=1}^{N} (a_i(y))_{x_i} = u \qquad \text{in } \mathbf{R}^N \times (0,\infty),$$
$$y(x,0) = y_0(x), \qquad x \in \mathbf{R}^N.$$

arise in the mathematical description of traffic control (M. Slemrod [1]) and in other problems of practical interest. For such a system we may prove a controllability result similar to Propositon 2.3. More precisely, it turns out that (Barbu [17]) the feedback control $u(t) = -\rho \operatorname{sign}^0 y(t)$ steers every initial data $y_0 \in \overline{D(A)} \cap L^\infty(\mathbf{R}^N)$ into the origin in a finite time $T \le \rho^{-1}\|y_0\|_{L^\infty(\mathbf{R}^N)}$. Here, $\operatorname{sign}^0 y = y/\|y\|_{L^1(\mathbf{R}^N)}$ if $y \neq 0$ and $\operatorname{sign}^0 0 = 0$.

4.3.5. *Nonlinear Hyperbolic Equations*

We are given two real Hilbert spaces V and H such that $V \subset H \subset V'$ and the inclusion mapping of V into H is continuous and densely defined. We have denoted by V' the dual of V and H is identified with its own dual. As usual, we denote by $\|\cdot\|$ and $|\cdot|$ the norms of V and H, respectively, and by $(\cdot,\cdot)$ the duality pairing between V and V' and the scalar product of H.

We shall consider here the second order Cauchy problem

$$\frac{d^2y}{dt^2} + Ay + B\left(\frac{dy}{dt}\right) \ni f(t) \qquad \text{in } (0,T),$$
$$y(0) = y_0, \qquad \frac{dy}{dt}(0) = y_1, \tag{3.112}$$

where A is a linear continuous and symmetric operator from V to V' and $B \subset V \times V'$ is maximal monotone. We assume further that

$$(Ay, y) + \alpha|y|^2 \ge \omega\|y\|^2 \qquad \forall y \in V, \tag{3.113}$$

where $\omega > 0$ and $\alpha \in \mathbf{R}$.

One principal motivation and model for Eq. (3.112) is the nonlinear hyperbolic boundary value problem

$$\frac{\partial^2 y}{\partial t^2} - \Delta y + \beta\left(\frac{\partial y}{\partial t}\right) \ni f(x,t) \qquad \text{in } \Omega \times (0,T),$$
$$y = 0 \qquad \text{in } \partial\Omega \times (0,T),$$
$$y(x,0) = y_0(x), \quad \frac{\partial y}{\partial t}(x,0) = y_1(x) \qquad \text{in } \Omega, \tag{3.114}$$

where β is a maximal monotone graph in $\mathbf{R} \times \mathbf{R}$ and Ω is a bounded open subset of $\mathbf{R}^N$ with a smooth boundary.

Regarding the general problem (3.112), we have the following existence result:

Theorem 3.6. *Let $f \in W^{1,1}([0,T]; H)$ and $y_0 \in V$, $y_1 \in D(B)$ be given such that*

$$\{Ay_0 + By_1\} \cap H \neq \emptyset. \tag{3.115}$$

Then there is a unique function $y \in W^{1,\infty}([0,T]; V) \cap W^{2,\infty}([0,T]; H)$ that satisfies

$$\frac{d^+}{dt}\frac{dy}{dt} + Ay(t) + B\left(\frac{d^+}{dt}y(t)\right) \ni f(t) \qquad \forall t \in [0,T),$$

$$y(0) = y_0, \qquad \frac{dy}{dt}(0) = y_1, \tag{3.116}$$

where $(d^+/dt)\, dy/dt$ is considered in the topology of H and $(d^+/dt)y$ in V.

Proof. Let $X = V \times H$ be the Hilbert space with the scalar product

$$\langle U_1, U_2 \rangle = (Au_1, u_2) + \alpha(u_1, u_2) + (v_1, v_2),$$

where $U_1 = [u_1, v_1]$, $U_2 = [u_2, v_2]$. In the space X, define the operator $\mathscr{A}: D(\mathscr{A}) \subset X \to X$ by

$$D(\mathscr{A}) = \{[u,v] \in V \times H; \{Au + Bv\} \cap H \neq \emptyset\},$$
$$\mathscr{A}[u,v] = [-v; \{Au + Bv\} \cap H] + \sigma[u,v], \qquad [u,v] \in D(\mathscr{A}), \tag{3.117}$$

where $\sigma = \sup\{\alpha(u,v)/((Au,u) + \alpha|u|^2 + |v|^2);\ u \in V, v \in H\}$.

We may write Eq. (112) as

$$\frac{dy}{dt} - z = 0 \quad \text{in } (0,T),$$

$$\frac{dz}{dt} + Ay + Bz = f.$$

Equivalently,

$$\frac{d}{dt}U(t) + \mathscr{A}U(t) - \sigma U(t) = F(t), \qquad t \in (0,T),$$

$$U(0) = U_0, \tag{3.118}$$

where

$$U(t) = [y(t), z(t)], \qquad F(t) = [0, f(t)], \qquad U_0 = [y_0, y_1].$$

It is easily seen that $\mathscr{A}$ is monotone in $X \times X$. Let us show that it is maximal monotone, i.e., $R(I + \mathscr{A}) = V \times H$ where I is the unity operator in $V \times H$. To this end, let $[g, h] \in V \times H$ be arbitrarily given. Then the equation $U + \mathscr{A}U = [g, h]$ can be written as

$$y - z + \sigma y = g,$$
$$z + Ay + Bz + \sigma z \ni h.$$

Substituting $y = (1 + \sigma)^{-1}(z + g)$ in the second equation of the previous system, we obtain

$$(1 + \sigma)z + (1 + \sigma)^{-1} Az + Bz \ni h - (1 + \sigma)^{-1} Ag.$$

Under our assumptions, the operator $z \xrightarrow{\Gamma} (1 + \sigma)z + (1 + \sigma)^{-1}Az$ is continuous, positive, and coercive from V to V'. Then $R(\Gamma + B) = V'$ (see Corollary 1.5 in Chapter 2), and so the previous equation has a solution $z \in D(B)$ and *a fortiori* $[g, h] \in R(I + \mathscr{A})$.

Then the conclusions of Theorem 3.6 follow by Theorem 1.2 because there is a unique solution $U \in W^{1,\infty}([0, T]; V \times H)$ to problem (3.118) satisfying

$$\frac{d^+}{dt}U(t) + \mathscr{A}U(t) - \sigma U(t) \ni F(t) \qquad \forall t \in [0, T),$$

i.e.,

$$\frac{d^+}{dt}y(t) = z(t) \qquad \forall t \in [0, T),$$

$$\frac{d^+}{dt}z(t) + Ay(t) + B(z(t)) \ni f(t) \qquad \forall t \in [0, T),$$

where $(d^+/dt)\, y(t)$ is in the topology of V whilst $(d^+/dt)z(t)$ is in the topology of H. ■

The operator B that arises in Eq. (3.112) might be multivalued. For instance, if $B = \partial\varphi$ where $\varphi: V \to \overline{\mathbf{R}}$ is a lower semicontinuous convex function, problem (3.112) reduces to a variational inequality of hyperbolic type.

In order to apply Theorem 3.6 to the hyperbolic problem (3.114), we take $V = H_0^1(\Omega)$, $H = L^2(\Omega)$, $V' = H^{-1}(\Omega)$, $A = -\Delta$, and $B: H_0^1(\Omega) \to$

$H^{-1}(\Omega)$ defined by $B = \partial\varphi$ where $\varphi: H_0^1(\Omega) \to \overline{\mathbf{R}}$ is the function

$$\varphi(y) = \int_\Omega j(y(x))\,dx \qquad \forall y \in H_0^1(\Omega), \qquad \beta = \partial j. \tag{3.119}$$

The operator B is an extension of the operator $(B_0 y)(x) = \{w \in L^2(\Omega);\ w(x) \in \beta(y(x))$ a.e. $x \in \Omega\}$ from $H_0^1(\Omega)$ to $H^{-1}(\Omega)$. More precisely (see H. Brézis [5]), if $\mu \in \partial\varphi(y) \in H^{-1}(\Omega)$, then μ is a bounded measure on Ω and $\mu = \mu_a\,dx + \mu_s$ where the absolutely continuous part $\mu_a \in L^1(\Omega)$ has the property that $\mu_a(x) \in \beta(y(x))$ a.e. $x \in \Omega$. If $D(\beta) = \mathbf{R}$, then $\mu \in L^1(\Omega)$ and $\mu(x) \in \beta(y(x))$ a.e. $x \in \Omega$. On the other hand, it follows by Lemma 2.2 in Chapter 2 that if $\mu \in H^{-1}(\Omega) \cap L^1(\Omega)$ is such that $\mu(x) \in \beta(y(x))$ a.e. $x \in \Omega$, then $\mu \in By$.

Then, applying Theorem 3.6, we get:

Corollary 3.5. *Let β be a maximal monotone graph in $\mathbf{R} \times \mathbf{R}$ and let $B = \partial\varphi$, where φ is defined by* (3.119). *Let $y_0 \in H_0^1(\Omega) \cap H^2(\Omega)$, $y_1 \in H_0^1(\Omega)$, and $f \in L^2(Q)$ be such that $\partial f/\partial t \in L^2(Q)$ and*

$$\mu_0(x) \in \beta(y_0(x)) \qquad \text{a.e. } x \in \Omega \qquad \text{for some } \mu_0 \in L^2(\Omega). \tag{3.120}$$

Then there is a unique function $y \in C([0,T];\ H_0^1(\Omega))$ such that

$$\begin{aligned} &\frac{\partial y}{\partial t} \in C([0,T];\ L^2(\Omega)) \cap C([0,T];\ H_0^1(\Omega)),\\ &\frac{\partial^2 y}{\partial t^2} \in L^\infty(0,T;\ L^2(\Omega)); \end{aligned} \tag{3.121}$$

$$\begin{aligned} &\frac{d^+}{dt}\,\frac{\partial y}{\partial t}(t) - \Delta y(t) + B\left(\frac{\partial}{\partial t}y(t)\right) \ni f(t) \qquad \forall t \in [0,T)\\ &y(x,0) = y_0(x), \qquad \frac{\partial y}{\partial t}(x,0) = y_1(x) \qquad \text{in } \Omega,\\ &y = 0 \qquad \text{in } \partial\Omega \times (0,T). \end{aligned} \tag{3.122}$$

Assume further that $D(\beta) = \mathbf{R}$. Then $\Delta y(t) \in L^1(\Omega)$ for all $t \in [0,T)$ and

$$\frac{d^+}{dt}\,\frac{dy}{dt}(t) - \Delta y(x,t) + \mu(x,t) = f(x,t), \qquad x \in \Omega,\ t \in [0,T), \tag{3.123}$$

where $\mu(x,t) \in \beta((\partial y/\partial t)(x,t))$ a.e. $x \in \Omega$.

(We note that condition (3.120) implies (3.115).)

Problems of the form (3.114) arise frequently in mechanics. For instance, if $\beta(r) = r|r|$ this equation models the behavior of elastic membrane with the resistance proportional to the velocity.

As another example, consider the unilateral hyperbolic problem

$$\frac{\partial^2 y}{\partial t^2} = \Delta y + f \quad \text{in } \left\{(x,t) \in Q;\ \frac{\partial y}{\partial t}(x,t) > \psi(x)\right\},$$

$$\frac{\partial^2 y}{\partial t^2} \geq \Delta y + f \quad \text{in } Q, \qquad \frac{\partial}{\partial t} y \geq \psi \quad \text{in } Q,$$

$$y = 0 \quad \text{in } \partial\Omega \times [0,T),$$

$$y(x,0) = y_0(x), \qquad \frac{\partial y}{\partial t}(x,0) = y_1(x) \quad \text{in } \Omega, \tag{3.124}$$

where $\psi \in H^2(\Omega)$ is such that $\psi \leq 0$ a.e. in $\partial\Omega$.

Clearly, we may write this variational inequality in the form (3.112), where $V = H_0^1(\Omega)$, $H = L^2(\Omega)$, $A = -\Delta$, and $B \subset H_0^1(\Omega) \times H^{-1}(\Omega)$ is defined by

$$Bu = \{w \in H^{-1}(\Omega);\ (w, u - v) \geq 0\ \forall v \in K\}$$

for all $u \in D(B) = K = \{u \in H_0^1(\Omega);\ u \geq \psi \text{ a.e. in } \Omega\}$.

Then, applying Theorem 3.6, we get:

Corollary 3.6. *Let $f, f_t \in L^2(Q)$ and $y_0 \in H_0^1(\Omega) \cap H^2(\Omega)$, $y_1 \in H_0^1(\Omega)$ be such that $y_1(x) \geq \psi(x)$ a.e. $x \in \Omega$. Then there is a unique function $y \in W^{1,\infty}([0,T]; H_0^1(\Omega))$ with $\partial y/\partial t \in W^{1,\infty}([0,T]; L^2(\Omega))$ satisfying*

$$\int_\Omega \left(\frac{d^+}{dt} \frac{\partial y}{\partial t}(x,t) \left(\frac{\partial y}{\partial t}(x,t) - u(x) \right) + \nabla y(x,t) \cdot \nabla \left(\frac{\partial y}{\partial t}(x,t) - u(x) \right) \right) dx$$

$$\leq \int_\Omega f(x,t) \left(\frac{\partial y}{\partial t}(x,t) - u(x) \right) dx \qquad \forall t \in [0,T), \qquad \forall u \in K,$$

$$y(x,0) = y_0(x), \quad \frac{\partial y}{\partial t}(x,0) = y_1(x), \qquad \forall x \in \Omega. \qquad \blacksquare \tag{3.125}$$

The Nonlinear Wave Equation. We shall study now the problem

$$\frac{\partial^2 y}{\partial t^2} - \Delta y + g(y) = f \quad \text{in } \Omega \times (0,T) = Q,$$
$$y(x,0) = y_0(x), \quad \frac{\partial y}{\partial t}(x,0) = y_1(x) \quad \text{in } \Omega,$$
$$y = 0 \quad \text{in } \partial\Omega \times (0,T) = \Sigma, \tag{3.126}$$

where Ω is a bounded and open subset of $\mathbf{R}^N$, with a sufficiently smooth boundary (of class C^2, for instance), and $g \in W^{1,\infty}(\mathbf{R})$ satisfies the following conditions:

(i) $|g'(r)| \le L(1 + |r|^p)$ a.e. $r \in \mathbf{R}$,

where $0 \le p \le 2/(N-2)$ if $N > 2$, and p is any positive number if $1 \le N \le 2$;

(ii) $rg(r) \ge 0 \quad \forall r \in \mathbf{R}$.

In the special case where $g(y) = \mu|y|^\rho y$, assumptions (i), (ii) are satisfied for $0 < \rho \le 2/(N-2)$ if $N > 2$, and for $\rho \ge 0$ if $N \le 2$. For $\rho = 2$ this is the Klein–Gordon equation, which arises in the quantum field theory (see Reed and Simon [1]).

In the sequel, we shall denote by ψ the primitive of g, i.e., $\psi(r) = \int_0^r g(t)\,dt$.

Theorem 3.7. *Let f, $\partial f/\partial t \in L^2(Q)$ and $y_0 \in H_0^1(\Omega) \cap H^2(\Omega)$, $y_1 \in H_0^1(\Omega)$ be such that $\psi(y_0) \in L^1(\Omega)$. Then under assumptions (i), (ii) there is a unique function y that satisfies*

$$y \in L^\infty(0,T; H_0^1(\Omega) \cap H^2(\Omega)) \cap C^1([0,T]; H_0^1(\Omega)),$$
$$\frac{\partial y}{\partial t} \in C([0,T]; H_0^1(\Omega)), \quad \frac{\partial^2 y}{\partial t^2} \in L^\infty(0,T; L^2(\Omega)),$$
$$\psi(y) \in L^\infty(0,T; L^1(\Omega)), \tag{3.127}$$

and

$$\frac{\partial^2 y}{\partial t^2} - \Delta y + g(y) = f \quad \text{a.e. in } Q,$$
$$y(x,0) = y_0(x), \quad \frac{\partial y}{\partial t}(x,0) = y_1(x) \quad \text{a.e. } x \in \Omega. \tag{3.128}$$

Proof. We shall write Eq. (3.126) as a first order differential equation in $X = H_0^1(\Omega) \times L^2(\Omega)$, i.e.,

$$\frac{dy}{dt} - z = 0, \quad \frac{dz}{dt} - \Delta y + g(y) = f \qquad \text{in } [0, T]. \tag{3.129}$$

Equivalently,

$$\begin{aligned} &\frac{d}{dt}U(t) + A_0 U(t) + GU(t) = F(t), \qquad t \in [0, T], \\ &\qquad\qquad\qquad\qquad U(0) = [Y_0, Y_1], \end{aligned} \tag{3.130}$$

where $U(t) = [y(t), z(t)]$, $G(U) = [0, g(y)]$, $A_0 U = [-u, -\Delta y]$, and $F(t) = [0, f(t)]$.

The space $X = H_0^1(\Omega) \times L^2(\Omega)$ is endowed with the usual norm:

$$\|U\|_X^2 = \|y\|_{H_0^1(\Omega)}^2 + \|z\|_{L^2(\Omega)}^2, \qquad U = [y, z].$$

We note first that the operator G is locally Lipschitz on X. Indeed, we have

$$\|G(y_1, z_1) - G(y_2, z_2)\|_X = \|g(y_1) - g(y_2)\|_{L^2(\Omega)}.$$

On the other hand,

$$\begin{aligned} |g(y_1) - g(y_2)| &\le \left|\int_0^1 g'(\lambda y_1 + (1 - \lambda) y_2)\, d\lambda\, (y_1 - y_2)\right| \\ &\le L|y_1 - y_2| \int_0^1 (1 + |\lambda(y_1 - y_2) + y_2|^\alpha)\, d\lambda \\ &\le C|y_1 - y_2|(\max(|y_1|^\alpha, |y_2|^\alpha) + 1) \\ &\qquad\qquad\qquad\qquad\qquad\qquad \forall y_1, y_2 \in \mathbf{R}. \end{aligned}$$

Hence, for any $z \in L^2(\Omega)$ and $y_i \in H_0^1(\Omega)$, $i = 1, 2$, we have

$$\begin{aligned} &\int_\Omega z(x)|g(y_1(x)) - g(y_2(x)|\, dx \\ &\qquad \le C \int_\Omega z(x)|y_1(x) - y_2(x)|(\max(|y_1(x)|^\alpha, |y_2(x)|^\alpha) + 1)\, dx \end{aligned}$$

and, therefore, by the Hölder inequality

$$\begin{aligned} &\int_\Omega z|g(y_1) - g(y_2)|\, dx \\ &\qquad \le C\|z\|_{L^2(\Omega)}\|y_1 - y_2\|_{L^\beta(\Omega)} \max(\|y_1\|_{L^{\alpha\delta}(\Omega)}^\alpha, \|y_2\|_{L^{\alpha\delta}(\Omega)}^\alpha) \\ &\qquad\quad + C\|z\|_{L^2(\Omega)}\|y_1 - y_2\|_{L^2(\Omega)}, \end{aligned}$$

where $1/\beta + 1/\delta + 1/2 = 1$. Now, we take in the latter inequality $\delta = N$ and $\beta = 2N/(N-2)$. We get

$$\|g(y_1) - g(y_2)\|_2 \le C\|y_1 - y_2\|_{2N/(N-2)} \max(\|y_1\|_{N\alpha}^{\alpha}, \|y_2\|_{N\alpha}^{\alpha}) + C\|y_1 - y_2\|_2 \qquad \forall y_1, y_2 \in H_0^1(\Omega).$$

Then, by the Sobolev imbedding theorem and assumption (i), we have

$$\|y_i\|_{N\alpha} \le C_i\|y_i\|_{H_0^1(\Omega)}, \qquad i = 1, 2,$$
$$\|y_1 - y_2\|_{2N/(N-2)} \le C_0\|y_1 - y_2\|_{H_0^1(\Omega)}.$$

(We have denoted by $\|\cdot\|_p$ the L^p norm.) This yields

$$\|g(y_1) - g(y_2)\|_2 \le C\|y_1 - y_2\|_{H_0^1(\Omega)}\big(\max\big(\|y_1\|_{H_0^1(\Omega)}^{\alpha}, \|y_2\|_{H_0^1(\Omega)}^{\alpha}\big) + 1\big)$$

and, therefore,

$$\|G(y_1, z_1) - G(y_2, z_2)\|_X \le C\|y_1 - y_2\|_{H_0^1(\Omega)}\big(1 + \max\big(\|y_1\|_{H_0^1(\Omega)}^{\alpha}, \|y_2\|_{H_0^1(\Omega)}^{\alpha}\big)\big) \qquad \forall y_1, y_2 \in H_0^1(\Omega), \quad (3.131)$$

as claimed. ■

Let $r > 0$ be arbitrary but fixed. Define the operator $\tilde{G}: X \to X$,

$$\tilde{G}(y, z) = \begin{cases} G(y, z) & \text{if } \|y\|_{H_0^1(\Omega)} \le r, \\ G\left(r\dfrac{y}{\|y\|_{H_0^1(\Omega)}}, z\right) & \text{if } \|y\|_{H_0^1(\Omega)} > r. \end{cases}$$

By (3.131) we see that the operator $\tilde{G}$ is Lipschitz on X. Hence, $A_0 + G$ is ω-m-accretive on X and by the standard existence theorem (Theorem 1.2) we conclude that the Cauchy problem

$$\frac{d}{dt}U(t) + A_0U(t) + \tilde{G}U(t) = F(t) \qquad \text{a.e. } t \in (0, T),$$
$$U(0) = [y_0, y_1], \qquad (3.132)$$

has a unique solution $U \in W^{1,\infty}([0,T]; X)$. This implies that there is a unique $y \in W^{1,\infty}([0,T]; H_0^1(\Omega))$ with $dy/dt \in W^{1,\infty}([0,T]; L^2(\Omega))$ such that

$$\frac{d^2y}{dt^2} - \Delta y + \tilde{g}(y) = f(t) \qquad \text{a.e. } t \in (0,T),$$

$$y(0) = y_0, \qquad \frac{dy}{dt}(0) = y_1 \qquad \text{in } \Omega, \tag{3.133}$$

where $\tilde{g}: H_0^1(\Omega) \to L^2(\Omega)$ is defined by

$$\tilde{g}(y) = \begin{cases} g(y) & \text{if } \|y\|_{H_0^1(\Omega)} \le r, \\ g\left(r\dfrac{y}{\|y\|_{H_0^1(\Omega)}}\right) & \text{if } \|y\|_{H_0^1(\Omega)} > r. \end{cases}$$

Choose r such that $\|y_0\|_{H_0^1(\Omega)} < r$. Then there is an interval $[0, T_r]$ such that $\|y(t)\|_{H_0^1(\Omega)} \le r$ for $t \in [0, T_r]$ and $\|y(t)\| > r$ for $t > T_r$. Then we have

$$\frac{\partial^2 y}{\partial t^2} - \Delta y + g(y) = f \qquad \text{in } \Omega \times (0, T_r),$$

and multiplying this by y_t and integrating on $\Omega \times (0,t)$, we get the energy equality

$$\|y_t(t)\|_2^2 + \|y(t)\|_{H_0^1(\Omega)}^2 + 2\int_\Omega \psi(y(x,t))\,dx$$

$$= \|y_1\|_2^2 + \|y_0\|_{H_0^1(\Omega)}^2 + 2\int_\Omega \psi(y_0(x))\,dx + 2\int_0^t \int_\Omega f y_t\,dx\,ds.$$

Since $\psi(y) \ge 0$ and $\psi(y_0) \in L^1(\Omega)$, by Gronwall's lemma we see that

$$\|y_t(t)\|_2 \le \left(\|y_1\|_2^2 + \|y_0\|_{H_0^1(\Omega)}^2 + 2\|\psi(y_0)\|_{L^1(\Omega)}\right)^{1/2}$$

$$+ \int_0^{T_r} \|f(s)\|_2\,ds$$

and, therefore,

$$\|y_t(t)\|_2^2 + \|y(t)\|_{H_0^1(\Omega)}^2 + 2\int_\Omega \psi(y(x,t))\,dx$$

$$\leq \|y_1\|_2^2 + \|y_0\|_{H_0^1(\Omega)}^2 + 2\int_\Omega \psi(y_0)\,dx + 2\left(\int_0^t \|f(s)\|_2^2\,ds\right)^{1/2}$$

$$\times\left(\left(\|y_1\|_2^2 + \|y_0\|_{H_0^1(\Omega)}^2 + 2\|\psi(y_0)\|_{L^1(\Omega)}\right)^{1/2} + \int_0^{T_r}\|f(s)\|_2\,ds\right).$$

The latter estimate shows that given $y_0 \in H_0^1(\Omega)$, $y_1 \in L^2(\Omega)$, $T > 0$, and $f \in L^2(Q_T)$, there is a sufficiently large r such that $\|y(t)\|_{H_0^1(\Omega)} \leq r$ for $t \in [0,T]$, i.e., $T = T_r$. We may infer, therefore, that for r large enough the function y found as the solution to (3.133) is in fact a solution to Eq. (3.128) satisfying all the requirements of Theorem 3.7.

The uniqueness of y satisfying (3.127), (3.128) is the consequence of the fact that such a function is the solution (along with $z = \partial y/\partial t$) to the ω-accretive differential equation (3.132).

By the previous proof, it follows that if one merely assumes that

$$y_0 \in H_0^1(\Omega), \qquad y_1 \in L^2(\Omega), \qquad \psi(y_0) \in L^1(\Omega),$$

then there is a unique function y,

$$y \in C([0,T]; H_0^1(\Omega)), \qquad \frac{\partial y}{\partial t} \in C([0,T]; L^2(\Omega)),$$

that satisfies Eq. (3.126) in a mild sense. However, if $\psi(y_0) \notin L^1(\Omega)$ or if one drops assumption (ii) then the solution to (3.126) exists only in a neighborhood of the origin.

The same method can be used to obtain existence for the nonlinear wave equation with nonhomogeneous Dirichlet conditions,

$$\begin{aligned} &\frac{\partial^2 y}{\partial t^2} - \Delta y + g(y) = f && \text{in } Q,\\ &y(x,0) = y_0(x), \quad \frac{\partial y}{\partial t}(x,t) = y_1(x) && \text{in } \Omega,\\ &y = u && \text{in } \Sigma. \end{aligned} \tag{3.134}$$

Bibliographical Notes and Remarks

Section 1. The main result of Section 1.1 is due to M. G. Crandall and L. C. Evans [1] (see also M. G. Crandall [2]), whilst Theorem 1.3 has been previously proved by M. G. Crandall and T. Liggett [1]. The existence and uniqueness of integral solutions for problem (1.1) (see Theorem 1.2) is due to Ph. Bénilan [1]. Theorems 1.4–1.7 were established in a particular case in Hilbert space by Y. Komura [1] (see also T. Kato [2]) and later extended to Banach spaces with uniformly convex duals by M. G. Crandall and A. Pazy [1], [2]. Note that the generation theorem, 1.3, remains true for ω-accretive operators satisfying the extended range condition (Kobayashi [1])

$$\liminf_{h \downarrow 0} \frac{1}{h} d(x, R(I + \lambda A)) = 0 \qquad \forall x \in \overline{D(A)},$$

where $d(x, K)$ is the distance from x to K (see also R. H. Martin [1]). These results have been partially extended to time-dependent accretive operators $A(t)$ in the works of T. Kato [2], M. G. Crandall and A. Pazy [3], N. Pavel [1], and K. Kobayashi *et al.* [1].

The basic properties of continuous semigroups of contractions have been established by Y. Komura [2], T. Kato [4], and M. G. Crandall and A. Pazy [1, 2]. For other significant results of this theory, we refer the reader to author's book [2]. The results of Section 1.5 are due to H. Brézis [1, 4]. These results were partially extended to the time-dependent case $(d/dt)y + \partial\varphi(t, y) \ni 0$ in the works of J. C. Peralba [1], J. Moreau [3], H. Attouch and A. Damlamian [1, 2], J. Watanabe [1], N. Kenmochi [1], and Otani [1]. For some extensions of Theorem 1.10 to the subgradient differential equation (1.74) with a nonconvex function φ we refer to E. DeGiorgi *et al.* [1] and M. Degiovanni, *et al.* [1].

Section 2. Theorem 2.1 is essentially due to Kobayashi and Miyadera [1]. A number of related results have been obtained previously by Miyadera and Oharu [1], Brézis and Pazy [1], J. Goldstein [1], and H. Brézis [6] (see Remark 2.1). In the case $A_n = \partial\varphi_n$, the convergence of (2.1) is equivalent to the convergence of φ_n in the sense of Mosco, and the previous approximation results become more specific (see H. Attouch [1, 2]).

Theorem 2.2 is due to Brézis and Pazy [2] (see also Miyadera and Oharu [1]). Theorem 2.3 along with other results of this type have been established in the previously cited paper of Brézis and Pazy [1]. For other related results we refer the reader to S. Reich [1, 2] and E. Schecter [1].

There is a large literature on numerical approximation of solutions to Eq. (1.1) but we simply refer the reader to the book [1] of J. W. Jerome and the bibliography given there.

Approximation schemes derived from the nonlinear Chernoff theorem or Lie–Trotter product formula were obtained in the works Y. Kobayashi [2], E. Magenes and C. Verdi [1], and E. Magenes [2]. Theorem 2.3 is due to H. Brézis [6]. For other significant results in this direction and recent references, we refer to Vrabie's book [1]. The general theory of nonlinear semigroups of contractions and evolutions generated by accretive mappings is treated in the books of V. Barbu [2], R. H. Martin [2], K. Deimling [1], and in the survey of M. G. Crandall [2]. For other results such as asymptotic behavior and existence of periodic and almost periodic solutions to problem (1.1), we refer the reader to the monographs of A. Haraux [1] and Gh. Moroşanu [1].

Section 3. The theory of nonlinear accretive differential equations was first applied to semilinear parabolic equations and parabolic variational inequalities by H. Brézis [2, 3]. In the L^1 setting, these problems were studied by Ph. Bénilan [1], Y. Konishi [1], Massey [1], and L. C. Evans [1] (Proposition 3.3).

There is an extensive literature on parabolic variational inequalities and the Stefan problem (see J. L. Lions [1], G. Duvaut and J. L. Lions [1], A. Friedman [1], C. M. Elliott and J. R. Ockendon [1], and A. M. Meirmanov [1] for significant results and complete references on this subject). Here we were primarily interested in the existence results that arise as direct consequences of the general theory developed previously, and we tried to put in perspective those models of free boundary problems which can be formulated as nonlinear differential equations of accretive type. The L^1-space semigroup approach to the nonlinear diffusion equation (3.79) was initiated by Ph. Bénilan [1] (see also Y. Konishi [1]), whilst the H^{-1} approach is due to H. Brézis [2] (Theorem 3.3). The smoothing effect of semigroup generated by the nonlinear diffusion operator in $L^1(\Omega)$ (Theorem 3.4) was discovered by Ph. Bénilan [2] and L. Véron [1] but the proof given here is essentially due to A. Pazy [1]. For other related contributions to existence and regularity of solutions to the porous medium equation, we refer to Bénilan, *et al.* [1], M. G. Crandall and M. Pierre [1], H. Brézis and A. Friedman [1]. Other properties of the semigroup generated by $-\Delta\beta$ in $L^1(\Omega)$ such as the existence of a finite extinction time were established by I. Diaz [2]. The null controllability of the system $y_t - \Delta\beta(y) \ni u$ is proved in V. Barbu [18]. We refer to O. Cârjă [1] for other controllability results of nonlinear systems of accretive type. The semigroup approach to the conservation law equation (3.10) (Theorem 3.5) is due to M. G. Crandall [1]. Theorem 3.6 along with other existence results for abstract hyperbolic equations have been established by H. Brézis [3] (see also the author's book [2] for other results of this type).

Chapter 5 | Optimal Control of Parabolic Variational Inequalities

In this chapter we will consider optimal control problems governed by semilinear parabolic equations and variational inequalities of parabolic type studied in the previous chapter. The main emphasis is put on first order necessary conditions of optimality. We will illustrate the breadth of applicability of these results on several problems involving the control of moving boundary in some physical problems.

5.1. Distributed Optimal Control Problems

5.1.1. Formulation of the Problem

Consider a general control process in $\Omega \subset \mathbf{R}^N$ governed by the abstract parabolic equation

$$\begin{aligned} y' + Ay + \beta(y - \psi) &\ni Bu + f \qquad \text{in } Q = \Omega \times (0,T), \\ y(0) &= y_0 \qquad \text{in } \Omega, \end{aligned} \tag{1.1}$$

and with the pay-off

$$G(y,u) = \int_0^T (g(t,y(t)) + h(u(t)))\,dt + \varphi_0(y(T)).$$

Here, $y'(t) = (dy/dt)(t)$ is the strong derivative of the function $y\colon Q \to \mathbf{R}$ viewed as a function of t from $[0,T]$ to $L^2(\Omega)$, and $f \in L^2(Q)$, $y_0 \in L^2(\Omega)$ are given functions. Throughout in the following, $\Omega \subset \mathbf{R}^N$ is a bounded open subset with C^2-boundary. As for the operators $A\colon D(A) \subset L^2(\Omega) \to$

$L^2(\Omega)$, $B: U \to L^2(\Omega)$ and the functions $\beta: \mathbf{R} \to 2^{\mathbf{R}}$, $g: [0, T] \times L^2(\Omega) \to \mathbf{R}$, $\varphi_0: L^2(\Omega) \to \mathbf{R}$, $h: U \to \overline{\mathbf{R}} =]-\infty, \infty]$, their properties are specified in hypotheses (i)–(iv) in the following.

Throughout this chapter, H will be the space $L^2(\Omega)$ endowed with the usual scalar product $(\cdot, \cdot)$ and norm $|\cdot|_2$. We are given also a real Hilbert space V that is dense in H, and

$$V \subset H \subset V'$$

algebraically and topologically. We shall denote by $\|\cdot\|$ the norm of V and by $(\cdot, \cdot)$ the pairing between V and V' that coincides with the scalar product of H on $H \times H$. The norm of V' will be denoted by $\|\cdot\|_*$.

The following hypotheses will be in effect throughout this chapter:

(i) *The injection of V into H is compact.*

(ii) *A is a linear, continuous, and symmetric operator from V to V' such that*

$$(Ay, y) \geq \omega\|y\|^2 + \alpha|y|_2^2 \qquad \forall y \in V,$$

where $\omega > 0$ and $\alpha \in \mathbf{R}$.

(iii) *β is a maximal monotone graph in $\mathbf{R} \times \mathbf{R}$ such that $0 \in D(\beta)$ and $\psi \in H^2(\Omega)$.*

There exists C independent of ε such that

$$(Ay, \beta_\varepsilon(y - \psi)) \geq -C(1 + |\beta_\varepsilon(y - \psi)|_2)(1 + |y|_2) \qquad \forall y \in D(A_H). \quad (1.2)$$

Moreover,

$$(Ay, \xi(y)) \geq -C(1 + |\xi(y)|_2)(1 + |y|_2) \qquad \forall y \in D(A_H), \quad (1.3)$$

for every monotonically increasing function $\xi \in C^1(\mathbf{R})$, $\xi(0) = 0$. Here, $\beta_\varepsilon = \varepsilon^{-1}(1 - (1 + \varepsilon\beta)^{-1})$ and $D(A_H) = \{y \in V;\ Ay \in H\}$.

(iv) *B is a linear continuous operator from a real Hilbert space U to H.*

The norm and the scalar product of U will be denoted by $|\cdot|_U$ and $\langle \cdot, \cdot \rangle$, respectively.

(v) *The function* $h: U \to \overline{\mathbf{R}}$ *is convex and lower semicontinuous (l.s.c.). There exist* $c_1 > 0$ *and* $c_2 \in \mathbf{R}$ *such that*

$$h(u) \geq c_1 |u|_U^2 - c_2 \qquad \forall u \in U.$$

(vi) *The function* $g: [0,T] \times H \to \mathbf{R}$ *is measurable in* t, $g(\cdot, 0) \in L^1(0,T)$, *and for every* $r > 0$ *there exists* $L_r > 0$ *independent of* t *such that*

$$|g(t,y) - g(t,z)| + |\varphi_0(y) - \varphi_0(z)| \leq L_r |y - z|_2$$
$$\forall t \in [0,T], |y|_2 + |z|_2 \leq r.$$

Instead of (vi), in most of the situations encountered subsequently one might alternatively assume that g is a normal continuous convex integrand and φ_0 is continuous and convex, i.e.:

(vi)$'$ *The function* $g: [0,T] \times H \to \mathbf{R}$ *is measurable in* t, *continuous and convex in* y, *and* $\varphi_0 : H \to \mathbf{R}$ *is continuous and convex. There are the functions* $\gamma \in L^2(0,T; H)$, $\delta \in L^1(0,T)$ *such that*

$$g(t,y) \geq (\gamma(t), y) + \delta(t) \qquad \forall t \in [0,T], y \in H.$$

As seen earlier in Chapter 4 (Section 3.2), we may equivalently write Eq. (1.1) as

$$y' + Ay + \partial\varphi(y) \ni f + Bu \qquad \text{in } (0,T),$$
$$y(0) = y_0 \tag{1.4}$$

where $\varphi(y) = \int_\Omega j(y(x) - \psi(x))\,dx \quad \forall y \in L^2(\Omega) = H, \quad \partial j = \beta$. Then, $D(\varphi) = \{y \in H;\ j(y - \psi) \in L^1(\Omega)\}$ and

$$\partial\varphi(y) = \{w \in L^2(\Omega);\ w(x) \in \beta(y(x) - \psi(x)) \text{ a.e. } x \in \Omega\}.$$

By Proposition 3.6, Chapter 4, for every $f \in L^2(Q)$, $u \in L^2(0,T; U)$, and every $y_0 \in D(\varphi) \cap V = \{y \in V;\ j(y - \psi) \in L^1(\Omega)\}$ the Cauchy problem (1.4) has a unique solution $y = y(t, y_0, u) \in W^{1,2}([0,T]; H) \cap C([0,T]; V)$, $Ay \in L^2(0,T; H)$. If $y_0 \in \overline{D(\varphi) \cap V} = \overline{D(\varphi)}$, then

$$y \in C([0,T]; H) \cap L^2(0,T; V), \qquad t^{1/2} y' \in L^2(0,T; H),$$

and

$$t^{1/2} Ay \in L^2(0,T; H).$$

Examples of equations of the form (1.1) satisfying the preceding hypotheses have been discussed in Chapters 3 and 4. For instance, if $V = H_0^1(\Omega)$, $A = -\Delta$, $\psi \equiv 0$, and β is a monotonically increasing continuous function, then the control system (1.1) reduces to a semilinear parabolic equation

$$
\begin{aligned}
&y_t(x,t) - \Delta y(x,t) - \beta(y(x,t)) = Bu + f \qquad \text{in } Q,\\
&y(x,0) = y_0(x) \qquad \text{in } \Omega,\\
&y(x,t) = 0 \qquad \text{in } \Sigma = \partial\Omega \times (0,T).
\end{aligned} \tag{1.5}
$$

If β is the multivalued graph,

$$
\beta(r) = \begin{cases} 0, & r > 0,\\ ]-\infty, 0], & r = 0,\\ \varnothing, & r < 0, \end{cases} \qquad \forall r \in \mathbf{R}, \tag{1.6}
$$

and $\psi \le 0$ in $\partial\Omega$ then assumptions (i)–(iii) are satisfied and problem (1.1) reduces to the obstacle parabolic problem

$$
\begin{aligned}
&y_t - \Delta y = Bu + f \qquad \text{a.e. in } \{(x,t);\, y(x,t) > \psi(x)\},\\
&y_t(x,t) = \max\{f(x,t) + (Bu)(x,t) + \Delta\psi(x), 0\}\\
&\qquad\qquad \text{a.e. in } \{(x,t);\, y(x,t) = \psi(x)\},\\
&y(x,t) \ge \psi(x) \qquad \text{a.e. in } Q,\\
&y = 0 \qquad \text{in } \Sigma,\\
&y(x,0) = y_0(x) \qquad \text{in } \Omega.
\end{aligned} \tag{1.7}
$$

Instead of $-\Delta$ we may consider of course a general second order elliptic differential operator

$$
A_0 y = -\sum_{i,j=1}^{N} (a_{ij} y_{x_i})_{x_j} + a_0 y,
$$

with the boundary conditions $\alpha_2\, \partial y/\partial\nu + \alpha_1 y = 0$ in $\partial\Omega$ (see Section 1.2 in Chapter 3, and Section 3.2 in Chapter 4).

Recalling that $A_H + \partial\varphi$ is maximal monotone in $H \times H$, we may write Eq. (1.4) as

$$
\begin{aligned}
&y'(t) + \partial\phi(y(t)) \ni Bu(t) + f(t) \qquad \text{a.e. } t \in (0,T),\\
&y(0) = y_0
\end{aligned} \tag{1.8}
$$

where

$$
\phi(y) = \tfrac{1}{2}(Ay, y) + \varphi(y), \qquad y \in V.
$$

The optimal control problem we shall study here is the following:

(P)

$$\text{Minimize}\left\{\int_0^T (g(t, y(t, y_0, u)) + h(u(t)))\, dt + \varphi_0(y(T, y_0, u));\right.$$

$$\left. u \in L^2(0, T; U)\right\},$$

where $y = y(t, y_0, u)$ is the solution (1.1) (equivalently (1.8)).

Proposition 1.1. *For every $y_0 \in \overline{D(\phi)} = \overline{D(\varphi) \cap V}$, problem* (P) *has at least one solution u.*

Proof. The proof is standard. However, we outline it for reader's convenience. Let d be the infimum in (P). It is readily seen by virtue of assumption (vi) or (vi)′ that $d > -\infty$. Hence, there is a sequence $\{u_n\} \subset L^2(0, T; U)$ such that

$$d \leq G(y_n, u_n) \leq d + n^{-1}, \tag{1.9}$$

where $y_n = y(t, y_0, u_n)$ and $G(y, u) = \int_0^T (g(t, y(t, y_0, u)) + h(u(t)))\, dt + \varphi_0(y(T, y_0, u))$.

By assumption (v), (vi) or (vi)′, we see that the u_n remain in a bounded subset of $L^2(0, T; U)$. Hence, there is $u^* \in L^2(0, T; U)$ such that on a subsequence, again denoted u_n,

$$u_n \to u^* \qquad \text{weakly in } L^2(0, T; U).$$

Now if take the scalar product of (1.8) (where $u = u_n$, $y = y_n$) with y_n and ty_n', we get the estimates (see the proof of Theorem 1.10 in Chapter 4)

$$\int_0^T \phi(y_n(t))\, dt + |y_n(t)|_2^2 + \int_0^T t|y_n'|^2\, dt + t\phi(y(t))$$

$$\leq C\left(|y_0|_2^2 + \int_0^T \left(|Bu_n(t)|_2^2 + |f(t)|_2^2\right) dt\right).$$

Hence, $\{y_n\}$ remains in a bounded subset of $L^2(0, T; V) \cap W^{1,2}([\eta, T]; H) \cap C([\eta, T]; V)$ for every $\eta \in (0, T)$. Then, by the Arzelà–Ascoli theorem we infer that, on a subsequence,

$$\begin{aligned}
y_n &\to y^* && \text{weakly in } L^2(0,T;V), \\
& && \text{strongly in every } C([\eta,T];H), \\
y_n' &\to (y^*)' && \text{weakly in every } L^2(\eta,T;H), \\
\partial\phi(y_n) &\to \xi && \text{weakly in every } L^2(\eta,T;H).
\end{aligned}$$

In particular, this implies that $y_n \to y^*$ strongly in $L^2(0,T;\ H)$ and

$$\xi(t) \in \partial\phi(y^*(t)) \qquad \text{a.e. in } (0,T),$$

because the operator $y \to \{w \in L^2(a,b;\ H);\ w(t) \in \partial\phi(y(t)) \text{ a.e. } t \in (a,b)\}$ is maximal monotone. Hence, $y^* = y(t, y_0, u^*)$. Next, we have, for $n \to \infty$,

$$\varphi_0(y_n(T)) \to \varphi_0(y^*(T)),$$
$$\int_0^T g(t, y_n(t))\, dt \to \int_0^T g(t, y^*(t))\, dt$$

if (vi) holds, and

$$\liminf_{n\to\infty} \int_0^T g(t, y_n(t))\, dt \geq \int_0^T g(t, y^*(t))\, dt$$

if (vi)′ holds. Finally, since the function $u \to \int_0^T h(u)\, dt$ is weakly lower semicontinuous (because it is convex and l.s.c.), we have

$$\liminf_{n\to\infty} \int_0^T h(u_n(t))\, dt \geq \int_0^T h(u^*(t))\, dt.$$

Then by (1.9) we see that $G(y^*, u^*) = d$, as desired. ■

To prove existence in problem (P) we have used the fact that every level subset $\{y\colon \phi(y) \leq \lambda\}$ is compact in H (as a consequence of hypothesis (i)). Thus, Proposition 1.1 remains true for every control system of the form (1.8) with this property. In particular, this applies to the following optimal control problem:

(P_1) *Minimize* $G(y,u)$ *on all* $y \in W^{1,2}([0,T];\ H) \cap L^2(0,T;\ H^2(\Omega))$ *and* $u \in L^2(0,T;\ U)$, *subject to*

$$\begin{aligned}
y_t - \Delta y &= Bu + f && \text{a.e. in } Q, \\
y(x,0) &= y_0(x) && \text{a.e. } x \in \Omega, \\
\frac{\partial y}{\partial \nu} + \beta(y) &\ni 0 && \text{a.e. in } \Sigma.
\end{aligned} \tag{1.10}$$

Here, $f \in L^2(Q)$, and β is a maximal monotone graph in $\mathbf{R} \times \mathbf{R}$, i.e., $\beta = \partial j$ where $j: \mathbf{R} \to \overline{\mathbf{R}}$ is a l.s.c. convex function.

As seen earlier, problem (1.10) can be written as (1.8) where

$$\phi(y) = \tfrac{1}{2} \int_\Omega |\nabla y|^2 \, dx + \int_{\partial\Omega} j(y) \, d\sigma$$

and $\partial\phi(y) = -\Delta y$, $D(\partial\phi) = \{y \in H^2(\Omega);\ \partial y/\partial\nu + \beta(y) \ni 0 \text{ a.e. in } \partial\Omega\}$.

Hence, if $y_0 \in D(\phi)$ under assumptions (iv)–(vi) (or (vi)′) problem (P_1) has at least one solution.

5.1.2. *The Approximating Control Process*

Let $(y^*, u^*) \in (W^{1,2}([0,T];\ H) \cap L^2(0,T;\ D(A_H))) \times L^2(0,T;\ U)$ be any optimal pair for problem (P), where $y_0 \in D(\varphi) \cap V$, i.e.,

$$y_0 \in V, \qquad j(y_0 - \psi) \in L^1(\Omega). \tag{1.11}$$

For every $\varepsilon > 0$, consider the adapted optimal control problem:

(P_ε) Minimize

$$G^\varepsilon(y,u) = \int_0^T \left(g^\varepsilon(t, y(t)) + h(u(t)) + \tfrac{1}{2}|u(t) - u^*(t)|_U^2\right) dt + \varphi_0^\varepsilon(y(T))$$

on all $y \in W^{1,2}([0,T];\ H) \cap L^2(0,T;\ D(A_H))$, $u \in L^2(0,T;\ H)$ *subject to*

$$y' + Ay + \beta^\varepsilon(y - \psi) = Bu + f \qquad \text{a.e. } t \in (0,T),$$
$$y(0) = y_0. \tag{1.12}$$

Here, $g^\varepsilon: [0,T] \times H \to \mathbf{R}$ and $\varphi_0^\varepsilon: H \to \mathbf{R}$ are defined by (see formula (2.79) in Chapter 2)

$$g^\varepsilon(t,y) = \int_{\mathbf{R}^n} g(t, P_n y - \varepsilon\Lambda_n \tau)\rho_n(\tau) \, d\tau, \tag{1.13}$$

$$\varphi_0^\varepsilon(y) = \int_{\mathbf{R}^n} \varphi_0(P_n y - \varepsilon\Lambda_n \tau)\rho_n(\tau) \, d\tau, \tag{1.14}$$

where $n = [\varepsilon^{-1}]$, ρ_n is a mollifier in $\mathbf{R}^n$ and $P_n : H \to X_n$, $\Lambda_n : \mathbf{R}^n \to X_n$ are given by

$$P_n u = \sum_{i=1}^{n} u_i e_i, \qquad u = \sum_{i=1}^{\infty} u_i e_i, \qquad \Lambda_n \tau = \sum_{i=1}^{n} \tau_i e_i$$

($(e_i)_{i=1}^{\infty}$ is an orthonormal basis in H).

The functions $\beta^{\varepsilon} : \mathbf{R} \to \mathbf{R}$ are continuously differentiable, monotonically increasing, $\beta^{\varepsilon}(0) = 0$ and they satisfy the following conditions:

(k) $\liminf\limits_{\varepsilon \to 0} j^{\varepsilon}(r_{\varepsilon}) \geq j(r) \qquad$ if $\lim\limits_{\varepsilon \to 0} r_{\varepsilon} = r$;

(kk) $\lim\limits_{\varepsilon \to 0} j^{\varepsilon}(r) = j(r) \qquad \forall r \in \mathbf{R}$;

(kkk) $|\beta^{\varepsilon}(r) - \beta_{\varepsilon}(r)| \leq C \qquad \forall \varepsilon > 0, r \in \mathbf{R}$;

where $j^{\varepsilon}(r) = \int_o^r \beta^{\varepsilon}(s)\,ds$, $\beta_{\varepsilon} = \varepsilon^{-1}(1 - (1 + \varepsilon\beta)^{-1})$ and $\beta = \partial j$. For instance, the functions β^{ε} defined by (see formula (2.71) in Chapter 3)

$$\beta^{\varepsilon}(r) = \int_{-\infty}^{\infty} \big(\beta_{\varepsilon}(r - \varepsilon^2(\theta) - \beta_{\varepsilon}(-\varepsilon^2\theta)\big)\rho(\theta)\,d\theta + \beta_{\varepsilon}(0), \quad (1.15)$$

where ρ is a C_0^{∞}-mollifier, satisfy these conditions.

If β is defined by

$$\beta(r) = \begin{cases} 0 & \text{if } r > 0, \\ ]-\infty, 0] & \text{if } r = 0, \end{cases} \quad (1.16)$$

we may take β^{ε} of the following form:

$$\beta^{\varepsilon}(r) = \begin{cases} \varepsilon^{-1} r + 2^{-1} & \text{for } r \leq -\varepsilon, \\ -(2\varepsilon^2)^{-1} r^2 & \text{for } -\varepsilon < r \leq 0, \\ 0 & \text{for } r > 0. \end{cases} \quad (1.17)$$

As noted earlier, we may equivalently write Eq. (1.12) as

$$y'(t) + Ay(t) + \nabla\varphi^{\varepsilon}(y(t)) = Bu(t) + f(t) \qquad \text{a.e. } t \in (0, T),$$
$$y(0) = y_0, \quad (1.18)$$

where

$$\varphi^{\varepsilon}(y) = \int_{\Omega} j^{\varepsilon}(y - \psi)\,dx \qquad \forall y \in L^2(\Omega).$$

It is readily seen that for $\varepsilon \to 0$ the solution $y_{\varepsilon}^u \in W^{1,2}([0,T]; H) \cap L^2(0,T; D(A_H))$ to problem (1.18) approximates the solution y^u to (1.1).

Moreover, we have:

Lemma 1.1. *Let $u_\varepsilon \to u$ weakly in $L^2(0,T;U)$ for $\varepsilon \to 0$. Then*

$$y_\varepsilon^{u_\varepsilon} \to y^u \qquad \begin{array}{l}\text{weakly in } W^{1,2}([0,T];H) \cap L^2(0,T;D(A_H)),\\ \text{strongly in } L^2(0,T;V) \cap C([0,T];H).\end{array}$$

Proof. We write Eq. (1.12) as

$$y' + Ay + \nabla\varphi_\varepsilon(y) = Bu + f + \nabla\varphi_\varepsilon(y) - \nabla\varphi^\varepsilon(y),$$

where $\nabla\varphi_\varepsilon(y) = (\partial\varphi)_\varepsilon(y) = \beta_\varepsilon(y + \psi)$ a.e. in Ω. Then, by hypothesis (1.2) and condition (kkk) it follows that

$$\|\nabla\varphi_\varepsilon(y_\varepsilon^u)\|_{L^2(0,T;H)} + \|Ay_\varepsilon^u\|_{L^2(0,T;H)} + \|y_\varepsilon^u\|_{W^{1,2}([0,T];H)} \le C(\|u\|_{L^2(0,T;U)} + 1),$$

where C is independent of u. Then by the Ascoli–Arzelà theorem and Aubin's compactness lemma we conclude that there is a subsequence, again denoted ε, such that

$$\begin{array}{rl} y_\varepsilon^{u_\varepsilon} \to y & \text{strongly in } C([0,T];H) \cap L^2(0,T;V),\\ Ay_\varepsilon^{u_\varepsilon} \to A_H y & \text{weakly in } L^2(0,T;H),\\ (y_\varepsilon^{u_\varepsilon})' \to y' & \text{weakly in } L^2(0,T;H),\\ \nabla\varphi^\varepsilon(y_\varepsilon^{u_\varepsilon}) \to \xi & \text{weakly in } L^2(0,T;H). \end{array}$$

By condition (kkk),

$$|j^\varepsilon(z)| \le j_\varepsilon(z)| + C|z| \qquad \forall z \in \mathbf{R},\ \varepsilon > 0.$$

Inasmuch as for $\varepsilon \to 0$, $j_\varepsilon(z) \to j(z)$ in $L^1(Q)$ (Theorem 2.2) in Chapter 2), we have

$$\lim_{\varepsilon\to 0} \int_Q j^\varepsilon(z(x,t) - \psi(x))\,dx\,dt = \int_Q j(z(x,t) - \psi(x))\,dx\,dt \qquad \forall z \in L^2(Q).$$

Similarly, by condition (kk) and the Fatou lemma,

$$\liminf_{\varepsilon\to 0} \int_Q j^\varepsilon(y_\varepsilon(x,t)) - \psi(x))\,dx\,dt \ge \int_Q j(y(x,t) - \psi(x))\,dx\,dt.$$

Then, letting ε tend to zero in the inequality

$$\int_Q j^\varepsilon(y_\varepsilon(x,t) - \psi(x))\,dx\,dt$$

$$\leq \int_Q j^\varepsilon(z(x,t) - \psi(x))\,dx\,dt + (\nabla\varphi^\varepsilon(y_\varepsilon(t)), y_\varepsilon(t) - z(t))\,dt$$

$$\forall z \in L^2(Q),$$

we get

$$\int_0^T \varphi(y(t))\,dt \leq \int_0^T \varphi(z(t))\,dt + \int_0^T (\xi(t), y(t) - z(t))\,dt$$

$$\forall z \in L^2(0,T;\,H) = L^2(Q),$$

and this yields (see Section 2.2 in Chapter 2)

$$\xi(t) \in \partial\varphi(y(t)) \qquad \text{a.e. } t \in (0,T).$$

Hence $y = y^u$, thereby completing the proof. ∎

By Proposition 1.1, for every $\varepsilon > 0$, problem (P_ε) has at least one optimal pair $(y_\varepsilon, u_\varepsilon)$.

Lemma 1.2. *For $\varepsilon \to 0$,*

$$u_\varepsilon \to u^* \qquad \text{strongly in } L^2(0,T;\,U),$$

$$y_\varepsilon \to y^* \qquad \text{strongly in } L^2(0,T;\,V) \cap C([0,T];\,H),$$

$$\text{weakly in } W^{1,2}([0,T];H) \cap L^2(0,T;\,D(A_H)), \tag{1.19}$$

$$\beta^\varepsilon(y_\varepsilon - \psi) \to \xi \qquad \text{weakly in } L^2(0,T;\,H), \tag{1.20}$$

where $\xi = f + Bu^* - (y^*)' - Ay^* \in \partial\beta(y^*)$ a.e. in Q.

Proof. For any $\varepsilon > 0$, we have

$$G^\varepsilon(y_\varepsilon, u_\varepsilon) \leq \varphi_0^\varepsilon(y_\varepsilon^{u^*}(T)) + \int_0^T \big(g^\varepsilon(t, y_\varepsilon^{u^*}(t)) + h(u^*(t))\big)\,dt.$$

(We recall that y_ε^u is the solution to Eq. (1.12).) By Lemma 1.1, $y_\varepsilon^{u^*} \to y^*$ in $C([0,T];\,H)$ and so, by Proposition 2.15 in Chapter 2,

$$g^\varepsilon(t, y_\varepsilon^{u^*}(t)) \to g(t, y^*(t)) \qquad \forall t \in [0,T].$$

Then, by the Lebesgue dominated convergence theorem,

$$\lim_{\varepsilon\to 0} \int_0^T g^\varepsilon(t, y_\varepsilon(t))\,dt = \int_0^T g(t, y^*(t))\,dt.$$

Similarly

$$\lim_{\varepsilon\to 0} \varphi_0^\varepsilon(y_\varepsilon^{u^*}(T)) = \varphi_0(y^*(T)),$$

whence

$$\limsup_{\varepsilon\to 0} G^\varepsilon(y_\varepsilon, u_\varepsilon) \le G(y^*, u^*). \tag{1.21}$$

On the other hand, since by assumption (v), $\{u_\varepsilon\}$ is bounded in $L^2(0,T;\, U)$, there exists $u_1 \in L^2(0,T;\, U)$ such that, on some subsequence $\varepsilon \to 0$,

$$u_\varepsilon \to u_1 \qquad \text{weakly in } L^2(0,T;U),$$

and so, by Lemma 1.1,

$$y_\varepsilon \to y_1 = y^{u_1} \qquad \text{strongly in } C([0,T];\, H).$$

Since the function $u \to \int_0^T h(u(t))\,dt$ is weakly lower semicontinuous on $L^2(0,T;\, U)$, we have

$$\liminf_{\varepsilon\to 0} G^\varepsilon(y_\varepsilon, u_\varepsilon) \ge G(y_1, u_1) \ge G(y^*, u^*)$$

and, by (1.21),

$$\lim_{\varepsilon\to 0} \int_0^T |u_\varepsilon - u^*|_U^2\,dt = 0.$$

Hence $y_1 = y^*, u_1 = u^*$, and (1.19), (1.20) follow by Lemma 1.1. ∎

Consider the Cauchy problem

$$\begin{aligned} p_\varepsilon' - Ap_\varepsilon - \dot\beta^\varepsilon(y_\varepsilon - \psi)p_\varepsilon &= \nabla g^\varepsilon(t, y_\varepsilon) \qquad \text{in } (0,T),\\ p_\varepsilon(T) &= -\nabla\varphi_0^\varepsilon(y_\varepsilon(T)), \end{aligned} \tag{1.22}$$

which has a unique solution $p_\varepsilon \in L^2(0,T;\, V) \cap C([0,T];\, H)$ with $p_\varepsilon' \in L^2(0,T;\, V')$ (see e.g., Theorem 1.9′ in Chapter 4). (Here, $\dot\beta^\varepsilon = (\beta^\varepsilon)'$.) On the other hand, since $(y_\varepsilon, u_\varepsilon)$ is optimal for problem (P_ε), we have

$$G^\varepsilon(y^{u_\varepsilon+\lambda v}, u_\varepsilon + \lambda v) \ge G^\varepsilon(y_\varepsilon, u_\varepsilon) \qquad \forall\lambda > 0,\ v \in L^2(0,T;V).$$

This yields

$$\int_0^T (h'(u_\varepsilon, v) + (\nabla g^\varepsilon(t, y_\varepsilon), z_\varepsilon) + \langle u_\varepsilon - u^*, v\rangle)\, dt$$
$$+ (\nabla \varphi_0^\varepsilon(y_\varepsilon(T)), z_\varepsilon(T)) \ge 0 \qquad \forall v \in L^2(0, T; U), \quad (1.23)$$

where $z_\varepsilon \in W^{1,2}([0,T];\ H) \cap L^2(0,T;\ D(A_H))$ is the solution to the linear equation

$$z' + Az + \dot{\beta}^\varepsilon(y_\varepsilon - \psi)z = Bv \qquad \text{a.e. } t \in (0,T),$$
$$z(0) = 0$$

and h' is the directional derivative of h (see formula (2.7) in Chapter 2). If multiply Eq. (1.22) by z_ε and integrate on $(0,T)$ we get, by (1.23),

$$\int_0^T (h'(u_\varepsilon, v) + \langle u_\varepsilon - u^*, v\rangle - \langle B^* p_\varepsilon, v\rangle)\, dt \ge 0 \qquad \forall v \in L^2(0,T; V),$$

and this yields (Proposition 2.8 in Chapter 2)

$$B^* p_\varepsilon(t) \in \partial h(u_\varepsilon(t)) + u_\varepsilon(t) - u^*(t) \qquad \text{a.e. } t \in (0,T). \quad (1.24)$$

Equations (1.22), (1.24) taken all together represent the Euler–Lagrange optimality conditions for problem (P_ε).

Lemma 1.3. *There is $C > 0$ independent of ε such that*

$$|p_\varepsilon(t)|_2^2 + \int_0^T \|p_\varepsilon(t)\|^2\, dt \ge C \qquad \forall \varepsilon > 0, t \in [0,T], \quad (1.25)$$

$$\int_Q |\mathrm{P}_\varepsilon \dot{\beta}^\varepsilon(y_\varepsilon - \psi)|\, dx\, dt \le C \qquad \forall \varepsilon > 0. \quad (1.26)$$

Proof. We take the scalar product of (1.22) with $p_\varepsilon(t)$ and integrate over $[t,T]$. Since $\dot{\beta}^\varepsilon \ge 0$, we get

$$\tfrac{1}{2}|p_\varepsilon(t)|_2^2 \le \tfrac{1}{2}|p_\varepsilon(T)|_2^2 - \omega \int_t^T \|p_\varepsilon(s)\|^2\, ds + \alpha \int_t^T |p_\varepsilon(s)|_2^2\, ds$$
$$+ \int_t^T |\nabla g^\varepsilon(s, y_\varepsilon(s))|_2 |p_\varepsilon(s)|_2\, ds, \qquad t \in [0,T].$$

On the other hand, by hypothesis (vi), Section 1.1, we have

$$\|\nabla g^\varepsilon(t, y_\varepsilon)\|_{L^\infty(0,T;\,H)} + |\nabla \varphi_0^\varepsilon(y_\varepsilon(T))|_2 \le C \qquad \forall \varepsilon > 0,$$

and so by Gronwall's lemma we arrive at estimate (1.25). To get (1.26), we multiply Eq. (1.22) by $\zeta(p_\varepsilon)$ and integrate on $Q = \Omega \times (0,T)$, where ζ is a smooth monotonically increasing approximation of the sign function such that $\zeta(0) = 0$. For instance,

$$\zeta = \zeta_\lambda(r) = \int_{-\infty}^{\infty} (\gamma_\lambda(r - \lambda\theta) - \gamma_\lambda(-\lambda\theta))\rho(\theta)\,d\theta,$$

where $\gamma_\lambda(r) = r|r|^{-1}$ for $|r| \geq \lambda$, $\gamma_\lambda(r) = \lambda^{-1}r$ for $|r| < \lambda$, and ρ is a C_0^∞-mollifier.

Then $(A\zeta(p_\varepsilon(t)), \zeta(p_\varepsilon(t)) \geq 0$ and, therefore,

$$\int_Q \dot{\beta}^\varepsilon(y_\varepsilon - \psi)\zeta(p_\varepsilon)p_\varepsilon\,dx\,dt \leq \int_Q |\nabla_y g^\varepsilon(t, y_\varepsilon)\zeta(p_\varepsilon)|\,dx\,dt + \|p_\varepsilon(T)\|_{L^1(\Omega)} \qquad \forall \varepsilon > 0.$$

Then, letting ζ tend to the sign function, we get estimate (1.26). ∎

Now, since $\{Ap_\varepsilon\}$ is bounded in $L^2(0,T; V')$ and $\{\dot{\beta}^\varepsilon(y_\varepsilon - \psi)p_\varepsilon\}$ is bounded in $L^1(0,T; L^1(\Omega))$, we may infer that $\{p_\varepsilon'\}$ is bounded in $L^1(0,T; L^1(\Omega) + V')$ and so, by the Sobolev imbedding theorem, $\{p_\varepsilon'\}$ is bounded in $L^1(0,T; Y^*)$, where $Y^* = (H^s(\Omega))' + V'$ $(s > N/2)$ is the dual of $Y = H^s(\Omega) \cap V$.

Since the injection of H into Y^* is compact and the set $\{p_\varepsilon(t)\}$ is bounded in H for every $t \in [0,T]$, by the Helly theorem we conclude that there is a function $p \in BV([0,T]; Y^*)$ such that, on a subsequence $\varepsilon_n \to 0$,

$$p_{\varepsilon_n}(t) \to p(t) \qquad \text{strongly in } Y^*, \qquad \forall t \in [0,T].$$

(Here, $BV([0,T]; Y^*)$ is the space of all Y^*-valued functions $p: [0,T] \to Y^*$ with bounded variation on $[0,T]$.) On the other hand, by estimate (1.25) we see that

$$p_{\varepsilon_n} \to p \qquad \text{weak star in } L^\infty(0,T; H), \text{weakly in } L^2(0,T; V). \tag{1.27}$$

Now since the injection of V into H is compact, for every $\lambda > 0$ there is $\delta(\lambda) > 0$ such that (see J. L. Lions [1], Chapter 1, Lemma 5.1),

$$|p_{\varepsilon_n}(t) - p(t)|_2 \leq \|p_{\varepsilon_n}(t) - p(t)\| + \delta(\lambda)\|p_{\varepsilon_n}(t) - p(t)\|_{Y^*} \qquad \forall t \in [0,T], \forall n \in \mathbf{N}^*.$$

This yields

$$p_{\varepsilon_n} \to p \qquad \text{strongly in } L^2(0,T; H), \tag{1.28}$$

and

$$p_{\varepsilon_n}(t) \to p(t) \qquad \text{weakly in } H, \forall t \in [0,T]. \tag{1.29}$$

Moreover, by estimate (1.26) we infer that there is $\mu \in (L^\infty(Q))^*$ such that, on a generalized subsequence λ of ε_n,

$$\dot{\beta}^\lambda(y_\lambda - \psi)p_\lambda \to \mu \qquad \text{weak star in } (L^\infty(Q))^*. \tag{1.30}$$

(We may also view μ as a bounded Radon measure on $\overline{Q}$.)

Now, since $\{\nabla\varphi_0^{\varepsilon_n}(y_{\varepsilon_n}(T))\}$ is bounded in H, we may assume by Proposition 2.15, Chapter 2, that

$$\nabla\varphi_0^{\varepsilon_n}\big(y_{\varepsilon_n}(T)\big) \to -p(T) \in \partial\varphi_0(y^*(T)) \qquad \text{weakly in } H,$$

and

$$\nabla_y g^{\varepsilon_n}(t, y_{\varepsilon_n}) \to \xi \qquad \text{weak star in } L^\infty(0,T;H).$$

Lemma 1.4. *We have*

$$\xi(t) \in \partial g(t, y^*(t)) \qquad \text{a.e. } t \in (0,T).$$

Proof. Going back to the proof of Proposition 2.15, we notice that

$$\begin{aligned}
&\delta^{-1}(g^\varepsilon(t, y_\varepsilon(t) + \delta z)) - g^\varepsilon(t, y_\varepsilon(t))) \\
&\quad = \delta^{-1}\int_{\mathbf{R}^n}(g(t, P_n y_\varepsilon(t) + \delta P_n z - \varepsilon\Lambda_n\tau) \\
&\quad\quad - g(t, P_n y_\varepsilon(t) - \varepsilon\Lambda_n\tau)\rho_n(\tau))\, d\tau,
\end{aligned}$$

and this yields

$$(\nabla g^\varepsilon(t, y_\varepsilon(t)), z(t)) \le \int_{\mathbf{R}^n} g^0(t, P_n y_\varepsilon(t) - \varepsilon\Lambda_n\tau, P_n z)\rho_n(\tau)\, d\tau$$

$$\forall z \in L^1(0,T;H).$$

Integrating over $[0,T]$ and letting $\varepsilon \to 0$, we get

$$\int_0^T(\xi(t), z(t))\, dt \le \int_0^T g^0(t, y^*(t), z(t))\, dt \qquad \forall z \in L^1(0,T;H),$$

and this implies the pointwise inequality

$$\xi(t), z) \le g^0(t, y^*(t), z) \qquad \forall z \in H, \text{a.e. } t \in (0,T),$$

and therefore $\xi(t) \in \partial g(t, y^*(t))$ a.e. $t \in (0,T)$, as claimed. ■

Now, letting $\varepsilon = \varepsilon_n \to 0$ in Eq. (1.22), we conclude that there is $p \in BV([0,T]; Y^*) \cap L^2(0,T; V) \cap L^\infty(0,T; H)$ that satisfies the equation

$$\begin{aligned} p' - Ap - \mu &\in \partial g(t, y^*) \qquad \text{in } (0,T), \\ p(T) &\in -\partial\varphi_0(y^*(T)), \end{aligned} \tag{1.31}$$

where p' is the derivative of p in the sense of vectorial distributions on $(0,T)$. Moreover, since the map $\partial h: U \to U$ is closed we see by (1.19), (1.24), and (1.27) that

$$B^*p(t) \in \partial h(u^*(t)) \qquad \text{a.e. } t \in (0,T). \tag{1.32}$$

Modifying p on a set of measure zero, we may also assume that the lateral limits $p(t+0)$ and $p(t-0)$ exist in $L^2_{\mathrm{w}}(\Omega)$ everywhere on $(0,T)$ (respectively, at $t = 0$ and $t = T$).

Summarizing, we have proved the following weak form of the maximum principle for problem (P):

Proposition 1.2. *Let (y^*, u^*) be an arbitrary optimal pair for problem* (P). *Then there exist the function $p \in L^\infty(0,T; H) \cap L^2(0,T; V) \cap BV([0,T]; Y^*)$ and a measure $\mu \in (L^\infty(Q))^*$ satisfying Eqs.* (1.31) *and* (1.32). ■

We shall call such a function p the *dual extremal arc* of problem (P). The properties of p as well as the optimality system (1.31) will be explicated for some particular cases.

5.1.3. Optimal Control of Semilinear Parabolic Equations

We shall study here problem (P) in the special case where $A = -\Delta$, $V = H_0^1(\Omega)$, $V' = H^{-1}(\Omega)$, $\psi \equiv 0$, and $\beta: \mathbf{R} \to \mathbf{R}$ is a locally Lipschitz, monotonically increasing function.

In other words, we will consider the case where the state equation is given by

$$\begin{aligned} \frac{\partial y}{\partial t} - \Delta y + \beta(y) &= Bu + f \qquad \text{in } Q = (\Omega \times (0,T), \\ y &= 0 \qquad \text{in } \Sigma = \partial\Omega \times (0,T), \\ y(x,0) &= y_0(x) \qquad \text{in } \Omega. \end{aligned} \tag{1.33}$$

Here, $y_0 \in H_0^1(\Omega)$, $f \in L^2(Q)$, and $B \in L(U, L^2(\Omega))$.

Theorem 1.1. *Let* $(y^*, u^*) \in W^{1,2}([0,T]; H) \cap L^2(0,T; (H_0^1(\Omega) \cap H^2(\Omega))) \times L^2(0,T; U)$ *be any optimal pair for problem* (P) *having* (1.33) *as state system, where* β *is locally Lipschitz and monotonically increasing. Then there are*

$$p \in BV([0,T]; Y^*) \cap L^2(0,T; H_0^1(\Omega)) \cap L^\infty(0,T; H)$$

and $\mu \in (L^\infty(Q))^*$ *such that* $p' - Ap - \mu \in L^\infty(0,T; H)$ *and*

$$p' - Ap - \mu \in \partial g(t, y^*) \qquad \text{a.e. in } Q, \tag{1.34}$$

$$\mu_a(x,t) \in p(x,t)\, \partial\beta(y^*(x,t)) \qquad \text{a.e. } (x,t) \in Q, \tag{1.35}$$

$$p(T) + \partial\varphi_0(y^*(T)) \ni 0 \qquad \text{in } \Omega, \tag{1.36}$$

$$B^*p(t) \in \partial h(u^*(t)) \qquad \text{a.e. } t \in (0,T). \tag{1.37}$$

Further assume that

$$0 \le \beta'(r) \le C(|\beta(r)| + |r| + 1) \qquad \text{a.e. } r \in \mathbf{R}. \tag{1.38}$$

Then $p \in AC([0,T]: Y^*) \cap C_w([0,T]; H) \cap C([0,T]; L^1(\Omega))$ *and* $\mu_a = \mu \in L^1(Q)$.

Here, μ_a is the absolutely continuous part of the measure μ and $\partial\beta$ is the generalized gradient of β.

Proof. Let $p \in L^\infty(0,T; H) \cap L^2(0,T; H_0^1(\Omega)) \cap \mathrm{BV}([0,T]; Y^*)$ be the function arising in Proposition 1.2 and let $\mu \in (L^\infty(Q))^*$ be the measure defined by (1.30). It remains to prove (1.35).

By the Egorov theorem, for each $\eta > 0$ there is a measurable subset $Q_\eta \subset Q$ such that $y^*, p \in L^\infty(Q_\eta)$, $m(Q \setminus Q_\eta) \le \eta$, and

$$y_{\varepsilon_n} \to y^*, \quad p_{\varepsilon_n} \to p \qquad \text{strongly in } L^\infty(Q_\eta),$$

$$\dot\beta^{\varepsilon_n}(y_{\varepsilon_n}) \to g_\eta \qquad \text{weak star in } L^\infty(Q_\eta).$$

By Lemma 2.4, Chapter 3, we have

$$g_\eta(x,t) \in \partial\beta(y^*(x,t)) \qquad \forall (x,t) \in Q_\eta.$$

This clearly implies that $\mu_a(x,t) \in p(x,t)\, \partial\beta(y^*(x,t))$ a.e. $(x,t) \in Q$ because

$$\dot\beta^\lambda(y_\lambda)p_\lambda \to \mu \qquad \text{weak star in } (L^\infty(Q))^*.$$

Now if β satisfies condition (1.38), then we have

$$0 \le \dot\beta^\varepsilon(r) \le C(|\beta^\varepsilon(r)| + |r| + 1) \qquad \text{a.e. } r \in \mathbf{R},$$

and so for any measurable subset E of Q we have

$$\int_E |\dot{\beta}^\varepsilon(y_\varepsilon)p_\varepsilon|\,dx\,dt$$
$$\leq C\left(\int_E |\beta^\varepsilon(y_\varepsilon)p_\varepsilon|\,dx\,dt + \int_E |p_\varepsilon|(1+|y_\varepsilon|)\,dx\,dt\right).$$

Since $\{\beta^\varepsilon)y_\varepsilon)\}$ is bounded in $L^2(Q)$ whilst p_ε and y_ε are strongly convergent, the last inequality implies that, for every $\delta > 0$, $\exists\omega(\delta)$ such that

$$\int_E |\dot{\beta}^\varepsilon(y_\varepsilon)p_\varepsilon|\,dx\,dt \leq \delta$$

if $m(E) < \omega(\delta)$. Then, by the Dunford–Pettis criterion, $\{\dot{\beta}^\varepsilon(y_\varepsilon)p_\varepsilon\}$ is weakly compact in $L^1(Q)$ and therefore $\mu \in L^1(Q)$, thereby completing the proof. ■

Remark 1.1. Assume now that $f \in L^q(Q)$, $B \in L(U, L^q(\Omega))$, and $y_0 \in W_0^{2-2/q,q}(\Omega)$, $j(y_0) \in L^1(\Omega)$ where $q > \max(N,2)$ and $j(y) = \int_0^y |\beta(s)|^{q-2}\beta(s)\,ds$. Then the solution y_ε to system

$$\frac{\partial y_\varepsilon}{\partial t} - \Delta y_\varepsilon + \beta^\varepsilon(y_\varepsilon) = Bu_\varepsilon + f \qquad \text{in } Q,$$
$$y_\varepsilon(0) = y_0 \qquad \text{in } \Omega, \qquad y_\varepsilon = 0 \qquad \text{in } \partial\Omega,$$

satisfies $\|\beta^\varepsilon(y_\varepsilon)\|_{L^q(Q)} \leq C$ (see Remark 3.1 in Chapter 4). Hence, $\{y_\varepsilon\}$ is bounded in $W_q^{2,1}(Q)$ and therefore it is compact in $C(\overline{Q})$. This implies that $\{\dot{\beta}^\varepsilon(y_\varepsilon)\}$ is bounded in $L^\infty(Q)$ and so $\mu \in L^2(Q)$. Therefore, the optimality system (1.34)–(1.36) becomes, in this case,

$$\frac{\partial p}{\partial t} + \Delta p - p\,\partial\beta(y^*) \ni 0 \qquad \text{a.e. in } Q$$
$$p(x,T) \in -\partial\varphi_0(y^*x,T)) \qquad \text{a.e. in } \Omega, \qquad p = 0 \qquad \text{in } \Sigma.$$

5.1.4 The Optimal Control of the Obstacle Problem

We shall consider now the case where β is defined by (1.16), i.e.,

$$\beta(r) = 0 \quad \text{if } r > 0, \qquad \beta(0) =]-\infty, 0], \qquad \beta(r) = \emptyset \quad \text{if } r < 0.$$

As seen earlier, in this case Eq. (1.1) reduces to the obstacle problem (1.7).

Theorem 1.2. *Let* (y^*, u^*) *be optimal in problem* (P), *where* $A = -\Delta$, $V = H_0^1(\Omega)$. *Then there is a function* $p \in L^2(0,T;\ H_0^1(\Omega)) \cap L^\infty(0,T;\ L^2(\Omega)) \cap BV([0,T];\ Y^*)$ *with* $p' - Ap \in (L^\infty(Q))^*$ *that satisfies the equations*

$$(p' + \Delta p)_a \in \partial g(t, y^*) \qquad \text{a.e. in } [y^* > \psi], \tag{1.39}$$

$$p(f + Bu^* + \Delta y^*) = 0 \qquad \text{a.e. in } [y^* = \psi], \tag{1.40}$$

$$p(T) + \partial\varphi_0(y^*(T)) \ni 0 \qquad \text{in } \Omega, \tag{1.41}$$

$$B^*p(t) \in \partial h(u^*(t)) \qquad \text{a.e. } t \in (0,T). \tag{1.42}$$

Here, p' is the derivative of $p: [0,T] \to L^2(\Omega)$ in the sense of distributions. Equation (1.39) should be understood of course in the following sense: There is an increasing family $\{Q_k\}_{k=1}^\infty$ of measurable subsets of Q such that $m(Q \setminus Q_k) \le k^{-1}$ and

$$\int_Q p(x,t)\gamma_t(x,t)\,dx\,dt - \int_Q \nabla p(x,t) \cdot \nabla\gamma(x,t)\,dx\,dt$$

$$- \int_\Omega p(x,t)\gamma(x,T)\,dx + \int_Q \gamma(x,t)\,\partial g(t, y^*(x,t))\,dx\,dt = 0$$

for all $\gamma \in L^2(0,T;\ H_0^1(\Omega)) \cap C(\overline{Q}) \cap C([0,T];\ Y)$ such that $\gamma_t \in L^2(0,T;\ H^{-1}(\Omega))$, $\gamma(x,0 = 0$ and $\operatorname{supp}\gamma \subset \{(x,t) \in Q;\ y^*(x,t) > \psi(x)\} \cap Q_k$. Equivalently,

$$p_t + \Delta p = \xi_a + \xi_s \qquad \text{in } [(x,t);\ y^*(x,t) > \psi(x)],$$

where $\xi_a \in L^1(Q)$, $\xi_a(x,t) \in \partial g(t, y^*(x,t))$ a.e. $(x,t) \in Q$, and ξ_s is a singular measure with respect to the Lebesgue measure on Q.

Proof of Theorem 1.2. The proof closely follows that of Theorem 2.2 in Chapter 3. However, for the sake of simplicity we will take here, in the approximating problem (P_ε), β^ε of the form (1.17). Let $\varepsilon = \varepsilon_n$ be the sequence that occurs in the proof of Proposition 1.2. We have

$$p_\varepsilon\beta^\varepsilon(y_\varepsilon - \psi) - p_\varepsilon(y_\varepsilon - \psi)\dot\beta^\varepsilon(y_\varepsilon - \psi)$$
$$= \begin{cases} 2^{-1}p_\varepsilon & \text{in } [y_\varepsilon - \psi \le -\varepsilon], \\ 2^{-1}(y_\varepsilon - \psi)^2 p_\varepsilon & \text{in } [-\varepsilon \le y_\varepsilon - \psi \le 0], \\ 0 & \text{in } [y_\varepsilon - \psi > 0]. \end{cases}$$

Since $\{p_\varepsilon\dot\beta^\varepsilon(y_\varepsilon - \psi)\}$ is bounded in L^1(Q), this yields

$$\|p_\varepsilon\beta^\varepsilon(y_\varepsilon - \psi) - p_\varepsilon(y_\varepsilon - \psi)\dot\beta^\varepsilon(y_\varepsilon - \psi)\|_{L^1(Q)} \le C\varepsilon \qquad \forall\varepsilon > 0. \tag{1.43}$$

On the other hand, since $\{\dot{\beta}^{\varepsilon}(y_\varepsilon - \psi)p_\varepsilon\}$ is bounded in $L^1(Q)$, we have

$$\frac{1}{\varepsilon}\int_{[y_\varepsilon \le \psi - \varepsilon]} |p_\varepsilon(x,t)|\,dx\,dt + \frac{1}{\varepsilon^2}\int_{[-\varepsilon \le y_\varepsilon - \psi \le 0]} |y_\varepsilon(x,t) - \psi(x)|\,|p_\varepsilon(x,t)|\,dx\,dt \le C, \quad (1.44)$$

whilst

$$\|p_\varepsilon \beta^{\varepsilon}(y_\varepsilon - \psi)\|_{L^1(Q)} = \frac{1}{\varepsilon}\int_{[y_\varepsilon - \psi \le -\varepsilon]} \left|p_\varepsilon\left(y_\varepsilon - \psi + \frac{\varepsilon}{2}\right)\right| dx\,dt + \frac{1}{2\varepsilon^2}\int_{[-\varepsilon \le y_\varepsilon - \psi \le 0]} |p_\varepsilon|(y_\varepsilon - \psi)^2\,dx\,dt. \quad (1.45)$$

Since

$$\int_{[y_\varepsilon - \psi \le -\varepsilon]} (y_\varepsilon - \psi)^2\,dx\,dt \le C\varepsilon^2$$

(because $\{\beta^{\varepsilon}(y_\varepsilon - \psi)\}$ is bounded in $L^2(Q)$), it follows by (1.44) and (1.45) that

$$p_\varepsilon \beta^{\varepsilon}(y_\varepsilon - \psi) \to 0 \quad \text{strongly in } L^1(Q) \quad (1.46)$$

and, by (1.43),

$$p_\varepsilon \beta^{\varepsilon}(y_\varepsilon - \psi)(y_\varepsilon - \psi) \to 0 \quad \text{strongly in } L^1(Q). \quad (1.47)$$

On the other hand, as seen earlier in the proof of Proposition 1.2,

$$p_\varepsilon \dot{\beta}^{\varepsilon}(y_\varepsilon - \psi) \to p(f + Bu^* - Ay^*) \quad \text{weakly in } L^1(Q).$$

Hence,

$$p(f + Bu^* - Ay^*) = 0 \quad \text{a.e. in } Q.$$

Next, by the Egorov theorem, for every $\eta > 0$ there is a measurable subset $Q_\eta \subset Q$ such that $m(Q \setminus Q_\eta) \le \eta$ (m is the Lebesgue measure), $y^* \in L^\infty(Q_\eta)$, and

$$y_\varepsilon \to y^* \quad \text{unformly on } Q_\eta.$$

This implies that (see (1.30))

$$\mu(y^* - \psi) = 0 \qquad \text{in } Q_\eta,$$

i.e.,

$$\int \mu_a(x,t)(y^*(x,t) - \psi(x))\varphi(x)\,dx\,dt + \mu_s((y^* - \psi)\varphi) = 0$$

for all $\varphi \in L^\infty(Q)$ such that supp $\varphi \subset Q_\eta$. This yields

$$(y^*(x,t) - \psi(x))\mu_a(x,t) = 0 \qquad \text{a.e. } (x,t) \in Q_\eta \cap Q_k,$$

where $\{Q_k\}_{k=1}^\infty$ is an increasing family of measurable sets $Q_k \subset Q$ such that $m(Q \setminus Q_k) \leq k^{-1}$ and $\mu_s = 0$ on Q_k.

Hence,

$$(y^* - \psi)\mu_a = 0 \qquad \text{a.e. in } Q,$$

and this clearly implies Eq. (1.39). ∎

Remark 1.2. If $f \in L^q(\Omega)$, $B \in L(U, L^q(\Omega))$, and $y_0 \in W_0^{2-2/q,q}(\Omega)$, $y_0 \geq \psi$ a.e. in Q, where $q > \max(N, 2)$, then as seen earlier (see Remark 1.1) $\{y_\varepsilon\}$ is bounded in $W_q^{2,1}(Q)$ and compact in $C(\overline{Q})$. Then, by (1.30) and (1.47), we conclude that

$$\mu(y^* - \psi) = 0$$

and so Eq. (1.39) becomes, in this case,

$$p_t + \Delta p \in \partial g(t, y^*) \qquad \text{in } [y^* > \psi].$$

The previous methods applies *mutatis-mutandis* to the optimal control problem (P_1).

Theorem 1.3. *Let (y^*, u^*) be optimal in problem (P_1), where β is locally Lipschitz and satisfies condition (1.38). Then there is $p \in AC([0,T];$ $(H^s(\Omega))') \cap C_w([0,T];\ L^2(\Omega)) \cap L^2(0,T;\ H^1(\Omega))$, $s > N/2$ such that $\partial p/\partial \nu \in L^1(\Sigma)$ and*

$$p_t + \Delta p \in \partial g(t, y^*) \qquad \text{a.e. in } Q, \tag{1.48}$$

$$p(x,T) + \partial\varphi_0(y^*(T))(x) \ni 0 \qquad \text{a.e. in } \Omega, \tag{1.49}$$

$$\frac{\partial p}{\partial \nu} + p\,\partial\beta(y^*) \ni 0 \qquad \text{a.e. in } \Sigma, \tag{1.50}$$

$$B^*p \in \partial h(u^*) \qquad \text{a.e. in } (0,T). \tag{1.51}$$

In the case of Signorini problem, i.e., where the state equation is

$$y_t - \Delta y = Bu \qquad \text{in } Q,$$
$$y(x,0) = y_0(x) \qquad \text{in } \Omega,$$
$$y \geq 0, \quad \frac{\partial y}{\partial \nu} \geq 0, \quad y\,\frac{\partial y}{\partial \nu} = 0 \qquad \text{in } \Sigma, \tag{1.52}$$

we have:

Theorem 1.4. *Let (y^*, u^*) be optimal. Then there is $p \in BV([0,T];$ $(H^s(\Omega))') \cap L^2(0,T;\ H^1(\Omega)) \cap L^\infty(0,T;\ L^2(\Omega))$ with $\partial p/\partial \nu \in (L^\infty(\Sigma))^*$ such that*

$$p_t + \Delta p \in \partial g(t, y^*) \qquad \text{in } Q, \tag{1.53}$$
$$p\,\frac{\partial y^*}{\partial \nu} = 0 \qquad \text{in } \Sigma, \tag{1.54}$$
$$y^* \left(\frac{\partial p}{\partial \nu}\right)_{\mathrm{a}} = 0 \qquad \text{in } \Sigma, \tag{1.55}$$
$$p(T) + \partial \varphi_0(y^*(T)) \ni 0 \qquad \text{a.e. in } \Omega, \tag{1.56}$$
$$B^* p \in \partial h(u^*). \tag{1.57}$$

Equations (1.52)–(1.57) taken all together represent a quasivariational inequality (see, e.g., J. L. Lions [4]). Equations (1.53), (1.55), and (1.56) should be interpreted in the following sense:

$$\int_Q (p\varphi_t - \nabla p \cdot \nabla \varphi)\,dx\,dt - \int_\Omega \partial \varphi_0(y^*(T))\,\varphi(x,T)\,dx$$
$$+ \int_Q \partial g(t, y^*)\,\varphi(x,t)\,dx\,dt = 0$$

for all $\varphi \in L^2(0,T;\ H^1(\Omega)) \cap W^{1,2}([0,T];\ L^2(\Omega))$ such that $\varphi(x,0) \equiv 0$ and $\varphi = 0$ in $\Sigma_k \cap [y^* = 0]$; Σ_k is a sequence of measurable subsets of Σ such that $m(\Sigma_k) \to 0$ as $k \to \infty$.

Proof of Theorem 1.3. Since the proof is essentially the same as that of Theorem 1.1, it will be sketched only.

Consider the approximating problem

$(\mathrm{P}_1^\varepsilon)$ Minimize $G^\varepsilon(y,u)$ on all $(y,u) \in W^{1,2}([0,T];\ L^2(\Omega)) \cap L^2(0,T;\ H^2(\Omega)) \cap L^2(0,T;\ U))$, subject to

$$\begin{aligned} y_t - \Delta y &= Bu + f \qquad \text{in } Q,\\ y(x,0) &= y_0(x) \qquad \text{in } \Omega,\\ \frac{\partial y}{\partial \nu} + \beta^\varepsilon(y) &= 0 \qquad \text{a.e. in } \Sigma, \end{aligned} \tag{1.58}$$

where $y_0 \in H^1(\Omega)$, $j(y_0) \in L^1(\Omega)$, β^ε satisfies (k)–(kkk) in Section 1.2, and (y^*, u^*) is an arbitrary optimal control for problem (P_1).

Arguing as in the proof of Lemma 1.1, it follows that if $u_\varepsilon \to u$ weakly in $L^2(0,T;\ U)$ then

$$\begin{aligned} y_\varepsilon^{u_\varepsilon} \to y^u \qquad & \text{strongly in } L^2(0,T;\ H^1(\Omega)) \cap C([0,T];\ L^2(\Omega)),\\ & \text{weakly in } W^{1,2}([0,T];\ L^2(\Omega)) \cap L^2(0,T;\ H^2(\Omega)). \end{aligned} \tag{1.59}$$

(Here, y^u is the solution to (1.58).) Moreover, $\beta^\varepsilon(y_\varepsilon) \to -\partial y^u/\partial \nu \in \beta(y^u)$ *weakly in* $L^2(\Sigma)$. Indeed, multiplying Eq. (1.58) by $y_\varepsilon = y_\varepsilon^{u_\varepsilon}$ and $-\Delta y_\varepsilon$, we get the estimates

$$|y_\varepsilon(t)|_2^2 + \int_0^t \|y_\varepsilon(s)\|_{H^1(\Omega)}^2\, ds \le C\left(1 + \int_0^T |u_\varepsilon(t)|_U^2\, dt\right)\|y_\varepsilon\|_{H^1(\Omega)}^2$$

$$\|y_\varepsilon(t)\|_{H^1(\Omega)}^2 + \int_0^t |\Delta y_\varepsilon(s)|_2^2\, ds$$

$$\le C\left(1 + \|y_0\|_{H^1(\Omega)}^2 + \int_0^T |u_\varepsilon(t)|_U^2\, dt\right).$$

Hence, $\{y_\varepsilon\}$ is bounded in $C([0,T];\ L^2(\Omega)) \cap L^\infty(0,T;\ H^1(\Omega)) \cap L^2(0,T;\ H^2(\Omega))$ and $\{\partial y_\varepsilon/\partial t\}$ is bounded in $L^2(0,T;\ L^2(\Omega))$. Thus, $\{y_\varepsilon\}$ is compact in $C([0,T];\ L^2(\Omega)) \cap L^2(0,T;\ H^1(\Omega))$ and weakly compact in $L^2(0,T;\ H^2(\Omega)) \cap W^{1,2}([0,T];\ L^2(\Omega))$. Then (1.59) follows by standard arguments.

Now, let (y^*, u^*) be optimal for problem (P_1). (The existence of such a pair follows by Proposition 1.1.) Then, arguing as in the proof of Lemma

1.2, it follows that

$$\begin{aligned}
u_\varepsilon &\to u^* && \text{strongly in } L^2(0,T;U),\\
y_\varepsilon &\to y^* && \text{strongly in } C([0,T];L^2(\Omega)) \cap L^2(0,T;H^1(\Omega)),\\
\beta^\varepsilon(y_\varepsilon) &\to -\frac{\partial y^*}{\partial \nu} && \text{weakly in } L^2(\Sigma),\\
y_\varepsilon &\to y^* && \text{weakly in } W^{1,2}([0,T];L^2(\Omega)) \cap L^2(0,T;H^2(\Omega)).
\end{aligned}$$

Now, let $p_\varepsilon \in L^2(0,T;\ H^1(\Omega)) \cap W^{1,2}([0,T];\ (H^1(\Omega))') \cap C([0,T];$ $L^2(\Omega))$ be the solution to boundary value problem

$$\begin{aligned}
\frac{\partial p_\varepsilon}{\partial t} + \Delta p_\varepsilon &= \nabla g^\varepsilon(t, y_\varepsilon) && \text{in } Q,\\
p_\varepsilon(x,T) + \partial\varphi_0(y_\varepsilon(T))(x) &= 0 && \text{in } \Omega,\\
\frac{\partial p_\varepsilon}{\partial \nu} + \dot\beta^\varepsilon(y_\varepsilon)p_\varepsilon &= 0 && \text{in } \Sigma.
\end{aligned}$$

(The existence of p_ε follows, for instance, by Theorem 1.9′ in Chapter 4, where $V = H^1(\Omega)$ and $A(t)\colon V \to V'$ is defined by

$$(A(t)u,v) = \int_\Omega \nabla u \cdot \nabla v\,dt + \int_{\partial\Omega} \dot\beta^\varepsilon(y_\varepsilon)uv\,d\sigma \qquad \forall u,v \in H^1(\Omega).$$

Then, reasoning as in the proof of Proposition 1.2, it follows that

$$B^*p_\varepsilon \in \partial h(u_\varepsilon) + u_\varepsilon - u^* \qquad \text{a.e. in } (0,T),$$

and

$$|p_\varepsilon(t)|_2^2 + \int_0^T \|p_\varepsilon(t)\|_{H^1(\Omega)}^2\,dt + \int_\Sigma |\dot\beta^\varepsilon(y_\varepsilon)p_\varepsilon|\,dx\,dt \le C \qquad \forall \varepsilon > 0.$$

Hence, $\{dp_\varepsilon/dt\}$ is bounded in $L^1(0,T;\ L^1(\Omega)) + L^1(0,T;\ (H^1(\Omega))') \subset$ $L^1(0,T;\ (H^s(\Omega))')$, where $s > N/2$. Then, by the Helly theorem, there is $p \in BV([0,T];\ (H^s(\Omega))')$ such that, on some subsequence, again denoted ε,

$$\begin{aligned}
p_\varepsilon(t) &\to p(t) && \text{strongly in } (H^s(\Omega))', \quad \forall t \in [0,T],\\
p_\varepsilon &\to p && \text{weakly in } L^2(0,T;H^1(\Omega)),\\
& && \text{weak star in } L^\infty(0,T;L^2(\Omega)).
\end{aligned} \tag{1.60}$$

This implies as in the proof of Proposition 1.2 that

$$p_\varepsilon \to p \qquad \text{strongly in } L^2(Q)$$

and

$$\dot{\beta}^\varepsilon(y_\varepsilon)p_\varepsilon \to \delta \qquad \text{weak star in } (L^\infty(\Sigma))^*.$$

Clearly, p satisfies the system

$$\begin{aligned} \frac{\partial p}{\partial t} + \Delta p &\in \partial g(t, y^*) && \text{in } Q, \\ p(x,T) + \partial\varphi_0(y^*(T))(x) &\ni 0 && \text{in } \Omega, \\ \frac{\partial p}{\partial \nu} + \delta &= 0 && \text{in } \Sigma, \\ B^*p &\in \partial h(u^*). \end{aligned}$$

Now, if β satisfies condition (3.18) it follows as in the proof of Theorem 1.1 that $\{\dot{\beta}^\varepsilon(y_\varepsilon)p_\varepsilon\}$ is weakly compact in $L^1(\Omega)$ and so $\delta \in L^1(\Sigma)$, whilst by Lemma 2.4 in chapter 3 we see that

$$\delta(x,t) \in \partial\beta(y^*(x,t))\, p(x,t) \qquad \text{a.e. } (x,t) \in \Sigma,$$

because $y_\varepsilon \to y^*$ strongly in $L^2(\Sigma)$ and $p_\varepsilon \to p$ weakly in $L^2(\Sigma)$. This completes the proof of Theorem 1.3. ■

Proof of Theorem 1.4. Arguing as in the proof of Theorem 1.2, we find (see (1.43)–(1.45)) that

$$\begin{aligned} p_\varepsilon \beta^\varepsilon(y_\varepsilon) &\to 0 && \text{strongly in } L^1(\Sigma), \\ p_\varepsilon \dot{\beta}^\varepsilon(y_\varepsilon) y_\varepsilon &\to 0 && \text{strongly in } L^1(\Sigma), \end{aligned}$$

The latter implies, as in the proof of Theorem 1.2, that $\delta_a y^* = 0$ in Σ. On the other hand, by (1.60) it follows that on a subsequence, again denoted ε, we have

$$\begin{aligned} p_\varepsilon &\to 0 && \text{a.e. in } \Sigma_1, \\ \beta^\varepsilon(y_\varepsilon) &\to 0 && \text{a.e. in } \Sigma_2, \end{aligned}$$

where $\Sigma = \Sigma_1 \cup \Sigma_2$. Since $\beta^\varepsilon(y_\varepsilon) \to -\partial y^*/\partial\nu$ weakly in $L^2(\Sigma)$ and $p_\varepsilon \to p$ weakly in $L^2(\Sigma)$, we conclude that

$$p\,\frac{\partial y^*}{\partial \nu} = 0 \qquad \text{a.e. in } \Sigma,$$

as claimed. ■

5.1.5. Optimal Control Problems with Infinite Horizon

Consider the problem

$$\inf\left\{\int_0^\infty (g(y(s)) + h(u(s)))\,ds;\ u \in L^2(\mathbf{R}^+;U)\right\} = \psi_\infty(y_0), \quad (1.61)$$

subject to

$$\frac{dy}{dt} + Ay + \partial\varphi(y) \ni Bu \qquad \text{a.e. } t > 0,$$

$$y(0) = y_0, \qquad (1.62)$$

where A is a linear continuous symmetric operator from V to V' satisfying the coercivity condition

$$(Ay, y) \geq \omega\|y\|^2, \qquad y \in V,$$

$B \in L(U, H)$, and $\varphi{:}H \to \overline{\mathbf{R}}$ is a l.s.c., convex function such that $0 \in \partial\varphi(0)$ and

$$(Ay, \partial\varphi_\varepsilon(y)) \geq -C\big(1 + |\partial\varphi_\varepsilon(y)|^2\big)(1 + |y|) \qquad \forall y \in D(A_H),$$

where $D(A_H) = \{y \in V;\ \ Ay \in H\}$, $\partial\varphi_\varepsilon(y) = \varepsilon^{-1}(y - (1 + \varepsilon\,\partial\varphi)^{-1}y)$. Here, U, V, and H are real Hilbert spaces such that $V \subset H \subset V'$ algebraically and topologically. We shall assume that the injection of V into H is compact and we shall use the standard notations for the norms in V, H, and U (see Section 1.1).

The functions $h{:}\, U \to \overline{\mathbf{R}}$ and $g{:}\, H \to \mathbf{R}$ satisfy the conditions:

(j) h is convex, lower semicontinuous, $h(0) = 0$, and

$$h(u) \geq \gamma|u|_U^2 \qquad \forall y \in U,$$

for some $\gamma > 0$.

(jj) g is locally Lipschitz, $g \geq 0$, and $g(0) = 0$.

We denote by ϕ the function $\frac{1}{2}(Ay, y) + \varphi(y)$ and recall that $\partial\phi = A_H + \partial\varphi$.

Proposition 1.3. *Assuming the preceding conditions, the function* $\psi_\infty{:}\,\overline{D(\phi)} \to \mathbf{R}$ *is locally Lipschitz, and for every* $y_0 \in \overline{D(\phi)}$ *the infimum defining* $\psi_\infty(y_0)$ *is attained.*

Proof. Let $y_0 \in \overline{D(\phi)}$ be arbitrary but fixed. If $y(t, y_0, u)$ is the solution to the Cauchy problem (1.62), then we have

$$|y(t, y_0, 0)| \le Ce^{-\omega t}|y_0| \qquad \forall t > 0,$$

and so $g(y(t, y_0, 0)) \in L^1(\mathbf{R})$. Hence, $\psi_\infty(y_0) < \infty$. Now let $u_n \in L^2(\mathbf{R}^+; U)$ be such that

$$\psi_\infty(y_0) \le \int_0^\infty (g(y_n(s)) + h(u_n(s)))\, ds \le \psi_\infty(y_0) + \frac{1}{n},$$

where $y_n(t) = y(t, y_0, u_n)$.

By (j), we see that $\{u_n\}$ is bounded in $L^2(\mathbf{R}^+;U)$ and so, by Eq. (1.62), we have

$$|y_n(t)| \le e^{-\omega t}|y_0| + \|B\| \int_0^t |u_n(s)|_U e^{-\omega(t-s)}\, ds \le C \qquad \forall t \ge 0,$$

$$\int_0^T t|y_n'(t)|^2\, dt + \int_0^T \|y_n(t)\|^2\, dt \le C_T \qquad \forall T > 0. \tag{1.63}$$

Hence, on a subsequence, again denoted $\{n\}$,

$$u_n \to u^* \qquad \text{weakly in } L^2(\mathbf{R}^+, U)$$
$$y_n(t) \to y(t, y_0, u^*) \qquad \text{strongly in } C([\delta, T]; H), \qquad \forall 0 < \delta < T,$$

and by the Fatou lemma,

$$\int_0^T (g(y(t, y_0, u^*)) + h(u^*(t)))\, dt \le \psi_\infty(y_0) \qquad \forall T > 0,$$

and this implies that

$$\int_0^\infty (g(y(t, y_0, u^*)) + h(u^*(t)))\, dt = \psi_\infty(y_0).$$

Now, for $y_0 \in B_r = \{y \in H; |y| \le r\}$, we have

$$\psi_\infty(y_0) \le \int_0^\infty g(y(t, y_0, 0))\, dt \le C_r$$

and so we may confine the infimum in (1.61) to those $u \in L^2(\mathbf{R}^+;U)$ for which

$$\gamma \int_0^\infty |u(t)|_U^2\, dt \le \int_0^\infty h(u(t))\, dt \le C_r.$$

Then, by estimate (1.63), we have

$$|y(t, y_0, u)| \leq re^{-\omega t} + \|B\|(C_r + C) \leq C_r^1 \qquad \forall t \geq 0, \ y_0 \in B_r.$$

Now, if $y_0, z_0 \in B_r$, we have

$$\psi_\infty(y_0) - \psi_\infty(z_0) \leq \int_0^\infty (g(y(t, y_0, v^*)) - g(y(t, z_0, v^*)))\, dx,$$

where $v^* \in L^2(\mathbf{R}^+; U)$ is such that

$$\psi_\infty(z_0) = \int_0^\infty (g(y(t, y_0, v^*)) + h(v^*(t)))\, dt.$$

Since g is locally Lipschitz and

$$|y(t, y_0, v^*) - y(t, z_0, v^*)| \leq e^{-\omega t}|y_0 - z_0| \qquad \forall t \geq 0,$$

we see that

$$\psi_\infty(y_0) - \psi_\infty(z_0) \leq L_r|y_0 - z_0| \qquad \forall y_0, z_0 \in B_r,$$

as desired. ■

In particular, it follows by Proposition 1.3 that for every $y_0 \in \overline{D(\phi)}$ problem (1.65) has at least one optimal pair $(y^*, u^*) \in C(\mathbf{R}^+; H) \times L^2(\mathbf{R}^+; U)$.

We notice that, for every $t > 0$,

$$\psi_\infty(y_0) = \inf\left\{\int_0^t (g(y(s, y_0, z)) + h(u(s)))\, ds + \psi_\infty(y(t, y_0, u)); \right.$$
$$\left. u \in L^2(0, t; U)\right\}. \tag{1.64}$$

Proposition 1.4. *In addition to (j), (jj), let us suppose that $F = \partial\varphi \in C^1(H)$, $N(B^*) = \{0\}$, and h is Gâteaux differentiable on U. Then if (y^*, u^*) is an optimal pair for problem (1.61) there is $p \in C([0, \infty); H) \cap L^2(\mathbf{R}^+; V) \cap L^\infty(\mathbf{R}^+; H)$ such that*

$$\frac{dp}{dt}(t) - Ap(t) - \nabla F(y^*(t))p(t) \in \partial g(y^*(t)) \qquad \text{a.e. } t > 0,$$

$$p(t) \in -\partial\psi_\infty(y^*(t)) \qquad \forall t \geq 0, \tag{1.65}$$

$$B^*p(t) = \partial h(u^*(t)) \qquad \forall t \geq 0. \tag{1.66}$$

Proof. As seen in the preceding, for every $t > 0$, (y^*, u^*) is optimal for problem (1.64). This implies by a standard device that there is $p^t \in C([0,t]; H) \cap W^{1,2}((0,t]; H) \cap L^2(0,t;V)$ such that

$$\frac{dp^t}{ds}(s) - Ap^t(s) - \nabla F(y^*(s))\, p^t(s) \in \partial g(y^*(s)) \qquad \text{a.e. } s \in (0,t),$$

$$p^t(t) \in -\partial\psi_\infty(y^*(t)),$$

$$B^*p^t(s) = \partial h(u^*(s)) \qquad \text{a.e. } s \in (0,t). \tag{1.67}$$

Since $N(B^*) = \{0\}$ and ∂h is single valued it follows that $p^t = p^{\hat{t}}$ on $[0,t]$ for $t \leq \hat{t}$. Now, let $p: [0,\infty) \to H$ be the function

$$p(s) = p^t(s) \qquad \text{for } 0 \leq s \leq t.$$

Since $y^* \in L^\infty(0,\infty: H)$, we see by (1.67) that

$$|p^t(s)|^2 + \int_0^t \|p^t(s)\|^2\, ds \leq C \qquad \forall s \in [0,t],$$

where C is independent of t. Hence, $p \in L^\infty(\mathbf{R}^+; H) \cap L^2(\mathbf{R}^+; V)$ satisfies Eqs. (1.65), (1.66), as claimed. ■

Remark 1.3. In order to avoid a tedious argument, we did not put the preceding result in its most general form. However, it is readily seen that it remains true for semilinear parabolic equations of the form (1.1) with β locally Lipschitz as well as for the obstacle problem.

5.2. Boundary Control of Parabolic Variational Inequalities

5.2.1. *The Obstacle Problem with Dirichlet Boundary Conditions*

We will consider here the controlled obstacle problem

$$\frac{\partial y}{\partial t} - \Delta y + \beta(y - \psi) \ni f \qquad \text{in } Q = \Omega \times (0,T),$$

$$y(x,0) = y_0(x) \qquad \text{in } \Omega,$$

$$y = u \qquad \text{in } \Sigma = \partial\Omega \times (0,T), \tag{2.1}$$

where $\beta: \mathbf{R} \to 2^{\mathbf{R}}$ is is the graph

$$\beta(r) = \begin{cases} 0, & r > 0, \\ ]-\infty, 0], & r = 0, \\ \varnothing, & r < 0, \end{cases}$$

$\psi \in C^2(\overline{\Omega})$, $\psi \leq 0$ in $\partial\Omega$, and Ω is a bounded open subset of $\mathbf{R}^N$ with a smooth boundary $\partial\Omega$ (of class C^2, for instance). Regarding the functions f, y_0, and u, we assume that

$$f \in L^p(Q), \qquad y_0 \in W_0^{2-2/p,p}(\Omega), \qquad y_0 \geq \psi \quad \text{in } \Omega, \tag{2.2}$$

$$u \in W_p^{2-1/p,1-1/2p}(\Sigma), \qquad u \geq \psi \text{ in } \Sigma \qquad u(x,0) = y_0(x), \qquad x \in \Omega, \tag{2.3}$$

where $p \geq 2$. Here, $W_p^{2l,l}(\Omega)$ is the usual Sobolev space on Σ (see, e.g., Ladyzhenskaya *et al.* [1], p. 96).

As seen earlier, problem (2.1) can be written equivalently as

$$\begin{aligned} &\frac{\partial y}{\partial t} - \Delta y \geq f, \quad y \geq \psi \qquad \text{in } Q, \\ &\frac{\partial y}{\partial t} - \Delta y = f \qquad \text{in } [y > \psi], \\ &\quad y(x,0) = y_0(x) \qquad \text{in } \Omega, \qquad y = u \text{ in } \Sigma. \end{aligned} \tag{2.4}$$

Regarding existence, we have:

Proposition 2.1. *Under the preceding assumptions, system* (2.1) *has a unique solution* $y \in W_p^{2,1}(Q)$ *and*

$$\|y\|_{W_p^{2,1}(Q)} \leq C\left(\|y_0\|_{W_0^{2-2/p,p}(\Omega)} + \|f\|_{L^p(Q)} + \|u\|_{W_p^{2-1/p,1-1/2p}(\Sigma)} + 1\right), \tag{2.5}$$

where C is independent of y_0, f, and u.

If $p > (N+2)/2$ then y is Hölder continuous in (x,t) and the map $(y_0, f, u) \to y$ is compact from $W_p^{2-1/p,p}(\Omega) \times L^p(\Omega) \times W_p^{2-1/p,1-1/2p}(\Sigma)$ to $C(\overline{Q})$.

Proof. The last part of the proposition is an immediate consequence of regularity properties of the elements of $W_p^{2,1}(Q)$ (Ladyzhenskaya *et al.* [1], p. 98).

For existence in (2.1), consider the approximating equation

$$\begin{aligned} \frac{\partial y}{\partial t} - \Delta y + \beta^{\varepsilon}(y - \psi) &= f && \text{in } Q, \\ y(x,0) &= y_0(x) && \text{in } \Omega, \\ y &= 0 && \text{in } \Sigma, \end{aligned} \tag{2.6}$$

where β^{ε} is defined by (1.17).

The existence for problem (2.6) follows from the following linear result (Ladyzhenskaya *et al.* [1], p. 388).

Lemma 2.1. *The boundary value problem*

$$\begin{aligned} y_t - \Delta y &= f && \text{in } Q, \\ y(x,0) &= y_0(x) && \text{in } \Omega, \\ y &= u && \text{in } \Sigma, \end{aligned} \tag{2.7}$$

has for every $f \in L^p(Q)$ and y_0, u satisfying (2.2), (2.3) *a unique solution $y \in W_p^{2,1}(Q)$, which satisfies the estimate*

$$\|y\|_{W_p^{2,1}(Q)} \leq C(\|y_0\|_{W_0^{2-2/p,p}(\Omega)} + \|u\|_{W^{2-1/p,1-1/2p}(\Sigma)} + \|f\|_{L^p(Q)}). \tag{2.8}$$

Now, let $\xi \in W_p^{2,1}(Q)$ be the solution to (2.7). Then the problem

$$\begin{aligned} (z_\varepsilon)_t - \Delta z_\varepsilon + \beta^{\varepsilon}(z_\varepsilon + \xi - \psi) &= f && \text{in } Q, \\ z_\varepsilon(x,0) &= 0 && \text{in } \Omega, \\ z_\varepsilon &= 0 && \text{in } \Sigma, \end{aligned} \tag{2.9}$$

has a unique solution $z_\varepsilon \in W_p^{2,1}(Q)$. Indeed, by the standard contraction principle it follows that (2.9) has a unique mild solution in $C([0,T]; L^p(\Omega))$. Hence, $\beta^{\varepsilon}(z_\varepsilon + \xi - \psi) \in L^p(Q)$ and by Lemma 2.1 we infer that $z_\varepsilon \in W_p^{2,1}(Q)$.

If we set $y_\varepsilon = z_\varepsilon + \xi$, we get

$$\begin{aligned} \frac{\partial}{\partial t} y_\varepsilon - \Delta y_\varepsilon + \beta^{\varepsilon}(y_\varepsilon - \psi) &= f && \text{in } Q, \\ y_\varepsilon(x,0) &= y_0(x) && \text{in } \Omega, \\ y_\varepsilon &= u && \text{in } \Sigma. \end{aligned} \tag{2.10}$$

We may write Eq. (2.10) as

$$\begin{aligned} \frac{\partial}{\partial t} y_\varepsilon - \Delta y_\varepsilon - \varepsilon^{-1}(y_\varepsilon - \psi)^- &= f + \nu_\varepsilon \quad \text{in } Q, \\ y_\varepsilon(x,0) &= y_0(x) \quad \text{in } \Omega, \\ y_\varepsilon &= u \quad \text{in } \Sigma, \end{aligned} \tag{2.11}$$

where

$$\nu_\varepsilon(x,t) = \begin{cases} \dfrac{1}{2} & \text{if } y_\varepsilon - \psi < -\varepsilon, \\ \dfrac{(y_\varepsilon - \psi)^2}{2\varepsilon^2} + \dfrac{1}{\varepsilon}(y_\varepsilon - \psi) & \text{if } \quad -\varepsilon \le y_\varepsilon - \psi \le 0, \\ 0 & \text{if } y_\varepsilon - \psi > 0. \end{cases} \tag{2.12}$$

We set $\delta_\varepsilon = -(1/\varepsilon)(y_\varepsilon - \psi)^-$. We have

$$-\int_\Omega \Delta(y_\varepsilon - \psi)|\delta_\varepsilon|^{p-2}\delta_\varepsilon \, dx \ge 0$$

and

$$\frac{\partial y_\varepsilon}{\partial t}|\delta_\varepsilon|^{p-2}\delta_\varepsilon = \frac{1}{p}\frac{\partial}{\partial t}|\delta_\varepsilon|^p \qquad \text{a.e. in } Q.$$

Multiplying Eq. (2.11) by $|\delta_\varepsilon|^{p-2}\delta_\varepsilon$ and integrating on Q, we get, after some calculation involving the Hölder inequality,

$$\int_\Omega |\delta_\varepsilon(x,t)|^p \, dx \le \int_0^t \int_\Omega |\delta_\varepsilon|^p \, dx\, ds + C(\|f\|^p_{L^p(Q)} + 1)$$

and so, by Gronwall's lemma,

$$\int_\Omega |\delta_\varepsilon(x,t)|^p \, dx \le C(\|f\|^p_{L^p(Q)} + 1) \qquad \forall t \in [0,T].$$

By Lemma 2.1 we have, therefore,

$$\|y_\varepsilon\|_{W_p^{2,1}(Q)} \le C(\|f\|_{L^p(Q)} + \|y_0\|_{W^{2-2/p,p}(\Omega)} + \|u\|_{W^{2-1/p,1-1/2p}(\Sigma)} + 1). \tag{2.13}$$

In particular, it follows that, for $\varepsilon \to 0$,

$$\begin{aligned} (y_\varepsilon - \psi)^- &\to 0 \quad \text{a.e. in } Q, \\ \nu_\varepsilon &\to 0 \quad \text{a.e. in } Q. \end{aligned} \tag{2.14}$$

Finally, by estimate (2.13) we see that, on a subsequence,

$$y_\varepsilon \to y \qquad \text{weakly in } W_p^{2,1}(Q)$$

and, in particular,

$$\begin{aligned} y_\varepsilon &\to y && \text{in } L^2(0,T;\, H_0^1(\Omega)),\\ \frac{dy_\varepsilon}{dt} &\to \frac{dy}{dt} && \text{weak ly in } L^2(Q),\\ \beta^\varepsilon(y_\varepsilon - \psi) &\to \eta && \text{weakstar in } L^\infty(0,T;\, L^p(\Omega)). \end{aligned} \tag{2.15}$$

We have, therefore,

$$\begin{aligned} &\frac{dy}{dt} - \Delta y + \eta = 0 && \text{in Q},\\ &y(0) = y_0, \quad y = u && \text{in } \Sigma \end{aligned} \tag{2.16}$$

This implies that

$$\int_Q \left(\frac{dy}{dt}(y - z) + \nabla y \cdot \nabla (y - z) \right) dx\,dt + \int_Q \eta(y - z)\, dx\,dt \geq 0 \tag{2.17}$$

for all $z \in K = \{z \in W_p^{2,1}(Q);\ z \geq 0\}$. Since

$$\begin{aligned} &\lim_{\varepsilon \to 0} \int_Q \beta^\varepsilon(y_\varepsilon - \psi)(y_\varepsilon - z)\, dx\,dt \\ &\quad = \int_Q \eta(x,t)(y(x,t) - z)\, dx\,dt \geq 0 \qquad \forall z \in K, \end{aligned}$$

We infer by (2.15) and (2.16) that y is the solution to Eq. (2.1), i.e., $\eta(x,t) \in \beta(y(x,t) - \psi(x))$ a.e. $(x,t) \in Q$. Estimate (2.5) is implied by (2.13) and (2.14), whilst the uniqueness is obvious. ∎

As seen earlier in Chapter 4 (Eq. (3.13)), perhaps the most important physical model for problem (2.1) is the one phase Stefan problem describing the melting process of a solid.

5.2.2. The Optimal Control Problem

We shall study here the optimal control problem:

$$\text{Minimize} \quad \int_0^T g(t,y))\, dt + \phi(u) + \varphi_0(y(T)) \tag{2.18}$$

on all $y \in W_p^{2,1}(Q)$ *and* $u \in W^{2-1/p,1-1/2p}(\Sigma)$, $p > (N+2)/2$, *subject to the state system* (2.1) *and to the constraints*

$$u \geq 0 \quad \text{in } \Sigma, \qquad u(x,0) = y_0(x) \quad \text{in } \Omega. \tag{2.19}$$

(For simplicity, we take $\psi \equiv 0$.)

Here, $g: [0,T] \times L^2(\Omega) \to \mathbf{R}^+$, $\varphi_0: L^2(\Omega) \to \mathbf{R}^+$ are locally Lipschitz and $\phi: W^{2-1/p,1-1/2p}(\Sigma) \to \overline{\mathbf{R}}$ is a lower semicontinuous convex function such that for all $u \in D(\phi)$, $u \geq 0$ a.e. in Σ, $u(x,0) = y_0(x)\ \forall x \in \partial\Omega$.

The latter assumption allows us to incorporate the control constraints (2.19) into the cost functional of problem (2.18).

We set $X_p = W^{2-1/p,1-1/2p}(\Sigma)$ and denote by $|||\cdot|||_p$ the natural norm of X_p.

Theorem 2.1. *Let* $(y^*, u^*) \in W_p^{2,1}(Q) \times X_p$ *be an optimal pair for problem* (2.18). *Then there is* $\xi \in L^2(Q)$, $p \in L^\infty(0,T;\ L^2(\Omega)) \cap L^2(0,T;\ H_0^1(\Omega)) \cap BV([0,T];\ H^{-s}(\Omega))$, $s > N/2$ *such that*

$$\frac{\partial}{\partial t}p + \Delta p \in (L^\infty(Q))^*, \qquad \frac{\partial p}{\partial \nu} \in X_p^*, \tag{2.20}$$

$$\left(\frac{\partial}{\partial t}p + \Delta p - \xi\right)y^* = 0, \qquad \xi \in \partial g(t, y^*) \qquad \text{a.e. in } Q, \tag{2.21}$$

$$\left(\frac{\partial}{\partial t}y^* - \Delta y^* - f\right)p = 0 \qquad \text{a.e. in } Q, \tag{2.22}$$

$$-\frac{\partial p}{\partial \nu} \in \partial\phi(u^*) \qquad \text{in } \Sigma. \tag{2.23}$$

We note that the product $y^*(p_t + \Delta p - \xi)$ makes sense because $y^* \in W_p^{2,1}(Q) \subset C(\overline{Q})$.

Proof. We shall use the standard method. Consider the approximating control problem:

Minimize

$$\int_0^T g^\varepsilon(t, y(t))\,dt + \phi(u) + \tfrac{1}{2}|||u - u^*|||_p^2 + \varphi_0^\varepsilon(y(T)), \tag{2.24}$$

subject to (2.1), (2.19).

Here, g^ε and φ_0^ε are defined as in the previous cases and, since the map $u \to y$ is compact from X_p to $C(\overline{Q})$, it follows by a standard device that problem (2.24) has for every $\varepsilon > 0$ at least one solution $(y_\varepsilon, u_\varepsilon) \in W_p^{2,1}(Q) \times X_p$. Moreover, using Lemma 2.1 and arguing as in the proof of Proposition 1.2, it follows that, for $\varepsilon \to 0$,

$$\begin{aligned} u_\varepsilon &\to u^* && \text{strongly in } X_p, \\ y_\varepsilon &\to y^* && \text{strongly in } W_p^{2,1}(Q) \subset C(\overline{Q}), \\ \beta^\varepsilon(y_\varepsilon) &\to f + \Delta y^* - y_t^* && \text{weakly in } L^p(Q). \end{aligned} \tag{2.25}$$

Now let $p_\varepsilon \in H^{2,1}(Q) \cap L^2(0,T; H_0^1(\Omega))$ be the solution to the boundary value problem

$$\begin{aligned} \frac{\partial p_\varepsilon}{\partial t} + \Delta p_\varepsilon - p_\varepsilon \dot\beta^\varepsilon(y_\varepsilon) &= \nabla g^\varepsilon(t, y_\varepsilon) && \text{in } Q, \\ p_\varepsilon &= 0 && \text{in } \Sigma, \\ p_\varepsilon(T) + \nabla\varphi_0^\varepsilon(y_\varepsilon(T)) &= 0 && \text{in } \Omega. \end{aligned} \tag{2.26}$$

Then, by a little calculation, we find that

$$\frac{\partial p_\varepsilon}{\partial \nu} + \partial\phi(u_\varepsilon) + F(u_\varepsilon - u^*) \ni 0 \qquad \text{in } \Sigma, \tag{2.27}$$

where $F: X_p \to X_p^*$ is the duality mapping of X_p and $\partial\phi: X_p \to X_p^*$ is the subdifferential of ϕ. Now multiplying Eq. (2.26) by p_ε and sign p_ε, we get the estimate

$$\|p_\varepsilon(t)\|_{L^2(\Omega)}^2 + \int_0^T \|p_\varepsilon(t)\|_{H_0^1(\Omega)}^2\, dt + \int_Q |\dot\beta^\varepsilon(y_\varepsilon) p_\varepsilon|\, dx\, dt \le C. \tag{2.28}$$

Hence, on a subsequence, we have (see the proof of Proposition 1.2)

$$\begin{aligned} p_\varepsilon &\to p && \text{strongly in } L^2(Q), \text{ weakly in } L^2(0,T; H_0^1(\Omega)), \\ & && \text{weak star in } L^\infty(0,T; L^2(\Omega)), \\ p_\varepsilon(t) &\to p(t) && \text{strongly in } H^{-s}(\Omega) \text{ for every } t \in [0,T], \end{aligned} \tag{2.29}$$

where $p \in BV([0,T]; H^{-s}(\Omega))$, $s > N/2$. Moreover, there is $\mu \in (L^\infty(Q))^*$ such that, on a generalized subsequence of $\{\varepsilon\}$,

$$\dot\beta^\varepsilon(y_\varepsilon) p_\varepsilon \to \mu \qquad \text{weak star (vaguely) in } (L^\infty(Q))^*.$$

Then, letting ε tend to zero in Eq. (2.26), we get

$$\begin{aligned} &\frac{\partial p}{\partial t} + \Delta p - \mu \in \partial g(t, y^*) && \text{in } Q, \\ &p(T) + \partial \varphi_0(y^*(T)) \ni 0 && \text{in } \Omega, \\ &p = 0 && \text{in } \Sigma. \end{aligned} \tag{2.30}$$

In other words, $\exists \xi \in L^2(Q)$, $\xi(x,t) \in \partial g(t, y^*)(x,t)$ a.e. $(x,t) \in Q$, such that

$$\int_Q \left(p \frac{\partial \varphi}{\partial t} + \nabla p \cdot \nabla \varphi \right) dx\,dt + \mu(\varphi) + \int_Q \xi \varphi \, dx\,dt$$
$$= \int_\Omega p(x,T) \varphi(x,T)\, dx$$

for all $\varphi \in L^2(0,T; H_0^1(\Omega)) \cap L^\infty(Q)$ such that $\partial\varphi/\partial t \in L^2(0,T; H^{-1}(\Omega))$ and $\varphi(x,0) \equiv 0$. Moreover, by (1.46) and (1.47), we have

$$p_\varepsilon \beta^\varepsilon(y_\varepsilon) \to p\left(f - \frac{\partial y^*}{\partial t} + \Delta y^* \right) = 0 \qquad \text{strongly in } L^1(Q),$$
$$p_\varepsilon \dot{\beta}^\varepsilon(y_\varepsilon) y_\varepsilon \to 0 \qquad \text{strongly in } L^1(Q).$$

Then, by (2.25), we infer that

$$\mu y^* = 0,$$

and Eqs. (2.21), (2.22) follow.

Now, let $\chi \in W_p^{2,1}(Q)$ be the solution to the boundary value problem

$$\begin{aligned} &\frac{\partial \chi}{\partial t} - \Delta \chi = 0 && \text{in } Q, \\ &\chi = w && \text{in } \Sigma, \\ &\chi(x,0) = \chi_0(x) && \text{in } \Omega, \end{aligned} \tag{2.31}$$

where $w \in X_p$, $\chi_0 \in W_p^{2-2/p}(\Omega)$ and $\chi_0(x) = w(x,0)$ a.e. $x \in \Omega$. Then, by Lemma 2.1,

$$\| \chi \|_{W^{2,1}{}_p(Q)} \le C\left(||| w |||_p + \| \chi_0 \|_{W_p^{2-2/p}(\Omega)} \right).$$

On the other hand, by the trace theorem (see, e.g., Ladyzhenskaya *et al.* [1]) we may choose $\chi_0 \in W_p^{2-2/p}(\Omega)$ such that $\chi_0 = w(\cdot, 0)$ in $\partial\Omega$ and

$$\| \chi_0 \|_{W_p^{2-2/p}(\Omega)} \le C \| w(0, \cdot) \|_{W_p^{2-4/p}(\Omega)} \le C\, ||| w |||_p .$$

With such a choice, we have

$$\|\chi\|_{W_p^{2,1}(Q)} \le C\|w\|_p \qquad \forall w \in X_p. \tag{2.32}$$

Now multiplying Eq. (2.30) by χ and integrating on Q, after some calculation we get

$$\left|\int_\Sigma w \frac{\partial p_\varepsilon}{\partial \nu}\, d\delta\, dt\right|$$

$$\le |\mu(\chi)| + \int_Q |\xi|\,|\chi|\, dx\, dt$$

$$+ \int_\Omega (|\chi_0|\,|p_\varepsilon(x,0)| + |p_\varepsilon(x,T)|\,|\chi(x,T)|)\, dx,$$

and by estimate (2.32) and the trace theorem, we get

$$\left|\int_\Sigma w \frac{\partial p_\varepsilon}{\partial \nu}\, d\sigma\, dt\right| \le C\|w\|_p \qquad \forall w \in X_p.$$

Hence, $\{\partial p_\varepsilon/\partial \nu\}$ is bounded in X^*_p (the dual of X_p) and, letting ε tend to zero in Eq. (2.25), it follows by (2.27) that

$$\frac{\partial p}{\partial \nu} + \partial\phi(u^*) \ni 0 \qquad \text{in } \Sigma,$$

as claimed. ■

We will consider now a variant of problem (2.18).

$$\text{Minimize} \quad \int_0^T g(t, y(t))\, dt + \varphi_0(y(T)) \tag{2.33}$$

on all $y \in W_p^{2,1}(Q)$, $u \in L^\infty(0,T)$, *and* $v \in W^{1,\infty}([0,T])$, *subject to the state system*

$$\frac{\partial y}{\partial t} - \Delta y = f_0 \qquad \text{in } \{(x,t) \in Q;\ y(x,t) > 0\},$$

$$y \ge 0, \quad \frac{\partial y}{\partial t} - \Delta y \ge f_0 \qquad \text{in } Q,$$

$$y = g_0(x)v(t) \qquad \text{in } \Sigma,$$

$$y(x,0) = 0, \qquad \text{in } x \in \Omega, \tag{2.34a}$$

$$v' + f(v) = u \qquad \text{a.e. } t \in (0,T),$$

$$v(0) = 0, \tag{2.34b}$$

and to the control constraints

$$0 \le u(t) \le M \qquad \text{a.e. } t \in (0,T),$$
$$\int_0^T u(t)\, dt = L. \tag{2.35}$$

Here, $f_0 \in L^p(Q)$, $g_0 \in W_0^{2-1/p}(\partial\Omega)$, $p > (N+2)/2$, and

$$g_0 \ge 0 \quad \text{in } \partial\Omega, \qquad 0 < MT < L. \tag{2.36}$$

The function $f: \mathbf{R} \to \mathbf{R}$ is Lipschitz and continuously differentiable.

Denote by $U_0 \subset L^\infty(0,T)$ the class of functions u satisfying the constraints (2.35).

To be more specific, we shall assume that

$$g(t,y) = \int_\Omega g^1(y(x))\, dx, \qquad \varphi_0(y) = \int_\Omega g^2(y(x))\, dx, \qquad y \in L^2(\Omega),$$

where $g^i: \mathbf{R} \to \mathbf{R}$, $i = 1,2$ are continuously differentiable functions satisfying the conditions

$$g^i(y) \ge -C(1+|y|) \qquad \forall y \in \mathbf{R},\ i = 1,2. \tag{2.37}$$

Theorem 2.2. *Let* (y^*, u^*) *be optimal for problem* (2.33). *Then there are* $p \in L^2(0,T;\ H_0^1(\Omega)) \cap BV([0,T];\ H^{-s}(\Omega)) \cap L^\infty(0,T;\ L^2(\Omega))$ *and* $q \in AC([0,T])$ *such that*

$$\frac{\partial p}{\partial t} + \Delta p = -g_y^1(y) \qquad \text{in } [y^* > 0],$$
$$p = 0 \qquad \text{in } [y^* = 0, f_0 \ne 0],$$
$$p = 0 \qquad \text{in } \Sigma, \qquad p(x,T) = -g_y^2(y^*(x,T)), \tag{2.38}$$

$$q'(t) - f'(u^*(t))q(t) = \int_t^T \int_{\partial\Omega} g_0 \frac{\partial p}{\partial \nu}\, dx\, ds \qquad \text{a.e. } t \in (0,T),$$
$$q(T) = 0, \tag{2.39}$$

$$u^*(t) = \begin{cases} 0 & \text{if } q(t) < \lambda, \\ M & \text{if } q(t) > \lambda, \end{cases} \qquad \text{a.e. } t \in (0,T), \tag{2.40}$$

where λ *is some real number.*

Proof. Since the proof is essentially the same as that of Theorem 2.1, it will be sketched only.

Consider the approximating penalized problem:

$$\text{Minimize} \quad \left\{ \int_Q g^1(y(x,t))\,dx\,dt + \int_\Omega g^2(y(x,T))\,dx + \tfrac{1}{2}\int_0^T (u(t) - u^*(t))^2\,dt \right\} \tag{2.41}$$

on all (y, u, v), subject to

$$\frac{\partial y}{\partial t} - \Delta y + \beta^\varepsilon(y) = f_0 \qquad \text{in } Q,$$

$$y = g_0 v \qquad \text{in } \Sigma, \qquad y(x,0) = 0 \qquad \text{in } \Omega,$$

and to constraints (2.34b), (2.35).

If (y^*, u^*, v^*) is optimal for problem (2.41), it follows as before that

$$u_\varepsilon \to u^* \qquad \text{strongly in } L^2(0,T),$$

$$y_\varepsilon \to y^* \qquad \text{strongly in } W_p^{2,1}(Q),$$

$$v_\varepsilon \to v^* \qquad \text{strongly in } W^{1,2}([0,T]).$$

On the other hand, problem (2.41) has the following optimality system:

$$\frac{\partial p_\varepsilon}{\partial t} + \Delta p_\varepsilon - \dot{\beta}^\varepsilon(y_\varepsilon)p_\varepsilon = g_y^1(y_\varepsilon) \qquad \text{in } Q,$$

$$p_\varepsilon = 0 \qquad \text{in } \Sigma,$$

$$p_\varepsilon(x,T) = -g_y^2(y_\varepsilon(x,T)), \qquad x \in \Omega, \tag{2.42}$$

$$q_\varepsilon'(t) - f'(v_\varepsilon(t))q_\varepsilon(t) = \int_t^T ds \int_{\partial\Omega} g_0(x)\,\frac{\partial p_\varepsilon}{\partial \nu}\,d\sigma,$$

$$q_\varepsilon(T) = 0, \tag{2.43}$$

$$q_\varepsilon(t) + u^*(t) - u_\varepsilon(t) \in \partial h(u_\varepsilon)(t) \qquad \text{a.e. } t \in (0,T), \tag{2.44}$$

where ∂h: $L^2(0,T) \to L^2(0,T)$ is the subdifferential of the indicator function h to the subset U_0, i.e.,

$$h(u) = \begin{cases} 0 & \text{if } u \in U_0, \\ \infty & \text{otherwise.} \end{cases}$$

Now, we may represent ∂h as (see, e.g., Section 2 in Chapter 2)

$$\partial h(u) = \partial h_0(u) + \partial h_1(u), \qquad u \in L^2(0,T),$$

where h_0 is the indicator function of $\{u \in L^2(0,T);\ 0 \le u \le M$ a.e. in $(0,T)\}$ and h_1 is the indicator function of $\{u \in L^2(0,T);\ \int_0^T u(t)\,dt = L\}$. As shown earlier, we have

$$\partial h_0(u) = \{w \in L^2(0,T);\ w(t) = 0 \text{ a.e. in } [t;\ 0 < u(t) < M],\\ w \ge 0 \text{ a.e. in } [t;\ u(t) = M],\ w \ge 0 \text{ a.e. in } [t;\ u(t) = 0]\}$$

and $\partial h_1(u) = \{\lambda \in \mathbf{R}\}$. Then, by (2.44), we see that

$$u_\varepsilon(t) = \begin{cases} 0 & \text{a.e. in } [t;\ q_\varepsilon(t) + u^*(t) - u_\varepsilon(t) < \lambda_\varepsilon], \\ M & \text{a.e. in } [t;\ q_\varepsilon(t) + u^*(t) - u_\varepsilon(t) > \lambda_\varepsilon], \end{cases} \tag{2.44$'$}$$

where $\lambda_\varepsilon \in \mathbf{R}$.

Now using the standard estimate for the solution p_ε to Eq. (2.42), it follows that, on a subsequence,

$$p_\varepsilon(t) \to p(t) \qquad \text{strongly in } H^{-s}(\Omega),\ \forall t \in [0,T],$$
$$p_\varepsilon \to p \qquad \text{strongly in } L^2(Q), \text{ weakly in } L^2(0,T;\ H_0^1(\Omega)),$$

where $s > N/2$ and $p \in BV([0,T];\ H^{-s}(\Omega))$. Moreover, we have

$$\dot\beta^\varepsilon(y_\varepsilon)p_\varepsilon \to \mu \qquad \text{weak star in } (L^\infty(Q))^*,$$
$$\frac{\partial p_\varepsilon}{\partial \nu} \to \frac{\partial p}{\partial \nu} \qquad \text{weakly in } X_p^*.$$

Then, letting ε tend to zero in the system (2.42)–(2.44), we see that p satisfies the system

$$\frac{\partial p}{\partial t} + \Delta p - \mu = g_y^1(y^*) \qquad \text{in } Q,$$
$$p = 0 \qquad \text{in } \Sigma,$$
$$p(x,T) = -g_y^2(y^*(x,T)) \qquad \text{in } \Omega, \tag{2.45}$$

$$q'(t) - f'(u^*(t))q(t) = \int_t^T \int_{\partial\Omega} g_0 \frac{\partial p}{\partial \nu}\, d\sigma\, ds \qquad \text{a.e. } t \in (0,T),$$
$$q(T) = 0. \tag{2.46}$$

Similarly, letting ε tend to zero in Eq. (2.44)$'$, we get

$$u^*(t) = \begin{cases} 0 & \text{a.e. } [t \in (0,T);\ q(t) < \lambda], \\ M & \text{a.e. } [t \in (0,T);\ q(t) > \lambda], \end{cases} \tag{2.47}$$

where $\lambda \in \mathbf{R}$. Moreover, as seen in the previous proof we have $\mu = 0$ in $\{(x,t) \in Q;\ y^*(x,t) > 0\}$ and $p = 0$ in $\{(x,t) \in Q;\ y^*(x,t) = 0,\ f_0 \neq 0\}$, thereby completing the proof. ■

In particular, it follows by Theorem 2.2 that if $\partial p/\partial \nu \geq 0$ in $\partial\Omega$ (this happens, for instance, if $g_y^1 \leq 0$, $g_y^2 \geq 0$, by virtue of the maximum principle for parabolic equations), then q is monotonically increasing and so every optimal control u^* for problem (2.32) has at most one switching point t_1. More will be said about this in Example 2 following.

We note also that Theorem 2.2 extends to control systems (2.33) with boundary conditions of the following form (see, e.g., Barbu and Barron [1]):

$$y = \sum_{i=1}^{m} g_i(x) v_i(t) \qquad \text{in } \Sigma$$

$$v' + f(v) = u \qquad \text{a.e. in } (0,T), \qquad v(0) = 0,$$

where $v = (v_1, \ldots, v_m)$, $u = (u_1, \ldots, u_n)$, $f = (f_1, \ldots, f_m)$, and $g_i \in W^{2-1/p}(\partial\Omega)$, $i = 1, \ldots, m$.

5.2.3. *Examples*

1. *Control of oxygen diffusion in an absorbing tissue*. The oxygen diffusion in absorbing tissue $\Omega \subset \mathbf{R}^3$ is governed by the obstacle problem (see, e.g., E. Magenes [1], and Elliott and Ockendon [1], p. 127)

$$\begin{aligned}
&\frac{\partial y}{\partial t} - \Delta y + 1 \geq 0, \quad y \geq 0 \qquad \text{in } Q = \Omega \times (0,T),\\
&\frac{\partial y}{\partial t} - \Delta y + 1 = 0 \qquad \text{in } \{(x,t) \in Q;\ y(x,t) > 0\},\\
&y(x,0) = y_0(x), \qquad x \in \Omega,\\
&y = u \qquad \text{in } \Sigma = \partial\Omega \times (0,T),
\end{aligned} \tag{2.48}$$

where y_0 is the initial distribution of oxygen concentration in the tissue and $u(t)$ is a prescribed concentration on $\partial\Omega$ at moment t, which satisfies the constraints $u(t) \in U_0\ \forall t \in [0,T]$,

$$U_0 = \{u \in X_p;\ u(x,0) = y_0(x)\ \forall x \in \partial\Omega,\ u \geq 0 \text{ in } \Sigma\}.$$

We associate with control system (2.48) the cost functional

$$\|u\|_p^2 + \int_\Omega (y(x,T)) - y^0(x))^2 \, dx, \tag{2.49a}$$

where $y^0 \in L^2(\Omega)$.

This is a problem of the form (2.18), where $g \equiv 0$ and

$$\phi(u) = \begin{cases} \|u\|_p^2 & \text{if } u \in U_0, \\ +\infty & \text{otherwise,} \end{cases} \qquad \forall u \in X_p,$$

$$\varphi_0(y) = \|y - y^0\|_{L^2(\Omega)}^2, \qquad y \in L^2(\Omega).$$

Then, $\partial\phi(u) = Fu + N_{U_0}(u)$ $\forall u \in X$, where $F: X_p \to X_p^*$ is the duality mapping of $X_p = W_p^{2-1/p, 1-1/2p}(\Sigma)$ and $N_{U_0} \subset X_p^*$ is the normal cone to U_0, i.e.,

$$N_{U_0}(u) = \{\eta \in X_p^*; (\eta, u - v) \geq 0 \ \forall v \in U_0\}.$$

If we take $v = u + \varphi$, there $\varphi \geq 0$, $\varphi \in C_0^\infty(\Sigma)$, we get $(\eta, \varphi) \leq 0$. Hence, η is a nonpositive Radon measure on Σ, i.e., $\eta \in M(\Sigma)$. Moreover, it is readily seen that $\eta = 0$ in $\{(x,t) \in \Sigma; u(x,t) > 0\}$. Then, by Theorem 2.1, the optimality system for this problem is

$$\begin{aligned} \frac{\partial p}{\partial t} + \Delta p &= 0 && \text{in } [y^* > 0], \\ p &= 0 && \text{in } [y^* = 0], \\ p &= 0 && \text{in } \Sigma, \end{aligned} \tag{2.49b}$$

whilst the optimal control u^* is given by

$$\begin{aligned} \frac{\partial p}{\partial \nu} + Fu^* &= 0 && \text{in } [u^* > 0], \\ u^* &= 0 && \text{in } \left[\frac{\partial p}{\partial \nu} > 0\right], \end{aligned}$$

because $(\eta, u^*) = 0$, $\eta \leq 0$ for all $\eta \in N_{U_0}(u^*)$. Equivalently,

$$u^* = -F^{-1}\left(\frac{\partial p}{\partial \nu}\right) \quad \text{in } \left[\frac{\partial p}{\partial \nu} < 0\right], \qquad u^* = 0 \quad \text{in } \left[\frac{\partial p}{\partial \nu} > 0\right]. \tag{2.50}$$

2. *Optimal control of the one phase Stefan problem.* Consider the melting process of a body of ice $\Omega \subset \mathbf{R}^3$ maintained at 0°C in contact with a region

of water on Γ_1 and at controlled temperature u on Γ_2. The boundary $\partial\Omega$ is composed of two disjoint parts Γ_1 and Γ_2, and $\overline{\Gamma}_1 \cap \overline{\Gamma}_2 = \emptyset$. Let $\theta(x,t)$ be the water temperature of point $x \in \Omega$ at time t. Initially, the water occupies the domain $\Omega_0 \subset \Omega$ at temperature θ_0 (see Fig. 4.1). If $t = \sigma(x)$ is the equation of the water–ice interface, then the temperature distribution θ satisfies the one phase Stefan problem

$$\begin{aligned}
\frac{\partial\theta}{\partial t} - \Delta\theta = 0 \qquad & \text{in } \{(x,t) \in Q;\ \sigma(x) < t < T\},\\
\theta = 0 \qquad & \text{in } \{(x,t) \in Q;\ \sigma(x) \geq t\},\\
\theta(x,0) = \theta_0(x) \qquad \forall x \in \Omega_0, \qquad & \theta(x,0) = 0 \qquad \forall x \in \Omega \setminus \Omega_0,
\end{aligned} \tag{2.51}$$

along with the Dirichlet boundary conditions

$$\begin{aligned}
\theta(x,t) &= g_0(x)u(t) \qquad && \text{in } \Sigma_1 = \Gamma_1 \times (0,T),\\
\theta(x,t) &= 0 && \text{in } \Sigma_2 = \Gamma_2 \times (0,T),
\end{aligned} \tag{2.52}$$

where

$$\begin{aligned}
& g_0 \in W_0^{2-1/p}(\Gamma_1), \quad g_0 \geq 0 \qquad \text{in } \Gamma_1,\\
& \theta_0 \in C(\overline{\Omega}_0), \qquad p > (N+2)/2, \qquad \theta_0(x) > 0 \qquad \forall x \in \overline{\Omega}_0.
\end{aligned} \tag{2.53}$$

We will consider here the following model optimization problem associated with the controlled melting process (2.51), (2.52):

$$\text{Maximize} \quad \int_Q \theta(x,t)\,dx\,dt \tag{2.54}$$

on all θ *and* u *subject to* (2.51), (2.52) *and to the control constraints* $u \in U_0$, *where*

$$U_0 = \left\{u \in L^\infty(0,T);\ 0 \leq u(t) \leq M,\ \int_0^T u(t)\,dt = L\right\}.$$

We will assume of course that $MT > L$.

As seen in Chapter 4, (Section 3.2), by the transformation

$$y(x,t) = \int_0^t \chi(x,s)\theta(x,s)\,ds,$$

where χ is the characteristic function of $Q_+ = \{(x,t);\ \sigma(x) < t \le T\}$, the system (2.51), (2.52) reduces to the controlled obstacle problem

$$\frac{\partial y}{\partial t} - \Delta y = f_0 \qquad \text{in } \{(x,t) \in Q;\ y(x,t) > 0\},$$

$$y \ge 0, \quad \frac{\partial y}{\partial t} - \Delta y \ge f_0 \qquad \text{in } Q,$$

$$y(x,0) = 0 \qquad \text{in } \Omega, \tag{2.55}$$

$$y(x,t) = g_0(x)v(t) \qquad \forall (x,t) \in \Sigma_1;$$

$$y(x,t) = 0 \qquad \forall (x,t) \in \Sigma_2, \tag{2.56}$$

$$v'(t) = u(t) \qquad \text{a.e. } t \in (0,T), \qquad v(0) = 0, \tag{2.57}$$

where $f_0 = \theta_0$ in Ω_0 and $f_0 = -\rho$ in $\Omega \setminus \Omega_0$.

In terms of y, problem (2.54) becomes:

$$\text{Maximize} \quad \int_\Omega y(x,T)\,dx \tag{2.58}$$

on all $(y,u) \in W_p^{2,1}(Q) \times U_0$ *satisfying* (2.55)–(2.57).

Applying Theorem 2.2, where $f \equiv 0$, $g^1 \equiv 0$, $g^2(y) = -y$, we find that every optimal control u^* is of the form (2.40), where

$$q'(t) = -\int_t^T \int_{\Gamma_1} g_0(x)\,\frac{\partial p}{\partial \nu}\,dx\,ds \qquad \forall t \in [0,T],$$

$$q(T) = 0, \tag{2.59}$$

and

$$\frac{\partial p}{\partial t} + \Delta p = 0 \qquad \text{in } \{(x,t) \in Q;\ y^*(x,t) > 0\},$$

$$p = 0 \qquad \text{in } \{(x,t) \in Q;\ y^*(x,t) = 0,\ f_0(x,t) \ne 0\},$$

$$p = 0 \qquad \text{in } \Sigma, \qquad p(x,T) = 1 \qquad \forall x \in \Omega. \tag{2.60}$$

Inasmuch as $\partial y^*/\partial t - \Delta y^* > 0$ in $\Omega_0 \times (0,T)$, we infer that $y^* > 0$ in $\Omega_0 \times (0,T)$.

Hence, p satisfies Eq. (2.60) in $\{(x,t) \in Q;\ x \in \Omega_0,\ t \in (0,T)\}$ and, by standard regularity results for parabolic Dirichlet problems (see for instance Chapter 4, Section 3.1), we know that $p \in C^{2,1}(\overline{\Omega}_0 \times [0,T))$ and $p \in C([0,T];\ L^2(\Omega))$. Then, by virtue of the maximum principle for linear parabolic operators (see, e.g., Porter and Weinberger [1], p. 170) we have

$$p > 0 \quad \text{in } \Omega_0 \times (0,T) \qquad \frac{\partial p}{\partial \nu} < 0 \quad \text{in } \Gamma_1 \times (0,T).$$

Then, by Eqs. (2.47) and (2.59), it follows that q is increasing and the optimal control u^* has one switch point t_0, i.e.,

$$u^*(t) = \begin{cases} 0, & t \in [0,t_0), \\ M, & t \in [t_0,T], \end{cases} \tag{2.61}$$

where $Mt_0 = L$.

We have therefore proved:

Corollary 2.1. *Under assumptions* (2.53) *the optimal control problem* (2.54) *has a unique solution* u^*, *given by* (2.61). ■

We will consider now the following problem:

> Given the surface $S_0 = \{(t,x);\ t = l(x)\}$, find $u \in U_0$ such that S_0 is as "close as possible" to the free boundary $S = \partial Q_+$ of problem (2.51).

This is an inverse Stefan problem in which the melting surface is known and the temperature on the surface Γ_1 has to be determined.

Let y^0 be a given smooth function on Q such that $y^0(x,t) = 0$ for $0 \le t \le l(x)$ and

$$\frac{\partial y^0}{\partial t} - \Delta y^0 \le f_0 \qquad \text{in } Q,$$
$$y^0 \le 0 \qquad \text{in } \Sigma, \qquad y^0(x,0) \le 0 \qquad \text{in } \Omega, \tag{2.62}$$

where f_0 is defined as before. Then the least square approach to this inverse problem (which in general is not well-posed) leads us to a problem of the form (2.33), i.e.:

$$\text{Minimize} \quad \int_Q (y(x,t) - y^0(x,t))^2\, dx\, dt \tag{2.63}$$

on all $(y,u) \in W_p^{2,1}(Q) \times U_0$ *satisfying* (2.55)–(2.57).

Then, by Theorem 2.2, every optimal control u^* of problem (2.63) is given by Eq. (2.40), where

$$q'(t) = -\int_t^T \int_{\Gamma_1} g_0 \frac{\partial p}{\partial \nu} \, d\sigma \, ds \qquad \text{a.e. } t \in (0,T), \qquad q(T) = 0,$$

and

$$\begin{aligned} \frac{\partial p}{\partial t} + \Delta p &= 2(y^* - y^0) \qquad && \text{in } [y^* > 0], \\ p &= 0 && \text{in } [y^* = 0;\ f_0 \neq 0], \\ p &= 0 && \text{in } \Sigma, \qquad p(x,T) = 0 \qquad \text{in } \Omega. \quad (2.64) \end{aligned}$$

By (2.55), (2.62), and the maximum principle, it follows that $y^* \geq y^0$ in Q. Then, by (2.64), we conclude (again by virtue of the maximum principle) that $\partial p / \partial \nu > 0$ in Σ_1 and so, by (2.40), we see that $u^* = M$ on $[0, t_0]$, $u^* = 0$ on $[t_0, T]$, where $Mt_0 = L$.

5.2.4. *The Obstacle Problem with Neumann Boundary Control*

The previous results remain true for optimal control problems with payoff

$$\int_0^T g(t, y(t) \, dt + h(u) + \varphi_0(y(T)) \tag{2.65}$$

and governed by the variational inequality (2.1), i.e.,

$$\begin{aligned} &\frac{\partial y}{\partial t} - \Delta y = f_0 \qquad \text{in } \{(x,t) \in Q;\ y(x,t) > 0\}, \\ &\frac{\partial y}{\partial t} - \Delta y \geq f_0, \quad y \geq 0 \qquad \text{in } Q = \Omega \times (0,T), \\ &\quad y(x,0) = y_0(x), \qquad x \in \Omega, \end{aligned} \tag{2.66}$$

with the boundary conditions

$$\frac{\partial y}{\partial \nu} + \alpha y = u \qquad \text{in } \Sigma_1, \qquad y = 0 \qquad \text{in } \Sigma_2, \tag{2.67}$$

$$\frac{dv}{dt} + \Lambda v = Bu \qquad \forall t \in (0,T), \qquad v(0) = 0. \tag{2.68}$$

Here, Ω is a bounded, open subset of $\mathbf{R}^n$ with a sufficiently smooth boundary $\partial\Omega = \Gamma_1 \cup \Gamma_2$, $\overline{\Gamma}_1 \cap \overline{\Gamma}_2 = \emptyset$, $\Sigma_i = \Gamma_i \times (0,T)$, $i = 1,2$; Λ is a linear continuous operator from $L^2(\Sigma_1)$ to itself, B is a linear continuous operator from a Hilbert space of controllers U to $L^2(\Sigma_1)$, $\alpha > 0$, and

$$f_0 \in W^{1,2}([0,T]; L^2(\Omega)), \tag{2.69}$$

$$y_0 \in H^2(\Omega), \qquad y_0 = 0 \quad \text{in } \Gamma_2, \qquad \frac{\partial y_0}{\partial \nu} + \alpha y_0 = 0 \quad \text{in } \Gamma_1, \qquad y_0 \geq 0. \tag{2.70}$$

Regarding the functions $g: [0,T] \times L^2(\Omega) \to \mathbf{R}$, $\varphi_0: L^2(\Omega) \to \mathbf{R}$, and $h: U \to \overline{\mathbf{R}}$ we will assume that:

(i) h is convex, lower semicontinuous and

$$h(u) \geq \gamma \|u\|_U^2 + C \qquad \forall u \in U, \tag{2.71}$$

for some $\gamma > 0$ and $C \in \mathbf{R}$,

(ii) g is measurable in t, $g(t,0) \in L^\infty(0,T)$, and there exists $C \in \mathbf{R}$ such that

$$g(t,y) + \varphi_0(y) \geq C(\|y\|_{L^2(\Omega)} + 1) \qquad \forall y \in L^2(\Omega). \tag{2.72}$$

For every $r > 0$, there exists $L_r > 0$ such that

$$|g(t,y) - g(t,z)| + |\varphi_0(y) - \varphi_0(z)| \leq L_r \|y - z\|_{L^2(\Omega)}$$

for all $t \in [0,T]$ and $\|y\|_{L^2(\Omega)} + \|z\|_{L^2(\Omega)} \leq r$.

Under assumptions (2.69), (2.70) the boundary value problem (2.66)–(2.68) has for every $u \in U$ a unique solution $y \in W^{1,2}([0,T]; V) \cap W^{1,\infty}([0,T]; H)$ (see Corollary 3.3, Chapter 4). Here, $V = \{y \in H^1(\Omega);\ y = 0 \text{ in } \Gamma_2\}$, $H = L^2(\Omega)$. Moreover, $y = \lim_{\varepsilon \to 0} y_\varepsilon$ strongly in $C([0,T]; H)$ and weakly in $W^{1,\infty}([0,T]; H) \cap W^{1,2}([0,T]; V)$, where y is the solution to the approximating equation

$$\begin{gathered}
\frac{\partial}{\partial t} y - \Delta y + \beta^\varepsilon(y) = f_0 \qquad \text{in } Q = \Omega \times (0,T), \\
y(x,0) = y_0(x), \\
\frac{\partial y}{\partial \nu} + \alpha y = v \qquad \text{in } \Sigma_1, \qquad y = 0 \qquad \text{in } \Sigma_2, \\
\frac{dv}{dt} + \Lambda v = Bu \qquad \text{in } (0,T), \qquad v(0) = 0,
\end{gathered} \tag{2.73}$$

where β^ε is defined by (1.17).

The following estimate holds:

$$\|y_t\|_{W^{1,\infty}([0,T];\, H) \cap W^{1,2}([0,T];\, V)} \le C(1 + \|u\|_U).$$

Then, by a standard device (see Proposition 1.1), it follows that optimal control problem (2.65) admits at least one optimal control u^*. Regarding the characterization of optimal controllers, we have:

Theorem 2.3. *Let* (y^*, u^*) *be an optimal pair for problem* (2.65), (2.66). *Then there exists* $p \in L^2(0,T;\ V) \cap L^\infty(0,T;\ L^2(\Omega)) \cap BV([0,T];\ (V \cap H^s(\Omega))')$, $s > N/2$, *such that* $\partial p/\partial t + \Delta p \in (L^\infty(Q))^*$ *and*

$$\left(\frac{\partial p}{\partial t} + \Delta p\right)_{\mathrm{a}} \in \partial g(t, y^*) \qquad \text{a.e. in } \{(x,t) \in Q;\ y^*(x,t) > 0\}, \tag{2.74}$$

$$p(T) + \partial\varphi_0(y^*(T)) \ni 0 \qquad \text{in } \Omega, \tag{2.75}$$

$$\frac{\partial p}{\partial \nu} + \alpha p = 0 \qquad \text{in } \Sigma_1, \qquad p = 0 \qquad \text{in } \Sigma_2, \tag{2.76}$$

$$p = 0 \qquad \text{a.e. in } \{(x,t) \in Q;\ y^*(x,t) = 0,\ f_0(x,t) \neq 0\}, \tag{2.77}$$

$$B^* \int_t^T e^{-\Lambda^*(s-t)} p(s)\, ds \in \partial h(u^*). \tag{2.78}$$

(Λ^* *is the adjoint of* Λ.)

If $N = 1$, *then* $y^* \in C(\overline{Q})$ *and Eq.* (2.74) *becomes*

$$\frac{\partial p}{\partial t} + \Delta p = \xi \in \partial g(t, y^*) \qquad \text{in } \{(x,t) \in Q;\ y^*(x,t) > 0\}, \tag{2.74$'$}$$

where $\xi \in L^2(Q)$.

Here, $BV([0,T];\ (V \cap H^s(\Omega))')$ is the space of functions with bounded variation from $[0,T]$ to $(V \cap H^s(\Omega))'$.

Proof. Since the proof is essentially the same as that of Theorem 2.1, it will be sketched only. Also, for the sake of simplicity we will assume that g and φ_0 are differentiable on $L^2(\Omega)$.

For every $\varepsilon > 0$, consider the approximating control problem:

Minimize

$$\int_0^T (g(t, y(t))\, dt + h(u) + \tfrac{1}{2}\|u - u^*\|_U^2 + \varphi_0(y(T)) \tag{2.79}$$

on all $(y, u) \in (W^{1,\infty}([0,T];\ H) \cap W^{1,2}([0,T];\ V)) \times L^2(\Sigma_1)$, *subject to Eq.* (2.72).

Let $(y_\varepsilon, u_\varepsilon)$ be a solution to problem (2.78).

By assumptions (i), (ii) we see that, for $\varepsilon \to 0$,

$$\begin{aligned} u_\varepsilon &\to u^* && \text{strongly in } U, \\ y_\varepsilon &\to y^* && \text{strongly in } C([0,T];\, H), \\ & && \text{weakly in } W^{1,\infty}([0,T];\, H) \cap W^{1,2}([0,T];\, V), \\ v_\varepsilon &\to v^* && \text{strongly in } W^{1,2}([0,T];\, L^2(\Gamma_1)), \end{aligned}$$

where

$$\frac{dv_\varepsilon}{dt} + \Lambda v_\varepsilon = B\mu_\varepsilon \qquad \text{a.e. in } (0,T), \qquad v_\varepsilon(0) = 0.$$

On the other hand, we have

$$\int_0^T \big(\nabla_y g(t, y_\varepsilon(t)), z(t)\big)\, dt + h'(u_\varepsilon, w) + \langle u_\varepsilon - u^*, w\rangle \geq 0 \qquad \forall w \in U,$$

where h' is the directional derivative of h, $(\cdot,\cdot)$ and $\langle \cdot,\cdot \rangle$ are the scalar products in $L^2(\Omega)$ and U, respectively, whilst z is the solution to

$$\begin{aligned} z_t - \Delta z + \dot\beta^\varepsilon(y_\varepsilon) z &= 0 && \text{in } Q, \\ z(x,0) &= 0 && \text{in } \Omega, \\ \frac{\partial z}{\partial \nu} + \alpha z &= V && \text{in } \Sigma_1, \qquad z = 0 \qquad \text{in } \Sigma_2 \\ \frac{dv}{dt} + \Lambda v &= Bw && \text{in } (0,T), \qquad v(0) = 0. \end{aligned}$$

Let $p_\varepsilon \in W^{1,2}([0,T];\, L^2(\Omega)) \cap L^2(0,T;\, V)$ be the solution to boundary value problem

$$\begin{aligned} &\frac{\partial}{\partial t} p_\varepsilon + \Delta p_\varepsilon - p_\varepsilon \dot\beta^\varepsilon(y_\varepsilon) = \nabla_y g(t, y_\varepsilon) && \text{in } Q, \\ &p_\varepsilon(x,T) + \nabla\varphi_0(y_\varepsilon(T))(x) = 0, && x \in \Omega, \\ &\frac{\partial p_\varepsilon}{\partial \nu} + \alpha p_\varepsilon = 0 \qquad \text{in } \Sigma_1, \qquad p_\varepsilon = 0 && \text{in } \Sigma_2. \end{aligned} \tag{2.80}$$

After some calculation, we get that

$$h'(u_\varepsilon, w) + \langle u_\varepsilon - u^*, w\rangle - \int_{\Gamma_1} p_\varepsilon(\sigma,t)\left(\int_0^t e^{-\Lambda(t-s)}(Bw)(s)\, ds\right) d\sigma\, dt \geq 0 \qquad \forall w \in U.$$

This yields

$$B^*\left(\int_t^T e^{-\Lambda^*(s-t)}p_\varepsilon(s)\,ds\right) + u_\varepsilon - u^* \in \partial h(u_\varepsilon).$$

Next, we multiply Eq. (2.80) by p_ε and sign p_ε and integrate on Q. We obtain the estimate

$$\|p_\varepsilon(t)\|^2_{L^2(\Omega)} + \int_0^T \|p_\varepsilon(t)\|^2_{H^1(\Omega)}\,dt + \int_Q |\dot\beta^\varepsilon(y_\varepsilon)p_\varepsilon|\,dx\,dt \le C.$$

Hence, $\{(p_\varepsilon)_t\}$ is bounded in $L^1(0,T;\ L^1(\Omega)) + L^2(0,T;\ V') \subset L^1(0,T;\ (H^s(\Omega)\cap V)')$ for $s > N/2$ (by Sobolev's imbedding theorem). Thus, on a subsequence, we have

$$p_\varepsilon \to p \qquad \text{weakly in } L^2(0,T;V), \qquad \text{weak star in } L^\infty(0,T;\ L^2(\Omega)),$$

and by the Helly theorem,

$$p_\varepsilon(t) \to p(t) \qquad \text{strongly in } (H^s(\Omega)\cap V)', \qquad \forall t \in [0,T].$$

Now, since the injection of V into $L^2(\Omega)$ is compact, for every $\lambda > 0$ we have

$$\begin{aligned}&\|p_\varepsilon(t) - p(t)\|_{L^2(\Omega)}\\ &\qquad \le \lambda\|p_\varepsilon(t) - p(t)\|_V + \delta(\lambda)\|p_\varepsilon(t) - p(t)\|_{H^s(\Omega)\cap V}.\end{aligned}$$

Hence, $p_\varepsilon \to p$ strongly in $L^2(Q)$ and $p_\varepsilon(t) \to p(t)$ weakly in $L^2(\Omega)$ $\forall t \in [0,T]$ and, on a generalized sequence,

$$\dot\beta^\varepsilon(y_\varepsilon)p_\varepsilon \to \mu \qquad \text{weak star in } (L^\infty(Q))^*.$$

Finally, arguing as in the previous proofs we see that (on a subsequence)

$$\begin{aligned}p_\varepsilon\beta^\varepsilon(y_\varepsilon) &\to p(f_0 - y_t^* + \Delta y^*) &&\text{strongly in } L^1(Q),\\ p_\varepsilon\dot\beta^\varepsilon(y_\varepsilon) &\to 0 &&\text{strongly in } L^1(Q).\end{aligned}$$

Combining the preceding relations, we conclude that p satisfies Eqs. (2.74)–(2.78). If $N = 1$, then it follows that $y_\varepsilon \to y^*$ in $C(\overline{Q})$ and so we infer that

$$\mu y^* = (p_t + \Delta p - \xi)\mu = 0 \qquad \text{in } Q,$$

where $\xi = \lim_{\varepsilon \to 0} \nabla_y g(t, y_\varepsilon)$ (in $L^2(Q)$) and μy^* stands for the product of μ with y^*. ■

Theorem 2.3 can be applied as in the previous example to optimal control of the one phase Stefan problem with boundary value conditions

$$\frac{\partial \theta}{\partial \nu} + \alpha(\theta - u) = 0 \quad \text{in } \Sigma_1, \qquad \theta = 0 \quad \text{in } \Sigma_2.$$

5.3. The Time-Optimal Control Problem

5.3.1. *The Formulation of the Problem*

Consider the control process described by the nonlinear Cauchy problem

$$\begin{aligned} y'(t) + My(t) &\ni u(t), \qquad t > 0, \\ y(0) &= y_0, \end{aligned} \tag{3.1}$$

where M is a maximal monotone mapping in a Hilbert space H with the norm $|\cdot|$ and scalar product $(\cdot, \cdot)$.

Then, as seen earlier, for every $y_0 \in \overline{D(M)}$ and $u \in L^1_{\text{loc}}(0, \infty; H)$ problem (3.1) has a unique mild solution $y = y(t, y_0, u) \in C([0, \infty); H)$. Denote by $\mathscr{U}$ the class of control functions u,

$$\mathscr{U} = \{u \in L^\infty(0, \infty; H); u(t) \in K \text{ a.e. } t > 0\}, \tag{3.2}$$

where K is a closed bounded and convex subset of H. Let $y_0, y_1 \in \overline{D(M)}$ be fixed. A control $u \in \mathscr{U}$ is called admissible if it steers y_0 to y_1 in a finite time T, i.e., $y(T, y_0, u) = y_1$.

If $K = \{u \in H; |u| \le \rho\}$, then we have:

Lemma 3.1. *Assume that $y_0 \in \overline{D(M)}$ and $y_1 \in D(M)$, $|M^0 y_1| < \rho$. Then there is at least one admissible control $u \in \mathscr{U}$.*

Proof. We shall argue as in the proof of Proposition 2.3 in Chapter 4. Consider the feedback law

$$u(t) = -\rho \operatorname{sgn}(y(t) - y_1).$$

Since the operator $My + \operatorname{sign}(y - y_1)$ is monotone in $H \times H$, the Cauchy problem

$$\frac{dy}{dt} + My + \rho \operatorname{sgn}(y - y_1) \ni 0,$$
$$y(0) = y_0,$$

has a unique mild solution $y \in C([0,\infty); H)$. If $y_0 \in D(M)$, then y is a.e. differentiable on $(0,\infty)$ and we have, therefore,

$$\frac{1}{2}\frac{d}{dt}|y(t) - y_1|^2 + \rho|y(t) - y_1| \leq |M^0 y_1||y(t) - y_1| \qquad \text{a.e. } t > 0,$$

because M is monotone. (Here, M^0 is the minimal section of M.) This yields

$$|y(t) - y_1| \leq (|M^0 y_1| - \rho)t + |y_0 - y_1| \qquad \forall t \geq 0.$$

Hence, $y(t) = y_1$ for $t \geq (\rho - |M^0 y_1|)^{-1}|y_0 - y_1|$. This clearly extends to all $y_0 \in \overline{D(M)}$. ■

The smallest time t for which $y(t, y_0, u) = y_1$ is called the *transition time* of the control u, and the infimum $T(y_0, y_1)$ of the transition times of all admissible controls $u \in \mathscr{U}$ is called *minimal time*, i.e.,

$$T(y_0, y_1) = \inf\{T;\ \exists u \in \mathscr{U} \text{ such that } y(T, y_0, u) = y_1\}. \tag{3.3}$$

A control $u \in \mathscr{U}$ for which $y(T(y_0, y_1), y_0, u) = 0$ (if any) is called a *time-optimal control* of system (3.1) and the pair $(y(t, y_0, u), u)$ is called a *time-optimal pair* of system (3.1)).

Proposition 3.1. *Let M be maximal monotone and let $S(t) = e^{-Mt}$, the semigroup generated by M on $\overline{D(M)}$, be compact for every $t > 0$. Then under conditions of Lemma* 3.1 *there exists at least one time-optimal control for system* (3.1).

Proof. Let $y_0 \in \overline{D(M)}$ be arbitrary but fixed. We know that $T^0 = T(y_0, y_1) < \infty$. Hence, there is a sequence $T_n \to T^0$ and u_n such that $y(T_n, y_0, u_n) = y_1$. Let $y_n = y(t, y_0, u_n)$ be the corresponding solution to Eq. (3.1). Without loss of generality, we may assume that $u_n \to u^*$ weak star in $L^\infty(0, T; H)$, where $T^0 < T_0 < \infty$ (we extend u_n be zero outside the interval $[0, T_n]$) and by Theorem 2.4, Chapter 4, we have, on a subsequence,

$$y_n(t) \to y(t, y_0, u^*).$$

This clearly implies that $y(T^0, y_0, u) = y_1$, and so u^* is a time-optimal control for system (3.1). Recall that if $M = \partial\phi$ where $\phi\colon H \to \overline{\mathbf{R}}$ is a lower semicontinuous convex function, then the assumptions of Proposition 3.1 hold if, for every $\lambda \in \mathbf{R}$, the level sets $\{x \in H;\ \phi(x) \leq \lambda\}$ are compact in H. ■

In the linear case, every time-optimal control is a bang-bang control and satisfies a maximum principle type result (Fattorini [1]). More precisely, if M is the generator of an analytic semigroup then every time-optimal control u^* for system (3.1), where $K = \{u \in H;\ |u| \leq \rho\}$, can be represented as

$$u^*(t) = \rho \operatorname{sgn} p(t) \qquad \text{a.e. } t > 0,$$

where p is the solution to adjoint equation

$$p' - M^*p = 0 \qquad \text{a.e. } t > 0,$$

and $\operatorname{sgn} p = p|p|^{-1}$ if $p \neq 0$, $\operatorname{sgn} 0 = \{w;\ |w| \leq 1\}$. (For other results of this type, we refer to Balakrishnan [1], J. L. Lions [2], and H. O. Fattorini [2].)

Next we shall prove a similar result for some classes of nonlinear accretive systems of the form (3.1).

5.3.2. The Time-Optimal Control Problem for Smooth Systems

We shall consider here the time-optimal problem for system (3.1) in the case where

$$M = A + F, \tag{3.4}$$

and:

(i) $-A$ is the infinitesimal generator of a C_0-semigroup of contractions e^{-At} that is analytic and compact;

(ii) $F\colon H \to H$ is continuously differentiable, monotone, and its Fréchet derivative F' is bounded on bounded subsets;

(iii) K is a closed, convex, and bounded subset of H and $\{p \in H;\ |p| \leq \gamma\} \subset K$ for some $\gamma > 0$.

In particular, it follows by (i), (ii) that $A + F$ is maximal monotone in $H \times H$.

Theorem 3.1. *Assume that* $y_0, y_1 \in D(A)$ *and that hypotheses* (i)–(iii) *are satisfied. Let* (y^*, u^*) *be any time-optimal pair corresponding to* y_0, y_1, *where* $|Ay_1 + Fy_1| < \gamma$. *Then*

$$u^*(t) \in \partial H_K(p(t)) \qquad \text{a.e. } t \in (0, T^*) \quad (3.5)$$

$$p'(t) - A^*p(t) - (F'(y^*))^*p = 0 \qquad \text{in } [0, T^*], \quad (3.6)$$

$$H_K(p(t)) - (Ay^*(t) + Fy^*(t), p(t)) = 1 \qquad \text{a.e. } t \in (0, T^*). \quad (3.7)$$

Here, $T^* = T(y_0, y_1)$ is the minimal time and H_K is the support function of K, i.e.,

$$H_K(p) = \sup\{(p, u);\ u \in K\} \qquad \forall p \in H.$$

Since, by (3.7), $p(t) \neq 0$ $\forall t \in [0, T^*]$ it follows by (3.5) that u^* is a bang-bang control, i.e.,

$$u^*(t) \in \operatorname{Fr} K \qquad \text{a.e. } t \in (0, T^*).$$

(∂H_K is the subdifferential of H_K.)

The solution p to Eq. (3.6) is considered of course in the mild sense and A^* denotes the dual of A.

The idea of proof is to approximate the time-optimal problem by the free time-optimal control problem

$$\min\Big\{T + \int_0^T \Big(h(u(t)) + \frac{\varepsilon}{2}|u(t)|^2\Big)\,dt + \frac{1}{2\varepsilon}|e^{-A\varepsilon}(y(T) - y_1)|^2$$
$$+ \frac{1}{2}\int_0^\infty dt\,\Big|\int_0^t (u(s) - u^*(s))\,ds\Big|^2, \quad (3.8)$$

where the minimum is taken over all $T > 0$ and $u \in L^2(0, T; U)$, $y \in C([0, T]; H)$ satisfying Eq. (3.1) with $M = A + F$.

Here, $h \colon H \to \mathbf{R}$ is the indicator function of K, i.e.,

$$h(u) = \begin{cases} 0 & \text{if } u \in K, \\ +\infty & \text{otherwise}. \end{cases}$$

It is readily seen that problem (3.8) has at least one solution $(y_\varepsilon, u_\varepsilon, T_\varepsilon)$.

Lemma 3.2. *Let* $(y_\varepsilon, u_\varepsilon, T_\varepsilon)$ *be optimal in problem* (3.8). *Then for* $\varepsilon \to 0$, *we have* $T_\varepsilon \to T^* = T(y_0, y_1)$ *and*

$$\int_0^t (u_\varepsilon - u^*)\, ds \to 0 \qquad \text{strongly in } L^2(0,\infty; H), \tag{3.9}$$

$$u_\varepsilon \to u^* \qquad \text{weak star in } L^\infty(0, T^*; H), \tag{3.10}$$

$$y_\varepsilon \to y^* \qquad \text{strongly in } C([0,T^*]; H),$$

$$\text{weakly in } W^{1,2}([0,T^*]; H). \tag{3.11}$$

Proof. We have

$$T_\varepsilon + \int_0^{T_\varepsilon} \left(\frac{\varepsilon}{2}|u_\varepsilon|^2 + h(u_\varepsilon) \right) dt + (2\varepsilon)^{-1} |e^{-A\varepsilon}(y_\varepsilon(T_\varepsilon) - y_1)|^2$$

$$+ \frac{1}{2} \int_0^\infty dt \left| \int_0^t (u_\varepsilon - u^*)\, ds \right|^2 \le T^* + \frac{\varepsilon}{2} \int_0^{T^*} |u^*|^2\, dt. \tag{3.12}$$

(We extend u_ε and u^* by 0 on $[T_\varepsilon, +\infty)$ and $[T^*, +\infty)$, respectively.) Hence, $\limsup_{\varepsilon \to 0} T_\varepsilon \le T^*$ and

$$|y_\varepsilon(T_\varepsilon) - y_1| \to 0 \qquad \text{as } \varepsilon \to 0.$$

Now, let $\varepsilon_n \to 0$ be such that $T_{\varepsilon_n} \to T_0$ and $u_{\varepsilon_n} \to u_0$ weak star in $L^\infty(0,\infty; H)$. Since $-A$ generates an analytic semigroup and $y_0 \in D(A)$, we have (see Theorem 4.6 in Chapter 1),

$$\|y_\varepsilon'\|_{L^2(0,T_0; H)} \le C \qquad \forall \varepsilon > 0.$$

Note also that

$$|y_\varepsilon(t)| \le C \qquad \forall t \in (0,\infty).$$

Now, since the semigroup e^{-At} is compact we deduce by the Arzelà–Ascoli theorem that $\{y_\varepsilon\}$ is compact in $C([0,T_0]; H)$. Hence, on a subsequence, again denoted ε_n, we have

$$y_{\varepsilon_n} \to \tilde{y} \qquad \text{strongly in } C([0,T_0]; H);\ \text{weakly in } W^{1,2}([0,T_0]; H),$$

where $\tilde{y}$ is the solution to (3.1) with $u = \tilde{u}$. Clearly, $\tilde{y}(T_0) = y_1$ and so u_0 is admissible. Hence $T_0 = T^*$ and by (3.12) we have also that

$$\int_0^t (u_{\varepsilon_n} - u^*)\, ds \to 0 \qquad \text{strongly in } L^2(0,\infty; H),$$

as claimed. ■

Lemma 3.3. *Let* $(y_\varepsilon, u_\varepsilon, T_\varepsilon)$ *be optimal in problem* (3.8). *Then there is* $p_\varepsilon \in W^{1,2}([0,T_\varepsilon); H) \cap C([0,T_\varepsilon]; H)$ *such that*

$$y_\varepsilon' + Ay_\varepsilon + Fy_\varepsilon = u_\varepsilon \qquad \text{a.e. } t \in (0, T_\varepsilon), \tag{3.13}$$

$$p_\varepsilon' - A^* p_\varepsilon - (F'(y_\varepsilon))^* p_\varepsilon = 0 \qquad \text{a.e. } t \in (0, T_\varepsilon), \tag{3.14}$$

$$y_\varepsilon(0) = y_1, \qquad p_\varepsilon(T_\varepsilon) = -\frac{1}{\varepsilon} e^{-A^*\varepsilon} e^{-A\varepsilon}(y_\varepsilon(T_\varepsilon) - y_1), \tag{3.15}$$

$$p_\varepsilon(t) \in \partial h(u_\varepsilon(t)) + \varepsilon u_\varepsilon(t) + \int_t^{T_\varepsilon} ds \int_0^s (u_\varepsilon(\tau) - u^*(\tau))\, d\tau \qquad \forall t \in [0, T_\varepsilon], \tag{3.16}$$

$$\int_0^t (u_\varepsilon(s) - u^*(s))\, ds = 0, \qquad u_\varepsilon(t) = u^*(t) \qquad \forall t \geq T_\varepsilon, \tag{3.17}$$

$$-(Ay_\varepsilon(T_\varepsilon) + Fy_\varepsilon(T_\varepsilon), p_\varepsilon(T_\varepsilon)) + H_K(p_\varepsilon(T_\varepsilon) - \varepsilon u_\varepsilon(T_\varepsilon)) + \frac{\varepsilon}{2}|u_\varepsilon(T_\varepsilon)|^2 = 1. \tag{3.18}$$

Here, $\partial h(u) = \{u \in H;\ (w, u - v) \geq 0\ \ \forall v,\ \|v\| \leq \rho\}$ and $H_K(p) = \sup\{(p, v);\ v \in K\}$ is the support function of K.

Proof. Since $(y_\varepsilon, u_\varepsilon, T_\varepsilon)$ is optimal, we have

$$\begin{aligned}
&\int_0^{T_\varepsilon} \Big(h(u_\varepsilon(t)) + \frac{\varepsilon}{2}|u_\varepsilon(t)|^2\Big)\, dt \\
&\quad + \frac{1}{2\varepsilon}|e^{-A\varepsilon}(y_\varepsilon(T_\varepsilon) - y_1)|^2 + \frac{1}{2}\int_0^\infty dt \left|\int_0^t (u_\varepsilon - u^*)\, ds\right|^2 \\
&\leq \int_0^{T_\varepsilon} \Big(h(u_\varepsilon(t) + \lambda v(t)) + \frac{\varepsilon}{2}|u_\varepsilon(t) + \lambda v(t)|^2\Big)\, dt \\
&\quad + \frac{1}{2\varepsilon}|e^{-A\varepsilon} y(T_\varepsilon, u_\varepsilon + \lambda v, y_0) - y_1|^2 \\
&\quad + \frac{1}{2}\int_0^\infty dt \left|\int_0^t (u_\varepsilon + \lambda v - u^*)\, ds\right|^2 \qquad \forall \lambda > 0,\ v \in L^\infty(0, \infty; H),
\end{aligned}$$

where $y(t, u, y_0)$ is the solution to system (3.1). Subtracting, dividing by λ, and letting λ tend to zero, we get

$$\int_0^{T_\varepsilon} (h'(u_\varepsilon(t), v(t)) + \varepsilon(u_\varepsilon(t), v(t)))\, dt - \int_0^{T_\varepsilon} (p_\varepsilon(t), v(t))\, dt$$
$$+ \int_0^\infty dt \left(\int_0^t (u_\varepsilon(s) - u^*(s))\, ds, \int_0^t v(\tau)\, d\tau \right) \geq 0$$
$$\forall v \in L^\infty(0, \infty; H),$$

where p_ε is the solution to (3.14), (3.15) and h' is the directional derivative of h. This yields

$$\int_0^{T_\varepsilon} (h'(u_\varepsilon(t), v(t)) + \varepsilon(u_\varepsilon(t), v(t)) - (p_\varepsilon(t), v(t)))\, dt$$
$$+ \int_0^\infty \left(v(\tau), \int_\tau^\infty dt \int_0^t (u_\varepsilon - u^*)\, ds \right) d\tau \geq 0 \qquad \forall v \in L^\infty(0, \infty; H),$$

which implies (3.16) and (3.17).

It remains to prove (3.18). We note first that

$$T_\varepsilon + (2\varepsilon)^{-1} |e^{-A\varepsilon}(y_\varepsilon(T_\varepsilon) - y_1)|^2 + \frac{\varepsilon}{2} \int_0^{T_\varepsilon} |u_\varepsilon(t)|^2\, dt$$
$$\leq T_\varepsilon - \lambda + (2\varepsilon)^{-1} |e^{-A\varepsilon}(y_\varepsilon(T_\varepsilon - \lambda) - y_1)|^2$$
$$+ \frac{\varepsilon}{2} \int_0^{T_\varepsilon - \lambda} |u_\varepsilon(t)|^2\, dt \qquad \forall 0 < \lambda < T_\varepsilon. \tag{3.19}$$

Since $(\varepsilon I + \partial h)^{-1}$ is Lipschitz on H and $p_\varepsilon \in W^{1,2}([0, T_\varepsilon]; H)$ (because $p_\varepsilon(T_\varepsilon) \in D(A^*)$ and the semigroup e^{-A^*t} is analytic), we see by Eq. (3.16) that u_ε is Hölder continuous on $[0, T_\varepsilon]$. Hence, $y_\varepsilon \in C^1([0, T_\varepsilon]; H)$ (see Theorem 4.5 in Chapter 1) and so we may pass to limit in (3.19), getting

$$-(y_\varepsilon'(T_\varepsilon), p_\varepsilon(T_\varepsilon)) + \frac{\varepsilon}{2} |u_\varepsilon(T_\varepsilon)|^2 \leq -1.$$

Equivalently,

$$(Ay_\varepsilon(T_\varepsilon) + Fy_\varepsilon(T_\varepsilon) - u_\varepsilon(T_\varepsilon), p_\varepsilon(T_\varepsilon)) + \frac{\varepsilon}{2} |u_\varepsilon(T_\varepsilon)|^2 \leq -1. \tag{3.20}$$

On the other hand, it follows by (3.16) that

$$-(u_\varepsilon(T_\varepsilon), p_\varepsilon(T_\varepsilon)) + \varepsilon |u_\varepsilon(T_\varepsilon)|^2 + (\partial h(u_\varepsilon(T_\varepsilon)), u_\varepsilon(T_\varepsilon)) = 0,$$

i.e.,

$$(u_\varepsilon(T_\varepsilon), p_\varepsilon(T_\varepsilon)) = \varepsilon|u_\varepsilon(T_\varepsilon)|^2 + h^*(\partial h(u_\varepsilon(T_\varepsilon))) = 0,$$

where $h^* = H_K$. Hence, by (3.16), we have

$$(u_\varepsilon(T_\varepsilon), p_\varepsilon(T_\varepsilon)) = \varepsilon|u_\varepsilon(T_\varepsilon)|^2 + H_K(p_\varepsilon(T_\varepsilon) - \varepsilon u_\varepsilon(T_\varepsilon)).$$

Substituting in (3.20) yields

$$(Ay_\varepsilon(T_\varepsilon) + Fy_\varepsilon(T_\varepsilon), p_\varepsilon(T_\varepsilon)) - \frac{\varepsilon}{2}|u_\varepsilon(T_\varepsilon)|^2 - H_K(p_\varepsilon(T_\varepsilon) - \varepsilon u_\varepsilon(T_\varepsilon)) \le -1.$$

On the other hand, we have

$$T_\varepsilon + (2\varepsilon)^{-1}|e^{-A\varepsilon}(y_\varepsilon(T_\varepsilon) - y_1)|^2 + \frac{\varepsilon}{2}\int_0^{T_\varepsilon}|u_\varepsilon(t)|^2\,dt$$
$$\le T_\varepsilon + \lambda + (2\varepsilon)^{-1}|e^{-A\varepsilon}(y_\varepsilon(T_\varepsilon + \lambda) - y_1)|^2 + \frac{\varepsilon}{2}\int_0^{T_\varepsilon+\lambda}|u_\varepsilon(t)|^2\,dt,$$

and this yields the opposite inequality

$$(Ay_\varepsilon(T_\varepsilon) + Fy_\varepsilon(T_\varepsilon), p_\varepsilon(T_\varepsilon)) - \frac{\varepsilon}{2}|u_\varepsilon(T_\varepsilon)|^2 - H_K(p_\varepsilon(T_\varepsilon) - \varepsilon u_\varepsilon(T_\varepsilon)) \ge -1,$$

as claimed. ■

Proof of Theorem 3.1. Since $0 \in \operatorname{int} K$, we have

$$H_K(p) \ge \gamma|p| \qquad \forall p \in H,$$

where $\gamma > 0$. Then, by Eq. (3.18), it follows that

$$\frac{\varepsilon}{2}|u_\varepsilon(T_\varepsilon)|^2 + \gamma|p_\varepsilon(T_\varepsilon) - \varepsilon u_\varepsilon(T_\varepsilon)|$$
$$\le 1 + |Fy_\varepsilon(T_\varepsilon) + Ay_1||p_\varepsilon(T_\varepsilon)| + C\varepsilon$$

because the operator $e^{-A^*\varepsilon}e^{-A\varepsilon}A$ is positive.

Since, as seen in Lemma 3.2, $y_\varepsilon(T_\varepsilon) \to y_1$ as $\varepsilon \to 0$ the preceding implies that

$$\frac{\varepsilon}{2}|u_\varepsilon(T_\varepsilon)|^2 + |p_\varepsilon(T_\varepsilon)| \le C \qquad \forall \varepsilon > 0.$$

Then, by the variation of constants formula

$$p_\varepsilon(t) = e^{-A^*(T_\varepsilon - t)} p_\varepsilon(T_\varepsilon) + \int_t^T e^{-A^*(s-t)} (F'(y_\varepsilon(s)))^* p_\varepsilon(s)\, ds$$

and the compactness of the semigroup e^{-A^*t}, we conclude that on a subsequence, again denoted ε, we have

$$p_\varepsilon(t) \to p(t) \qquad \text{strongly in } H, \forall t \in [0, T^*], \tag{3.21}$$

Where p is the solution to Eq. (3.6). Then, letting ε tend to zero in (3.16), we see that

$$p(t) \in \partial h(u^*(t)) \qquad \text{a.e. } t \in (0, T^*),$$

which is equivalent to (3.5).

It remains to prove (3.7). To this end, we note first that by Eqs. (3.13), (3.14) we have

$$\frac{d}{dt}(Ay_\varepsilon + Fy_\varepsilon, p_\varepsilon) = (u_\varepsilon, p'_\varepsilon) \qquad \text{a.e. } t \in (0, T_\varepsilon). \tag{3.22}$$

We set

$$v_\varepsilon(t) = \int_t^{T_\varepsilon} ds \int_0^s (u_\varepsilon(\tau) - u^*(\tau))\, d\tau \qquad \forall t \in [0, T_\varepsilon].$$

Then we may write Eq. (3.16) as

$$u_\varepsilon(t) = \nabla h_\varepsilon^*(p_\varepsilon(t) - v_\varepsilon(t)) \qquad \forall t \in [0, T_\varepsilon],$$

where $h_\varepsilon^*(p) = \sup\{(p, u) - (\varepsilon/2)|u|^2;\ u \in K\}$. Inasmuch as

$$\frac{d}{dt} h_\varepsilon^*(p_\varepsilon(t) - v_\varepsilon(t)) = (\nabla h_\varepsilon^*(p_\varepsilon(t) - v_\varepsilon(t)), p'_\varepsilon(t) - v'_\varepsilon(t)) \qquad \text{a.e. } t \in (0, T_\varepsilon)$$

it follows by (3.22) that

$$\frac{d}{dt}((Ay_\varepsilon(t) + Fy_\varepsilon(t), p_\varepsilon(t)) - h_\varepsilon^*(p_\varepsilon(t) - v_\varepsilon(t))) = (u_\varepsilon(t), v'_\varepsilon(t)) \qquad \text{a.e. } t \in (0, T_\varepsilon)$$

and, integrating from 0 to T_ε, we get

$$\begin{aligned}(Ay_\varepsilon(T_\varepsilon) &+ Fy_\varepsilon(T_\varepsilon), p_\varepsilon(T_\varepsilon)) - h_\varepsilon^*(p_\varepsilon(T_\varepsilon)) \\ &= (Ay_\varepsilon(t) + Fy_\varepsilon(t), p_\varepsilon(t)) + h_\varepsilon^*(p_\varepsilon(t) - v_\varepsilon(t)) \\ &\quad + \int_t^T (u_\varepsilon(s), v'_\varepsilon(s))\, ds \qquad \forall t \in [0, T_\varepsilon].\end{aligned}$$

Then, letting ε tend to zero, it follows by Lemma 3.2, (3.18), and (3.21) that

$$(Ay^*(t) + Fy^*(t), p(t)) - h^*(p(t))$$
$$= (Ay^*(T^*) + Fy^*(T^*), p(T^*)) - H_K(p(T^*)) = -1 \qquad \forall t \in [0, T^*],$$

thereby completing the proof. ■

5.3.3. The Time-Optimal Control Problem for Semilinear Parabolic Equations

We shall study here the time-optimal control problem in the case where $H = L^2(\Omega)$, $K = \{u \in L^2(\Omega); |u(x)| \leq \rho$ a.e. $x \in \Omega\}$, $My = -\Delta y + \beta(y)$ $\forall y \in D(M) = \{y \in H^2(\Omega) \cap H_0^1(\Omega); \exists w \in L^2(\Omega)$ such that $w(x) \in \beta(y(x))$ a.e. $x \in \Omega\}$. Here, Ω is a bounded and open subset of $\mathbf{R}^N$ with a sufficiently smooth boundary (of class $C^{1,1}$, for instance) and β is a maximal monotone graph in $\mathbf{R} \times \mathbf{R}$ such that $0 \in \beta(0)$.

In other words, we shall study the problem:

(P) $\inf\{T; \exists u \in \mathscr{U}_\rho, y(T, y_0, u) = y_1\}$,

where $y(t, y_0, u)$ is the solution to semilinear parabolic boundary value problem

$$\frac{\partial y}{\partial t} - \Delta y + \beta(y) \ni u \quad \text{in } \Omega \times (0, \infty),$$
$$y(x, 0) = y_0(x) \quad x \in \Omega,$$
$$y = 0 \quad \text{in } \partial\Omega \times (0, \infty), \tag{3.23}$$

and

$$\mathscr{U}_\rho = \{u \in L^\infty(\Omega \times (0, \infty)); |u(x,t)| \leq \rho \text{ a.e. } (x,t) \in \Omega \times (0, \infty)\}.$$

Regarding y_0 and y_1, we shall assume that

$$y_0, y_1 \in D(M) \cap L^\infty(\Omega) \quad \text{and} \quad \|M^0 y_1\|_{L^\infty(\Omega)} < \rho, \tag{3.24}$$

where $(M^0 y)(x) = \inf\{|w|; w \in -\Delta y(x) + \beta(y(x))\}$, $x \in \Omega$.

It turns out that under the preceding assumptions problem (P) has at least one solution (T^*, y^*, u^*). This follows as in the proof of Lemma 3.1, using Lemma 3.4 following.

Lemma 3.4. *Under assumptions* (3.24), *there is at least one admissible control* $u \in \mathscr{U}_\rho$ *for problem* (P).

Proof. We note that Lemma 3.1 is inapplicable here since int $K = \emptyset$. However, we shall use the same method to prove the existence of an admissible control (see also Proposition 3.5 in Chapter 4). Namely, consider the feedback control

$$u(x,t) = -\rho \operatorname{sign}(y(x,t) - y_1(x)) \qquad \forall (x,t) \in \Omega \times (0,\infty),$$

where sign $r = r/|r|$ if $r \neq 0$, sign $0 = [-1,1]$. Then the boundary value problem

$$\frac{\partial y}{\partial t} - \Delta y + \beta(y) + \rho \operatorname{sign}(y - y_1) \ni 0 \qquad \text{in } \Omega \times (0,\infty),$$
$$y(x,0) = y_0(x), \qquad x \in \Omega,$$
$$y = 0 \qquad \text{in } \partial\Omega \times (0,\infty),$$

has a unique solution $y \in W^{1,2}([0,T];\ L^2(\Omega)) \cap L^2(0,T;\ H_0^1(\Omega) \cap H^2(\Omega))$ for every $T > 0$ because by Theorem 2.4, in Chapter 2, the operator

$$\tilde{M}y = My + \rho \operatorname{sign}(y - y_1), \qquad y \in D(M),$$

is maximal monotone in $L^2(\Omega)$. As a matter of fact, $\tilde{M} = \partial\phi$ where

$$\phi(y) = \frac{1}{2}\int_\Omega |\nabla y|^2\,dx + \int_\Omega (j(y) + \rho|y - y_1|)\,dx, \qquad y \in L^2(\Omega),$$

where $\partial j = \beta$.

Consider the function $w(x,t) = \|y_0 - y_1\|_{L^\infty(\Omega)} - (\rho - \mu)t$, where $\mu = \|M^0 y_1\|_{L^\infty(\Omega)}$. We have

$$\frac{\partial w}{\partial t} - \Delta w + \rho \operatorname{sign} w \ni \mu \qquad \text{in } \Omega \times \left(0, (\rho - \mu)^{-1}\|y_0 - y_1\|_{L^\infty(\Omega)}\right),$$
$$w(x,0) = \|y_0 - y_1\|_{L^\infty(\Omega)} \qquad \text{in } \Omega,$$
$$w \geq 0 \qquad \text{in } \partial\Omega \times \left(0, (\rho - \mu)^{-1}\|y_0 - y_1\|_{L^\infty(\Omega)}\right),$$

and by the maximum principle we see that

$$y(x,t) - y_1(x) \le w(x,t)$$
$$\forall (x,t) \in \Omega \times \left(0, (\rho - \mu)^{-1} \|y_0 - y_1\|_{L^\infty(\Omega)}\right).$$

Hence, $y(x,t) - y_1(x) \le \|y_0 - y_1\|_{L^\infty(\Omega)} - (\rho - \mu)t$, and by a symmetric argument it follows that

$$y(x,t) - y_1(x) \ge -\|y_0 - y_1\|_{L^\infty(\Omega)} + (\rho - \mu)t$$

for $x \in \Omega$, $0 \le t \le (\rho - \mu)^{-1}\|y_0 - y_1\|_{L^\infty(\Omega)}$. Hence, $y(x,t) = y_1(x)$ for $x \in \Omega$ and $t \ge (\rho - \mu)^{-1}\|y_0 - y_1\|_{L^\infty(\Omega)}$, as claimed. ■

Now we shall formulate a maximum principle type result for problem (P). We shall assume that y_0, y_1 satisfy (3.24) and

$$y_0 \in W_0^{2-2/q,q}(\Omega), \qquad q > \max(N,2). \tag{3.25}$$

Theorem 3.2. *Let (y^*, u^*, T^*) be optimal in problem* (P). *Then*

$$u^*(x,t) \in \rho \operatorname{sign} p(x,t) \qquad \text{a.e. } (x,t) \in \Omega \times (0,T^*), \tag{3.26}$$

where $p \in L^2(0,T^*;\ W_0^{1,q'}(\Omega)) \cap BV([0,T^*];\ H^{-s}(\Omega) + W^{-1,q}(\Omega))$, $s > N/2$, satisfies the system

$$\frac{\partial}{\partial t}p + \Delta p - \nu = 0 \qquad \text{in } \Omega \times (0,T^*), \tag{3.27}$$

$$-\int_\Omega (\nabla y^*(x,t) \cdot \nabla p(x,t) + \xi(x,t)p(x,t))\,dx$$
$$+ \rho \int_\Omega |p(x,t)|\,dx = 1 \qquad \text{a.e. } t \in (0,T^*). \tag{3.28}$$

Here $\xi \in L^2(\Omega \times (0,T^*))$ is such that $\xi(x,t) \in \beta(y^*(x,t))$ a.e. $(x,t) \in \Omega \times (0,T^*)$ *and $\nu \in (L^\infty(\Omega \times (0,T^*)))^*$.*

In particular, it follows by Eq. (3.28) that $p(t,\cdot) \not\equiv 0$ a.e. $t \in (0,T^*)$ and so for almost every $t \in (0,T^*)$ there is $\Omega_t \subset \Omega$, $m(\Omega_t) > 0$ such that $|u^*(x,t)| = \rho$ a.e. $x \in \Omega_t$.

Proof of Theorem 3.2. Let (y^*, u^*, T^*) be any optimal pair for problem (P). Proceeding as in the proof of Theorem 3.1, consider the approximating control problem

$$\min\left\{T + \int_0^T \left(h(u(t)) + \frac{\varepsilon}{2}|u(t)|_2^2\right) dt + \eta(\varepsilon)|y(T) - y_1|_2^2 \right.$$
$$\left. + \frac{1}{2}\int_0^\infty dt \left|\int_0^t (u(s) - u^*(s))\, ds\right|_2^2 ;\ u \in L^2(0,\infty; L^2(\Omega))\right\}, \quad (3.29)$$

where $y \in W^{1,2}([0,T]; L^2(\Omega)) \cap L^2(0,T; H_0^1(\Omega) \cap H^2(\Omega))\ \forall T > 0$ is the solution to system

$$\begin{aligned} &\frac{\partial y}{\partial t} + \Delta y + \beta^\varepsilon(y) = u && \text{in } \Omega \times (0,\infty), \\ &y(x,0) = y_0(x) && \text{in } \Omega, \\ &y = 0 && \text{in } \partial\Omega \times (0,\infty). \end{aligned} \quad (3.30)$$

Here, $\eta(\varepsilon) > 0$, $\eta(\varepsilon) \overset{\varepsilon\to 0}{\to} \infty$, $|\cdot|_2$ is the L^2-norm on Ω and β^ε is a smooth approximation of β satisfying conditions (k)–(kkk) in Section 1.2. The function $h\colon L^2(\Omega) \to \overline{\mathbf{R}}$ is the indicator function of $K_0 = \{u \in L^2(\Omega);\ |u(x)| \le \rho \text{ a.e. } x \in \Omega\}$.

Let $(y_\varepsilon, u_\varepsilon, T_\varepsilon)$ be optimal in problem (3.29). Then, by Lemma 3.3, there is $p_\varepsilon \in C([0,T_\varepsilon]; L^2(\Omega)) \cap L^2(0,T_\varepsilon; H_0^1(\Omega) \cap H^2(\Omega))$ such that

$$\begin{aligned} &\frac{\partial y_\varepsilon}{\partial t} - \Delta y_\varepsilon + \beta^\varepsilon(y_\varepsilon) = u_\varepsilon \qquad \text{in } Q_\varepsilon = \Omega \times (0,T_\varepsilon), \\ &y_\varepsilon(x,0) = y_0(x) \quad \text{in } \Omega, \qquad y_\varepsilon = 0 \quad \text{in } \Sigma_\varepsilon = \partial\Omega \times (0,T_\varepsilon), \end{aligned} \quad (3.31)$$

$$\begin{aligned} &\frac{\partial p_\varepsilon}{\partial t} + \Delta p_\varepsilon - \dot{\beta}^\varepsilon(y_\varepsilon)p_\varepsilon = 0 \qquad \text{in } Q_\varepsilon, \\ &p_\varepsilon(x,T_\varepsilon) = -2\eta(\varepsilon)(y_\varepsilon(x,T) - y_1) \quad \text{in } \Omega, \qquad p_\varepsilon = 0 \quad \text{in } \Sigma_\varepsilon, \end{aligned} \quad (3.32)$$

$$\begin{aligned} &p_\varepsilon(t) \in \partial h(u_\varepsilon(t)) + \varepsilon u_\varepsilon(t) + \int_t^{T_\varepsilon} ds \int_0^s (u_\varepsilon - u^*)\, d\tau, \\ &\int_0^t (u_\varepsilon - u^*)\, ds = 0, \qquad u_\varepsilon = u^* \quad \text{on } [T_\varepsilon, +\infty). \end{aligned} \quad (3.33)$$

$$\begin{aligned} &-\int_\Omega (\nabla y_\varepsilon(x,T_\varepsilon) \cdot \nabla p_\varepsilon(x,T_\varepsilon) - \beta^\varepsilon(y_\varepsilon(x,T_\varepsilon))p_\varepsilon(x,T_\varepsilon))\, dx \\ &+ \rho\int_\Omega |p_\varepsilon(x,T_\varepsilon) - \varepsilon u_\varepsilon(x,T_\varepsilon)|\, dx + \frac{\varepsilon}{2}|u_\varepsilon(T_\varepsilon)|_2^2 = 1. \end{aligned} \quad (3.34)$$

Now, arguing as in the proof of Lemma 3.2, we see that $T_\varepsilon \to T^* = T(y_0, y_1)$, $y_\varepsilon(T_\varepsilon) \to y_1$ in $L^2(\Omega)$, and

$$
\begin{aligned}
u_\varepsilon &\to u^* && \text{weak star in } L^\infty(0,T^*;\, L^2(\Omega)),\\
\int_0^t (u_\varepsilon - u^*)\, ds &\to 0 && \text{strongly in } L^2(0,\infty;\, L^2(\Omega)),\\
y_\varepsilon &\to y^* && \text{strongly in } C([0,T^*];\, L^2(\Omega)),
\end{aligned}
$$

weakly in $L^2(0,T^*;\, H_0^1(\Omega) \cap H^2(\Omega)) \cap W^{1,2}([0,T^*];\, L^2(\Omega))$.

$$(3.35)$$

Now, multiplying Eq. (3.32) by sign p_ε and integrating on Q_ε, we get

$$\|p_\varepsilon(t)\|_{L^1(\Omega)} + \int_{Q_\varepsilon} |\dot\beta^\varepsilon(y_\varepsilon) p_\varepsilon|\, dx\, dt \le \|p_\varepsilon(T_\varepsilon)\|_{L^1(\Omega)} \le C \quad (3.35)'$$

because by Eq. (3.34) and the monotonicity of β^ε it follows that $\{p_\varepsilon(T_\varepsilon)\}$ is bounded in $L^1(\Omega)$. To obtain further estimates on p_ε, we consider the boundary value problem

$$\frac{\partial v}{\partial t} - \Delta v = \sum_{i=1}^N (h_i)_{x_i} \qquad \text{in } Q^* = \Omega \times (0,T^*), \tag{3.36}$$

where $h_i \in L^2(0,T^*;\, L^q(\Omega))$, $i = 1,\dots,N$, $q > 2$. Problem (3.36) has a unique solution $v \in L^\infty(0,T^*;\, H_0^1(\Omega))$ with $\partial v/\partial t \in L^2(0,T^*;\, H^{-1}(\Omega))$ (see, e.g., Theorem 1.9 in Chapter 4). Moreover, if $q > N$ then $v \in L^\infty(Q^*)$ and

$$\|v\|_{L^\infty(Q^*)} \le C \sum_{i=1}^N \|h_i\|_{L^2(0,T^*;\, L^q(\Omega))}$$

(see Ladyzhenskaya *et al.* [1], p. 213).

Now, if we multiply Eq. (3.32) by v and integrate on Q^*, we get the inequality

$$\left| \sum_{i=1}^N \int_{Q^*} (p_\varepsilon)_{x_i} h_i\, dx \right| \le C \sum_{i=1}^N \|h_i\|_{L^2(0,T^*;\, L^q(\Omega))}$$

and, since $h = (h_1,\dots,h_N)$ is arbitrary in $L^2(0,T;\, L^q(\Omega))$,

$$\|p_\varepsilon\|_{L^2(0,T^*;\, W_0^{1,q'}(\Omega))} \le C \qquad \forall \varepsilon > 0, \tag{3.37}$$

where $1/q + 1/q' = 1$. Hence, $\{\partial p_\varepsilon/\partial t\}$ is bounded in $L^1(0, T^*; H^{-s}(\Omega) + W^{-1,q}(\Omega))$, where $s > N/2$. (If extend u_ε by 0 on $[T_\varepsilon, +\infty)$ we may assume that p_ε are defined on $[0, T^*]$.) Then, according to the Helly theorem, there is $p \in BV([0, T^*]; H^{-s}(\Omega) + W^{-1,q}(\Omega)) \cap L^2(0, T^*; W_0^{1,q'}(\Omega))$ such that, for some $\{\varepsilon_n\} \to 0$,

$$p_{\varepsilon_n}(t) \to p(t) \qquad \text{strongly in } H^{-s}(\Omega) + W^{-1,q}(\Omega), \qquad \forall t \in [0, T^*],$$
$$p_{\varepsilon_n} \to p \qquad \text{weakly in } L^2(0, T^*; W_0^{1,q'}(\Omega)).$$

On the other hand, since the injection of $W_0^{1,q'}(\Omega)$ into $L^{q'}(\Omega)$ is compact, for every $\delta > 0$ there is $\eta(\delta) > 0$ such that (see J. L. Lions [1], p. 71)

$$\begin{aligned}\|p_{\varepsilon_n}(t) - p(t)\|_{L^{q'}(\Omega)} \le{}& \delta \|p_{\varepsilon_n}(t) - p(t)\|_{W_0^{1,q'}(\Omega)} \\ &+ \eta(\delta)\|p_{\varepsilon_n}(t) - p(t)\|_{H^{-s}(\Omega)+W^{-1,q}(\Omega)} \\ &\qquad \forall t \in [0, T^*].\end{aligned}$$

This implies that

$$p_{\varepsilon_n} \to p \qquad \text{strongly in } L^2(0, T^*; L^{q'}(\Omega)). \tag{3.38}$$

Now, letting ε_n tend to zero in Eq. (3.32), we see that

$$\frac{\partial p}{\partial t} + \Delta p - \nu = 0 \qquad \text{in } Q^*,$$

where $\nu = \mathrm{w} - \lim \dot{\beta}^\varepsilon(y_\varepsilon)p_\varepsilon$ on some generalized sequence $\{\varepsilon\}$. Moreover, by (3.33) we get (3.26). If we multiply Eq. (3.31) by $\beta^\varepsilon(y_\varepsilon)|\beta^\varepsilon(y_\varepsilon)|^{q-2}$ and integrate on $Q_\varepsilon = \Omega \times (0, T_\varepsilon)$, we see that $\{\beta^\varepsilon(y_\varepsilon)\}$ is bounded in $L^q(Q^*)$. This implies that $\{y_\varepsilon\}$ is bounded in $W_q^{2,1}(Q^*)$ and so on a subsequence, again denoted ε_n,

$$\Delta y_{\varepsilon_n} - \beta^{\varepsilon_n}(y_{\varepsilon_n}) \to \Delta y^* - \xi \qquad \text{weakly in } L^q(Q^*), \tag{3.39}$$
$$y_{\varepsilon_n} \to y^* \qquad \text{in } C(\overline{Q}^*),$$
$$\xi(x,t) \in \beta(y^*(x,t)) \qquad \text{a.e. } (x,t) \in Q^*. \tag{3.40}$$

Now, multiplying Eq. (3.31) by p_ε', (3.32) by y_ε', and subtracting the results, we get, as in the proof of Theorem 3.1,

$$\begin{aligned}&\int_\Omega (-\Delta y_\varepsilon(x, T_\varepsilon) + \beta^\varepsilon(y_\varepsilon(x, T_\varepsilon)))p_\varepsilon(x, T_\varepsilon)\,dx - h_\varepsilon^*(p_\varepsilon(T_\varepsilon)) \\ &\quad = \int_\Omega (-\Delta y_\varepsilon(x,t) + \beta^\varepsilon(y_\varepsilon(x,t)))p_\varepsilon(x,t)\,dx - h_\varepsilon^*(p_\varepsilon(t) - v_\varepsilon(t)) \\ &\qquad + \int_t^T \int_\Omega u_\varepsilon(x,s)(v_\varepsilon)_t(x,s))\,dx\,ds = 0 \qquad \forall t \in [0, T_\varepsilon],\end{aligned}$$

where

$$v_\varepsilon(x,t) = \int_t^T ds \int_0^s (u_\varepsilon(\tau) - u^*(\tau))\, d\tau$$

and $h_\varepsilon^*(p) = \sup\{(p,u)_2 - (\varepsilon/2)\|u\|_2^2;\ u \in K_0\}$. Then, by (3.34), (3.35), and (3.39), it follows that

$$\begin{aligned}
\lim_{\varepsilon\to 0} \int_\Omega (-\Delta y_\varepsilon(x,t) + \beta^\varepsilon(y_\varepsilon(x,t))) p_\varepsilon(x,t)\, dx \\
&= \int_\Omega (-\Delta y^*(x,t) + \xi(x,t)) p(x,t)\, dx \\
&= \rho \int_\Omega |p(x,t)|\, dx - 1 \qquad \text{a.e. } t \in (0,T^*),
\end{aligned}$$

as claimed. ■

Now we shall consider some particular cases. The first one is that where

$$\beta(r) = \begin{cases} 0 & \text{for } r \geq 0, \\ \mathbf{R}^- & \text{for } r = 0. \end{cases}$$

As seen earlier, in this case the control system (3.33) reduces to the obstacle controlled problem

$$\begin{aligned}
\frac{\partial y}{\partial t} - \Delta y = u \qquad & \text{in } \{(x,t);\ y(x,t) > 0\}, \\
\frac{\partial y}{\partial t} - \Delta y \geq u, \quad y \geq 0 \qquad & \text{in } \Omega \times (0,\infty), \\
y(x,0) = y_0(x) \qquad \text{in } \Omega, \qquad y = 0 \qquad & \text{in } \partial\Omega \times (0,\infty).
\end{aligned} \tag{3.41}$$

We take β^ε in the following form (see (1.17)):

$$\beta^\varepsilon(r) = \begin{cases} \varepsilon^{-1} r + 2^{-1} & \text{for } r \leq -\varepsilon, \\ -(2\varepsilon^2)^{-1} r^2 & \text{for } -\varepsilon < r \leq 0, \\ 0 & \text{for } r > 0, \end{cases} \tag{3.42}$$

and set

$$\begin{aligned}
Q_\varepsilon^1 &= \{(x,t) \in Q_\varepsilon;\ y_\varepsilon(x,t) \leq -\varepsilon\}, \\
Q_\varepsilon^2 &= \{(x,t) \in Q_\varepsilon;\ -\varepsilon < y_\varepsilon(x,t) \leq 0\}.
\end{aligned}$$

Then we have

$$p_\varepsilon \beta^\varepsilon(y_\varepsilon) = \left(p_\varepsilon \dot\beta^\varepsilon(y_\varepsilon) y_\varepsilon + 2^{-1} p_\varepsilon\right)\chi_\varepsilon^1 + 2^{-1} p_\varepsilon \dot\beta^\varepsilon(y_\varepsilon) y_\varepsilon \chi_\varepsilon^2 \qquad \text{a.e. } (x,t) \in Q_\varepsilon, \quad (3.43)$$

where χ_ε^i, $i = 1, 2$, is the characteristic function of Q_ε^i.

Since $\{p_\varepsilon \dot\beta^\varepsilon(y_\varepsilon)\}$ is bounded in $L^1(Q_\varepsilon)$, $\{y_\varepsilon\}$ in $C(\overline{Q})$, and $\{\beta^\varepsilon(y_\varepsilon)\}$ is bounded in $L^q(Q_\varepsilon)$, it follows by (3.43) that, for some $\varepsilon_n \to 0$,

$$p_{\varepsilon_n} \beta^{\varepsilon_n}(y_{\varepsilon_n}) \to 0 \qquad \text{a.e. in } Q^*, \qquad (3.44)$$

whilst by (3.39), (3.40) we have

$$p_{\varepsilon_n} \beta^{\varepsilon_n}(y_{\varepsilon_n}) \to p(\Delta y^* - y_t^*) \in p\beta(y^*) \qquad \text{weakly in } L^1(Q^*).$$

Hence,

$$p_{\varepsilon_n} \beta^{\varepsilon_n}(y_{\varepsilon_n}) \to 0 \qquad \text{strongly in } L^1(Q^*),$$

and so

$$p\left(u^* + \frac{\partial y^*}{\partial t} + \Delta y^*\right) = 0 \qquad \text{a.e. in } Q^*.$$

Now, using (3.42) once again, we see that

$$p_\varepsilon\left(\beta^\varepsilon(y_\varepsilon) - \dot\beta^\varepsilon(y_\varepsilon) y_\varepsilon\right) \to 0 \qquad \text{strongly in } L^1(Q^*),$$

because $m(Q_\varepsilon^1), m(Q_\varepsilon^2) \to 0$ as $\varepsilon \to 0$. Hence, on a subsequence,

$$p_{\varepsilon_n} \dot\beta^{\varepsilon_n}(y_{\varepsilon_n}) y_{\varepsilon_n} \to 0 \qquad \text{strongly in } L^1(Q^*).$$

Since as seen earlier $y_{\varepsilon_n} \to y^*$ in $C(\overline{Q})$, this implies that $\nu y^* = 0$ in Q^*, i.e.,

$$\left(\frac{\partial p}{\partial t} + \Delta p\right) y^* = 0 \qquad \text{in } Q^*.$$

We have therefore proved the following theorem:

Theorem 3.3. *Let $y_0 \in W_0^{2-2/q,q}(\Omega) \cap H^2(\Omega) \cap L^\infty(\Omega)$, $q > \max(N, 2)$ be such that $y_0 \geq 0$ in Ω, and let $y_1 \in H_0^1(\Omega) \cap H^2(\Omega) \cap L^\infty(\Omega)$ be such that*

$$y_1 \geq 0 \qquad \text{in } \Omega, \qquad \|\Delta y_1\|_{L^\infty(\Omega)} < \rho.$$

Let (y^*, u^*) be any optimal pair for the time-optimal problem associated with system (3.41). Then there is $p \in L^2(0, T^*; W_0^{1,q'}(\Omega)) \cap BV([0,T]; H^{-s}(\Omega)$

$+ W^{-1,q}(\Omega))$ *such that* $(\partial/\partial t)p + \Delta p \in M(\overline{Q}^*)$ *and*

$$\frac{\partial p}{\partial t} + \Delta p = 0 \qquad \text{in } \{(x,t) \in Q^*;\ y^*(x,t) > 0\}, \tag{3.45}$$

$$p = 0 \qquad \text{in } \{(x,t) \in Q^*;\ y^*(x,t) = 0\}, \tag{3.46}$$

$$u^*(x,t) \in \rho \operatorname{sign} p(x,t) \qquad \text{a.e. } (x,t) \in Q^*, \tag{3.47}$$

$$\rho \| p(t) \|_{L^1(\Omega)} + \int_\Omega \Delta y^*(x,t)\, p(x,t)\, dx = 1 \qquad \text{a.e. } t \in (0,T^*). \tag{3.48}$$

■

Here, $M(\overline{Q}^*)$ is the space of bounded Radon measures on $\overline{Q}^*$.

This theorem clearly extends to the time-optimal problem for the variational inequality

$$\frac{\partial y}{\partial t} - \Delta y = u \qquad \text{in } \{y > \psi\},$$

$$\frac{\partial y}{\partial t} - \Delta y \geq u, \quad y \geq \psi \qquad \text{in } \Omega \times (0,\infty),$$

$$y(x,0) = y_0(x), \quad y = 0 \qquad \text{in } \partial\Omega \times (0,\infty),$$

where $\psi \in C^2(\overline{\Omega})$ is a given function such that $\psi \leq 0$ in $\partial\Omega$.

Now we shall consider the special case where $y_1 \equiv 0$. If we take, in the approximating problem (3.29), $\eta(\varepsilon) = \varepsilon^{-1/2}$, multiply Eq. (3.31) by $\partial y_\varepsilon/\partial t$, and integrate with respect to x, we get

$$\int_\Omega j^\varepsilon(y_\varepsilon(x,t))\, dx \leq C \qquad \forall \varepsilon > 0,\ t \in [0, T_\varepsilon],$$

where C is independent of ε and t. We recall that $j^\varepsilon(r) = \int_0^r \beta^\varepsilon(s)\, ds$ and so, by (3.42), it follows that

$$\int_\Omega j_\varepsilon^2(x, T_\varepsilon)\, dx \leq C\varepsilon \qquad \forall \varepsilon > 0.$$

If multiply Eq. (3.32) by p_ε^+ and integrate on $\Omega \times (t, T_\varepsilon)$, we get

$$\tfrac{1}{2} \int_\Omega |p_\varepsilon^+(x,t)|^2\, dx + \int_t^T \int_\Omega |\nabla p_\varepsilon^+(x,s)|^2\, dx\, ds$$

$$\leq \varepsilon^{-1/2} \int_\Omega |y_\varepsilon(x,T_\varepsilon)|^2\, dx \leq C\varepsilon^{1/2}.$$

Hence

$$
\begin{aligned}
p &\le 0 \qquad \text{in } Q^*,\\
\frac{\partial p}{\partial t} + \Delta p &= 0 \qquad \text{in } \{(x,t) \in Q^*;\ y^*(x,t) > 0\},\\
p &= 0 \qquad \text{in } \{(x,t) \in Q^*;\ y^*(x,t) = 0\},\\
u^*(x,t) &\in \rho \operatorname{sign} p(x,t) \qquad \text{in } Q^*.
\end{aligned}
\tag{3.49}
$$

If the open set $E = \{(x,t) \in Q^*;\ y^*(x,t) > 0\}$ is connected, then by the maximum principle we conclude that $p < 0$ in E, and so

$$u^*(x,t) = -1 \qquad \text{in } \{(x,t) \in Q^*;\ y^*(x,t) > 0\}. \tag{3.50}$$

We have obtained, therefore, a feedback representation for the time-optimal control u^*. In general, it follows by (3.50) that $u^* = -1$ in at least one component of the noncoincidence set $\{(x,t);\ y^*(x,t) > 0\}$.

We shall consider now the case where β is a monotonically increasing locally Lipschitz function on $\mathbf{R}$. Then we may take β^ε defined by the formula (2.74) in Chapter 3. By (3.40) we see that $\{\dot\beta^\varepsilon(y_\varepsilon)\}$ is bounded in $L^\infty(Q^*)$, and so extracting a further subsequence if necessary we may assume that

$$\dot\beta^{\varepsilon_n}(y_{\varepsilon_n}) \to g \qquad \text{weak star in } L^\infty(Q^*),$$

where $g(x,t) \in \partial\beta(y^*(x,t))$ a.e. $(x,t) \in Q^*$ (see Lemma 2.5 in Chapter 3). Then, by (3.38), we infer that $\nu \in L^{q'}(Q^*)$ and $\nu(x,t) \in \partial\beta(y^*(x,t))\, p(x,t)$ a.e. $(x,t) \in Q^*$, where $\partial\beta$ is the generalized gradient of β.

Then, by Theorem 3.2, we have:

Theorem 3.4. *Let y_0, y_1 satisfy* (3.24), (3.25) *and let β be monotonically increasing and locally Lipschitz on* $\mathbf{R}$. *Then if (y^*, u^*, T^*) is optimal for problem* (P) *there are $p \in L^2(0,T^*;\ W_0^{1,q'}(\Omega)) \cap BV([0,T^*];\ H^{-s}(\Omega) + W^{-1,q}(\Omega))$, $s > N/2$, and $\eta \in L^\infty(Q^*)$ such that*

$$u^*(x,t) \in \rho \operatorname{sign} p(x,t) \qquad \text{a.e. } (x,t) \in Q^*, \tag{3.51}$$

$$
\begin{aligned}
&\frac{\partial p}{\partial t} + \Delta p - \eta p = 0 \qquad \text{in } Q^*,\\
&\eta(x,t) \in \partial\beta(y^*(x,t)) \qquad \text{a.e. } (x,t) \in Q^*,
\end{aligned}
\tag{3.52}
$$

$$
\begin{aligned}
&-\int_\Omega (\nabla y^*(x,t) \cdot \nabla p(x,t) + \xi(x,t) p(x,t))\, dx\\
&\qquad + \rho \int_\Omega |p(x,t)|\, dx = 1 \qquad \text{a.e. } t \in (0,T^*),
\end{aligned}
\tag{3.53}
$$

where $\xi \in L^2(Q^*)$, $\xi(x,t) \in \beta(y^*(x,t))$ a.e. $(x,t) \in Q^*$. ∎

Remark 3.1. If in problem (P) we replace the set $\mathcal{U}_\rho$ by

$$\tilde{\mathcal{U}}_\rho = \left\{u \in L^\infty(0,\infty; L^2(\Omega));\ \|u(t)\|_{L^2(\Omega)} \le \rho \text{ a.e. } t \ge 0\right\},$$

then Theorems 3.2–3.4 remain true except that Eqs. (3.28), (3.48), and (3.53) are replaced by

$$-\int_\Omega (\nabla y^*(x,t) \cdot \nabla p(x,t) + \xi(x,t)p(x,t))\,dx$$
$$+ \rho\|p(t)\|_{L^2(\Omega)} = 1 \qquad \text{a.e. } t \in (0,T^*),$$

respectively

$$-\int_\Omega \nabla y^*(x,t) \cdot \nabla p(x,t)\,dx + \rho\|p(t)\|_{L^2(\Omega)} = 1 \qquad \text{a.e. } t \in (0,T^*),$$

in the case of the obstacle problem.

5.3.4. *Approximating Time-Optimal Control by Infinite Horizon Controllers*

Though the results of this section remain true for more general time-optimal problems, we confine ourselves to parabolic systems of the form (3.23). More precisely, we shall consider the time optimal control problem:

(P_1) $\inf = \{T;\ \exists u \in \mathcal{U},\ y(T, y_0, u) = 0\}$

where

$$\mathcal{U} = \{u \in L^\infty(0,\infty; L^2(\Omega));\ u(t) \in K \text{ a.e. } t > 0\}. \tag{3.54}$$

K is either the set $\{u \in L^\infty(\Omega);\ |u(x)| \le \rho$ a.e. $x \in \Omega\}$ or $\{u \in L^2(\Omega);\ \|u\|_2 \le \rho\}$, and $y = y(t, y_0, u)$ is the solution to system (3.23).

Here, β is a maximal monotone graph in $\mathbf{R} \times \mathbf{R}$ such that $0 \in \beta(0)$. We shall assume throughout this section that

$$y_0 \in H_0^1(\Omega), \qquad j(y_0) \in L^1(\Omega) \quad (\partial j = \beta) \tag{3.55}$$

if $K = \{u;\ \|u\|_2 \le \rho\}$, and $y_0 \in L^\infty(\Omega) \cap H_0^1(\Omega)$, $\beta(y_0) \in L^\infty(\Omega)$ if $K = \{u;\ \|u\|_{L^\infty(\Omega)} \le \rho\}$. As seen earlier (Lemmas 3.1 and 3.4), problem (P_1) admits at least one optimal pair (y^*, u^*).

We shall approximate problem (P_1) by the following family of infinite horizon optimal control problems:

$$(\mathrm{P}^{\varepsilon}) \qquad \text{Minimize} \quad \int_0^{\infty} (g^{\varepsilon}(y(t)) + h_{\varepsilon}(u(t)))\, dt$$

on all $u \in L^2_{\text{loc}}(\mathbf{R}^+; L^2(\Omega))$ *and* $y \in W^{1,2}_{\text{loc}}([0,\infty); L^2(\Omega)) \cap L^2_{\text{loc}}(\mathbf{R}^+; H^1_0(\Omega) \cap H^2(\Omega))$, *subject to*

$$\begin{aligned} \frac{\partial y}{\partial t} - \Delta y + \beta^{\varepsilon}(y) &= u \qquad \text{in } \Omega \times \mathbf{R}^+, \\ y(x,0) &= y_0(x) \qquad \text{in } \Omega, \\ y &= 0 \qquad \text{in } \partial\Omega \times \mathbf{R}^+. \end{aligned} \tag{3.56}$$

Here, β^{ε} is a smooth approximation of β, i.e., $\beta^{\varepsilon} \in C^2(\mathbf{R})$, $\dot{\beta}^{\varepsilon} \geq 0$, $\beta^{\varepsilon}(0) = 0$, $\dot{\beta}^{\varepsilon} \in L^{\infty}(\mathbf{R})$, and these satisfy assumptions (k)–(kkk) in Section 1.2. The functions $g^{\varepsilon}: L^2(\Omega) \to \mathbf{R}$ and $h_{\varepsilon}: L^2(\Omega) \to \mathbf{R}$ are defined by

$$h_{\varepsilon}(u) = \inf\left\{\frac{|u - v|_2^2}{2\varepsilon}; v \in K\right\}, \qquad u \in L^2(\Omega), \tag{3.57}$$

and

$$g^{\varepsilon}(y) = \pi\left(\frac{|y|_2^2}{\varepsilon^{1/4}}\right), \qquad y \in L^2(\Omega), \tag{3.58}$$

where $\pi \in C^1(\mathbf{R}^+)$ is such that $\pi' \geq 0$, $0 \leq \pi \leq 1$ in $\mathbf{R}^+$, and

$$\pi(y) = \begin{cases} 1 & \text{for } y \geq 2, \\ 0 & \text{for } 0 \leq y \leq 1. \end{cases}$$

Lemma 3.5. *For all ε sufficiently small, problem* $(\mathrm{P}^{\varepsilon})$ *admits at least one solution* $(y_{\varepsilon}, u_{\varepsilon})$.

Proof. It is readily seen that there exists at least one admissible pair (y, u) in problem $(\mathrm{P}^{\varepsilon})$. For instance, we may take u as in the proof of Lemma 3.1 and 3.4. Hence, there are the sequences u_n, y_n satisfying system (3.56) and such that

$$d \leq \int_0^{\infty} (g^{\varepsilon}(y_n) + h_{\varepsilon}(u_n))\, dt \leq d + n^{-1},$$

where d is the infimum in $(\mathrm{P}^{\varepsilon})$.

Then, by the definition of h^{ε}, we see that the u_n remain in a bounded subset of $L^2_{\text{loc}}(\mathbf{R}^+; L^2(\Omega))$. Hence, on a subsequence,

$$u_n \to u \qquad \text{weakly in } L^2_{\text{loc}}(\mathbf{R}^+; L^2(\Omega)),$$

and by Eq. (3.56) we see that the y_n remain in a bounded subset of $W^{1,2}_{\text{loc}}([0,\infty); L^2(\Omega)) \cap L^2_{\text{loc}}(\mathbf{R}^+; H^1_0(\Omega) \cap H^2(\Omega))$. Hence, we may assume that, for every $T > 0$,

$$\begin{aligned} y_n \to y = y(t, y_0, u) \qquad & \text{strongly in } L^2(0,T; L^2(\Omega)) \cap L^2(0,T; H^1_0(\Omega)), \\ & \text{weakly in } L^2(0,T; H^2(\Omega)), \end{aligned}$$

and by the Fatou lemma,

$$\liminf_{n\to\infty} \int_0^\infty g^\varepsilon(y_n)\, dt \geq \int_0^\infty g^\varepsilon(y)\, dt,$$

and

$$\liminf_{n\to\infty} \int_0^\infty h_\varepsilon(u_n)\, dt \geq \int_0^\infty h_\varepsilon(u)\, dt,$$

because the function $u \to \int_0^\infty h_\varepsilon(u)\, dt$ is convex and l.s.c. on $L^2_{\text{loc}}(\mathbf{R}^+; L^2(\Omega))$. Hence,

$$\int_0^\infty (g^\varepsilon(y) + h_\varepsilon(u))\, dt = d,$$

as desired. ■

Theorem 3.5. *Let $(y_\varepsilon, u_\varepsilon)$ be optimal in problem* (P^ε). *Then, on a subsequence, $\varepsilon \to 0$,*

$$\begin{aligned} u_\varepsilon \to u^* \qquad & \text{weak star in } L^\infty(0,T^*; L^2(\Omega)), \\ y_\varepsilon \to y^* \qquad & \text{weakly in } W^{1,2}([0,T^*]; L^2(\Omega)) \cap L^2(0,T^*; H^2(\Omega)), \\ & \text{strongly in } C([0,T^*]; L^2(\Omega)) \cap L^2(0,T^*; H^1_0(\Omega)), \end{aligned} \tag{3.59}$$

where T^* is the minimal time and (y^*, u^*) is an optimal pair for problem (P_1).

Proof. Let $(y_1^*, u_1^*) \in W^{1,2}([0,T^*]; L^2(\Omega)) \cap L^2(0,T^*; L^2(\Omega))$ be any optimal pair in problem (P_1). (We have already noted that such a pair exists.) We extend u_1^* and y_1^* by 0 on $[T^*, +\infty)$ and note that y_1^*, u_1^* is a solution to (3.23) on $\Omega \times (0,\infty)$. Now, let $\tilde{y}_\varepsilon$ be the solution to Eq. (3.56) for $u = u_1^*$. Since $h_\varepsilon(u_1^*(t)) = 0$ a.e. $t > 0$, we have

$$\int_0^\infty (g^\varepsilon(y_\varepsilon(t)) + h_\varepsilon(u_\varepsilon(t)))\, dt \leq \int_0^\infty g^\varepsilon(\tilde{y}_\varepsilon(t))\, dt$$

whereas, by Lemma 1.1,

$$|\tilde{y}_\varepsilon(t) - y_1^*(t)|_2 \leq C\varepsilon^{1/2} \qquad \forall t \in [0, T^*]. \tag{3.60}$$

On the other hand, by Eq. (3.56) we have the estimate

$$|\tilde{y}_\varepsilon(t)|_2 \le e^{-\omega(t-T^*)}|\tilde{y}_\varepsilon(T^*)|_2 \qquad \forall t \ge T^*$$

(because $u_1^* = 0$ on $[T^*, \infty)$), and along with (3.60) this yields

$$|\tilde{y}_\varepsilon(t)|_2 \le C\varepsilon^{1/2} \qquad \forall t \ge T^*.$$

Then, by the definition (3.58) of g^ε, it follows that, for all ε sufficiently small,

$$\int_0^\infty g^\varepsilon(\tilde{y}_\varepsilon(t))\, dt = \int_0^{T^*} g^\varepsilon(\tilde{y}_\varepsilon(t))\, dt \le \int_0^{T^*} g^\varepsilon(y_1^*(t))\, dt + C\varepsilon^{1/4},$$

and so

$$\limsup_{\varepsilon \to 0} \int_0^\infty (g^\varepsilon(y_\varepsilon(t)) + h_\varepsilon(u_\varepsilon(t)))\, dt \le T^*. \tag{3.61}$$

On the other hand, since $\{u_\varepsilon\}$ is bounded in $L^2_{\text{loc}}(\mathbf{R}^+; L^2(\Omega))$ it follows by Lemma 1.1 that there exists $u^* \in L^2_{\text{loc}}(\mathbf{R}^+; L^2(\Omega))$ such that, for every $T > 0$,

$$\begin{aligned} u_{\varepsilon_n} &\to u^* && \text{weakly in } L^2(0,T; L^2(\Omega)), \\ y_{\varepsilon_n} &\to y^* && \text{weakly in } W^{1,2}([0,T]; L^2(\Omega)) \cap L^2(0,T; H^2(\Omega)), \\ & && \text{strongly in } C([0,T]; L^2(\Omega)) \cap L^2(0,T; H^1(\Omega)), \end{aligned}$$

where $y^* = y(t, y_0, u^*)$.

We shall prove that u^* is a time-optimal control. We note first that, by (3.58) and (3.61), it follows that the Lebesgue measure of the set $\{t > 0; |y_\varepsilon(t)|_2^2 \ge 2\varepsilon^{1/4}\}$ is smaller than T^*. Thus, there are $\varepsilon_n \to 0$ and $t_n \in [0, 2T^*]$ such that

$$|y_{\varepsilon_n}(t_n)|_2^2 \le 2\varepsilon_n^{1/4} \qquad \forall n. \tag{3.62}$$

Extracting a further subsequence, we may assume that $t_n \to T_0$.

On the other hand, since $\{\partial y_{\varepsilon_n}/\partial t\}$ is bounded in every $L^2(0,T; L^2(\Omega))$, we have

$$|y_{\varepsilon_n}(t) - y_{\varepsilon_n}(t_n)|_2 \le C|t - t_n|^{1/2} \qquad \forall t \in [0, T_0].$$

Then, by (3.62), we conclude that $y^*(T_0) = 0$. Let $\tilde{T} = \inf\{T;\ y^*(T) = 0\}$. We will prove that $\tilde{T} = T^*$. To this end, for every $\varepsilon > 0$ consider the set $E_\varepsilon = \{t \in [0, \tilde{T}];\ |y_\varepsilon(t)|_2^2 \ge 2\varepsilon^{1/4}\}$. By (3.61), we see that

$$\limsup_{\varepsilon \to 0} m(E_\varepsilon) \le T^* \le \tilde{T},$$

where m denotes the Lebesgue measure. On the other hand,

$\limsup_{\varepsilon \to 0} m(E_\varepsilon) = \tilde{T}$, for otherwise there would exist $\delta > 0$ and $\varepsilon_n \to 0$ such that $m(E_{\varepsilon_n}) \le \tilde{T} - \delta$ $\forall n$. In other words, there would exist a sequence of measurable sets $A_n \subset [0, \tilde{T}]$ such that $m(A_n) \ge \delta$ and $|y_{\varepsilon_n}(t)|_2^2 \le 2\varepsilon_n^{1/4}$ $\forall t \in A_n$. Clearly, this would imply that

$$|y^*(t)|_2 \le (2\varepsilon_n^{1/4})^{1/2} + \nu_n \qquad \forall t \in A_n,$$

where $\nu_n \to 0$ as $n \to \infty$.

On the other hand, since $y^*(t) \ne 0$ $\forall t \in [0, \tilde{T}]$, we have

$$\lim_{n \to \infty} m\left\{t \in [0, \tilde{T}];\ |y^*(t)|^2 \le (2\varepsilon_n^{1/4})^{1/2} + \nu_n\right\} = 0.$$

The contradiction we have arrived at shows that indeed $\limsup_{\varepsilon \to 0} m(E_\varepsilon) = \tilde{T}$ and therefore $\tilde{T} = T^*$, as claimed. This completes the proof. ∎

Now, if $(y_\varepsilon, u_\varepsilon)$ is an optimal pair for problem (P^ε), by Proposition 1.3 there is $p_\varepsilon \in C([0,\infty); L^2(\Omega)) \cap W_{\mathrm{loc}}^{1,2}([0,\infty); L^2(\Omega)) \cap L_{\mathrm{loc}}^2(\mathbf{R}^+; H_0^1(\Omega) \cap H^2(\Omega))$ such that

$$\frac{\partial p_\varepsilon}{\partial t} + \Delta p_\varepsilon - \dot{\beta}^\varepsilon(y_\varepsilon) p_\varepsilon = G_\varepsilon(y_\varepsilon) \qquad \text{in } \Omega \times (0,\infty), \tag{3.63}$$

$$p_\varepsilon(t) = \nabla h_\varepsilon(u_\varepsilon(t)) \qquad \forall t \ge 0, \tag{3.64}$$

$$p_\varepsilon(t) \in -\partial\varphi_\varepsilon(y_\varepsilon(t)) \qquad \forall t \ge 0,$$

where $\varphi_\varepsilon(y_0)$ is the value of (P^ε) at $y_0 \in L^2(\Omega)$ and

$$G_\varepsilon(y) = 2y\varepsilon^{-1/4}\pi'\left(\frac{|y|_2^2}{\varepsilon^{1/4}}\right), \qquad y \in L^2(\Omega).$$

To be more specific, let us assume that $K = \{u \in L^2(\Omega);\ |u|_2 \le \rho\}$.

We may pass to limit in system (3.63), (3.64) to get that the optimal pair (y^*, u^*) given by Theorem 3.5 satisfies a maximum principle–type system. Indeed, by (3.63) and (3.56) we get

$$\frac{d}{dt}\left((p_\varepsilon, -\Delta y_\varepsilon + \beta^\varepsilon(y_\varepsilon)) - \rho|p_\varepsilon(t)|_2 - \frac{\varepsilon}{2}|p_\varepsilon(t)|_2^2 + g^\varepsilon(y_\varepsilon)\right) = 0$$

and, therefore,

$$\begin{aligned}(p_\varepsilon(t), -\Delta y_\varepsilon(t) + \beta^\varepsilon(y_\varepsilon(t)) - \rho|p_\varepsilon(t)|_2 \\ - \frac{\varepsilon}{2}|p_\varepsilon(t)|_2^2 + g^\varepsilon(y_\varepsilon(t)) \equiv C.\end{aligned}$$

Since $p_\varepsilon \in L^2(\mathbf{R}^+;\ L^2(\Omega))$, $g^\varepsilon(y_\varepsilon) \in L^1(\mathbf{R}^+)$, and

$$(p_\varepsilon(t_n), -\Delta y_\varepsilon(t_n) + \beta^\varepsilon(y_\varepsilon(t_n))) \to 0 \qquad \text{for some } t_n \to \infty,$$

we find that

$$\rho|p_\varepsilon(t)|_2 + \frac{\varepsilon}{2}|p_\varepsilon(t)|_2^2$$
$$= (p_\varepsilon(t), -\Delta y_\varepsilon(t) + \beta^\varepsilon(y_\varepsilon(t))) + g^\varepsilon(y_\varepsilon(t)) \qquad \forall t \geq 0. \quad (3.65)$$

On the other hand, a little calculation reveals that

$$\frac{d}{dt}(p_\varepsilon(t), -\Delta y_\varepsilon(t) + \beta^\varepsilon(y_\varepsilon(t))) \geq 0,$$

and so

$$(p_\varepsilon(t), -\Delta y_\varepsilon(t) + \beta^\varepsilon(y_\varepsilon(t))) \leq 0 \qquad \forall t \geq 0.$$

This implies that

$$\rho|p_\varepsilon(t)|_2 \leq 1 \qquad \forall t \geq 0. \tag{3.66}$$

Noticing that $G^\varepsilon(y_\varepsilon(t)) = 0$ for $|y_\varepsilon(t)|_2^2 \geq 2\varepsilon^{1/4}$, it follows by estimate (3.66) and Eq. (3.63) that $\{p_\varepsilon\}$ is bounded in $L^\infty(0, T^* - \delta;\ L^2(\Omega)) \cap L^2(0, T^* - \delta;\ H_0^1(\Omega))$ for every $\delta > 0$. Then, by using a standard device we find as in previous proofs that, on a subsequence $\varepsilon_n \to 0$,

$$p_{\varepsilon_n} \to p \qquad \text{strongly in } L^2(0, T^*;\ L^2(\Omega)),$$

$$p_{\varepsilon_n}(t) \to p(t) \qquad \text{strongly in } H^{-s}(\Omega), \text{ weakly in } L^2(\Omega) \text{ for } t \in [0, T^*),$$

where $p \in L^\infty(0, T^*;\ L^2(\Omega)) \cap L^2(0, T^*;\ H_0^1(\Omega)) \cap BV([0, T^*];\ H^{-s}(\Omega))$ satisfies the equations

$$\frac{\partial p}{\partial t} + \Delta p - \nu = 0 \qquad \text{in } \Omega \times (0, T^*), \tag{3.67}$$

$$u^*(t) = \rho \operatorname{sgn} p(t) \qquad \text{a.e. } t \in (0, T^*), \tag{3.68}$$

$$\rho|p(t)|_2 - (p(t), -\Delta y^*(t) + \beta(y^*(t))) = 1 \qquad \text{a.e. } t \in (0, T^*), \tag{3.69}$$

where $\nu \in (L^\infty(\Omega \times (0, T^*))^*$.

In particular, it follows by (3.68) and (3.69) that u^* is a bang-bang control, i.e., $|u^*(t)|_2 = \rho$ a.e. $t \in (0, T^*)$.

For special choices of β (for instance, β locally Lipschitz or a maximal monotone graph of the form (1.6)), we may deduce Theorems 3.3 and 3.4

from the preceding optimality system (see Remark 3.1). We refer the reader to author's book [7] for other results in this direction.

5.4. Approximating Optimal Control Problems via the Fractional Steps Method

5.4.1. *The Description of the Approximating Scheme*

We will return now to the optimal control problem (P) in Section 1.1, i.e.,

$$\text{Minimize} \quad \int_0^T (g(t, y(t)) + h(u(t)))\, dt + \varphi_0(y(T)) \tag{4.1}$$

on all $(y, u) \in C([0,T]; H) \cap L^2(0,T; U)$, *subject to the state system*

$$\begin{aligned} y'(t) + Ay(t) + Fy(t) &\ni (Bu)(t) + f(t) \qquad \text{a.e. } t \in (0,T), \\ y(0) &= y_0, \end{aligned} \tag{4.2}$$

in a real Hilbert space H.

Here, $B \in L(L^2(0,T; U), L^2(0,T; H))$, $g:[0,T] \times H \to \mathbf{R}$, $\varphi_0: H \to \mathbf{R}$, $h: U \to \overline{\mathbf{R}}$ satisfy assumptions (v), (vi) in Section 1.1, and U is a real Hilbert space. The operator $A: V \to V'$ is linear, continuous, symmetric, and coercive, i.e.,

$$(Ay, y) \geq \omega \|y\|^2 \qquad \forall y \in V$$

whilst $F = \partial\varphi: H \to H$, where $\varphi: H \to \overline{\mathbf{R}}$ is a l.s.c., convex function such that

$$\begin{aligned} (Ay, F_\lambda y) &\geq 0 \qquad \forall y \in D(A_H),\ \lambda > 0, \\ F_\lambda &= \lambda^{-1}\left(I - (I + \lambda F)^{-1}\right) = \partial\varphi_\lambda. \end{aligned} \tag{4.3}$$

We will assume further that the projection operator P of H onto $K = \overline{D(F)}$ maps V into itself and

$$(APy, Py) \leq (Ay, y) \qquad \forall y \in V. \tag{4.4}$$

Here, V is as usually a real Hilbert space compactly, continuously, and densely imbedded in H, with the norm denoted $\|\cdot\|$.

We will assume, finally, that

$$y_0 \in D(\varphi) \cap V, \qquad f \in L^2(0,T; H). \tag{4.5}$$

As seen earlier, assumptions (4.3), (4.5) imply that $A + F$ is maximal monotone. More precisely, $A + F = \partial\phi$, where

$$\phi(y) = \tfrac{1}{2}(Ay, y) + \varphi(y) \qquad \forall y \in V.$$

Then, the Cauchy problem has for every $u \in L^2(0,T; U)$ a unique solution $y^u \in W^{1,2}([0,T]; H) \cap L^2(0,T; D(A_H))$. Since the map $u \to y^u$ is compact from $L^2(0,T; U)$ to $C([0,T]; H)$, problem (4.1) has a solution. Here, we will approximate problem (4.1) by the following one:

$$\text{Minimize} \quad \int_0^T (g(t, y(t)) + h(u(t)))\, dt + \varphi_0(y(T)) \tag{4.6}$$

on all $y\colon [0,T] \to H$, $u \in L^2(0,T; U)$, *subject to*

$$y'(t) + Ay(t) = (Bu)(t) + f(t) \qquad \text{a.e. } t \in (i\varepsilon, (i+1)\varepsilon),$$
$$y_+(i\varepsilon) = w_i(\varepsilon) \qquad \text{for } i = 1, \dots, n-1, \qquad y_+(0) = y_0, \tag{4.7}$$
$$w_i' + Fw_i \ni 0 \qquad \text{in } (0, \varepsilon), \qquad \varepsilon = T/n,$$
$$w_i(0) = Py_-(i\varepsilon) \qquad \text{for } i = 1, 2, \dots, n-1. \tag{4.8}$$

Here, $y_-(i\varepsilon)$ and $y_+(i\varepsilon)$ are respectively the left and right limits of y at $i\varepsilon$.

Since, by assumption (4.3), $e^{-Ft}V \subset V$ for all $t > 0$, it is readily seen that problem (4.7), (4.8) has a unique solution $y\colon [0,T] \to H$, which is piecewise continuous and belongs to $W^{1,2}([i\varepsilon, (i+1)\varepsilon]; H) \cap L^2(i\varepsilon, (i+1)\varepsilon; V)$ on every interval $[i\varepsilon, (i+1)\varepsilon]$. Then, by a standard device, it follows that the optimal control problem (4.6) has for every $\varepsilon > 0$ at least one solution u_ε^*. We set

$$\Psi(u) = \int_0^T (g(y^u(t)) + h(u(t)))\, dt + \varphi_0(y^u(T))$$

and

$$\Psi_\varepsilon(u) = \int_0^T (g(y_\varepsilon^u(t)) + h(u(t)))\, dt + \varphi_0(y_\varepsilon^u(T)),$$

where y_ε^u is the solution to system (4.7), (4.8). Then, in terms of Ψ and Ψ_ε, we may rewrite problem (4.1) and (4.6) as

$$\min\{\Psi(u);\, u \in L^2(0,T; U)\}, \tag{4.9}$$

respectively,

$$\min\{\Psi_\varepsilon(u);\, u \in L^2(0,T; U)\}. \tag{4.10}$$

The main result of this section is the following convergence theorem.

Theorem 4.1. *Assume that beside the preceding hypotheses at least one of the following assumptions holds*:

(i) *B is compact from $L^2(0,T;\, U)$ to $L^2(0,T;\, H)$*;

(ii) *$F = \partial I_C$, where C is a closed convex subset of H.*

Then

$$\lim_{\varepsilon \to 0} \left(\inf\{\Psi_\varepsilon(u);\, u \in L^2(0,T;\, U)\}\right) = \inf\{\Psi(u);\, u \in L^2(0,T;\, U)\}, \tag{4.11}$$

and if $\{u_\varepsilon^*\}$ is a sequence of optimal controls for problem (4.6) *then*

$$\Psi(u_\varepsilon^*) \to \inf\{\Psi(u);\, u \in L^2(0,T;\, U)\}. \tag{4.12}$$

Moreover, every weak limit point of $\{u_\varepsilon^*\}$ for $\varepsilon \to 0$ is an optimal control of problem (4.1).

It is apparently clear that conceptually and practically the decoupled problem (4.6) is simpler than the original problem (4.1).

Now we shall briefly present some typical situations to which Theorem 4.1 is applicable.

1. Consider the distributed control system

$$\begin{aligned} &\frac{\partial y}{\partial t} - \Delta y + \beta(y) \ni f(x,t) + \sum_{i=1}^{m} v_i(t) a_i(x), \\ &\qquad\qquad\qquad\qquad\qquad\qquad (x,t) \in Q = \Omega \times (0,T), \\ &y(x,0) = y_0(x) \qquad \text{in } \Omega, \\ &y + \alpha \frac{\partial y}{\partial \nu} = 0 \qquad \text{in } \Sigma = \partial\Omega \times (0,T), \end{aligned} \tag{4.13}$$

$$\begin{aligned} &\frac{dv}{dt} + Dv = B_0 u \qquad \text{a.e. } t \in (0,T), \\ &v(0) = v_0, \qquad v = (v_1, \ldots, v_m), \end{aligned} \tag{4.14}$$

in an open domain $\Omega \subset \mathbf{R}^N$ with a sufficiently smooth boundary. Here, $\alpha \geq 0$, $\beta: \mathbf{R} \to \mathbf{R}$ is a maximal monotone graph (eventually, multivalued) such that $D(\beta) = \mathbf{R}$ and $0 \in \beta(0)$, D is a Lipschitz mapping from $\mathbf{R}^m$ to itself, $B_0 \in L(\mathbf{R}^p, \mathbf{R}^m)$, $a_i \in L^\infty(\Omega)$ for $i = 1, \ldots, m$, and

$$f \in L^2(Q), \qquad y_0 \in H^1(\Omega), \qquad j(y_0) \in L^1(\Omega),$$

where $\beta = \partial j$.

We may apply Theorem 4.1, where $U = \mathbf{R}^p$, $V = H^1(\Omega)$, $H = L^2(\Omega)$, $A{:}V \to V'$ is defined by

$$(Ay, z) = \int_\Omega \nabla y \cdot \nabla z \, dx + \frac{1}{\alpha} \int_{\partial\Omega} yz \, dx \qquad \forall y, z \in H^1(\Omega)$$

(respectively, $Ay = -\Delta y$ and $V = H_0^1(\Omega)$ if $\alpha = 0$), and

$$(Fy)(x) = \{w \in L^2(\Omega);\, w(x) \in \beta(y(x)) \text{ a.e.} x \in \Omega\},$$

$$(Bu)(t, x) = \sum_{i=1}^{m} a_i(x) v_i(t), \qquad u \in L^2(0, T; U).$$

Assumption (i) is obviously satisfied by virtue of the Arzelá theorem. We leave it to the reader to write the iterative scheme and to formulate the approximating problem (4.6) in the present situation.

2. Consider the optimal control problem (4.1) governed by the free boundary problem (the obstacle problem)

$$\begin{aligned} \frac{\partial y}{\partial t} - \Delta y &\geq u && \text{in } Q \\ \left(\frac{\partial y}{\partial t} - \Delta y - u\right) y &= 0 && \text{in } Q, \\ y(x,0) &= y_0(x) && \text{in } \Omega, \\ y &= 0 && \text{in } \Sigma, \end{aligned} \tag{4.15}$$

where $y_0 \in H_0^1(\Omega)$, $y_0 \geq 0$ a.e. in Ω.

As seen earlier, this control system is of the form (4.2), where $H = L^2(\Omega)$, $V = H_0^1(\Omega)$, $A = -\Delta$, and $F = \partial I_C$ where $C = \{y \in H_0^1(\Omega);\ y(x) \geq 0$ a.e. $x \in \Omega\}$. We note that in this case $(Py)(x) = y^+(x) = \max\{y(x), 0\}$, a.e. $x \in \Omega$, and assumptions (4.3) (4.4) are clearly satisfied since

$$\|\nabla Py\|_{L^2(\Omega)} \leq \|\nabla y\|_{L^2(\Omega)} \qquad \forall y \in H_0^1(\Omega).$$

Since $e^{-Ft}y = y^+$ $\forall t \geq 0$, and all $y \in L^2(\Omega)$, system (4.7), (4.8) becomes, in this case,

$$\begin{aligned} \frac{\partial y}{\partial t} - \Delta y &= u && \text{in } Q_\varepsilon^i = \Omega \times (i\varepsilon, (i+1)\varepsilon), \\ y &= 0 && \text{in } \Sigma_\varepsilon^i = \partial\Omega \times (i\varepsilon, (i+1)\varepsilon), \\ y(x,0) &= y_0(x) && \text{in } \Omega, \\ y_+(x, i\varepsilon) &= \max\{y_-(x, i\varepsilon), 0\} && \text{a.e. } x \in \Omega. \end{aligned} \tag{4.16}$$

Arguing as in the proof of Theorem 1.2, we get for the corresponding problem (4.6) the following optimality system (assume that g is Gâteaux differentiable and $\varphi_0 \equiv 0$):

$$\frac{\partial p}{\partial t} + \Delta p = \nabla_y g(t, y_\varepsilon^*) \qquad \text{in } Q_\varepsilon^i, \forall i,$$

$$p_-(x,(i+1)\varepsilon) = p_+(x,(i+1)\varepsilon)$$

$$\text{in } \{x; (y_\varepsilon^*)_+ (x,(i+1)\varepsilon) > 0\},$$

$$p_-(x,(i+1)\varepsilon) = 0 \qquad \text{in } \{x; (y_\varepsilon^*) + (x,(i+1)\varepsilon) = 0\},$$

$$p_-(x,T) = 0 \qquad \text{in } \Omega, \tag{4.17}$$

$$p \in \partial h(u^*) \qquad \text{a.e. } t \in (0,T). \tag{4.18}$$

This system can be solved numerically by a gradient type algorithm and the numerical tests performed by V. Arnăutu (see Barbu [11]) show that a large amount of computing time is saved using this scheme.

5.4.2. *The Convergence of the Scheme*

We will prove Theorem 4.1 here. The main ingredient of the proof is Proposition 4.1, which also has an interest in itself.

Proposition 4.1. *Under the assumptions of Theorem* 4.1, *if* $\{u_{\varepsilon_n}\}$ *is weakly convergent to* u *as* $\varepsilon_n \to 0$, *then*

$$y_{\varepsilon_n}^{u_{\varepsilon_n}} \to y^u(t) \qquad \text{strongly in } H, \forall t \in [0,T]. \tag{4.19}$$

We recall that y_ε^u *is the solution to system* (4.7), (4.8).

Let us postpone for the time being the proof of Proposition 4.1 and derive now Theorem 4.1.

Let u_ε^* be an optimal controller for problem (4.6) and let y_ε^* be the corresponding solution to system (4.7), (4.8). By assumption (v) in Section 1.1, $\{u_\varepsilon\}$ is bounded in $L^2(0,T;\, U)$ and so, on a subsequence $\varepsilon_n \to 0$ as $n \to \infty$,

$$u_{\varepsilon_n} \to u^* \qquad \text{weakly in } L^2(0,T;U), \tag{4.20}$$

whilst by Lemma 4.1,

$$y_{\varepsilon_n}^*(t) \to y^{u^*}(t) \qquad \text{strongly in } H, \forall t \in [0,T]. \tag{4.21}$$

(We set $y^*_{\varepsilon_n} = y^{u_{\varepsilon_n}}_{\varepsilon_n}$.) This clearly implies that

$$g(y^*_{\varepsilon_n}) \to g(y^{u^*}) \qquad \text{in } L^1(0,T), \tag{4.22}$$

and since $u \to \int_0^T h(u)$ is weakly lower semicontinuous, we have

$$\liminf_{n\to\infty} \int_0^T h(u_{\varepsilon_n})\,dt \geq \int_0^T h(u^*)\,dt. \tag{4.23}$$

On the other hand, we have

$$\Psi_{\varepsilon_n}(u_{\varepsilon_n}) \leq \Psi_{\varepsilon_n}(\tilde{u}^*) \qquad \forall n,$$

where $\tilde{u}^*$ is optimal in problem (4.1). Letting n tend to ∞, we get by (4.21)–(4.23) that

$$\Psi(u^*) \leq \liminf_{n\to\infty} \Psi_{\varepsilon_n}(u_{\varepsilon_n}) \leq \Psi(\tilde{u}^*) \tag{4.24}$$

and, therefore,

$$\Psi(u^*) = \lim_{n\to\infty} \Psi_{\varepsilon_n}(u_{\varepsilon_n}) = \Psi(\tilde{u}^*) = \inf\{\Psi(u);\, u \in L^2(0,T;U)\},$$

i.e., u^* is an optimal control for problem (4.1).

To prove (4.12), we set $\tilde{y}_\varepsilon = y^{u^*_\varepsilon}$ and note that, by (4.20) and the Arzelá–Ascoli theorem,

$$\tilde{y}_{\varepsilon_n} \to y^* = y^{u^*} \qquad \text{strongly in } C([0,T];H).$$

Hence,

$$\int_0^T g(\tilde{y}_{\varepsilon_n})\,dt + \varphi_0(\tilde{y}_{\varepsilon_n}(T)) \to \int_0^T g(y^*)\,dt + \varphi_0(y^*(T)),$$

whilst by (4.24) we see that

$$\int_0^T h(u_{\varepsilon_n}(t))\,dt \to \int_0^T h(u^*(t))\,dt.$$

Therefore, $\Psi(u^*_{\varepsilon_n}) \to \Psi(u^*)$, as claimed. This completes the proof. ∎

To prove Proposition 4.1 under hypothesis (i) we shall establish first a Lie–Trotter product formula for the nonhomogeneous Cauchy problem

$$\begin{aligned} y^* + Ay + Fy &= q, \qquad t \geq 0, \\ y(0) &= x, \end{aligned} \tag{4.25}$$

where $q \in L^1(\mathbf{R}^+; H)$, $x \in \overline{D(A) \cap D(F)} = \overline{D(F)} = K$, and A, F satisfy assumptions (4.3), (4.4). As mentioned earlier (Remark 1.4 in Chapter 4) we may write (4.25) as an autonomous differential equation

$$\frac{d}{dt}S(t)(x,q) + \mathscr{A}S(t)(x,q) = 0 \qquad t \geq 0, \tag{4.26}$$

in $X = H \times L^1(\mathbf{R}^+; H)$, where

$$\mathscr{A}(x,q) = \{(A+F)x - q(0), -q'\} \qquad \forall (x,q) \in D(\mathscr{A}),$$
$$D(\mathscr{A}) = \{(x,q) \in X;\ x \in D(A) \cap D(F), q \in W^{1,1}([0,\infty); H)\},$$

and $S(t)(x,q) = [y(t), q_t]$, $q_t(s) = q(t+s)$, $t, s \geq 0$ (i.e., $S(t)$ is the semigroup generated by $\mathscr{A}$ on X). On $\mathscr{K} = K \times L^1(\mathbf{R}^+; H)$ consider the semigroups of contraction $S_1(t)$ and $S_2(t)$ defined by

$$S_1(t)(x,q) = [z(t), q_t], \qquad t \geq 0, \tag{4.27}$$

$$S_2(t)(x,q) = [e^{-Ft}x, q], \qquad t \geq 0, \tag{4.28}$$

where

$$z' + Az = q \qquad \text{in } \mathbf{R}^+ = (0,\infty), \qquad z(0) = x,$$

and $w = e^{-Ft}x$ is the solution to

$$w' + Fw = 0 \qquad \text{in } \mathbf{R}^+, \qquad w(0) = x.$$

It is easily seen that $S_1(t)$ and $S_2(t)$ are generated by the operators $\mathscr{A}_1$ and $\mathscr{A}_2$:

$$\mathscr{A}_1(x,q) = [Ax - q(0), -q'], \qquad \mathscr{A}_2(x,q) = [Fx, 0].$$

Now let $P: H \to K$ be the projection on $K = \overline{D(F)}$, and let $Q: X \to \mathscr{K} = K \times L^1(\mathbf{R}^+; H)$ be the operator

$$Q(x,q) = [Px, q], \qquad x \in K, q \in L^1(\mathbf{R}^+; H).$$

Lemma 4.1. *For all $(x,q) \in \mathscr{K}$, we have*

$$\lim_{n\to\infty} \left(QS_1\left(\frac{t}{n}\right)S_2\left(\frac{t}{n}\right)\right)^n (x,q) = S(t)(x,q), \tag{4.29}$$

uniformly on compact intervals.

Proof. We will use the nonlinear Chernoff theorem (Theorem 2.2 in Chapter 4). To this end, consider the family $\{\Gamma(t)\}_{t\geq 0}$ of nonexpansive operators
on $\mathscr{K}$,

$$\Gamma(t) = QS_1(t)S_2(t), \qquad t \geq 0,$$

and set

$$X_t = (I + \lambda t^{-1}(I - \Gamma(t)))^{-1}(x, q), \qquad t \geq 0. \tag{4.30}$$

Then, according to Chernoff theorem, to prove (4.29) it suffices to show that

$$\lim_{t\to 0} X_t = (I + \lambda A)^{-1}(x, q), \qquad \forall\lambda > 0. \tag{4.31}$$

According to (4.27), (4.28), we may rewrite (4.30) as

$$(t + \lambda)x^t - \lambda P\left(e^{-At}e^{-Ft}x^t + \int_0^t e^{-A(t-s)}y^t(s)\,ds\right) = tx, \tag{4.32}$$

$$(t + \lambda)y^t(s) - \lambda y^t(s + t) = tq(s) \qquad \forall s \geq 0, \tag{4.33}$$

where $X_t = (x^t, y^t) \in X$ and e^{-At} is the semigroup generated on H by $-A_H$.

Inasmuch as the operators $(I + \lambda\mathscr{A})^{-1}$ and $(I + \lambda t^{-1}(I - \Gamma(t)))^{-1}$ are nonexpansive, without loss of generality we may assume that $x \in D(A) \cap D(F)$ and $q, q' \in L^2(\mathbf{R}^+; V) \cap L^1(\mathbf{R}^+; V)$ (the general case follows by density). Then, by Eq. (4.33), we see that $y^t \in L^1(\mathbf{R}^+; V) \cap L^2(\mathbf{R}^+; V)$, y^t is V-absolutely continuous on compact intervals, and

$$\|y^t\|_{L^i(\mathbf{R}^+; V)} \leq \|q\|_{L^i(\mathbf{R}^+; V)}, \qquad i = 1, 2,$$

$$\left\|\frac{dy^t}{ds}\right\|_{L^i(\mathbf{R}^+; V)} \leq \left\|\frac{dq}{ds}\right\|_{L^i(\mathbf{R}^+; V)}, \qquad i = 1, 2.$$

Hence, $\{y^t\}_{t>0}$ is compact in $L^1(\mathbf{R}^+; H) \cap C(\mathbf{R}^+; H)$ and, therefore,

$$y^t \to w \qquad \text{strongly in } H\text{, uniformly in } s \text{ on compacta.} \tag{4.34}$$

Since, by (4.33),

$$\int_\mu^\infty |y^t(s)|\,ds \leq \int_\mu^\infty |q(s)|\,ds \qquad \forall\mu > 0,$$

we may conclude, therefore, that

$$y^t \to w \qquad \text{strongly in } L^1(\mathbf{R}^+; H).$$

On the other hand, for each $\psi \in W^{1,2}([0,\infty); H)$, $\psi(0) = 0$, we have

$$\lim_{t\to 0} \int_0^\infty \left(\frac{y^t(s+t) - y^t(s)}{t}, \psi(s) \right) ds = -\int_0^\infty (w(s), \psi'(s))\, ds$$

and, therefore, for $t \to 0$,

$$\frac{1}{t}(y^t(s+t) - y^t(s)) \to w' \qquad \text{weakly in } L^2(\mathbf{R}^+; H).$$

Then, letting t tend to zero in (4.33), we see that

$$w - w' = q \qquad \text{a.e. in } \mathbf{R}^+.$$

Next, by (4.34), it follows that

$$\frac{1}{t}\int_0^t e^{-A(t-s)} y^t(s)\, ds \to w(0) \qquad \text{strongly in } H \text{ as } t \to 0.$$

To complete the proof it remains to be shown that, for $t \to 0$,

$$x^t \to x_\lambda^0 \qquad \text{strongly in } H, \tag{4.35}$$

where x_λ^0 is the solution to the equation

$$x_\lambda^0 + \lambda(Ax_\lambda^0 + Fx_\lambda^0) = x + \lambda w(0). \tag{4.36}$$

To this aim, we set $q^t = t^{-1}\int_0^t e^{-A(t-s)} y^t(s)\, ds$. Noticing that $P = (I + \partial I_K)^{-1}$, we may equivalently write (4.32) as

$$t^{-1}(x^t - e^{-At}x^t) + \lambda t^{-1} e^{-At}(x^t - e^{-Ft}x^t) \\ + \lambda t^{-1}\, \partial I_K(\lambda^{-1}(t+\lambda)x^t - t\lambda^{-1}x) \ni x + \lambda q^t - x^t. \tag{4.37}$$

Let $z^t(s) = e^{-As}x^t$ and let u be arbitrary but fixed in V. Multiplying the equation $z' + Az = 0$ by $z^t - u$ and integrating over $[0, t]$, we get

$$(e^{-At}x^t - x^t, x^t - u) + \tfrac{1}{2}|e^{-At}x^t - x^t|^2 \\ + \int_0^t \big(\varphi_1(z^t(s)) - \varphi_1(u)\big)\, ds \le 0, \tag{4.38}$$

where $\varphi_1(y) = \frac{1}{2}(Ay, y)$, $y \in V$.

Similarly,

$$(e^{-Ft}x^t - x^t, x^t - u) + \tfrac{1}{2}|e^{-Ft}x^t - x^t|^2 \\ + \int_0^t (\varphi(e^{-Fs}x^t) - \varphi(u))\, ds \le 0. \tag{4.39}$$

Hence,

$$
\begin{aligned}
& t^{-1}(e^{-Ft}x^t - x^t, e^{-At}(x^t - u)) + (2t)^{-1}|e^{-Ft}x^t - x^t|^2 \\
& \quad + t^{-1}\int_0^t \big(\varphi(e^{-Fs}x^t) - \varphi(u) + \varphi_1(e^{-As}x^t) - \varphi_1(u)\big)\, ds \\
& \quad \le t^{-1}(e^{-Ft}x^t - x^t, e^{-At}x^t - x^t) + t^{-1}|e^{-Ft}x^t - x^t|\,|e^{-At}u - u|.
\end{aligned} \tag{4.40}
$$

Now multiply (4.37) by $x^t - e^{-Ft}x^t + t\lambda^{-1}(x^t - x)$, and use the accretivity of e^{-At} along with the definition of ∂I_K to get

$$
\begin{aligned}
& t^{-1}(x^t - e^{-At}x^t, x^t - e^{-Ft}x^t) \\
& \quad \le -(x^t - e^{-At}x^t, x^t - x) - (x^t - e^{-Ft}x^t, e^{-At}(x^t - x)) \\
& \qquad + (x + \lambda q^t - x^t, x^t - e^{-Ft}x^t) + t\lambda^{-1}(x + \lambda q^t - x^t, x^t - x).
\end{aligned}
$$

Combining this with (4.38) and (4.40) yields

$$
\begin{aligned}
& \lambda(t^{-1}(e^{-At}x^t - x^t), x^t - u) + \lambda(t^{-1}e^{-At}(e^{-Ft}x^t - x^t), x^t - u) \\
& \quad + \lambda t^{-1}\int_0^t \big(\varphi(e^{-Fs}x^t) + \varphi_1(e^{-As}x^t)\big)\, ds \\
& \quad \le \lambda(\varphi(u) + \varphi_1(u)) + (2t)^{-1}\lambda|e^{-At}u - u|^2 \\
& \qquad + |x^t - e^{-Ft}x^t|(2|x^t - x| + |x + \lambda q^t - x^t|) \\
& \qquad + t\lambda^{-1}|x + \lambda q^t - x^t|\,|x^t - x|.
\end{aligned} \tag{4.41}
$$

Next, by (4.32), we have

$$x^t = t(\lambda + t)^{-1}x + \lambda(\lambda + t)^{-1}P(e^{-At}e^{-Ft}x^t + tq^t),$$

whilst by assumption (4.3) it follows that

$$\varphi_1(e^{-Ft}x) \le \varphi_1(x) \qquad \forall t \ge 0, \forall x \in V.$$

This yields

$$\|x^t\| \le \|x\| + \lambda\|q^t\| \le C \qquad \forall t > 0,$$

because as previously seen $\{q^t\}$ is bounded in $L^\infty(0, 1; V)$.

We may conclude, therefore, that $\{x^t\}$ is a compact subset of H and so on a subsequence, again denoted $\{t\}$, we have

$$x^t \to x_\lambda^0 \qquad \text{strongly in } H. \tag{4.42}$$

Since the functions φ and φ_1 are lower semicontinuous, by the Fatou lemma we have

$$\liminf_{t\to 0} t^{-1}\int_0^t \big(\varphi(e^{-Fs}x^t) + \varphi_1(e^{-As}x^t)\big)\,ds \ge \varphi(x_\lambda^0) + \varphi_1(x_\lambda^0). \quad (4.43)$$

Next, from (4.42), we see that (see also Theorem 1.10 in Chapter 4)

$$t^{-1}|u - e^{-At}u|^2 + |e^{-At}x^t - x^t| + |e^{-Ft}x^t - x^t| \to 0 \qquad \text{as } t \to 0. \quad (4.44)$$

Now, coming back to Eq. (4.37), it follows by the definition of ∂I_K that

$$\begin{aligned}
&\lambda t^{-1}\big(\partial I_K(\lambda^{-1}(t+\lambda)x^t - t\lambda^{-1}x), x^t - u\big)\\
&\quad\ge -\big(\partial I_K(\lambda^{-1}(t+\lambda)x^t - t\lambda^{-1}x), x^t - x\big)\\
&\quad= -\lambda^{-1}(e^{-At}x^t - x^t, x^t - x) - (e^{-Ft}x^t - x^t, e^{-At}(x^t - x))\\
&\qquad - t\lambda^{-1}(x + \lambda q^t - x^t, x^t - x),
\end{aligned}$$

and by (4.44) we infer that

$$\liminf_{t\to 0} t^{-1}\big(\partial I_K(\lambda^{-1}(t+\lambda)x^t - t\lambda^{-1}x), x^t - u\big) \ge 0.$$

Along with (4.37), (4.41), and (4.43), this yields

$$-\lim_{t\to 0}\,(x + \lambda q^t - x^t, x^t - u) \le \lambda(\phi(u) - \phi(x^0)),$$

where $\phi = \varphi + \varphi_1$. Hence,

$$(x_\lambda^0 - \lambda w(0) - x, x^0 - u) \le \lambda(\phi(u) - \phi(x^0)).$$

Since $\partial\phi = \partial\varphi + \partial\varphi_1$ and u is arbitrary in H, this implies that

$$x + \lambda w(0) - x_\lambda^0 \in \lambda(A + F)x_\lambda^0,$$

i.e., x_λ^0 is the solution to Eq. (4.36), as desired. ∎

Proof of Proposition 4.1. We shall assume first that hypothesis (i) holds. Let $\{u_\varepsilon\}$ be weakly convergent to u in $L^2(0,T;U)$ and let $y_\varepsilon = y_\varepsilon^{u_\varepsilon}$ be the corresponding solution to (4.7), (4.8). Then $\{Bu_\varepsilon\}$ is strongly convergent to Bu in $L^2(0,T;H)$, and if we extend Bu_ε and Bu by 0 on (T,∞), then by Lemma 4.1 the sequence $\tilde{y}_\varepsilon$ defined by

$$[\tilde{y}_\varepsilon(t), \tilde{u}_\varepsilon(t)] = (QS_1(\varepsilon)S_2(\varepsilon))^i(y_0, f + Bu_\varepsilon), \qquad t \in [i\varepsilon, (i+1)\varepsilon], \quad (4.45)$$

or, equivalently

$$\tilde{y}_\varepsilon(t) = P(y_\varepsilon)_-(i\varepsilon) \qquad \forall t \in [i\varepsilon, (i+1)\varepsilon], \tag{4.46}$$

is strongly convergent to $y^u(t)$ for every $t \in [0, T]$.

On the other hand we have, by (4.7) and (4.8),

$$|(y_\varepsilon^u)_-(i\varepsilon) - (y_\varepsilon^u)_+(i\varepsilon)| = |(y_\varepsilon^u)_-(i\varepsilon)_- e^{-F\varepsilon}P(y_\varepsilon^u)_-(i\varepsilon)| \tag{4.47}$$

and

$$(y_\varepsilon^u)_-(i\varepsilon) = e^{-A\varepsilon}e^{-F\varepsilon}P(y_\varepsilon^u)_-((i-1)\varepsilon) + \int_{(i-1)\varepsilon}^{i\varepsilon} e^{-A(i\varepsilon - s)}(Bu + f)\,ds. \tag{4.48}$$

Note also that

$$\int_{i\varepsilon}^{t} |(y_\varepsilon^u)'|^2\,ds + (Ay_\varepsilon^u, y_\varepsilon^u(t))$$
$$\leq (A(y_\varepsilon^u)_+(i\varepsilon), (y_\varepsilon^u)_+(i\varepsilon)) + \int_{i\varepsilon}^{t} \left(|Bu|^2 + |f|^2\right) ds$$

for $t \in [i\varepsilon, (i+1)\varepsilon]$.

Since, by virtue of assumptions (4.3), (4.4),

$$\begin{aligned}(A(y_\varepsilon^u)_+(i\varepsilon), (y_\varepsilon^u)_+(i\varepsilon)) &= \left(Ae^{-F\varepsilon}P(y_\varepsilon^u)_-(i\varepsilon), e^{-F\varepsilon}P(y_\varepsilon)_-(i\varepsilon)\right)\\ &\leq (AP(y_\varepsilon^u)_-(i\varepsilon), P(y^u)_-(i\varepsilon))\\ &\leq (A(y_\varepsilon^u)_-(i\varepsilon), (y^u)_-(i\varepsilon)),\end{aligned}$$

we get

$$\sum_{i=0}^{N-1} \int_{i\varepsilon}^{(i+1)\varepsilon} |(y_\varepsilon^u)'|^2\,ds + (Ay_\varepsilon^u(t), y_\varepsilon^u(t))$$
$$\leq (Ay_0, y_0) + \int_0^T \left(|Bu|^2 + |f|^2\right) dt, \tag{4.49}$$

and in particular it follows that $\{y_\varepsilon\}$ is bounded in $L^\infty(0, T; V)$ and $\{(y_\varepsilon)_-(i\varepsilon)\}$ is compact in H.

On the other hand, since $e^{-A\varepsilon}D(F) \subset D(F)$ $\forall \varepsilon > 0$ (by (4.3)), we see by (4.48) that

$$|(y_\varepsilon)_-(i\varepsilon) - P(y_\varepsilon)_-(i\varepsilon)|$$
$$\leq \int_{(i-1)\varepsilon}^{i\varepsilon} |Bu_\varepsilon + f|\,ds \leq C\varepsilon^{1/2} \qquad \forall i. \tag{4.50}$$

Along with (4.47), this yields

$$\lim_{\varepsilon\to 0} |(y_\varepsilon)_-(i\varepsilon) - (y_\varepsilon)_+(i\varepsilon)| = 0$$

because e^{-Ft} is continuous in t and $\{(y_\varepsilon)_-(i\varepsilon)\}$ is compact in H.

Now, using once again Eq. (4.7), we see that

$$\begin{aligned}|y_\varepsilon(t) - (y_\varepsilon)_+(i\varepsilon)|^2 &\\ \leq C\left(\varepsilon|(y_\varepsilon)_+(i\varepsilon)|^2 + \int_{i\varepsilon}^t (|Bu_\varepsilon|^2 + |f|^2)\, ds\right)& \\ \forall t \in [i\varepsilon, (i+1)\varepsilon],&\end{aligned}$$

and along with (4.50) this implies that

$$y_\varepsilon(t) \to y^u(t) \qquad \text{strongly in } H, \forall t \in [0, T],$$

as claimed.

Now we shall assume that hypothesis (ii) holds, i.e., $F = \partial I_C$. Then, $e^{-tF}P = P = (I + \lambda\, \partial I_C)^{-1}$ for all $t > 0$, $K = C$, and the system (4.7), (4.8) becomes

$$\begin{aligned} y'(t) + Ay(t) &= (Bu)(t) + f(t) \qquad \text{a.e. } t \in (i\varepsilon, (i+1)\varepsilon), \\ y_+(i\varepsilon) &= Py_-(i\varepsilon). \end{aligned} \tag{4.51}$$

Let $u_{\varepsilon_n} \to u$ be weakly convergent in $L^2(0, T; U)$. For simplicity, we set $u_{\varepsilon_n} = u_n$, $\varepsilon_n = \varepsilon$ and $y_{\varepsilon_n}^{u_{\varepsilon_n}} = y_n$. By the estimate (4.49), we have

$$\begin{aligned}\sum_{i=0}^{N-1} \int_{i\varepsilon}^{(i+1)\varepsilon} (|y_n'(t)|^2 + \|y_n(t)\|^2)\, dt& \\ \leq C\left(\|y_0\|^2 + \int_0^T (|u_n|_U^2 + |f|^2)\, dx\right)&\end{aligned} \tag{4.52}$$

and, therefore,

$$\int_0^T |Ay_n(t)|^2\, dt \leq C \qquad \forall n. \tag{4.53}$$

On the other hand, we have

$$\begin{aligned}\sum_{i=0}^{N-1} |(y_n)_+(i\varepsilon) - (y_n)_+(i\varepsilon)|& \\ \leq \sum_{i=0}^{N-1} |P(y_n)_-(i\varepsilon) - e^{-A\varepsilon}P(y_n)_-((i-1)\varepsilon)|& \\ + \int_0^T (|Bu_n(s)| + |f(s)|)\, ds.&\end{aligned} \tag{4.54}$$

Since $e^{-At}C \subset C$ $\forall t \geq 0$, we have

$$\begin{aligned}
&|P(y_n)_-(i\varepsilon) - e^{-A\varepsilon}P(y_n)_-((i-1)\varepsilon)| \\
&\quad = |P(y_n)_-(i\varepsilon) - Pe^{-A\varepsilon}P(y_n)_-((i-1)\varepsilon)| \\
&\quad \leq |(y_n)_-(i\varepsilon) - e^{-A\varepsilon}P(y_n)_-((i-1)\varepsilon)| \\
&\quad \leq \int_{(i-1)\varepsilon}^{i\varepsilon} (|Bu_n| + |f|)\, ds.
\end{aligned}$$

Substituting this in (4.54), we get the estimate

$$\sum_{i=0}^{N} |(y_n)_+(i\varepsilon) - (y_n)_-(i\varepsilon)| \leq 2\int_0^T (|Bu_n| + |f|)\, dt \leq C.$$

Along with (4.52), this yields

$$\bigvee_0^T y_n + \|y_n(t)\| \leq C \qquad \forall n, t \in [0,T],$$

where $\bigvee_0^T y_n$ stands for the variation of $y_n: [0,T] \to H$. Since the injection of V into H is compact we conclude, by virtue of the infinite dimensional Helly theorem, that on a subsequence, again denoted y_n,

$$y_n(t) \to y(t) \qquad \text{strongly in } H, \forall t \in [0,T]. \tag{4.55}$$

By estimate (4.53), it follows that $Ay \in L^2(0,T;\, H)$.

Now, let $z \in C$ be arbitrary but fixed, and let $t \in [k\varepsilon, (k+1)\varepsilon]$, $s \in [i\varepsilon, (i+1)\varepsilon]$, $i < k$, be two points on the interval $[0,T]$. By Eq. (4.51) we get

$$\begin{aligned}
&\tfrac{1}{2}\big(|y_n(t) - z|^2 - |(y_n)_+(k\varepsilon) - z|^2\big) \\
&\quad + \tfrac{1}{2}\sum_{j=1}^{k} \big(|(y_n)_-(j\varepsilon) - z|^2 - |(y_n)_+((j-1)\varepsilon) - z|^2\big) \\
&\quad + \tfrac{1}{2}\big(|(y_n)_-((i+1)\varepsilon) - z|^2 - |y_n(s) - z|^2\big) \\
&\quad = \int_s^t (Bu_n + f - Ay_n, y_n - z)\, d\tau.
\end{aligned}$$

This yields

$$\begin{aligned}
\tfrac{1}{2}|y_n(t) - z|^2 = {}& \int_s^t (Bu_n + f - Ay_n, y_n - z)\, d\tau + \tfrac{1}{2}|y_n(s) - z|^2 \\
&+ \tfrac{1}{2}\sum_{j=i+2}^{k+1} (|(y_n)_+(j\varepsilon) - z|^2 - |(y_n)_-(j\varepsilon) - z|^2.
\end{aligned}$$

On the other hand, we have

$$\begin{aligned}\tfrac{1}{2}\big(|(y_n)_+(j\varepsilon) - z|^2 - |(y_n)_-(j\varepsilon) - z|^2\big) &\le ((y_n)_+(j\varepsilon) - (y_n)_-(j\varepsilon), (y_n)_+(j\varepsilon) - z) \\ &= (P(y_n)_-(j\varepsilon) - (y_n)_-(j\varepsilon), P(y_n)_-(j\varepsilon) - z) \le 0,\end{aligned}$$

because $I - P = \lambda(\partial I_C)_\lambda \in \lambda\, \partial I_C P \ \forall \lambda > 0$. Hence,

$$\tfrac{1}{2}\big(|y_n(t) - z|^2 - |y_n(s) - z|^2\big) \le \int_s^t (Bu_n + f - Ay_n, y_n - z)\, d\tau,$$

and letting n tend to $+\infty$ we get

$$\tfrac{1}{2}\big(|y(t) - z|^2 - |y(s) - z|^2\big) \le \int_s^t (Bu + f - Ay, y - z)\, d\tau \tag{4.56}$$

for all $0 \le s \le t \le T$.

Before proceeding further, let us observe that $y(t) \in C$ a.e. $t \in (0, T)$. Indeed, we have

$$y_n(t) = e^{-A(t - i\varepsilon)} P(y_n)_-(i\varepsilon) + \int_{i\varepsilon}^t e^{-A(t-s)}(Bu_n + f)\, ds \qquad \forall t \in [i\varepsilon, (i+1)\varepsilon],$$

and, therefore,

$$|y_n(t) - Py_n(t)| \le \left|\int_{i\varepsilon}^t e^{-A(t-s)}(Bu_n + f)\, ds\right| \le C\varepsilon^{1/2},$$

because $e^{-At}C \subset C \ \forall t \ge 0$. Hence, $y(t) = Py(t)$, as claimed. Now, in (4.56) take $z = y(s)$. By Gronwall's lemma, we get

$$|y(t) - y(s)| \le \int_s^t |Bu + f + Ay|\, d\tau \qquad \text{for } 0 \le s \le t \le T,$$

and therefore the function $y: [0, T] \to H$ is absolutely continuous and almost everywhere differentiable. On the other hand, by (4.56) we have

$$\begin{aligned}&(y(t) - y(s), y(s) - z) \\ &\quad\le \int_s^t ((Bu)(\tau) + f(\tau) - Ay(\tau), y(\tau) - z)\, d\tau.\end{aligned}$$

Then, dividing by $t - s$ and letting s tend to t, we see that

$$(y'(t) + Ay(t) - (Bu)(t) - f(t), y(t) - z) \leq 0 \qquad \text{a.e. } t \in (0,T),$$

for all $z \in C$. Hence,

$$y'(t) + Ay(t) + \partial I_C(y(t)) \ni (Bu)(t) + f(t) \qquad \text{a.e. } t \in (0,\varepsilon T),$$

i.e., $y = y^u$, as claimed. This completes the proof of Proposition 4.1. ■

Bibliographical Notes and Remarks

Section 1. Theorems 1.1–1.4 along with other related results were established in the author's work [4, 5, 7]. In a particular case, Theorem 1.2 has been previously given by Ch. Saguez [2]. D. Tiba [1] (see also [3] and [4]) has obtained similar results for optimal control problems governed by hyperbolic equations of the form $y_{tt} - \Delta y + \gamma(y_t) \ni Bu$, nonlinear parabolic equations in divergent form, and the nonlinear diffusion equation $y_t - \Delta\beta(y) \ni Bu$ (see also D. Tiba and Zhou Meike [1]). In this context, we also mention the work of S. Aniţa [1] on optimal control of a free boundary problem that models the dynamics of population.

By similar methods and a sharp analysis of optimality system, A. Friedman [3] has obtained the exact description of the optimal controller for the obstacle problem with constraints of the form $\{u \in L^\infty(Q);\ 0 \leq u \leq M, \int_Q u\,dx\,dt = L\}$. Periodic optimal control problems for the two phase Stefan problem were studied by Friedman *et al.* [2]. Theorems 1.1 and 1.2, were extended by Zheng-Xu He [1] to state constraints problems of the form (P) (see also D. Tiba [3]). Numerical schemes for problems of this type were studied by V. Arnăutu [1]. We mention in this context the works of I. Pawlow [1], and M. Niezgodka and I. Pawlow [1].

Section 2. In a slightly different form, Theorems 2.1–2.3 were established first in the author's work [5, 7, 9] (see Friedman [3, 4] for the one phase Stefan problem; see also Ch. Moreno and Ch. Saguez [1]). There is an extensive literature on the inverse Stefan problem and optimal control of moving surfaces, and we refer the reader to the survey of K. H. Hoffmann and M. Niezgodka [1] for references and significant results. A different approach to the inverse Stefan problem that consists of reducing it to a linear optimal control problem in a noncylindrical domain was used in the works of V. Barbu [16], and V. Barbu, G. DaPrato, and J. P. Zolesio [1] (see also V. Arnăutu [2]). The control of the moving boundary of the two phase Stefan problem,

$$y_t - \Delta\beta(y) \ni f \quad \text{in } Q, \qquad \frac{\partial y}{\partial \nu} = 0 \quad \text{in } \Sigma_1, \qquad \frac{\partial y}{\partial \nu} + \alpha y = u \quad \text{in } \Sigma_2,$$

where β is the enthalpy function (see Section 3.3 in Chapter 4) has important industrial applications and was studied by the methods developed here by several authors, including Ch. Saguez [3], D. Tiba [4], D. Tiba and Zhou Meike [1], V. Arnăutu and V. Barbu [1], and D. Tiba and P. Neittaanamäki [1]. The optimal control of the moving boundary of a process modeling growth of a crystal was discussed in the work of Th. Seidman [1, 2].

Section 3. The main results of this section (Theorems 3.1 and 3.4) were established in the author's work [10, 12]. The approach presented in Section 3.3 was first used in the author's work [7, 9] to get first order necessary conditions of optimality for the time-optimal control problem. A different approach involving the Eckeland variational principle was developed by H. O. Fattorini [3].

Section 4. The contents of this section closely follows the author's work [14]. For other related results, we refer to the author's work [15, 19].

Chapter 6 | Optimal Control in Real Time

In this chapter we will be concerned with the feedback representation of optimal controllers to problems studied in the previous chapter. We will see that under quite general conditions such a control is a feedback control of the form $u = \partial h^*(-B^* \partial_y \psi(t, y))$, where ψ is a generalized solution to a certain Hamilton–Jacobi equation associated with the given problem (the dynamic programming equation).

6.1. Optimal Feedback Controllers

6.1.1. Closed Loop Systems

Consider a general control process

$$y'(t) + My(t) \ni Bu(t) + f(t), \qquad t \in [0, T],$$
$$y(0) = y_0, \tag{1.1}$$

in a real Hilbert space H, where $M = \partial\phi$, $\phi: H \to \overline{\mathbf{R}}$ is a lower semicontinuous convex function, $\partial\phi: H \to 2^H$ is the subdifferential of ϕ, $B \in L(U, H)$, and U is another real Hilbert space. Here, $y_0 \in \overline{D(\phi)}$ and $f \in L^2(0, T: H)$.

The control function $u: [0, T] \to U$ is said to be a *feedback control* if it can be represented as a function of the present state of the system (1.1), i.e.,

$$u(t) \in \Lambda(t, y(t)) \qquad \text{a.e. } t \in (0, T), \tag{1.2}$$

where $\Lambda: [0, T] \times H \to U$ is a multivalued mapping. Of course, some continuity and measurability assumptions on Λ are in order. A map $\Lambda: H \to U$ is said to be upper semicontinuous at y from H to U_w if for every weakly open subset D of U satisfying $\Lambda(y) \subset D$ there exists a neighborhood $B(y, \delta)$ of y such that $\Lambda(B(y, \delta)) \subset D$. The multivalued map $\Lambda(t)$: $[0, T] \to U$ is said to be measurable if for each closed subset C of U the set $\{t \in [0, T];\ \Lambda(t) \cap C \neq \emptyset\}$ is Lebesgue measurable. It is easily seen that if $R(\Lambda)$ is bounded in U then Λ is upper semicontinuous from H to U_w, and with weakly closed values if and only if Λ is closed in $H \times U_w$. (Here, U_w is the space U endowed with the weak topology.)

If in the state system (1.1) we replace the control u by the feedback control (1.2), we obtain the *closed loop system*

$$y' + My - B\Lambda(t, y) \ni f(t), \qquad t \in (0, T),$$
$$y(0) = y_0. \tag{1.3}$$

We say that the feedback control Λ is *compatible* with system (1.1) if (1.2) has at least one local solution.

In general, the Cauchy problem (1.2) is not well-posed unless we impose further conditions on the feedback law Λ.

We mention in this direction the following result due to Attouch and Damlamian [1]. More general results of this type can be found in Vrabie's book [1] (see also the monographs of Aubin and Cellina [1] and Filipov [1] for a complete treatment of nonmonotone differential inclusions in $\mathbf{R}^N$).

Proposition 1.1. *Let H be a separable Hilbert space and let $\Lambda_0 = B\Lambda$: $[0, T] \times \overline{D(M)} \to H_w$ be upper semicontinuous in y, measurable in t, and with compact convex values. Assume further that*

(a) *For each $y^0 \in \overline{D(M)}$ there exist $r > 0$ and $h_0 \in L^2(0, T; \mathbf{R}^+)$ such that*

$$\sup\{\|v\|_U;\ v \in \Lambda_0(t, y)\} \le h_0(t) \qquad \text{a.e. } t \in (0, T), \|y - y_0\| \le r$$

and every level set $\{y \in H;\ \phi(y) \le \lambda\}$ is compact in H. (1.4)

Then for each $y_0 \in \overline{D(M)}$ there is $0 < T_0 < T$ such that the Cauchy problem (1.3) *has at least one strong solution y on $[0, T_0]$ that satisfies*

$$y \in C([0, T_0]; H), \qquad y(t) \in D(M) \qquad \text{a.e. } t \in (0, T_0),$$
$$t^{1/2}\frac{dy}{dt} \in L^2(0, T_0; H), \qquad \phi(y) \in L^1(0, T), \qquad t^{1/2}My \in L^2(0, T; H). \tag{1.5}$$

If $y_0 \in D(\phi)$, then $y \in W^{1,2}([0, T_0]; H)$ and $\phi(y) \in AC([0, T_0]; H)$.

This means that there exists one measurable selection $\lambda(t)$ of $B\Lambda(t, y(t))$ such that $\lambda \in L^2(0, T; H)$ and

$$\frac{dy}{dt}(t) + My(t) \ni \lambda(t) + f(t) \qquad \text{a.e. } t \in (0, T).$$

Proof. Denote by $\Psi: L^1(0, T_0; H) \to L^2(0, T_0; H)$ the operator

$$(\Psi z)(t) = y(t) \qquad \text{a.e. } t \in (0, T_0),$$

where y is the solution to the Cauchy problem

$$y'(t) + My(t) = z(t) + f(t) \qquad \text{a.e. } t \in (0, T_0),$$
$$y(0) = y_0. \tag{1.6}$$

Let $D \subset L^2(0, T_0; H) \times L^2(0, T_0; H)$ be the multivalued mapping

$$D = \{[y, v] \in L^2(0, T_0; H) \times L^2(0, T_0; H);\ y(t) \in \overline{D(M)},$$
$$|y(t) - y_0| \le r \text{ a.e. } t \in (0, T_0),$$
$$v(t) \in B\Lambda(t, y(t)) \text{ a.e. } t \in (0, T_0)\}.$$

We have:

Lemma 1.1. *D is upper semicontinuous and with compact convex values from $L^2(0, T_0; H)$ into $L^2_w(0, T_0; H)$. Moreover, $Dy \neq \emptyset$ for all $y \in L^2(0, T_0; H)$, $y(t) \in \overline{D(M)}$, $|y(t) - y_0| \le r$ a.e. $t \in (0, T_0)$.*

(Here, $L^2_w(0, T_0; H)$ is the space $L^2(0, T_0; H)$ endowed with the weak topology.)

Proof. Let $y_n \to y$ strongly in $L^2(0, T_0; H)$ and $v_n \in Dy_n$, $v_n \to v$ weakly in $L^2(0, T_0; H)$. We have, therefore,

$$v_n(t) \in B\Lambda(t, y_n(t)) \qquad \text{a.e. } t \in (0, T).$$

By Mazur's theorem, there is $\{w_m\}$, a finite combination of the v_n, $n \ge m$, such that $w_m \to v$ strongly in $L^2(0, T_0; H)$ as $m \to \infty$. Thus, there is a measurable subset $I \subset (0, T)$ such that $m(I) = T$ and on a subsequence, again denoted n, we have

$$y_n(t) \to y(t) \qquad \text{strongly in } H, \forall t \in I,$$
$$y_n(t) \in D(M), \quad v_n(t) \in B\Lambda(t, y_n(t)), \qquad \forall t \in I,$$
$$w_m(t) \to v(t) \qquad \text{strongly in } H, \forall t \in I.$$

Since $B\Lambda(t, \cdot)$ is upper semicontinuous to U_w, for every weakly neighborhood $\mathscr{V}$ of $B\Lambda(t, y(t))$ there is a neighborhood $\mathscr{U}$ of $y(t)$ such that $B\Lambda(t, x) \subset \mathscr{V}$ for all $x \in \mathscr{U}$. This clearly implies that $v(t) \in B\Lambda(t, y(t))$ $\forall t \in I$.

Now let $y \in L^2(0, T_0; H)$ be such that $y(t) \in \overline{D(A)}$, $|y(t) - y_0| \leq r$ a.e. $t \in (0, T_0)$. Then, the multivalued mapping $t \to B\Lambda(t, y(t))$ is clearly measurable and so, according to a well-known selection result due to C. Castaign (see, e.g., C. Castaign and M. Valadier [1]) it has a measurable selection, which by condition (1.4) is in $L^2(0, T_0; H)$. ■

Now we come back to operator Ψ previously defined.

For any $\delta \in L^2(0, T_0; H)$, denote by x_δ the set $\{z \in L^2(0, T_0; H);$ $|z(t)| \leq \delta(t)$ a.e. $t \in (0, T)\}$.

Lemma 1.2. *The operator Ψ defined on X_δ is continuous from $L^2_w(0, T_0;$ $H)$ to $L^2(0, T_0; H)$.*

Proof. Let $z_n \in X$ be weakly convergent to z in $L^2(0, T_0; H)$ and denote by y_n the corresponding solutions to (1.6). We have the estimates (see Section 1.5, Chapter 4)

$$\int_0^{T_0} t|y_n'(t)|^2\, dt + t\phi(y_n(t)) + \int_0^{T_0} \phi(y_n(t))\, dt \leq C,$$

and by the Arzelà–Ascoli theorem we infer that $\{y_n\}$ is compact in $C([\varepsilon, T_0]; H)$ for every $\varepsilon > 0$. Hence, on a subsequence, we have

$$y_n(t) \to y(t) \qquad \text{strongly in } H, \forall t \in [0, T_0],$$

uniformly on every interval $[\varepsilon, T_0]$, $\varepsilon > 0$. By the Lebesgue dominated convergence theorem, it follows that $y \in L^2(0, T_0; H)$ and $y_n \to y$ strongly in $L^2(0, T_0; H)$. By standard arguments, this implies that $y = \Psi z$ is the solution to (1.6), as claimed. ■

Proof of Proposition 1.1 (continued). We may write problem (1.3) as

$$y \in \Psi D(y), \quad |y(t) - y_0| \leq r \qquad \text{a.e. } t \in (0, T_0)$$

or, equivalently,

$$w \in D\Psi w \qquad w \in X_\delta, \tag{1.7}$$

where $\delta = \|B\|_{L(U, H)} h_0$ (see assumption (3.3)).

By Lemmas 1.1 and 1.2, the operator $D\Psi$ is upper semicontinuous on $L^2_w(0, T_0; H)$, has compact convex values and X_δ is a compact subset of $L^2_w(0, T_0; H)$. Moreover, $D\Psi$ maps X_δ into itself if T_0 is sufficiently small. Indeed, for $z \in X_\delta$ we have, by Eq. (1.6),

$$\begin{aligned} &\tfrac{1}{2}\left(|y(t) - y_0|^2\right)' + \phi(y(t)) - \phi(y_0) \\ &\qquad \le (w(t) + f(t), y(t) - y_0) \qquad \text{a.e. } t \in (0, T_0), \end{aligned}$$

and this yields

$$|y(t) - y_0| \le C\left(t + \int_0^t (\delta(s) + |f(s)|)^2\, ds\right) \qquad \forall t \in [0, T_0].$$

(We assume first that $y_0 \in D(\phi)$.) Hence, for T_0 sufficiently small we have

$$|\Psi w(t) - y_0| \le r \qquad \forall t \in [0, T_0],$$

and so by assumption (1.4) it follows that $D\Psi w \in X_\delta$, as claimed.

If $y_0 \in \overline{D(\phi)} = \overline{D(M)}$, the same conclusion follows by density. Then, by the Kakutani theorem (see Theorem 2.2 in Chapter 1) in the space $L^2_w(0, T_0; H)$, we infer that the operator $D\Psi$ has at least one fixed point $w \in X_\delta$. Equivalently, there is $y = \Psi w$ such that

$$w = Dy \qquad \text{in } L^2(0, T_0; H).$$

By the definitions of D and Ψ we see that y is a solution to the Cauchy problem (1.3).

We note that the conditions (1.5) follow by Theorem 1.10 in Chapter 4. ■

Consider now the optimal control problem:

$$\text{Minimize} \quad \left\{\int_0^T (g(t, y(t)) + h(u(t)))\, dt + \varphi_0(y(T))\right\}, \tag{1.8}$$

subject to

$$\begin{aligned} y' = \partial\phi(y) &\ni Bu + f \qquad \text{a.e. } t \in (0, T), \\ y(0) &= y_0, \end{aligned} \tag{1.9}$$

where the functions $g: [0, T] \times H \to \mathbf{R}$, $\varphi_0: H \to \mathbf{R}$ and $h: U \to \overline{\mathbf{R}}$ satisfy conditions (v), (vi) from Section 1.1 in Chapter 5.

An optimal control u for problem (1.8), (1.9) that is represented in the feedback form (1.2) is called *optimal feedback control*.

We shall see in the sequel that for a quite general class of optimal control problems of the form (1.8) every optimal control can be represented as a feedback control, and the synthesis function Λ can be easily described in terms of the solutions to the optimality system associated with (1.8), (1.9).

To be more specific, we will consider the optimal control problem (1.8) governed by the semilinear parabolic equation

$$\begin{aligned} \frac{\partial y}{\partial t} - \Delta y + \beta(y) &= Bu + f && \text{in } Q = \Omega \times (0,T), \\ y &= 0 && \text{in } \Sigma = \partial\Omega \times (0,T), \\ y(x,0) &= y_0(x) && \text{in } \Omega_0. \end{aligned} \tag{1.10}$$

where β is a locally Lipschitz real valued function that satisfies the condition

$$0 \le \beta'(r) \le C(|\beta(r)| + |r| + 1) \qquad \text{a.e. } r \in \mathbf{R}. \tag{1.11}$$

As seen earlier this is a problem of the form (1.9), where $H = L^2(\Omega)$ and $\partial\phi(y) = -\Delta y + \beta(y)$ $\forall y \in D(\phi) = \{z \in H_0^1(\Omega) \cap H^2(\Omega);\ \beta(z) \in L^2(\Omega)\}$. It is readily seen that $D(\partial\phi)$ is dense in $L^2(\Omega)$. Indeed, for any $z \in H_0^1(\Omega)$ we have $(1 + \varepsilon\beta)^{-1}z \in D(\partial\phi)$ (we may assume $\beta(0) = 0$) and

$$\begin{aligned} (1 + \varepsilon\beta)^{-1} z(x) &\to z(x) && \text{a.e. } x \in \Omega \text{ for } \varepsilon \to 0, \\ |(1 + \varepsilon\beta)^{-1} z(x)| &\le |z(x)| && \text{a.e. } x \in (\Omega). \end{aligned}$$

Hence, $(1 + \varepsilon\beta)^{-1}z \to z$ strongly in $L^2(\Omega)$ as $\varepsilon \to 0$.

Define the map $\Gamma: [0,T] \times H \to H$ by

$$\Gamma(t, z_0) = \{p^t(t)\}, \qquad t \in [0,T],\ z_0 \in L^2(\Omega), \tag{1.12}$$

where (y^t, u^t, p^t) satisfy the system

$$\begin{aligned} &\frac{\partial}{\partial s} y^t - \Delta y^t + \beta(y^t) = Bu^t + f && \text{in } Q^t = \Omega \times (t,T), \\ &\frac{\partial}{\partial s} p^t + \Delta p^t - \partial\beta(y^t)p^t \in \partial g(s, y^t) && \text{in } Q^t, \\ &y^t(x,t) = z_0(x), \quad p^t(x,T) \ni -\partial\varphi_0(y(x,T)) && \text{in } \Omega, \\ &y^t = p^t = 0 \quad \text{in } \Sigma^t = \partial\Omega \times (t,T), \end{aligned} \tag{1.13}$$

$$B^* p^t(s) \in \partial h(u^t(s)) \qquad \text{a.e. } s \in (t,T), \tag{1.14}$$

and $(y^t, u^t) \in C([t,T];\ H) \times L^2(t,T;\ U)$ is a solution to optimization problem

$$\text{Minimize} \quad \left\{ \int_t^T (g(s, y(s)) + h(u(s)))\, ds + \varphi_0(y(T)), \right. \tag{1.15}$$

subject to $u \in L^2(t,T;\ U)$ and

$$\begin{aligned} &\frac{\partial y}{\partial s} - \Delta y + \beta(y) = Bu + f \qquad \text{in } Q^t, \\ &y(x,t) = z_0(x) \qquad \text{in } \Omega, \qquad y = 0 \qquad \text{in } \Sigma^t. \end{aligned} \tag{1.16}$$

As seen in Proposition 1.1 and Theorem 1.1 in Chapter 5, for every $(t, z_0) \in [0,T] \times L^2(\Omega)$ there are y^t, u^t, and $p^t \in AC([t,T];\ Y^*) \cap C_w([t,T];\ L^2(\Omega)) \cap L^2(t,T;\ H_0^1(\Omega))$ satisfying Eqs. (1.13), (1.14). Hence, Γ is well-defined on $[0,T] \times L^2(\Omega)$.

Theorem 1.1. *Every optimal control u^* for problem* (1.8), (1.10) *has the feedback representation*

$$u^*(t) \in \partial h^*(B^*\Gamma(t, y^*(t))) \qquad \text{a.e. } t \in (0,T), \tag{1.17}$$

where y^* is the corresponding optimal state.

Theorem 1.1 amounts to saying that every optimal control of problem (1.8), (1.10) is an optimal feedback control with the synthesis function

$$\Lambda(t,y) = \partial h^*(B^*\Gamma(t,y)) \qquad \forall (t,y) \in [0,T] \times L^2(\Omega). \tag{1.18}$$

Proof. Let (y^*, u^*) be any optimal pair for problem (1.8), (1.10). Then, obviously, (y^*, u^*) is also optimal for the problem:

$$\begin{aligned} &\text{Minimize} \quad \left\{ \int_t^T (g(s, y(s)) + h(u(s)))\, ds + \varphi_0(y(T)), \right. \qquad \text{subject to} \\ &\frac{\partial}{\partial s} y - \Delta y + \beta(y) = Bu + f \qquad \text{in } \Omega^t, \\ &y(x,t) = y^*(x,t) \qquad \text{in } \Omega, \qquad y = 0 \qquad \text{in } \Sigma^t, \end{aligned} \tag{1.19}$$

and so by Theorem 1.1 in Chapter 5 we infer that there are $p^t \in AC([t,T];Y^*) \cap C_w([t,T];\ L^2(\Omega)) \cap L^2(t,T;\ H_0^1(\Omega))$ such that

$$\begin{aligned} &\frac{\partial}{\partial s} p^t + \Delta p^t - \partial\beta(y^*) p^t \in \partial g(s, y^*) \qquad \text{in } \varphi^t, \\ &p^t(x,t) \in -\partial\varphi_0(y^*(x,T)), \qquad x \in \Omega, \qquad p^t = 0 \qquad \text{in } \Sigma^t, \end{aligned} \tag{1.20}$$

$$u^*(s) \in \partial h^*(B^* p^t(s)) \qquad \text{a.e. } s \in (t,T). \tag{1.21}$$

The function u^* being measurable, it is a.e. approximately continuous on $[0, T]$. This means that for almost all $t \in (0, T)$ there is a measurable set $E_t \subset (0, T)$ such that t is a density point for E_t and $u^*|_{E_t}$ is continuous at t.

Let us denote by E^t the set of all $s \in [t, T]$ for which (1.21) holds. Since, for almost all $t \in (0, T)$, t is a density point for $E^t \cap E_t$ there is a sequence $s_n \to t$ such that

$$u^*(s_n) \to u^*(t) \qquad \text{and} \qquad u^*(s_n) \in \partial h^*(B^* p^t(s_n)).$$

Since p^t is weakly continuous on $[t, T]$, we have

$$B^* p^t(s_n) \to B^* p^t(t) \qquad \text{weakly in } L^2(\Omega),$$

and ∂h^* being weakly–strongly closed in $U \times U$ we conclude that $u^*(t) \in \partial h^*(B^* p^t(t))$ a.e. $t \in (0, T)$. In other words, we have shown that

$$u^*(t) \in \partial h^*(B^* \Gamma(t, y^*(t))) \qquad \text{a.e. } t \in (0, T),$$

thereby completing the proof. ■

Regarding the properties of the *synthesis function* Λ, we have:

Proposition 1.2. *For each $t \in [0, T]$, the map $\Gamma(t, \cdot)$: $L^2(\Omega) \to L^2_{\mathrm{w}}(\Omega)$ is upper semicontinuous and bounded on bounded subsets. For each $y_0 \in L^2(\Omega)$, $\Gamma(\cdot, y_0)$: $[0, T] \to L^2(\Omega)$ is measurable.*

Proof. It is easily seen that $\Gamma(t, \cdot)$ is bounded on every bounded subset. We assume that Γ is not upper semicontinuous from $L^2(\Omega)$ to $L^2_{\mathrm{w}}(\Omega)$, and argue from this to a contradiction. This would imply that there are $y_0 \in L^2(\Omega)$, $y_n^0 \to y_0$ strongly in $L^2(\Omega)$, and $\eta_n^0 \in \Gamma(t, y_n^0)$ such that $\eta_n^0 \notin D$ for all n. Here, D is a weakly open subset of $L^2(\Omega)$ and

$$\Gamma(t, y_0) \subset D. \tag{1.22}$$

By the definition of Γ there are (y_n^t, u_n^t, p_n^t) satisfying the system (1.13), (1.14), and $p_n^t(t) = \eta_0^n$, $y_n^t(t) = y_0^n$. By (1.15, (1.16) it is easily seen that $\{y_n^t\}$ is bounded in $C([t, T]; L^2(\Omega))$ and $\{t^{1/2}\, dy_n^t/ds\}$ is bounded in $L^2(t, T; L^2(\Omega))$. Thus, on a subsequence, again denoted $\{n\}$, we have

$$
\begin{aligned}
u_n^t &\to u^t && \text{weakly in } L^2(t, T; U),\\
y_n^t(s) &\to y^t(s) && \text{strongly in } H, \forall s \in [t, T],\\
y_n^t &\to y^t && \text{weakly in } W^{1,2}([\delta, T]; H)\ \forall \delta \in (0, T),
\end{aligned}
$$

where y^t is the solution to (1.13) with the Cauchy condition $y^t(t) = y_0$. Then, by the estimates proved in Proposition 1.2, Chapter 5, we know that

(see also Theorem 1.1 in Chapter 5)

$$p_n^t \to p^t \qquad \text{strongly in } L^2(t,T;\, L^2(\Omega)),$$
$$\qquad\qquad \text{weak star in } L^\infty(t,T;\, L^2(\Omega)),$$

and

$$p_n^t(s) \to p^t(s) \qquad \text{strongly in } Y^*, \forall s \in [t,T],$$

where p^t is the solution to (1.13), (1.14).

Since $\{p_n^t(t)\}$ is bounded in $L^2(\Omega)$ and $p_n^t(t) \to p^t(t)$ strongly in $Y^* \supset L^2(\Omega)$, we conclude that

$$p_n^t(t) \to p^t(t) \qquad \text{weakly in } L^2(\Omega), \forall t \in [0,T].$$

Hence, $\eta_n^0 \to p^t(t) \in \Gamma(t, y_0)$ weakly in $L^2(\Omega)$, which by virtue of (1.22) implies that $\eta_n^0 \in D$ for all n sufficiently large. The contradiction at which we have arrived completes the proof. ■

Let $y_0 \in L^2(\Omega)$ be arbitrary but fixed. Then by a similar argument it follows that the multivalued function $t \to \Gamma(t, y_0) = p^t(t)$ is weakly upper semicontinuous from $[0, T]$ to $L^2(\Omega)$, which clearly implies that $(\Gamma(\cdot, y_0))^{-1}(C)$ is a closed subset of $[0, T]$ for each closed subset C of $L^2(\Omega)$, i.e., $\Gamma(\cdot, y_0)$ is measurable, as claimed.

Theorem 1.1 provides a simple method to compute the synthesis function Λ associated with problem (1.8), (1.9) by decoupling the corresponding optimality system. However, since in general the multivalued mapping (1.18) is neither upper semicontinuous nor has convex values, the existence result given by Proposition 1.1 is not applicable to closed loop system (1.2) and so it is not clear if the corresponding feedback control (1.17) is compatible with system (1.10).

However, if $h(u) = \frac{1}{2}|u|_U^2$ we may replace u by the relaxed feedback control

$$u = B^* \overline{\operatorname{conv} \Gamma(t, y)},$$

which by virtue of Propositions 1.1 and 1.2 can be implemented into system (1.10).

Let us observe that if, in the problem (1.8, (1.10), $g(t, \cdot\,)$, φ_0, and β are continuously differentiable with Lipschitz derivatives and if ∂h^* is Lipschitz, then by standard fixed point arguments it follows that for $|t - T| \le \delta$ sufficiently small, the system (1.13), (1.14) has a unique solution (y^t, p^t). Hence, for T sufficiently small, $\Gamma(t, \cdot\,)$ is single valued and continuous on a given bounded subset of $L^2(\Omega)$. Then the feedback control (1.17) is continuous, and so by Proposition 1.1 it is compatible with the system (1.10). Now, if we approximate g, φ_0, and h by g^ε, $(\varphi_0)^\varepsilon$, β^ε, and $h_\varepsilon(u) = h(u) + \varepsilon|u|_U^2$, respectively, we may construct an approximating feedback control compatible with the system (1.10) on a sufficiently small interval of time.

Remark 1.1. If β is merely locally Lipschitz or in the case of the obstacle problem, we may define the map Γ as

$$\Gamma(t, z_0) = p^t(t + 0),$$

where (y^t, p^t) satisfy the corresponding optimality system on $[t, T]$, and $p^t(t + 0) = \mathrm{w} - \lim_{s \to t, s > t} p^t(s)$ in $L^2(\Omega)$.

6.1.2. *The Optimal Value Function*

We shall consider here problem (P) studied in Chapter 5, i.e.,

$$\text{(P)} \qquad \text{Minimize} \int_0^T (g(y(t)) + h(u(t)))\,dt + \varphi_0(y(T))$$

on all $(y, u) \in C([0, T]; H) \times L^2(0, T; U)$, *subject to*

$$\frac{dy}{dt}(t) + Ay(t) + \partial\varphi(y(t)) \ni Bu(t) \qquad \text{a.e. } t \in (0, T),$$
$$y(0) = y_0, \tag{1.23}$$

where A is a linear, continuous, and symmetric operator from V to V' satisfying the coercivity condition

$$(Ay, y) \geq \omega\|y\|^2 \qquad \forall y \in V, \tag{1.24}$$

and $\varphi: H \to \overline{\mathbf{R}}$ is a lower semicontinuous convex function such that

$$(Ay, \partial\varphi_\varepsilon(y)) \geq -C\big(1 + |\partial\varphi_\varepsilon(y)|^2\big)(1 + |y|) \qquad \forall y \in D(A_H), \tag{1.25}$$

where $D(A_H) = \{y \in V;\ Ay \in H\}$ and $\partial\varphi_\varepsilon = \varepsilon^{-1}(I - (I + \varepsilon\,\partial\varphi)^{-1}$, $\varepsilon > 0$.

Here, H and V are real Hilbert spaces such that $V \subset H \subset V'$ algebraically and topologically, and the injection of V into H is compact.

We shall denote as usual by $(\cdot, \cdot)$ the scalar product of H, and also the pairing between V and V', and by $|\cdot|$ (respectively, $\|\cdot\|$) the norm of H (respectively, V).

As seen earlier, the operator

$$\partial\phi = A_H + \partial\varphi, \qquad A_H u = Au \cap H,$$

is maximal monotone in $H \times H$, and

$$\phi(y) = \tfrac{1}{2}(Ay, y) + \varphi(y), \qquad \overline{D(\phi)} = \overline{D(\varphi)}.$$

Regarding the functions $g: H \to \mathbf{R}$, $\varphi_0: H \to \mathbf{R}$, and the operator $B: U \to H$, we shall assume that hypotheses (iv), (v), and (vi) or (vi)′ of Section 1.1 in Chapter 5 are satisfied.

As noted earlier, for every $y_0 \in \overline{D(\phi)}$ the Cauchy problem (1.23) has a unique solution $y \in C([0,T]; H) \cap L^2(0,T; V)$ with $t^{1/2}\, dy/ds \in L^2(0,T; H)$, $t^{1/2}A_H y \in L^2(0,T; H)$. For $0 < t < s < T$ and $y_0 \in \overline{D(\phi)}$, denote by $y(s,t,y_0)$ the solution to the Cauchy problem

$$\frac{dy}{ds}(s) + \partial\phi(y(s)) \ni Bu(s) \qquad \text{a.e. } s \in (t,T), \qquad y(t) = y_0. \quad (1.26)$$

The function $\psi: [0.T] \times \overline{D(\phi)} \to \mathbf{R}$,

$$\psi(t,y_0) = \inf\left\{\int_t^T (g(y(s,t,y_0,u)) + h(u(s)))\, ds \right.$$
$$\left. + \varphi_0(y(T,t,y_0,u));\ u \in L^2(t,T;U)\right\} \quad (1.27)$$

is called the *optimal value function* of problem (P).

Proposition 1.3. *For every $(t,y_0) \in [0,T] \times \overline{D(\phi)}$ the infimum defining $\psi(t,y_0)$ is attained. For each $t \in [0,T]$, $\psi(t,\cdot)$ is locally Lipschitz on $\overline{D(\phi)}$, and for each $y_0 \in D(\partial\phi)$, $\psi(\cdot,y_0)$ is Lipschitz on $[0,T]$. Moreover, ψ is continuous in t and y on $[0,T] \times \overline{D(\phi)}$.*

Proof. By Proposition 1.1 in Chapter 5 we know that the infimum defining $\psi(t,y_0)$ is attained. Since $\partial\phi$ is monotone, we have

$$|y(s,t,y_0,u) - y(s,t,z_0,u)| \le |y_0 - z_0| \qquad \forall s \in [t,T]. \quad (1.28)$$

Now if multiply Eq. (1.23) by $y(s) - y^0$, where $y^0 \in D(\partial\phi)$, and integrate on $[t, s]$, we get

$$|y(s,t,y_0,u)| \le |y_0| + C\left(\int_t^s |u(\tau)|_U \, d\tau + 1\right) \qquad \forall s \in [t,T], \quad (1.29)$$

where C is independent of t and s.

Now let $y_0 \in \overline{D(\phi)}$ be such that $|y_0| \le r$. We have

$$\begin{aligned} \psi(t,y_0) &\le \int_t^T (g(y(s,t,y_0,0)) + h(0))\, ds + \varphi_0(y(T,t,y_0,0)) \\ &\le C_r. \end{aligned} \quad (1.30)$$

(We may assume that $0 \in D(h)$.) By virtue of assumption (v) we may therefore restrict the optimization problem (1.24) to the class $\mathscr{M}_r$ of $u \in L^2(t,T;\, U)$ satisfying the condition

$$\int_t^T |u(\tau)|_U^2 \, d\tau \le C_r^1 .$$

Then, by estimates (1.28), (1.29), it follows that for $u \in \mathscr{M}_r$ the function $y_0 \to \int_t^T g(y(s,t,y_0,u)\, ds + \varphi_0(y,T,t,y_0,u))$ is Lipschitz on $B_r = \{y \in \overline{D(\partial\phi)};\ |y| \le r\}$ with the Lipschitz constant independent of $u \in \mathscr{M}_r$. Since

$$\psi(t,y_0) - \psi(t,z_0) \le \int_t^T (g(y(s,t,y_0,u)) - g(y(s,t,z_0,u)))\, ds$$

for $y_0, z_0 \in B_r$ and some $u \in \mathscr{M}_r$ such that

$$\psi(t,z_0) = \int_t^T (g(y(s,t,z_0,u)) + h(u))\, ds + \varphi_0(y(T,t,z_0,u)),$$

we obtain that

$$|\psi(t,y_0) - \psi(t,z_0)| \le L_r |y_0 - z_0| \qquad \forall y_0, z_0 \in B_r,$$

as claimed.

On the other hand, for every $y_0 \in D(\partial\phi)$ we have

$$\begin{aligned} &|y(s,\tilde{t},y_0,u) - y(s,t,y_0,u)| \\ &\quad \le |y(t,\tilde{t},y_0,u) - y_0 + \int_{\tilde{t}}^t |Bu(\tau)| d\tau + |\partial\phi(y_0)|\,|t - \tilde{t}|. \end{aligned} \quad (1.31)$$

Now let us observe that without any loss of generality we may assume that, for $y_0 \in B_r$, $\mathscr{M}_r \subset \{u \in L^\infty(t,T;\, U);\ |u(\tau)|_U \le C_r\}$, where C_r is independent of t. Indeed, if u^t is a minimum point for the functional (1.27) then by

(1.29) we see that the corresponding optimal state y^t has an independent bound in $C([t,T]; H)$. On the other hand,

$$u^t(s) \in \partial h^*(B^*p^t(s)) \qquad \text{a.e. } s \in (t,T),$$

where $p^t \in L^\infty(t,T; H)$ and

$$|p^t(s)| \le C_r^2 \qquad \forall s \in [t,T].$$

This follows as in the proof of Proposition 1.2 in Chapter 5, approximating problem (1.27) by a family of smooth control problems and taking the limit of the corresponding optimality system,

$$\frac{dy_\varepsilon}{ds} + Ay + \nabla\varphi^\varepsilon(y_\varepsilon) = Bu_\varepsilon \qquad \text{in } (t,T),$$

$$\frac{dp_\varepsilon}{ds} - Ap_\varepsilon - \nabla^2\varphi^\varepsilon(y_\varepsilon)p_\varepsilon = \nabla g^\varepsilon(y_\varepsilon) \qquad \text{in } (t,T),$$

$$y_\varepsilon(t) = y_0, \qquad p_\varepsilon(T) = -\nabla(\varphi_0)^\varepsilon(y_\varepsilon(T)).$$

Since, by assumption (v), ∂h^* is bounded on bounded subsets, we conclude that

$$|u^t(s)|_U \le C_r^3 \qquad \forall y_0 \in Br,\ s \in [t,T],$$

and so by (1.31) we have

$$|y(s,\tilde{t},y_0,u) - y(s,t,y_0,u)| \le C_r^4|t - \tilde{t}| \qquad \forall u \in \mathscr{M}_r, \qquad (1.32)$$

for all $t \le \tilde{t} \le s \le T$. Now let $u_t \in L^2(t,T; H)$ and $y_t = y(s,t,y_0,u_t)$ be such that

$$\psi(t,y_0) = \int_t^T (g(y_t(s)) + h(u_t(s)))\,ds + \varphi_0(y_t(T)),$$

and let $v(s) = u_0$ for $s \in [\tilde{t},t]$, and $v(s) = u_t(s)$ for $s \in [t,T]$, where $h(u_0) < \infty$. We have

$$\begin{aligned}\psi(\tilde{t},y_0) - \psi(t,y_0) \le& \int_{\tilde{t}}^t (g(y(s,\tilde{t},y_0,v)) + h(u_0))\,ds\\ &+ \int_t^T (g(y(s,\tilde{t},y_0,v)) - g(y(s,t,y_0,v)))\,ds\\ &+ \varphi_0(y(T,\tilde{t},y_0,v)) - \varphi_0(y(T,t,y_0,v)),\end{aligned}$$

which along with the estimate (1.32) yields

$$|\psi(\tilde{t},y_0) - \psi(t,y_0)| \le L|t - \tilde{t}| \qquad \forall t,\tilde{t} \in [0,T].$$

(We note that L depends on r ($y_0 \in Br \cap D(\partial\phi)$).) ■

Now we shall prove the *dynamic programming principle* for problem (P).

Lemma 1.3. *For all $t \in [0,T]$, $s \in [t,T]$, and $y_0 \in \overline{D(\phi)}$, we have*

$$\psi(t,y_0) = \inf\left\{\int_t^s (g(y(\tau,t,y_0,u)) + h(u(\tau)))\,d\tau \right.$$
$$\left. + \psi(s, y(s,t,y_0,u)); u \in L^2(t,s;U)\right\}. \tag{1.33}$$

Proof. Let $(y,u) \in C([t,T]; H) \times L^2(t,T; U)$ be such that $y = y(s,t,y_0,u)$ and

$$\psi(t,y_0) = \int_t^T (g(y(\tau)) + h(u(\tau)))\,d\tau + \varphi_0(y(T))$$
$$= \int_t^s (g(y(\tau)) + h(u(\tau)))\,d\tau$$
$$+ \int_s^T (g(y(\tau)) + h(u(\tau)))\,d\tau + \varphi_0(y(T)).$$

This yields

$$\psi(t,y_0) \geq \psi(s,y(s)) + \int_t^s (g(y(\tau)) + h(u(\tau)))\,d\tau.$$

Hence,

$$\psi(t,y_0) \geq \inf\left\{\int_t^s (g(y(\tau,t,y_0,u)) + h(u(\tau)))\,d\tau \right.$$
$$\left. + \psi(s, y(s,t,y_0,u)); u \in L^2(t,T;U)\right\}.$$

On the other hand, for all $u \in L^2(t,T; U)$ and $y(s) = y(s,t,y_0,u)$, we have

$$\psi(t,y_0) \leq \int_t^s (g(y(\tau)) + h(u(\tau)))\,d\tau$$
$$+ \int_s^T (g(y(\tau)) + h(u(\tau))\,d\tau + \varphi_0(y(T)).$$

We may choose y and u in a such a way that

$$\psi(s,y(s)) = \int_0^T (g(y(\tau)) + h(u(\tau)))\,d\tau + \varphi_0(y(T)),$$

and so

$$\psi(t, y_0) \leq \inf\left\{\int_t^s (g(y(\tau, t, y_0, u)) + h(u(\tau)))\, d\tau \right.$$
$$\left. + \psi(s, y(s, t, y_0, u)); u \in L^2(t, T; U)\right\},$$

thereby completing the proof. (The continuity of ψ is obvious.) ■

6.1.3. *The Dynamic Programming Equation*

As in the classical theory of calculus of variations, we associate with the problem (P) the Hamilton–Jacobi equation

$$\psi_t(t, y) - h^*(-B^*\psi_y(t, y)) - (\partial\phi(y), \psi_y(t, y)) + g(y) = 0 \quad \text{in } [0, T] \times H$$
$$\psi(T, y) = \varphi_0(y) \qquad \forall y \in H. \tag{1.34}$$

where ψ_t and ψ_y represent the partial derivative of ψ with respect to t and y, whilst h^* is the conjugate of h.

Equation (1.34) is called the *dynamic programming equation* corresponding to problem (P).

In general, this equation does not have a solution in the classical sense even if the space H is finite dimensional and $\partial\phi$, g, and φ_0 are smooth. However, under supplementary conditions on problem (P) the optimal value function ψ satisfies Eq. (1.34) in some generalized sense.

Let us denote by $D_y^+\psi(t, y)$ the *superdifferential* of $\psi(t, \cdot)$ at y, i.e., the set of all $\eta \in H$ such that

$$\limsup_{z \to y} \{(\psi(t, z) - \psi(t, y) - (\eta, z - y))|z - y|^{-1}\} \leq 0. \tag{1.35}$$

Similarly, $D_y^-\psi(t, y)$, the *subdifferential* of $\psi(t, \cdot)$ at y, is defined as the set of all $w \in H$ such that

$$\liminf_{z \to y} \{(\psi(t, z) - \psi(t, y) - (w, z - y))|z - y|^{-1}\} \geq 0. \tag{1.35$'$}$$

We will assume now that:

(a) A is a linear, continuous and symmetric operator from V to V' satisfying condition (1.24);

(b) $V \subset H \subset V'$ algebraically and topologically; the injection of V into H is compact;

(c) The operator $F = \partial\varphi: H \to H$ is Fréchet differentiable on H with locally Lipschitz Fréchet differential ∇F;

(d) The functions g and φ_0 are continuously differentiable on H with locally Lipschitz derivatives ∇g and $\nabla\varphi_0$, respectively.

(e) The function $h: U \to \overline{\mathbf{R}}$ is convex, lower semicontinuous, and

$$h(u) \geq \alpha|u|_U^2 + \gamma \qquad \forall u \in U,$$

where $\alpha > 0$. Moreover, the conjugate function h^* is differentiable with locally Lipschitz derivative ∇h^*.

Theorem 1.2. *Let assumptions* (a)–(e) *be satisfied. Then the optimal value function* $\psi: [0,T] \times H \to H$ *is continuous, locally Lipschitz in* y *for every* $t \in [0,T]$, *Lipschitz in* t *for every* $y \in D(A_H)$, $D_y^+ \psi(t,y) \neq \emptyset$ $\forall (t,y) \in [0,T] \times H$, *and*

$$\begin{aligned} \psi_t(t,y) - h^*(-B^*\eta) - (A_H y + Fy, \eta) + g(y) &= 0 \\ &\text{a.e. } t \in (0,T), \forall y \in D(A_H), \\ \psi(T,y) = \varphi_0(y) \qquad \forall y \in H, & \qquad (1.36) \end{aligned}$$

where $\eta \in D_y^+ \psi(t,y)$.

More precisely, Eq. (1.36) holds for all (t,y) for which $\psi(\cdot,y)$ is differentiable at t.

Roughly speaking, Theorem 1.2 amounts to saying that ψ satisfies the Hamilton–Jacobi equation (1.34) in the following weak sense:

$$\begin{aligned} \psi_t(t,y) - h^*(-B^* D_y^+ \psi(t,y) - \big(A_H y + Fy, D_y^+ \psi(t,y)\big) + g(y) = 0 \\ \text{a.e. } t \in (0,T),\ y \in D(A_H). \quad (1.36)' \end{aligned}$$

Regarding the optimal feedback controllers, we have:

Theorem 1.3. *Under assumptions* (a)–(e), *every optimal control* u^* *of problem* (P) *is expressed as a function of the optimal state* y^* *by the feedback law*

$$u^*(t) \in \nabla h^*(-B^* D_y^+ \psi(t, y^*(t)) \qquad \forall t \in [0,T]. \qquad (1.37)$$

Proof of Theorem 1.2. Let $(t, y_0) \in [0,T] \times D(A_H)$ be arbitrary but fixed. Then, as seen in Chapter 5 (Proposition 1.1), there are $y^t \in C([t,T]; H)$

and $u^t \in L^2(t, T; U)$ such that

$$\psi(t, y_0) = \int_t^T (g(y^t(s)) + h(u^t(s)))\, ds + \varphi_0(y^t(T)) \tag{1.38}$$

and

$$\frac{d}{ds} y^t(s) + A_H y^t(s) + F y^t(s) = B u^t(s) \qquad \text{a.e. } s \in (t, T),$$
$$y^t(t) = y_0 . \tag{1.39}$$

Then, by the maximum principle, there is $p^t \in C([t, T]; H)$ such that

$$\frac{d}{ds} p^t(s) - A_H p^t(s) - \nabla F(y^t(s)) p^t(s) = \nabla g(y^t(s)), \qquad s \in [t, T],$$
$$p^t(T) = -\nabla \varphi_0(y^t(T)), \tag{1.40}$$
$$u^t(s) = \nabla h^*(B^* p^t(s)) \qquad \forall s \in (t, T). \tag{1.41}$$

Since $y^t \in W^{1,2}([t, T]; H)$, the functions $s \to \nabla g(y^t(s))$ and $s \to \nabla F(y^t(s))$ are Hölder continuous. Moreover, inasmuch as $p^t \in W^{1,2}([t, \delta]; H)$ for every $t < \delta < T$, we conclude that $s \to \nabla g(y^t(s)) + \nabla F(y^t(s))\, p^t(s)$ is Hölder continuous on every interval $[t, \delta] \subset [t, T]$, and so p^t is a classical solution to Eq. (1.40) on every $[t, \delta]$ (see Theorem 4.5 in Chapter 1).

Finally, since by (1.41) u^t is Hölder continuous on every $[t, \delta]$, we infer that y^t is a classical solution (i.e., a C^1-solution) to Eq. (1.39) on $[t, T)$.

Now, if multiply Eq. (1.39) by dp^t/ds, Eq. (1.40) by dy^t/ds, and subtract the results, we get

$$\frac{d}{ds}[(Ay^t(s), p^t(s)) + (Fy^t(s), p^t(s)) + g(y^t(s)) - h^*(B^* p^t(s))]$$
$$= 0 \qquad \forall s \in (t, T),$$

because, by virtue of Eq. (1.41),

$$\frac{d}{ds} h^*(B^* p^t(s)) = \left(B u^t(s), \frac{dp^t}{ds}(s)\right) \qquad \forall s \in (t, T).$$

Hence,

$$(Ay^t(s), p^t(s)) + (Fy^t(s), p^t(s)) + g(y^t(s)) - h^*(B^* p^t(s)) = \delta_t$$
$$\forall s \in (t, T). \tag{1.42}$$

(In the sequel we shall simply write $A = A_H$.)

We set

$$\Gamma(t, y_0) = \{-p^t(t);\ p^t \text{ satisfies Eqs. (1.40), (1.41) along with some optimal pair } (y^t, u^t)\}.$$

We have:

Lemma 1.4. *For every* $(t, y_0) \in [0, T] \times H$, *we have*

$$\Gamma(t, y_0) \subset D_y^+ \psi(t, y_0). \tag{1.43}$$

Proof. For any $x \in H$ we have

$$\begin{aligned} &\psi(t, x) - \psi(t, y_0) \\ &\quad \le \int_t^T \big(g(y_x(s)) - g(y^t(s))\big)\, ds + \varphi_0(y_x(T)) - \varphi_0(y^t(T)), \end{aligned} \tag{1.44}$$

where

$$\begin{aligned} y_x'(s) + Ay_x(s) + Fy_x(s) &= Bu^t(s), \qquad s \in (t, T), \\ y_x(t) &= x. \end{aligned}$$

Clearly, we have

$$|y_x(s) - y^t(s)| \le C|x - y_0| \qquad \forall s \in [t, T). \tag{1.45}$$

Now let w be the solution to the equation

$$\begin{aligned} w' + Aw + \nabla F(y^t)w &= 0 \qquad \text{in } [t, T), \\ w(t) &= x - y_0. \end{aligned} \tag{1.46}$$

It is easily seen that if $|y_0 - x| \le \delta(\varepsilon)$ then

$$|y_x(s) - y^t(s) - w(s)| \le \varepsilon|y_0 - x| \qquad \forall s \in [t, T), \tag{1.47}$$

and so by (1.44) we have

$$\begin{aligned} &\psi(t, x) - \psi(t, y_0) \\ &\quad \le \int_t^T (\nabla g(y^t(s)), w(s))\, ds + \big(\nabla\varphi_0(y^t(T)), w(T)\big) + \varepsilon|x - y_0| \\ &\qquad\qquad \text{for } |x - y_0| \le \delta(\varepsilon). \end{aligned} \tag{1.48}$$

Now we take the scalar product of Eq. (1.40) with w and integrate on $[t, T]$. We get

$$\psi(t, x) - \psi(t, y_0) \le -(p^t(t), w(t)) + \varepsilon|x - y_0| \qquad \text{for } |x - y_0| \le \delta(\varepsilon).$$

Hence,

$$-p^t(t) \in D_y^+ \psi(t, x_0),$$

as claimed. ■

Remark 1.2. It is easily seen that for $T - t$ sufficiently small, the system (1.39)–(1.41) has a unique smooth solution (y^t, p^t) and so $\Gamma(t, \cdot)$ is single valued. Moreover, arguing as in the previous proof, it follows that for every $R > 0$ there is $\eta(R) > 0$ such that $\psi(t, \cdot) \in C^1(B_R)$ and

$$\nabla_y \psi(t, y_0) = \Gamma(t, y_0) \qquad \forall t \in [T - \eta(R), T],$$

for all $y_0 \in B_R = \{y \in H; |y| \le R\}$.

Proof of Theorem 1.2 (continued). We shall assume first that $\nabla\varphi_0(y) \in D(A)$ for all $y \in H$. Then, by Eq. (1.40), we see that $p^t \in C^1([t,T];\ H)$ and $y^t \in C^1([t,T];\ H)$. Now let $y_0 \in D(A)$ be arbitrary but fixed and let $t \in [0,T]$ be such that $\psi(\cdot, y_0)$ is differentiable at t. By the definition of ψ, we have

$$\psi(t, y_0) = \inf\left\{\int_0^{T-t} (g(y(s)) + h(u(s)))\, ds + \varphi_0(y(T-t));\right.$$
$$\left. y' + Ay + Fy = Bu \text{ in } [0, T-t],\ y(0) = y_0\right\}.$$

We set $z^t(s) = y^t(t+s)$, $v^t(s) = u^t(t+s)$, $q^t(s) = p^t(t+s)$ for $s \in [0, T - t]$, where (y^t, p^t) is defined by (1.38)–(1.41). We have

$$\psi(t+\varepsilon, y_0) - \psi(t, y_0) \le -\int_{T-t-\varepsilon}^{T-t} (g(z^t(s)) + h(v^t(s)))\, ds$$
$$+ \varphi_0(z^t(T-t-\varepsilon)) - \varphi_0(z^t(T-t)).$$

This yields

$$\frac{\partial\psi}{\partial t}(t, y_0) \le -g(y^t(T)) - h(u^t(T))$$
$$-\big(\nabla\varphi_0(y^t(T)), Ay^t(T) + Fy^t(T) - Bu^t(T)\big),$$

and by (1.40) it follows that

$$\frac{\partial\psi}{\partial t}(t, y_0)$$
$$\le -g(y^t(T)) - (p^t(T), Ay^t(T) + Fy^t(T)) + h^*(-B^*p^t(T)).$$

On the other hand, we have

$$\begin{aligned}\psi(t-\varepsilon, y_0) - \psi(t, y_0) &\le \int_{T-t}^{T-t+\varepsilon} (g(\tilde{z}(s)) + h(\tilde{v}(s)))\, ds \\ &\quad + \varphi_0(\tilde{z}(T-t+\varepsilon)) - \varphi_0(\tilde{z}(T-t)),\end{aligned}$$

where $\tilde{z}(s) = z^t(s)$ on $[0, T-t]$, $\tilde{v}(s) = v^t(s)$ on $[0, T-t]$, and

$$\begin{aligned}\tilde{z}' + A\tilde{z} + F\tilde{z} &= Bv^t \qquad \text{in } [0, T-t+\varepsilon], \\ \tilde{z}(0) &= y_0.\end{aligned}$$

This yields

$$\begin{aligned}&-\frac{\partial \psi}{\partial t}(t, y_0) \\ &\quad \le g(y^t(T)) + h(u^t(T)) \\ &\qquad + \big(\nabla\varphi_0(y^t(T)), Bu^t(T) - Fy^t(T) - Ay^t(T)\big) \\ &\quad = g(y^t(T)) + (p^t(T), Ay^t(T) + Fy^t(T)) - h^*(-B^*p^t(T)).\end{aligned}$$

Then, by (1.42) and Lemma 1.4, it follows that

$$\frac{\partial \psi}{\partial t}(t, y_0) - h^*(-B^*\eta) - (\eta, Ay_0 + Fy_0) + g(y_0) = 0,$$

where $\eta = -p^t(0) \in D_y^+ \psi(t, y_0)$.

In the general case (i.e., if $\nabla\varphi_0(y) \notin D(A_H)$), consider the sequence $\{\varphi_0^\varepsilon\}$,

$$\varphi_0^\varepsilon(y) = \varphi_0(e^{-A_H \varepsilon} y) \qquad \forall \varepsilon > 0,\ y \in H,$$

and denote by ψ^ε the corresponding optimal value function (1.27). Let $y_\varepsilon^t, u_\varepsilon^t, p_\varepsilon^t$ satisfy the system (1.39)–(1.41) and

$$\psi^\varepsilon(t, y\ ) = \int_t^T \big(g(y_\varepsilon^t) + h(u_\varepsilon^t)\big)\, ds + \varphi_0^\varepsilon(y_\varepsilon^t(T)).$$

According to the first part of the proof, we have

$$\begin{aligned}\frac{\partial \psi^\varepsilon}{\partial t}(t, y_0) - h^*(B^*p_\varepsilon^t(t)) + (Ay_0 + Fy_0, p_\varepsilon^t(t)) + g^\varepsilon(y_0) &= 0 \\ &\qquad \text{a.e. } t \in (0, T), \\ \psi^\varepsilon(T, y_0) = \varphi_0^\varepsilon(y_0). &\qquad (1.49)\end{aligned}$$

By assumption (e) it follows that $\{u_\varepsilon^t\}$ is bounded in $L^2(t,T;U)$ and so on a subsequence, again denoted $\{\varepsilon\}$, we have

$$\begin{aligned} u_\varepsilon^t &\to \tilde{u}^t \qquad \text{weakly in } L^2(t,T;U),\\ y_\varepsilon^t &\to \tilde{y}^t \qquad \text{strongly in } C([t,T];H),\\ p_\varepsilon^t &\to \tilde{p}^t \qquad \text{strongly in } C([t,T];H), \end{aligned}$$

where $(\tilde{y}^t, \tilde{p}^t, \tilde{u}^t)$ satisfy Eqs. (1.39)–(1.41). It is also clear that

$$\psi^\varepsilon(t,y_0) \to \psi(t,y_0) \qquad \text{as } \varepsilon \to 0.$$

Then, letting ε tend to zero in (1.49), we see that

$$\lim_{\varepsilon\to 0} \frac{\partial\psi_\varepsilon}{\partial t}(t,y_0) - h^*(B^*\tilde{p}^t(t)) + (Ay_0 + Fy_0, \tilde{p}^t(t)) + g(y_0) = 0.$$

Since, as seen in Lemma 1.4, $\tilde{p}^t(t) \in -D_y^+\psi(t,y_0)$ we infer that

$$\frac{\partial\psi}{\partial t}(t,y_0) - h^*(-B^*\eta) - (Ay_0 + Fy_0, \eta) + g(y_0) = 0$$

$$\text{a.e. } t \in (0,T),$$

for some $\eta \in D_y^+\psi(t,y_0)$.

This completes the proof of Theorem 1.2. ■

Remark 1.3. Under the assumptions of Theorem 1.2, for every $R > 0$ there is $T = T(R)$ such that ψ is a classical solution to Eq. (1.36) on the domain $(B_R \cap D(A_H)) \times (0,T)$ (see Remark 1.2).

Proof of Theorem 1.3. Let (y^*, u^*) be any optimal pair in problem (P). Then, by Lemma 1.3, we see that for every $t \in (0,T)$, (y^*,u^*) is also optimal for the problem

$$\inf\left\{\int_t^T (g(y(s)) + h(u(s)))\,ds + \varphi_0(y(T));\ y' + Ay + Fy = Bu \right.$$
$$\left. \text{in } [t,T],\ y(t) = y^*(t),\ u \in L^2(t,T;U)\right\}.$$

This means that

$$u^*(s) = \nabla h^*(B^*p^t(s)) \qquad \forall s \in [t,T],$$

where p^t is a solution to the system (1.40) where $y^t = y^*$. Then, by Lemma 1.4, we conclude that $u^*(t) \in \nabla h^*(-B^*D_y^+\psi(t,y^*(t)))$ $\forall t \in [0,T]$, as claimed. ■

Now we shall recall the concept of *viscosity solution* for the Hamilton–Jacobi equation (1.34) (M. G. Crandall and P. L. Lions [3]).

Let $\varphi \in C([0,T] \times H)$. Then, y is a viscosity solution to (1.34) on $[0,T] \times H$ if for every continuously differentiable function $\chi: [0,T] \times H \to \mathbf{R}$ and $\theta: H \to \mathbf{R}$ satisfying:

(i) χ is weakly sequentially lower semicontinuous and $\nabla\chi$, $A\nabla\chi$ are continuous;

(ii) θ is radial, nondecreasing, and continuously differentiable on H.

if $(t_0, y_0) \in (0,T) \times H$ is a local maximum (respectively, minimum) of $\varphi - \chi - \theta$ (respectively, $\varphi + \chi + \theta$), we have

$$\chi_t(t_0, y_0) - h^*\big(-B^*(\chi_y(t_0, y_0) + \nabla\theta(y_0))\big) - \big(A\chi_y(t_0, y_0), y_0\big) \\ -\big(Fy_0, \chi_y(t_0, y_0) + \nabla\theta(y_0)\big) + g(y_0) \geq 0 \tag{1.50}$$

(respectively,

$$\chi_t(t_0, y_0) + h^*(-B^*(\chi_y(t_0, y_0,) + \nabla\theta(y_0))) - \big(A\chi_y(t_0, y_0), y_0\big) \\ -\big(\chi_y(t_0, y_0) + \nabla\theta(y_0), Fy_0\big) - g(y_0) \geq 0.) \tag{1.51}$$

Proposition 1.4. *Under assumptions* (a)–(e), *the optimal value function* ψ *is a viscosity solution to Eq.* (1.34).

Proof. Let (t_0, y_0) be a local maximum for $\psi - \chi - \theta$, i.e.,

$$\psi(t_0, y_0) - \chi(t_0, y_0) - \theta(y_0) \geq \psi(t,y) - \chi(t,y) - \theta(y) \qquad \forall (t,y) \in \mathscr{V},$$

where $\mathscr{V}$ is a neighborhood of (t_0, y_0). This yields, by virtue of Lemma 1.4,

$$\chi(t_0, y_0) - \chi(t, y(t)) + \theta(y_0) - \theta(y(t)) \\ \leq \psi(t_0, y_0) - \psi(t, y(t)) \leq \int_{t_0}^{t} (g(y(s)) + h(u(s)))\, ds,$$

where y, u satisfy Eq. (1.23) on $[t_0, T]$, with $y(t_0) = y_0$. Then, for u smooth, we have

$$-\chi_t(t_0, y_0) - \big(\chi_y(t_0, y_0), Bu(t_0) - Ay_0 - Fy_0\big) \\ -(\nabla\theta(y_0), Bu(t_0) - Ay_0 - Fy_0) \leq g(y_0) + h(u(t_0)). \tag{1.52}$$

For every $u_0 \in D(h)$, we may choose $u \in W^{1,2}([t_0, T]; U)$ such that $h(u) \in C([t_0, T])$. For instance, we may take u to be the solution to the Cauchy problem

$$u' + \partial h(u) \ni 0 \qquad \text{a.e. in } (t_0, T), \qquad u(t_0) = u_0 .$$

Then, by (1.52), it follows that

$$\begin{aligned}\chi_t(t_0, y_0) &- \big(A\chi_y(t_0, y_0), y_0\big) - \big(Fy_0, \chi_y(t_0, y_0) + \nabla\theta(y_0)\big) + g(y_0)\\ &\geq h^*\big(-B^*\big(\chi_y(t_0, y_0) + \nabla\theta(y_0)\big)\big),\end{aligned}$$

because θ is radial nonincreasing, i.e., $\theta(x) = \omega(|x|)$, $\omega' \geq 0$, and so $(Ay_0, \nabla\theta(y_0)) \geq 0$.

Assume now that $\varphi + \chi + \theta$ has a local minimum at (t_0, y_0). We have, therefore, by Lemma 1.4

$$\begin{aligned}\chi(t_0, y_0) &- \chi(t, y(t)) + \theta(y_0) - \theta(y(t))\\ &\leq \psi(t, y(t)) - \psi(t_0, y_0) = -\int_{t_0}^{t} (g(y(s)) + h(u(s)))\, ds,\end{aligned}$$

where $[y, u]$ is optimal for the problem (1.27) on the interval $[t_0, T]$.

This yields

$$\begin{aligned}-\chi_t(t_0, y_0) &- \big(\chi_y(t_0, y_0)\big) + \nabla\theta(y_0), Bu(t_0) - Ay_0 - Fy_0)\\ &\leq -g(y_0) - h(u(t_0)).\end{aligned}$$

(We note that since (y, u) satisfies a system of the form (1.39)–(1.41), u is smooth.)

Hence,

$$\begin{aligned}\chi_t(t_0, y_0) &- \big(A\chi_y(t_0, y_0), y_0\big) - \big(\chi_y(t_0, y_0) + \nabla\theta(y_0), Fy_0\big) - g(y_0)\\ &\geq h(u(t_0)) - \big(B^*\big(\chi_y(t_0, y_0) + \nabla\theta(y_0)\big), u(t_0)\big)\\ &\geq -h^*\big(-B^*\big(\chi_y(t_0, y_0) + \nabla\theta(y_0)\big)\big),\end{aligned}$$

as claimed. ■

It follows by the uniqueness results in Crandall and Lions [3] that ψ is the unique viscosity solution of Eq. (1.34). In the general case of unbounded operators $\partial\phi$, the Cauchy problem (1.34) is still well-posed in a certain class of generalized solutions introduced by D. Tătaru [1] (see also [3]).

As a matter of fact such a problem can be approximated by a family of smooth optimal control problems of the form

$$(\mathrm{P}_\varepsilon) \quad \text{Minimize} \quad \int_0^T \left(g^\varepsilon(y(t)) + h(u(t)) + \varepsilon |u(t)|_U^2\right) dt + \varphi_0^\varepsilon(y(T)),$$

subject to

$$\frac{dy}{dt} + Ay + \partial\varphi^\varepsilon(y) = Bu \qquad \text{in } [0,T],$$
$$y(0) = y_0,$$

where g^ε, φ_0^ε, and φ^ε are smooth approximations of g, φ_0, and φ, respectively.

Moreover, the optimal value function ψ^ε of problem (P_ε) satisfies Eq. (1.34) in the sense of Theorem 1.2 and $\psi^\varepsilon \to \psi$ on $[0,T] \times \overline{D(\phi)}$ as $\varepsilon \to 0$. In this context, we may view ψ as a generalized solution to Eq. (1.34). More will be said about this equation in Section 2.

Now let us come back to the case $H = L^2(\Omega)$, $A = -\Delta$, $V = H_0^1(\Omega)$, and

$$\partial\varphi(y)(x) = \beta(y(x)) \qquad \forall y \in L^2(\Omega), \text{a.e. } x \in \Omega, \qquad (1.53)$$

where $\beta: \mathbf{R} \to \mathbf{R}$ is a locally Lipschitz function. Though Theorems 1.2 and 1.3 are not applicable to the present situation we have, however:

Theorem 1.4. *If β satisfies condition* (1.11) *then every optimal control u^* is expressed as function of the corresponding optimal state y^* by the feedback law*

$$u^*(t) \in \partial h^*(-B^* \partial\psi(t, y^*(t))) \qquad \text{a.e. } t \in (0,T), \qquad (1.54)$$

where $\psi: [0,T] \times L^2(\Omega) \to \mathbf{R}$ is the optimal value function and $\partial\psi$ is the Clarke generalized gradient of $\psi(t, \cdot\,)$.

Proof. According to Lemma 1.4, for every $t \in [0,T]$ the optimal pair (y^*, u^*) is on the interval $[0,t]$ optimal for the problem

$$\inf\left\{\int_0^t (g(y(s,0,y_0,u)) + h(u(s)))\, ds + \psi(t, y(t,0,y_0,u));\right.$$
$$\left. u \in L^2(0,t;U)\right\}.$$

Then, by Theorem 1.1 in Chapter 5, for every $t \in [0,T]$ there is $p^t \in AC([0,t]; Y^*) \cap C_w([0,t]; H)$ that satisfies the equations

$$B^*p^t(s) \in \partial h(u^*(s)), \tag{1.55}$$

$$p^t(t) \in -\partial\psi(t, y^*(t)). \tag{1.56}$$

Since the function u^* is approximately continuous on $[0,T]$, arguing as in the proof of Theorem 1.1 it follows by Eq. (1.55) that

$$u^*(t) \in \partial h^*(B^*p^t(t)) \qquad \text{a.e. } t \in (0,T),$$

which implies (1.54), as desired. ■

Remark 1.4. If, under assumptions of Theorem 1.4, h is Gâteaux differentiable and $R(B)$ (the range of B) is dense in H then any dual extremal arc associated with (y^*, u^*) satisfies the equation

$$p(t) \in -\partial\psi(t, y^*(t)) \qquad \text{a.e. } t \in (0,T). \tag{1.57}$$

Indeed, by Eq. (1.55) we have $p(s) = p^t(s)$ $\forall s \in [0,t]$, which by virtue of (1.56) implies (1.57). (See C. Popa [2] for a similar result in the case of the obstacle control problem and Clarke and Vinter [1] for a related finite dimensional result.)

6.1.4. *Feedback Controllers for the Optimal Time Control Problem*

We shall consider here the time-optimal control problem (P) studied in Chapter 5 (Section 3.3) with the state system (3.23) and the control constraints $u \in \tilde{\mathscr{U}}_\rho = \{v \in L^\infty(0,\infty; L^2(\Omega)); |v(t)|_2 \le \rho \text{ a.e. } t > 0\}$. We shall assume here that $\beta: \mathbf{R} \to \mathbf{R}$ is locally Lipschitz and

$$0 \le \beta'(r) \le C(1 + |\beta(r)| + |r|) \qquad \text{a.e. } r \in \mathbf{R}. \tag{1.58}$$

Denote by φ the minimal time function

$$\varphi(y_0) = \inf\left\{T; \exists u \in \tilde{\mathscr{U}}_\rho \text{ such that } y(T, y_0, u) = 0\right\}.$$

As seen in Lemma 3.1, $\varphi(y_0) < \infty$ $\forall y_0 \in L^2(\Omega)$ and

$$\varphi(y_0) \le \varphi(z_0) + \rho|y_0 - z_0|_2 \qquad \forall y_0, z_0 \in L^2(\Omega).$$

Hence,

$$|\varphi(y_0) - \varphi(z_0)| \le \rho|y_0 - z_0|_2 \qquad \forall y_0, z_0 \in L^2(\Omega).$$

Also, it is readily seen that (the dynamic programming principle)

$$\varphi(y_0) = \inf\left\{t + \varphi(y(t, y_0, u));\, u \in \tilde{\mathscr{U}}_\rho\right\} \qquad \forall t > 0,\ y_0 \in L^2(\Omega). \tag{1.59}$$

Here, $y(t, y_0, u)$ is the solution to system (3.23), i.e.,

$$\begin{aligned} \frac{dy}{dt} - \Delta y + \beta(y) &= Bu && \text{in } \Omega \times \mathbf{R}^+, \\ y(x,0) &= y_0(x) && \text{in } \Omega, \\ y &= 0 && \text{in } \partial\Omega \times \mathbf{R}^+. \end{aligned} \tag{1.60}$$

This implies that every time-optimal pair (y^*, u^*) is also optimal in problem (1.58), and so by Theorem 1.1 in Chapter 5 for every $t \in [0, T^*)$ (T^* is the minimal time) there is $p^t \in AC([0,t];\, Y^*) \cap C_w([0,t];\, L^2(\Omega))$ $\cap\, C([0,t];\, L^1(\Omega))$ such that

$$\begin{aligned} \frac{dp^t}{ds} + \Delta p^t - p^t\partial\beta(y) &\ni 0 && \text{in } \Omega \times (0,t), \\ p^t(t) &\in -\partial\varphi(y^*(t)) && \text{in } \Omega, \\ u^*(s) &\in \rho \operatorname{sgn} p^t(s) && \text{a.e. } s \in (0,t), \end{aligned} \tag{1.61}$$

where $\operatorname{sgn} p^t = p^t/|p^t|_2$ if $p^t \neq 0$, $\operatorname{sgn} 0 = \{w \in L^2(\Omega);\ |w|_2 \le 1\}$. Then, arguing as in the proof of Theorem 1.1, we conclude that

$$u^*(t) \in -\rho \operatorname{sgn} \xi(t), \qquad \xi(t) \in \partial\varphi(y^*(t)) \quad \text{a.e. } t \in (0, T^*), \tag{1.62}$$

where $\partial\varphi$ is the generalized gradient (in the sense of Clarke) of $\varphi: L^2(\Omega) \to \mathbf{R}$.

We have proved, therefore:

Theorem 1.5. *Under the assumption* (1.58), *any time-optimal control u^* of system* (1.60) *admits the feedback representation* (1.62). ■

It turns out that, at least formally, the minimal time function φ is the solution to the Bellman equation associated with the given problem, i.e.,

$$\begin{aligned} (D\varphi(y), My) + \rho|D\varphi(y)|_2 &= 1, \qquad y \in L^2(\Omega), \\ \varphi(0) &= 0, \end{aligned} \tag{1.63}$$

where $My = -\Delta y + \beta(y)\ \forall y \in D(M) = \{y \in H_0^1(\Omega) \cap H^2(\Omega);\ \beta(y) \in L^2(\Omega)\}$. Indeed, coming back to approximating problem (P_ε) in Section 3.3, Chapter 5, we see by Eqs. (3.64), (3.65) there that

$$u_\varepsilon(t) = \rho \operatorname{sgn} p_\varepsilon(t) + \varepsilon p_\varepsilon(t) \qquad \text{a.e. } t > 0,$$

and therefore

$$u_\varepsilon(t) \in -\rho \operatorname{sgn} \partial\varphi_\varepsilon(y_\varepsilon(t)) - \varepsilon\, \partial\varphi_\varepsilon(y_\varepsilon(t)) \qquad \text{a.e. } t > 0. \quad (1.64)$$

Then, by (3.65), it follows that φ_ε is the solution to the stationary Hamilton–Jacobi equation

$$(\partial\varphi_\varepsilon(y), My) + \rho|\partial\varphi_\varepsilon(y)|_2 + \frac{\varepsilon}{2}|\partial\varphi_\varepsilon(y)|_2^2 = g^\varepsilon(y) \qquad \forall y \in L^2(\Omega),$$
$$\varphi(0) = 0, \qquad (1.65)$$

i.e., $\exists \eta_\varepsilon(y) \in \partial\varphi_\varepsilon(y)$ such that

$$(\eta_\varepsilon(y), My) + \left(\rho + \frac{\varepsilon}{2}\right)|\eta_\varepsilon(y)|_2^2 = g^\varepsilon(y) \qquad \forall y \in L^2(\Omega).$$

Since by Theorem 3.5, Chapter 5, $\varphi_\varepsilon \to \varphi$ as $\varepsilon \to 0$, $|\partial\varphi_\varepsilon(y)|_2 \leq \rho^{-1}$, and $g^\varepsilon \to 1$ as $\varepsilon \to 0$, we may view φ_ε as an approximating solution to Eq. (1.63) and so φ itself is a generalized solution to Eq. (1.63). As a matter of fact, it turns out that φ is a viscosity solution to Eq. (1.63) and that under a supplementary growth condition on β, it is unique in the class of weakly continuous functions on $L^2(\Omega)$ (see Barbu [20], Tătaru [2]).

6.2. A Semigroup Approach to the Dynamic Programming Equation

6.2.1. *Variational and Mild Solutions to the Dynamic Programming Equation*

We shall study here the Hamilton–Jacobi equation (1.34) in a more general context, namely in the case when $\partial\phi$ is replaced by a general maximal monotone operator $M \subset H \times H$. By the substitution $\varphi(t, y) = \psi(T - t, y)$ we reduce this problem to the forward Cauchy problem

$$\varphi_t(t, y) + h^*\big(-B^*\varphi_y(t, y)\big) + \big(My, \varphi_y(t, y)\big) = g(y),$$
$$\varphi(0, y) = \varphi_0(y). \qquad (2.1)$$

As seen earlier, there is a close connection between Eq. (2.1) and the optimal control problem

$$\inf\left\{\int_0^t (g(y(s)) + h(u(s)))\,ds + \varphi_0(y(t));\ u \in L^1(0,t;U)\right\} = \varphi(t,x), \tag{2.2}$$

where $y = y(s, x, u) \in C([0,t];\ H)$ is the mild solution to the Cauchy problem

$$\frac{dy}{ds} + My \ni Bu \qquad \text{in } (0,t), \qquad y(0) = x \in \overline{D(M)}. \tag{2.3}$$

The following hypotheses will be in effect throughout this section:

(i) H and U are real Hilbert spaces with the norms denoted $|\cdot|$ and $|\cdot|_U$, respectively. $B \in L(U,H)$ and M is a maximal monotone subset of $H \times H$ with the domain $D(M)$.

(ii) $h\colon U \to \overline{\mathbf{R}}$ is a lower semicontinuous convex function such that

$$\lim_{|u|_U \to \infty} h(u)/|u|_U = \infty. \tag{2.4}$$

Denote by h^* the conjugate function of h, i.e., $h^*(p) = \sup\{\langle p, u\rangle - h(u);\ u \in U\}$. We have denoted by $(\cdot,\cdot)$ the scalar product of H and by $\langle\cdot,\cdot\rangle$ the scalar product of U. B^* is the dual operator of B.

Given a metric space X we shall denote by $BUC(X)$ the space of all bounded and uniformly continuous real valued functions on X endowed with the usual sup norm:

$$\|f\|_b = \sup\{|f(x)|;\ x \in X\}, \qquad f \in BUC(X).$$

By Lip(X), we shall denote the space of all Lipschitz functions $f\colon X \to \mathbf{R}$.

In the following, the space X will be the closure $\overline{D(M)}$ of $D(M)$ in H.

If $g, \varphi_0 \in BUC(\overline{D(M)})$, then the function φ is well-defined on $[0,\infty) \times \overline{D(M)}$. We set

$$(S(t)\varphi_0)(x) = \varphi(t,x), \qquad t \geq 0,\ x \in \overline{D(M)}, \tag{2.5}$$

and call it *variational solution* to Hamilton–Jacobi equation (2.1). We have

Lemma 2.1. *Let $g \in BUC(\overline{D(M)})$. Then $S(t)\varphi_0 \in BUC(\overline{D(M)})$ for all $t \geq 0$, $\varphi_0 \in BUC(\overline{D(M)})$, and*

$$S(t)S(s)\varphi_0 = S(t+s)\varphi_0 \qquad \forall t, s \geq 0,\ \varphi_0 \in BUC\big(\overline{D(M)}\big). \tag{2.6}$$

$$\|S(t)\varphi_0 - S(t)\psi_0\|_b \leq \|\varphi_0 - \psi_0\|_b \qquad \forall t \geq 0,\ \varphi_0, \psi_0 \in BUC\big(\overline{D(M)}\big). \tag{2.6$'$}$$

In other words, S(t) is a semigroup of contractions on the Banach space $Y = BUC(\overline{D(M)})$.

Proof. Obviously, $S(t)\varphi_0$ is bounded on $\overline{D(M)}$ for every $t \geq 0$. For $x, x_0 \in \overline{D(M)}$ we have, for any $\varepsilon > 0$,

$$\begin{aligned} S(t)\varphi_0(x) &- S(t)\varphi_0(x_0) \\ &\leq \int_0^t (g(y(s,x,u)) - g(y(s,x_0,u)))\,ds \\ &\quad + \varphi_0(y(t,x,u)) - \varphi_0(y(t,x_0,u)) + \varepsilon, \end{aligned}$$

where $u \in L^1(0,t;U)$ is such that

$$\int_0^t g(y(s,x_0,u)) + h(u(s)))\,ds + \varphi_0(y(t,x_0,u)) \leq S(t)\varphi_0(x) + \varepsilon$$

and $y(s,x,u)$ is the mild solution to (2.3).

Recalling that (see Theorem 1.1, Chapter 4)

$$|y(t,x_0,u) - y(t,x,(v)| \leq |x_0 - x| + \int_0^t |B(u-v)|_U\,ds,$$

we conclude that $S(t)\varphi_0$ is uniformly continuous on $\overline{D(M)}$.

The semigroup property (2.6) is an immediate consequence of the optimality principle, whilst (2.6)$'$ follows by the obvious inequality

$$(S(t)\varphi_0)(x) - (S(t)\psi_0)(x) \leq \varphi_0(y(t,x,u)) - \psi_0(t,x,u) + \varepsilon, \tag{2.7}$$

where ε is arbitrary and suitable chosen. ■

The main result of this section (Theorem 2.1 following) amounts to saying that the semigroup $S(t)$ is generated on a certain subset of Y by an m-accretive operator.

The generator $\mathscr{A}$ of $S(t)$ is constructed as follows.

For any $f \in BUC(\overline{D(M)})$ and $\lambda > 0$, define the function

$$(R(\lambda)f)(x) = \inf\left\{\int_0^\infty e^{-\lambda t}(f(y(t)) + h(u(t)))\,dt;\right.$$

$$\left. u \in L^1_{\text{loc}}(\mathbf{R}^+; U),\ y' + My \ni Bu \text{ in } \mathbf{R}^+,\ y(0) = x\right\}. \quad (2.8)$$

It turns out that $R(\lambda)$ is a pseudo-resolvent in $Y = BUC(\overline{D(M)})$. More precisely, we have:

Lemma 2.2. *For every $\lambda > 0$, $R(\lambda)$ maps Y into itself and*

$$R(\lambda)f = R(\mu)((\mu - \lambda)R(\lambda)f + f) \qquad 0 < \lambda \le \mu < \infty. \quad (2.9)$$

Moreover, if $f \in \operatorname{Lip}(\overline{D(M)})$ then $R(\lambda)f \in \operatorname{Lip}(\overline{D(M)})$.

Proof. Let $u_0 \in U$ be such that $h(u_0) < \infty$ and let $y_0(t) = y(t, x, u_0)$ be the corresponding solution to (2.3). We have the obvious inequality

$$(R(\lambda)f)(x) \le \int_0^\infty e^{-\lambda t}(f(y_0(t)) + h(u_0(t)))\,dt \le C \qquad \forall x \in \overline{D(M)}.$$

Hence, $\sup\{(R(\lambda)f)(x);\ x \in \overline{D(M)}\} < +\infty$. Note also that $\inf\{(R(\lambda)f)(x);\ x \in \overline{D(M)}\} > -\infty$ because otherwise there would exist the sequences x_n, u_n such that

$$\int_0^\infty e^{-\lambda t}(f(y(t, x_n, u_n)) + h(u_n(t)))\,dt \to \infty.$$

This would imply that $\int_0^\infty e^{-\lambda t}h(u_n(t))\,dt \to -\infty$ as $n \to \infty$, which by virtue of the convexity of h and assumption (2.4) leads to a contradiction. Hence, $R(\lambda)f$ is bounded on $\overline{D(M)}$. To prove that $R(\lambda)f$ is uniformly continuous consider x_1, x_2 arbitrary but fixed in $\overline{D(M)}$. For every $\varepsilon > 0$, there are u_ε and $\bar{u}_\varepsilon$ such that

$$(R(\lambda)f)(x_1) \ge \int_0^\infty e^{-\lambda t}(f(y_\varepsilon(t)) + h(u_\varepsilon(t)))\,dt - \varepsilon,$$

$$(R(\lambda)f)(x_2) \ge \int_0^\infty e^{-\lambda t}(f(\bar{y}_\varepsilon(t)) + h(\bar{u}_\varepsilon(t)))\,dt - \varepsilon,$$

where $y_\varepsilon = y(t, x_1, u_\varepsilon)$, $\bar{y}_\varepsilon = y(t, x_2, \bar{u}_\varepsilon)$. We have

$$(R(\lambda)f)(x_1) - (R(\lambda)f)(x_2) \le \int_0^\infty e^{-\lambda t}(f(\bar{z}_\varepsilon(t)) - f(\bar{y}_\varepsilon(t)))\,dt + \varepsilon,$$

$$(R(\lambda)f)(x_2) - (R(\lambda)f)(x_1) \le \int_0^\infty e^{-\lambda t}(f(z_\varepsilon(t)) - f(y_\varepsilon(t)))\,dt + \varepsilon,$$

where $z_\varepsilon = y(t, x_2, u_\varepsilon)$ and $\bar{z}_\varepsilon = y(t, x_1, \tilde{u}_\varepsilon)$.

Since $|\bar{z}_\varepsilon(t) - \bar{y}_\varepsilon(t)| \le |x_1 - x_2|$ and $|y_\varepsilon(t) - z_\varepsilon(t)| \le |x_1 - x_2|$, we conclude that

$$|R(\lambda)f)(x_1) - (R(\lambda)f)(x_2)| \le \delta(\varepsilon) \qquad \text{if} |x_1 - x_2| \le \varepsilon.$$

By a similar argument, it follows that

$$R(\lambda)f \in \mathrm{Lip}(\overline{D(M)} \qquad \text{if } f \in \mathrm{Lip}(\overline{D(M)}).$$

Now by definition of $R(\lambda)$ we have, for all $\lambda, \mu > 0$,

$$\begin{aligned} &R(\mu)((\mu - \lambda)R(\lambda)f + f)(x) \\ &\quad = \inf\left\{\int_0^\infty e^{-\mu t}(f(y(t)) + h(u(t))\right. \\ &\qquad +(\mu - \lambda)\inf\left(\int_0^\infty e^{-\lambda s}(f(z(s) + h(v(s)))\,ds\right. \\ &\qquad \left. z' + Mz \ni Bv,\ z(0) = y(t),\ v \in L^1_{\mathrm{loc}}(\mathbf{R}^+; U)\right), \\ &\qquad \left. y' + My \ni Bu,\ y(0) = x\right\}. \end{aligned}$$

Therefore,

$$\begin{aligned} &R(\mu)((\mu - \lambda)R(\lambda)f + f)(x) \\ &\quad \le \int_0^\infty e^{-\mu t}(f(y(t)) + h(u(t))\,dt \\ &\qquad +(\mu - \lambda)\int_0^\infty e^{-\lambda s}(f(y(t+s)) + h(u(t+s)))\,ds \end{aligned}$$

for all (y, u) satisfying Eq. (2.3).

This yields

$$R(\mu)((\mu - \lambda)R(\lambda)f + f)(x) \le \int_0^\infty e^{-\lambda t}(f(y(t)) + h(u(t)))\,dt,$$

i.e.,

$$R(\mu)((\mu-\lambda)R(\lambda)f+f)(x) \le (R(\lambda)f)(x) \qquad \forall x \in D(M).$$

On the other hand, we have

$$(R(\lambda)f)(x) \le \int_0^t e^{-\lambda s}(f(y^*(s)) + h(u^*(s)))\,ds$$
$$+ \int_0^\infty e^{-\lambda s}(f(z^*(s-t)) + h(v^*(s-t)))\,ds \quad (2.10)$$

where (y^*, v^*) and (z^*, v^*) satisfy Eq. (2.3) with the initial conditions $y^*(0) = x$, $z^*(0) = y^*(t)$ and are chosen in a such a way that

$$R(\mu)((\mu-\lambda)R(\lambda)f+f)(x)$$
$$\ge \int_0^\infty e^{-\mu t}(f(y^*(t)) + h(u^*(t)))\,dt + (\mu-\lambda)\int_0^\infty e^{-\mu t}\,dt$$
$$+ \int_0^\infty e^{-\lambda s}(f(z^*(s)) + h(v^*(s)))\,ds - \varepsilon. \quad (2.11)$$

Now, if we multiply (2.10) by $e^{-(\mu-\lambda)t}$ and integrate on $(0,\infty)$, we find after some calculation involving (2.11) that

$$(R(\lambda)f)(x) \le R(\mu)((\mu-\lambda)R(\lambda)f+f)(x) + \varepsilon \qquad \forall \varepsilon > 0,$$

which completes the proof. ■

Now let $\mathscr{A}: D(\mathscr{A}) \subset Y \to Y$ be the operator (eventually, multivalued) defined by

$$\mathscr{A}R(1)f = f - R(1)f \qquad \forall f \in Y = BUC(\overline{D(M)}),$$
$$D(\mathscr{A}) = \{\varphi = R(1)f;\ f \in Y\}. \quad (2.12)$$

If the operator A is continuous on H and f, h^* are smooth, then it is readily seen that $\varphi = R(1)f$ satisfies the stationary Hamilton-Jacobi equation

$$\lambda\varphi(y) + (My, \eta(y)) + h^*(-B^*\eta(y)) = f(y),$$
$$\eta(y) \in \partial\varphi(y), \qquad \forall y \in D(M).$$

(This follows easily by the optimality system associated with the infinite horizon optimal control problem (2.8).) Thus, we may view the operator $\mathscr{A}$ as an extension on Y of the operator $y \to (My, \varphi_y(y)) + h^*(-B^*\varphi_y(y))$. If

M is linear, then we have (see Barbu and DaPrato [1])

$$\mathscr{A}\varphi = \{w \in Y; w(x) = (\eta(x), Mx) + h^*(-B^*\eta(x)) \ \forall x \in D(M), \\ \eta(x) \in \partial\varphi(x)\} \qquad \forall \varphi \in D(\mathscr{A}).$$

Lemma 2.3. *The operator* $\mathscr{A}$ *is* m-*accretive in* $Y \times Y$ *and*

$$(\lambda I + \mathscr{A})^{-1} = R(\lambda) \qquad \forall \lambda \in (0,1]. \tag{2.13}$$

(I is the unity operator in Y.)

Proof. By Eqs. (2.9), (2.12) we see that

$$(\lambda I + \mathscr{A})^{-1} = R(\lambda) \qquad \text{for } 0 < \lambda \leq 1, \tag{2.14}$$

whilst by definition of $R(\lambda)$ it is readily seen that

$$\|R(\lambda)f - R(\lambda)g\|_{\mathrm{b}} \leq \lambda^{-1}\|f - g\|_{\mathrm{b}} \qquad \forall \lambda > 0, f, g \in Y.$$

Hence, $\mathscr{A}$ is m-accretive (see Section 3.1 in Chapter 2). Moreover, by the resolvent equation

$$(\lambda I + \mathscr{A})^{-1}f = (\mu I + \mathscr{A})^{-1}((\mu - \lambda)(\lambda I + \mathscr{A})^{-1}f + f),$$

we infer that (2.13) holds for all $\lambda > 0$. ■

According to the Crandall–Liggett generation theorem (Theorem 1.3 in Chapter 4), for every $\varphi_0 \in \overline{D(\mathscr{A})}$ and $g \in Y$ the Cauchy problem

$$\frac{d\varphi}{dt} + \mathscr{A}\varphi \ni g \qquad \text{in } [0, \infty), \\ \varphi(0) = \varphi_0, \tag{2.15}$$

has a unique mild solution $\varphi \in C([0,\infty); Y)$, given by the exponential formula

$$\varphi(t) = \lim_{n \to \infty} J^n\left(\frac{t}{n}\right)\varphi_0 \qquad \forall t \geq 0, \tag{2.16}$$

where $J(\lambda)\varphi_0 = (I + \lambda\mathscr{A})^{-1}(\varphi_0 + \lambda g) = (I + \lambda\mathscr{A}_g)^{-1}\varphi_0$, where $\mathscr{A}_g\varphi = \mathscr{A}\varphi + g$.

Equivalently, $\varphi(t) = \lim_{\varepsilon \to 0} \varphi_\varepsilon(t)$ in every $C([0,T]; Y)$, where φ_ε is the solution to difference equations

$$\frac{1}{\varepsilon}(\varphi_\varepsilon(t) - \varphi_\varepsilon(t - \varepsilon)) + \mathscr{A}\varphi_\varepsilon(t) \ni g \qquad \forall t \geq \varepsilon, \\ \varphi_\varepsilon(t) = \varphi_0 \qquad \text{for } -\varepsilon \leq t \leq 0. \tag{2.17}$$

Moreover, the map $T(t): \overline{D(\mathscr{A})} \to \overline{D(\mathscr{A})}$ defined by

$$T(t)\varphi_0 = \varphi(t) \qquad \forall \varphi_0 \in \overline{D(\mathscr{A})}, t \geq 0, \tag{2.18}$$

is a continuous semigroup of nonlinear contractions of $\overline{D(\mathscr{A})}$. The function $t \to T(t)\varphi_0$ is called *mild solution* to Eq. (2.1).

6.2.2. The Equivalence of Variational and Mild Solutions

Coming back to the semigroup $S(t)$ defined by formula (2.5), one might suspect that

$$S(t) = T(t) \qquad \text{on } \overline{D(\mathscr{A})}.$$

Indeed, we have:

Theorem 2.1. *Assume that hypotheses* (i), (ii) *hold and that* $g \in$ Lip{$(\overline{D(M)})$. *Then*

$$S(t)\varphi_0 = T(t)\varphi_0 \qquad \forall t \geq 0, \forall \varphi_0 \in \overline{D(\mathscr{A})}. \tag{2.19}$$

Moreover, the operator $\mathscr{A}$ *is single valued and for every* $\varphi_0 \in D(\mathscr{A})$ *one has*

$$\lim_{t \downarrow 0} \frac{1}{t}((S(t)\varphi_0)(x) - \varphi_0(x)) = -(\mathscr{A}\varphi_0)(x) + g(x) \qquad \forall x \in D(M). \tag{2.20}$$

Before proving Theorem 2.1 we shall give a precise description of the closure $\overline{D(\mathscr{A})}$ of $D(\mathscr{A})$ in Y.

We shall denote by Z the set of all $\varphi \in Y$ having the property that the function $t \to S^M(t)\varphi = \varphi(e^{-Mt}x)$ defined from $[0, \infty)$ to Y is continuous into the origin. In other words, Z is the domain of the C_0-semigroup

$$S^M(t): [0, \infty) \to Y.$$

Proposition 2.1. *Under the preceding assumptions* $Z = \overline{D(\mathscr{A})}$.

Proof. We set

$$\mathscr{D} = \Big\{\varphi \in B \cup C\big(\overline{D(M)}\big) \cap \text{Lip}\big(\overline{D(M)}\big); |\varphi(e^{-Mt}x) - \varphi(x)| \leq L_\varphi t \\ \forall t \geq 0, x \in \overline{D(M)}\Big\} \tag{2.21}$$

or, equivalently,

$$\mathscr{D} = \left\{ \varphi \in BUC\big(\overline{D(M)}\big) \cap \operatorname{Lip}\big(\overline{D(M)}\big);\ |\varphi(y(t,x,u)) - \varphi(x)| \right.$$
$$\left. \leq L_{\varphi}\left(t + \int_0^t |u(s)|_U \, ds\right) \forall t \geq 0,\ u \in L^1(\mathbf{R}^+; U),\ x \in \overline{D(M)} \right\},$$
$$(2.21)'$$

where $y(t, x, u)$ is the solution to (2.3). It is readily seen that $Z = \overline{\mathscr{D}}$ (the closure of $\mathscr{D}$ in $BUC(\overline{D(M)})$). Indeed, the space $BUC(D(M)) \cap \operatorname{Lip}(\overline{D(M)})$ is dense in $B \cup C(\overline{D(M)})$ (see, e.g., Lasry and Lions [1]) and by the same argument it follows that $Z \cap \operatorname{Lip}(\overline{D(M)})$ is dense in Z. Now it is readily seen that for every $\varphi \in Z \cap \operatorname{Lip}(\overline{D(M)})$,

$$\varphi_\varepsilon(x) = \varepsilon^{-1} \int_0^\varepsilon \varphi(e^{-Mt}x)\, dt \overset{\varepsilon \to 0}{\to} \varphi \qquad \text{in } Y,$$

and

$$\varphi_\varepsilon \in \Delta.$$

This implies that $Z = \overline{\mathscr{D}}$, as claimed. Hence, to prove Proposition 2.1 it suffices to show that $\overline{\mathscr{D}} = \overline{D(\mathscr{A})}$. Toward this aim, we shall prove first that

$$(I + \mu\mathscr{A})^{-1} BUC\big(\overline{D(M)}\big) \cap \operatorname{Lip}\big(\overline{D(M)}\big) \subset \mathscr{D} \qquad \forall \mu > 0. \quad (2.22)$$

Let $\mu > 0$ and $f \in BUC(\overline{D(M)}) \cap \operatorname{Lip}(\overline{D(M)})$. We set $\varphi = (I + \mu\mathscr{A})^{-1} f = R(\mu^{-1})(\mu^{-1} f)$. For every $\varepsilon > 0$, there are $(y_\varepsilon, u_\varepsilon) \in C([0,T];\ H) \cap L^1(0,T;\ U)$ for any $T > 0$ such that $y_\varepsilon' + My_\varepsilon \ni Bu_\varepsilon$ in $[0,\infty)$, $y_\varepsilon(0) = x$, and

$$|\varphi(x) - \int_0^\infty e^{-t/\mu}(f(y_\varepsilon(t))\mu^{-1} + h(u_\varepsilon(t)))\, dt| \leq \varepsilon. \qquad (2.23)$$

Let (y, u) be any pair of functions satisfying Eq. (2.3). We have

$$|y(t) - y_\varepsilon(t)| \leq \|B\| \int_0^t (|u_\varepsilon(s)|_U + |u(s)|_U)\, ds$$

and, therefore,

$$|\varphi(y_\varepsilon(t)) - \varphi(y(t))| \leq C \int_0^t (|u_\varepsilon(s)|_U + |u(s)|_U)\, ds \qquad \forall t > 0, \quad (2.24)$$

because φ is Lipschitz (Lemma 2.3). (Here and everywhere in the following we shall denote by C several positive constants independent of ε, u, and t.) On the other hand, by the optimality principle we have

$$(R(\lambda)f)(x) = \inf\left\{\int_0^t e^{-\lambda s}(f(y(s)) + h(u(s)))\,ds + e^{-\lambda t}(R(\lambda)f)(y(t));\right.$$
$$\left. y' + My \ni Bu \text{ in } (0,t), y(0) = x\right\} \qquad \forall t > 0.$$

In view of (2.23), this yields

$$|\varphi(x) - e^{-t/\mu}\varphi(y_\varepsilon(t)) - \int_0^t e^{-s/\mu}(\frac{1}{\mu}f(y_\varepsilon(s)) + h(s))\,ds| \leq \varepsilon$$

and, therefore,

$$|\varphi(x) - e^{-t/\mu}\varphi(y_\varepsilon(t))| \leq C\left(t + \left|\int_0^t h(u_\varepsilon(s))\,ds\right|\right) + \varepsilon \qquad \forall t \geq 0. \tag{2.25}$$

On the other hand, by (2.24) we have

$$\int_0^t e^{-s/\mu}\left(\frac{1}{\mu}f(y_\varepsilon(s)) + h(u_\varepsilon(s))\right)ds + e^{-t/\mu}\varphi(y_\varepsilon(t))$$
$$\leq \int_0^t e^{-s/\mu}\left(\frac{1}{\mu}f(y_0) + h(u_0)\right)ds + e^{-t/\mu}\varphi(y_0(t)) + \varepsilon,$$

where $u_0 \in U$ is such that $h(u_0) < \infty$ and $y_0 = y(t, y, u_0)$. This yields

$$\int_0^t h(u_\varepsilon(s))\,ds \leq C\left(t + \int_0^t |u_\varepsilon(s)|_U\,ds\right) + \varepsilon \qquad \forall t > 0,$$

because h is bounded from below by an affine function. Then, by assumption (2.4), we see that

$$\left|\int_0^t h(u_\varepsilon(s))\,ds\right| + \int_0^t |u_\varepsilon(s)|_U\,ds \leq Ct + \varepsilon \qquad \forall t \geq 0.$$

Substituting this into (2.24) and using (2.25), we get

$$|\varphi(y(t)) - \varphi(x)| \leq C\left(t + \int_0^t |u(s)|_U\,ds\right) + \varepsilon \qquad \forall \varepsilon > 0, t > 0,$$

where C is independent of ε, u, and t. Hence, $\varphi \in \mathscr{D}$.

Since the space $BUC(\overline{D(M)}) \cap \mathrm{Lip}(\overline{D(M)})$ is dense in $BUC(\overline{D(M)})$ and the operator $(I + \mu\mathscr{A})^{-1}$ is nonexpansive on $Y = BUC(\overline{D(M)})$, we conclude that

$$(I + \mu\mathscr{A})^{-1}f \in \overline{\mathscr{D}} \qquad \forall f \in Y.$$

On the other hand, we have (see Proposition 3.2 in Chapter 2), for $\mu \to 0$,

$$(I + \mu\mathscr{A})^{-1} f \to f \qquad \text{in } Y,$$

for every $f \in \overline{D(\mathscr{A})}$. Along with (2.22), this implies that $\overline{D(\mathscr{A})} \subset \overline{\mathscr{D}}$.

To prove that $\overline{\mathscr{D}} \subset \overline{D(\mathscr{A})}$ consider f, an arbitrary element of $\mathscr{D}$, and set

$$f_\mu = (I + \mu\mathscr{A})^{-1} f = R(\mu^{-1})(\mu^{-1} f), \qquad \mu > 0.$$

We shall prove that $f_\mu \to f$ in Y as $\mu \to 0$. Since $f_\mu \in D(\mathscr{A})$ for all $\mu > 0$ this clearly will imply that $f \in \overline{D(\mathscr{A})}$, thereby completing the proof.

For every $\mu > 0$ and all $x \in \overline{D(M)}$, there are $(y_\mu, u_\mu) \in C([0,\infty); H) \cap L^1_{\text{loc}}(0,\infty); U)$, $y_\mu = y(t, x, u_\mu)$ such that

$$f_\mu(x) \geq \int_0^\infty e^{-t/\mu}\left(\frac{1}{\mu} f(y_\mu(t)) + h(u_\mu(t))\right) dt - \mu$$

and

$$f_\mu(x) \leq \int_0^\infty e^{-t/\mu}\left(\frac{1}{\mu} f(y(t,x,u)) + h(u(t))\right) dt$$

for all $u \in L^1_{\text{loc}}(0,\infty; U$. In particular, for $u \equiv u_0 \equiv$ constant and $y_0 = y(t, x, u_0)$ we have

$$\begin{aligned} f_\mu(x) - f(x) &\leq \frac{1}{\mu}\int_0^\infty e^{-t/\mu}(f(y_0(t)) - f(x) + \mu h(u_0))\, dt \\ &\leq \frac{C}{\mu}\int_0^\infty t e^{-t/\mu}\, dt + \mu|h(u_0)| = \mu(C + |h(u_0)|) \end{aligned}$$

$$\forall \mu > 0, \quad (2.26)$$

because $f \in \mathscr{D}$. Similarly,

$$\begin{aligned} &f(x) - f_\mu(x) \\ &\quad\leq \mu + \frac{1}{\mu}\int_0^\infty e^{-t/\mu}(f(x) - f(y_\mu(t)))\, dt - \int_0^\infty e^{-t/\mu} h(u_\mu(t))\, dt \\ &\quad\leq \mu + \frac{C}{\mu}\int_0^\infty t e^{-t/\mu}\, dt + \frac{C}{\mu}\int_0^\infty e^{-t/\mu}\left(\int_0^t |u_\mu(s)|_U\, ds\right) dt \\ &\qquad - \int_0^\infty e^{-t/\mu} h(u_\mu(t))\, dt \\ &\quad\leq C\left(\mu + \int_0^\infty e^{-t/\mu}|u(t)|_U\, dt\right). \end{aligned}$$

On the other hand, by (2.26) we see that

$$\int_0^\infty e^{-t/\mu} h\big(u_\mu(t)\big)\,dt \le C\mu + \frac{1}{\mu}\int_0^\infty e^{-t/\mu}\big(f(x) - f(y_\mu(t))\big)\,dt$$

$$\le C\left(\mu + \int_0^\infty e^{-t/\mu}|u_\mu(t)|_U\,dt\right) \qquad \forall \mu > 0.$$

Then, by assumption (2.4), we see that for every $\delta > 0$ there is $N(\delta)$ such that

$$N\int_{[|u_\mu(t)|_U \ge \delta]} e^{-t/\mu}|u_\mu(t)|_U\,dt$$

$$\le C\mu + \int_{[|u_\mu(t)|_U \ge \delta]} e^{-t/\mu}|u_\mu(t)|_U\,dt + C\mu\delta.$$

Hence,

$$\int_{[|u_\mu(t)|_U \ge \delta]} e^{-t/\mu}|u_\mu(t)|_U\,dt \le C\mu(\delta + 1)(N(\delta) - 1)^{-1}.$$

This implies that $\int_0^\infty e^{-t/\mu} h(u_\mu(t))\,dt \to 0$ as $\mu \to 0$, and so by (2.26), (2.27) we conclude that $f_\mu \to f$ in Y as $\mu \to 0$, as claimed. ■

Proof of Theorem 2.1. We shall prove first that $S(t)$ maps $\overline{D(\mathscr{A})} = \overline{\mathscr{D}}$ into itself. To this end, we fix φ_0 in $\mathscr{D}$. By the optimality principle we have, for all $0 \le s \le t < \infty$,

$$(S(t)\varphi_0)(x) = \inf\left\{\int_0^s (g(y(\tau)) + h(u(\tau)))\,d\tau + (S(t-s)\varphi_0)(y(s));\right.$$

$$\left. y' + My \ni Bu \text{ in } [0,s),\ y(0) = x,\ u \in L^1(0,t;U)\right\}.$$

Hence,

$$(S(t)\varphi_0)(x) - (S(t)\varphi_0)(y(s))$$

$$\le \int_0^s (g(y_0(\tau)) + h(u_0)))\,d\tau + \|S(s)\varphi_0 - \varphi_0\|_{\mathrm{b}}, \qquad (2.28)$$

where $u_0 \equiv$ constant, $h(u_0) < \infty$, and $y_0 = y(t, x, u_0)$. Moreover, there exist $y_\varepsilon, u_\varepsilon$ satisfying Eq. (2.3) and such that

$$(S(t)\varphi_0)(x) - (S(t)\varphi_0(y_\varepsilon(s)) \geq \int_0^s (g(y_\varepsilon(\tau)) + h(u_\varepsilon(\tau)))\, d\tau - \|S(s)\varphi_0 - \varphi_0\|_b - \varepsilon, \quad (2.29)$$

$$\int_0^s (g(y_\varepsilon(\tau)) + h(u_\varepsilon(\tau)))\, d\tau + (S(t-s)\varphi_0)(y_\varepsilon(s)) \leq (S(t)\varphi_0)(x) + \varepsilon \leq \int_0^s (g(y_0(\tau)) + h(u_0)))\, d\tau + (S(t-s)\varphi_0)(y_0(s)) + \varepsilon,$$

and yields

$$\int_0^s h(u_\varepsilon(\tau))\, d\tau \leq C\left(\varepsilon + s + \int_0^s |u_\varepsilon(\tau)|_U\, d\tau\right), \qquad 0 \leq s \leq t.$$

Then, using once again assumption (2.4), we get

$$\int_0^s |h(u_\varepsilon(\tau)|\, d\tau + \int_0^s |u_\varepsilon(\tau)|_U\, d\tau \leq Cs \qquad \forall s \in (0, t).$$

Now, coming back to inequalities (2.28), (2.29), we get

$$|(S(t)\varphi_0)(x) - (S(t)\varphi_0)(y(s))| \leq C\left(s + \int_0^s |u_\varepsilon(\tau)|_U\, d\tau\right) + \|S(s)\varphi_0 - \varphi_0\|_b \qquad \forall s \in (0, t), \quad (2.30)$$

where (y, u) is an arbitrary pair of functions satisfying (2.3).

On the other hand, by the definition of $S(t)$ it follows by similar argument that

$$|(S(s)\varphi_0)(x) - \varphi_0(x)| \leq Cs, \qquad \forall s > 0,\ x \in \overline{D(\mathscr{A})},$$

which along with (2.30) implies that $S(t)\varphi_0 \in \mathscr{D}$, as claimed.

Now we shall apply the nonlinear Chernoff theorem (Theorem 2.2 in Chapter 4) on the space Y where $C = \overline{\mathscr{D}}$, $F(t) = S(t)$, and $A = \mathscr{A}_0$, $\mathscr{A}_0 \varphi = \mathscr{A}\varphi + g$. To this end, we shall prove that for every $\mu > 0$,

$$\lim_{\varepsilon \downarrow 0} \left(I + \frac{\mu}{\varepsilon}(I - S(\varepsilon)) \right)^{-1} \varphi_0$$
$$= (I + \mu \mathscr{A}_0)^{-1} \varphi_0 = (I + \mu \mathscr{A})^{-1}(\varphi_0 + \mu g) \qquad \forall \varphi_0 \in \overline{\mathscr{D}}. \quad (2.31)$$

Let us postpone for the time being the proof of (2.31).

If (2.31) holds, then we have

$$T(t)\varphi_0 = \lim_{n \to \infty} \left(S\left(\frac{t}{n} \right) \right)^n \varphi_0 = S(t)\varphi_0 \qquad \forall \varphi_0 \in \overline{\mathscr{D}}, t \geq 0, \quad (2.32)$$

i.e., $T(t) \equiv S(t)$ on $\overline{\mathscr{D}} = \overline{D(\mathscr{A})}$.

Now, by definition of $S(t)$, we have

$$(S(t)\varphi_0)(x) - \varphi_0(x)$$
$$\leq \int_0^t (g(y(s)) + h(u(s)))\, ds + \varphi_0(y(t)) - \varphi_0(x) \quad (2.33)$$

for all (y, u) satisfying Eq. (2.3). If $\varphi_0 \in D(\mathscr{A})$, then there is $f \in Y\}$ *such that*

$$\varphi_0(x) = (R(1)f)(x)$$
$$= \inf\left\{ \int_0^t e^{-s}(f(y(s)) + h(u(s)))\, ds + e^{-t}\varphi_0(y(t)); \right.$$
$$\left. y' + My \ni Bu,\ y(0) = x,\ u \in L^1_{\text{loc}}(0, \infty; U) \right\}.$$

Then, by (2.33), *for every* $\varepsilon > 0$ *there exist* $u_\varepsilon \in L^1_{\text{loc}}(\mathbf{R}^+; U)$, $y_\varepsilon = y(t, x, u_\varepsilon)$ such that

$$(S(t)\varphi_0)(x) - \varphi_0(x)$$
$$\leq \int_0^t (1 - e^{-s}) h(u_\varepsilon(s))\, ds + \int_0^t g(y_\varepsilon(s))\, ds$$
$$-(e^{-t} - 1)\varphi_0(y_\varepsilon(t)) - \int_0^t e^{-s} f(y_\varepsilon(s))\, ds + \varepsilon, \qquad t \geq 0,$$
$$(2.34)$$

i.e.,

$$\int_0^t e^{-s}h(u_\varepsilon(s))\,ds + \int_0^t e^{-s}f(y_\varepsilon(s))\,ds + e^{-t}\varphi_0(y_\varepsilon(t))$$
$$\leq \varphi_0(x) + \varepsilon$$

This yields

$$\int_0^t e^{-t}h(u_\varepsilon(s))\,ds \leq C\left(t + \int_0^t |u_\varepsilon(s)|\,ds\right) + \varepsilon \qquad \forall x \in D(M),$$

because by Eq. (2.3) we have

$$|y_\varepsilon(t) - x| \leq t|Mx| + \int_0^t |Bu_\varepsilon(s)|\,ds \qquad \forall t \geq 0.$$

We may conclude, therefore, that

$$\int_0^t (|h(u_\varepsilon(s))| + |u_\varepsilon(s)|_U)\,ds \leq C(t + \varepsilon) \qquad \forall t \geq 0,$$

where C is independent of ε and t.

We take $\varepsilon = t^2$. Then, by (2.34), we have

$$\limsup_{t \downarrow 0} \frac{(S(t)\varphi_0)(x) - \varphi_0(x)}{t} \leq g(x) + \varphi_0(x) - f(x). \quad (2.35)$$

Similarly, we have

$$\varphi_0(x) - (S(t)\varphi_0)(x)$$
$$\leq \int_0^t e^{-s}(f(z_\varepsilon(s)) + h(v_\varepsilon(s)))\,ds + e^{-t}\varphi_0(z_\varepsilon(t))$$
$$- \int_0^t (g(z_\varepsilon(s)) + h(v_\varepsilon(s)))\,ds - \varphi_0(z_\varepsilon(t)) + \varepsilon, \quad (2.36)$$

where

$$z_\varepsilon' + Mz_\varepsilon \ni Bv_\varepsilon \qquad \text{in } (0,\infty), \qquad z_\varepsilon(0) = x,$$

and

$$\int_0^t (g(z_\varepsilon(s)) + h(v_\varepsilon(s)))\,ds + \varphi_0(z_\varepsilon(t)) \le (S(t)\varphi_0)(x) + \varepsilon$$

Then, arguing as before, we see that

$$\int_0^t (|h(v_\varepsilon(s))| + |v_\varepsilon(s)|_U)\,ds \le C(t+\varepsilon) \qquad \forall x \in D(M),$$

and so by (2.36) we have

$$\limsup_{t \downarrow 0} \frac{1}{t}(\varphi_0(x) - (S(t)\varphi_0)(x)) \le f(x) - \varphi_0(x) - g(x) \qquad \forall x \in D(M),$$

which along with (2.35) implies that

$$\lim_{t \downarrow 0} \frac{1}{t}((S(t)\varphi_0)(x) - \varphi_0(x)) = \varphi_0(x) - f(x) + g(x) \qquad \forall x \in D(M),$$

for $\varphi_0 = R(1)f$ and all $f \in Y$. Hence, $\mathscr{A}$ is single valued and

$$\lim_{t \downarrow 0} \frac{1}{t}(S(t)\varphi_0)(x) - \varphi_0(x)) = -(\mathscr{A}\varphi_0)(x) + g(x)$$

for all $x \in D(M)$ and all $\varphi_0 \in D(\mathscr{A})$, as desired.

To complete the proof, it remains to verify (2.31). Since $(I + \mu\mathscr{A})^{-1}$ and $(I + (\mu/\varepsilon)(I - S(\varepsilon)))^{-1}$ are nonexpansive on $\overline{\mathscr{D}} = \overline{D(\mathscr{A})}$, it suffices to prove (2.31) for $\varphi_0 \in \mathscr{D}$. We set

$$\varphi_\varepsilon = \left(I + \frac{\mu}{\varepsilon}(I - S(\varepsilon))\right)^{-1} \varphi_0, \qquad \varphi = (I + \mu\mathscr{A})^{-1}(\varphi_0 + \mu g).$$

We have

$$\varphi_\varepsilon = \frac{\varepsilon}{\varepsilon + \mu}\varphi_0 + \frac{\mu}{\varepsilon + \mu} S(\varepsilon)\varphi_\varepsilon. \tag{2.37}$$

Equivalently,

$$\varphi_\varepsilon(x) = \frac{\varepsilon}{\varepsilon+\mu}\varphi_0(x) + \frac{\mu}{\varepsilon+\mu}\inf\left\{\int_0^\varepsilon (g(y(t)) + h(u(t)))\,dt + \varphi_\varepsilon(y_\varepsilon(\varepsilon));\right.$$
$$\left. y' + My \ni Bu,\ y(0) = x,\ u \in L^1(0,\varepsilon;U)\right\}. \tag{2.38}$$

Recall that by the definition of $S(t)$ we have

$$\|S(\varepsilon)\varphi_0\|_{\mathrm{Lip}(\overline{D(M)})} \le \varepsilon\|g\|_{\mathrm{Lip}(\overline{D(M)})} + \|\varphi_0\|_{\mathrm{Lip}(\overline{D(M)})}$$

and

$$\|S(\varepsilon)\varphi_0\|_{\mathrm{b}} \le \varepsilon(\|g\|_{\mathrm{b}} + C) + \|\varphi_0\|_{\mathrm{b}} \qquad \forall g \in \mathrm{Lip}(\overline{D(M)}),$$

where $\|\cdot\|_{\mathrm{Lip}(\overline{D(M)})}$ is the Lipschitz norm on $\mathrm{Lip}(\overline{D(M)})$. Then, by (2.38), we get the estimates

$$\|\varphi_\varepsilon\|_{\mathrm{Lip}(\overline{D(M)})} \le \|\varphi_0\|_{\mathrm{Lip}(\overline{D(M)})} + \mu\varepsilon\|g\|_{\mathrm{Lip}(\overline{D(M)})},$$
$$\|\varphi_\varepsilon\|_{\mathrm{b}} \le C \qquad \forall \varepsilon > 0. \tag{2.39}$$

Now let $(y_\varepsilon, u_\varepsilon) \in C([0,\varepsilon];\ H) \times L^1(0,\varepsilon;\ U)$ be such that $y_\varepsilon = y(t, x, u_\varepsilon)$, i.e.,

$$y_\varepsilon' + My_\varepsilon \ni Bu_\varepsilon \qquad \text{in } (0,\infty), \qquad y_\varepsilon(0) = x,$$

and

$$\frac{\mu}{\varepsilon+\mu}\left(\int_0^\varepsilon (g(y_\varepsilon) + h(u_\varepsilon))\,dt + \varphi_\varepsilon(y_\varepsilon(\varepsilon))\right) + \frac{\varepsilon}{\varepsilon+\mu}\varphi_0(x)$$
$$\le \varphi_\varepsilon(x) + \varepsilon^2.$$

This yields

$$\frac{\mu}{\varepsilon+\mu}\int_0^\varepsilon h(u_\varepsilon)\,dt$$
$$\le \varphi_\varepsilon(y_0(\varepsilon)) - \varphi_\varepsilon(y_\varepsilon(\varepsilon)) + \frac{\mu}{\varepsilon+\mu}\int_0^\varepsilon (g(y_0(t)) - g(y_\varepsilon(t)))\,dt + C\varepsilon, \tag{2.40}$$

where $y_0 = y(t, x, u_0)$, $h(u_0) < \infty$.

Then, by the estimate (2.39), we have

$$\int_0^{\varepsilon} h(u_\varepsilon(t))\,dt \le C\left(\varepsilon + \int_0^{\varepsilon} |u_\varepsilon(t)|_U\,dt\right)$$

and, using once again assumption (2.4), we get

$$\int_0^{\varepsilon} (|h(u_\varepsilon(t))| + |u_\varepsilon(t)|_U)\,dt \le C\varepsilon \qquad \forall \varepsilon > 0, \tag{2.41}$$

where C is independent of ε. Since $\varphi_0 \in \mathscr{D}$, this implies that

$$|\varphi_0(y(t)) - \varphi_0(x)| \le C\varepsilon \qquad \forall x \in \overline{D(M)},\, t \in (0, \varepsilon). \tag{2.42}$$

We set $\varphi = (I + \mu\mathscr{A})^{-1}(\varphi_0 + \mu g)$. As noted earlier, we have

$$\begin{aligned}\varphi(x) = \inf\Bigg\{&\int_0^{\varepsilon} e^{-t/\mu}(g(y(t)) + \mu^{-1}\varphi_0(y(t)) + h(u(t)))\,dt \\ &+ e^{-\varepsilon/\mu}\varphi(y(\varepsilon));\ y' + My \ni Bu \text{ in } (0, \varepsilon),\ y(0) = x, \\ &u \in L^1(0, \varepsilon; U)\Bigg\},\end{aligned}$$

and so there exists a pair $(z_\varepsilon, v_\varepsilon)$ satisfying the Cauchy problem (2.3) on $(0, \varepsilon)$ and such that

$$\begin{aligned}\varphi(x) \ge &\int_0^{\varepsilon} e^{-t/\mu}(g(z_\varepsilon(t)) + \mu^{-1}\varphi_0(z_\varepsilon(t)) + h(v_\varepsilon(t)))\,dt \\ &+ e^{-\varepsilon/\mu}\varphi(z_\varepsilon(\varepsilon)) - \varepsilon^2.\end{aligned} \tag{2.43}$$

Then, arguing as before, we see that

$$\int_0^{\varepsilon} (|h(v_\varepsilon(t))| + |v_\varepsilon(t)|_U)\,dt \le C\varepsilon \qquad \forall \varepsilon > 0 \tag{2.44}$$

and, therefore,

$$|\varphi_0(z_\varepsilon(t)) - \varphi_0(x)| \le C\varepsilon \qquad \forall t \in (0, \varepsilon),\, x \in \overline{D(M)}. \tag{2.45}$$

Now, by (2.42), (2.43), it follows that

$$\begin{aligned}
&\varphi_\varepsilon(x) - \varphi(x) \\
&\quad \le \frac{\varepsilon}{\varepsilon+\mu}\varphi_0(x) + \frac{\mu}{\varepsilon+\mu}\int_0^\varepsilon (g(z_\varepsilon) + h(v_\varepsilon))\,dt \\
&\qquad + \frac{\mu}{\varepsilon+\mu}\varphi_\varepsilon(z_\varepsilon(\varepsilon)) - e^{-\varepsilon/\mu}\varphi(z_\varepsilon(\varepsilon)) \\
&\qquad - \int_0^\varepsilon e^{-t/\mu}(g(z_\varepsilon(t)) + \mu^{-1}\varphi_0(z_\varepsilon(t)) + h(v_\varepsilon(t)))\,dt + \varepsilon,
\end{aligned} \tag{2.46}$$

whilst by (2.41) and (2.43) we have

$$\begin{aligned}
&\varphi(x) - \varphi_\varepsilon(x) \\
&\quad \le \int_0^\varepsilon e^{-t/\mu}(g(y_\varepsilon) + \mu^{-1}\varphi_0(y_\varepsilon) + h(u_\varepsilon))\,dt + e^{-\varepsilon/\mu}\varphi(y_\varepsilon(\varepsilon)) \\
&\qquad - \frac{\varepsilon}{\varepsilon+\mu}\int_0^\varepsilon (g(y_\varepsilon) + h(u_\varepsilon))\,dt - \frac{\varepsilon}{\varepsilon+\mu}\varphi_0(x) + \varepsilon.
\end{aligned} \tag{2.47}$$

Using the estimates (2.39), (2.41), (2.42), (2.44), and (2.45), in inequalities (2.46), (2.47), we see that

$$\|\varphi_\varepsilon - \varphi\|_b \le \delta(\varepsilon) \to 0 \qquad \text{as } \varepsilon \to 0,$$

which completes the proof of (2.31). ■

Remark 2.1. Theorem 2.1 is related to existence of Lie generators for nonlinear semigroups of contractions (J. W. Neuberger [1, 2]). Indeed, if $g \equiv 0$, $B = 0$, and

$$h(u) = \begin{cases} 0 & \text{if } u = 0, \\ \infty & \text{otherwise}, \end{cases}$$

then $S(t) \equiv S^M(t)$, i.e.,

$$(S(t)\varphi)(x) = \varphi(e^{-Mt}x) \qquad \forall x \in D(M),$$

for all $\varphi \in Y = BUC(\overline{D(M)})$.

Then, by Theorem 2.1, it follows in particular that this semigroup is generated on Z by a single valued m-accretive operator.

6.2.3. *Approximation of the Dynamic Programming Equation*

We shall consider here the forward Hamilton–Jacobi equation (2.1), where $M = A + \partial\varphi$ and $A \in L(V, V')$, $\varphi: H \to \overline{\mathbf{R}}$, g, h, and φ_0 satisfy the assumptions of Section 1.2. As mentioned earlier, the function

$$\psi(t,x) = \inf\left\{\int_0^t (g(y(t)) + h(u(t)))\,dt + \varphi_0(y(t));\right.$$
$$y' + Ay + \partial\varphi(y) \ni Bu \text{ a.e. in } (0,t),$$
$$\left. y(0) = x,\, u \in L^1(0,t;U)\right\}, \tag{2.48}$$

called a *variational solution* to Eq. (2.1), i.e.,

$$\psi_t(t,x) + h^*(-B^*\psi_x(t,x)) + (Ax + \partial\varphi(t,x), \psi_x(t,x)) = g(x),$$
$$(t,x) \in (0,T) \times H,$$
$$\psi(0,x) = \varphi_0(x), \tag{2.49}$$

is under certain regularity assumptions on φ, a strong solution in some generalized sense (Theorem 1.2). It should be said that due to its complexity Eq. (2.49) is hard to solve or approximate by standard methods. Here, we will briefly discuss the Lie–Trotter scheme (the method of fractional steps) for a such a equation. Formally, a such a scheme for Eq. (2.49) is defined by

$$\psi_t^\varepsilon(t,x) + h^*(-B^*\psi_x^\varepsilon(t,x)) + (Ax, \psi_x^\varepsilon(t,x)) = 0$$
$$\text{in } [i\varepsilon, (i+1)\varepsilon] \times \overline{D(\varphi)},$$
$$\psi_+^\varepsilon(i\varepsilon, x) = \psi_+^\varepsilon(i\varepsilon, e^{-\varepsilon\,\partial\varphi}x) + \varepsilon g(e^{-\varepsilon\,\partial\varphi}x), \qquad i = 1, \ldots, N-1,$$
$$\psi^\varepsilon(0,x) = \varphi_0(x) \qquad \forall x \in \overline{D(\varphi)}, \tag{2.50}$$

where $N\varepsilon = T$.

In general, we do not know whether this scheme is convergent for $\varepsilon \to 0$. However, this happens in some significant situations and in particular if $\varphi = I_C$, where C is a closed subset of H such that

$$(APy, Py) \le (Ay, y) \qquad \forall y \in V,$$
$$(I + \lambda A_H)^{-1} C \subset C \qquad \forall \lambda > 0. \tag{2.51}$$

(Here, P is the projection operator on C.) We recall that $e^{-t\,\partial\varphi} = P$ $\forall t \ge 0$.

As seen earlier, this corresponds to an optimal control problem governed by the variational inequality

$$(y'(t) + Ay(t) - Bu(t), y(t) - z) \leq 0 \qquad \text{a.e. } t \in (0,T), \forall z \in C,$$
$$y(0) = y_0 .$$

Theorem 2.2. *Under the preceding assumptions*

$$\psi^{\varepsilon}(t,x) \overset{\varepsilon \to 0}{\to} \psi(t,x) \qquad \forall (t,x) \in [0,T] \times C, \tag{2.52}$$

where ψ *is the variational solution to* (2.48). ■

Theorem 2.2 is a direct consequence of convergence Theorem, 4.1, in Chapter 5 but we omit the details (see the author's paper [15]).

It should be observed that Eq. (2.50) is structurally simpler than the original equation and its solution φ can be explicitly written. Indeed, we have

$$\psi^{\varepsilon}(t,x) = \inf\left\{\int_{i\varepsilon}^{t} h(u)\,ds + \psi_{-}^{\varepsilon}(i\varepsilon, Py(t)) + \varepsilon g(y(t));\right.$$
$$\left. y' + Ay = Bu \text{ in } (i\varepsilon, t),\ y(0) = x,\ u \in L^1(0,t;U)\right\},$$
$$t \in [i\varepsilon, (i+1)\varepsilon], \tag{2.53}$$

and in view of the maximum principle (optimality system) we have

$$\psi^{\varepsilon}(t,x) = \int_{i\varepsilon}^{t} h(u_{\varepsilon}^{*})\,ds + \psi_{-}^{\varepsilon}(i\varepsilon, Py_{\varepsilon}^{*}(t)) + \varepsilon g(Ty_{\varepsilon}^{*}(t)),$$

where

$$\begin{aligned} (y_{\varepsilon}^{*})' + Ay_{\varepsilon}^{*} &= Bu_{\varepsilon}^{*} && \text{in } (i\varepsilon, t), \\ p_{\varepsilon}' - Ap_{\varepsilon} &= 0 && \text{in } (i\varepsilon, t), \\ y_{\varepsilon}^{*}(i\varepsilon) &= x, && \\ u_{\varepsilon}^{*}(s) &\in \partial h^{*}(B^{*}p_{\varepsilon}(s)) && \text{a.e. } s \in (i\varepsilon, t). \end{aligned}$$

Hence,

$$u_\varepsilon^*(s) \in \partial h^*(B^* e^{-(t-s)A} p) \qquad \text{a.e. } s \in (i\varepsilon, t),$$

for some $p \in H$. Substituting this into (2.53), we get

$$\begin{aligned}\psi^\varepsilon(t,x) = \inf_{p \in H} \bigg\{ & \int_{i\varepsilon}^t h(\partial h^*(B^* e^{-(t-s)A} p))\, ds + \varepsilon g(e^{-(t-i\varepsilon)A} x) \\ & + \psi_-^\varepsilon \bigg(i\varepsilon, \bigg(P \bigg(e^{-(t-i\varepsilon)A} x \\ & + \int_{i\varepsilon}^t e^{-(t-s)A} B\, \partial h^*(B^* e^{-(t-s)A} p)\, ds \bigg)\bigg)\bigg) \\ & \text{for } t \in [i\varepsilon, (i+1)\varepsilon],\ x \in C. \qquad (2.54)\end{aligned}$$

Moreover,

$$u = \partial h^*\big(-B^* \psi_y^\varepsilon(T-t, y)\big) \qquad \forall t \in [0,T],$$

is a suboptimal feedback control for problem (P) in Section 1.2.

We will illustrate this on the following example: $H = L^2(\Omega)$, $V = H_0^1(\Omega)$, $A = -\Delta$, $C = \{y \in H_0^1(\Omega);\ y(x) \geq 0 \text{ a.e. } x \in \Omega\}$, $B \equiv I$, $h(u) = \frac{1}{2}\int_\Omega u^2\, dx$, $g \equiv 0$, and $\varphi_0(y) = \frac{1}{2}\int_\Omega (y - y^0)^2\, dx$. This corresponds to following optimal control problem:

$$\text{Minimize} \quad \frac{1}{2}\int_Q u^2\, dx\, dt + \frac{1}{2}\int_\Omega \left(y(x,T) - y^0(x)\right)^2 dx$$

on all $u \in L^2(Q)$, $Q = \Omega \times (0,T)$ *and* $y \in L^2(0,T;\ H_0^1(\Omega) \cap H^2(\Omega))$, $y_t \in L^2(Q)$, *subject to*

$$\frac{\partial y}{\partial t} - \Delta y = u \qquad \text{in } \{(x,t) \in Q;\ y(x,t) > 0\},$$

$$\frac{\partial y}{\partial t} - \Delta y \geq u, \quad y \geq 0 \qquad \text{in } Q,$$

$$y(x,0) = y_0(x) \qquad \text{in } \Omega.$$

In this case, Eq. (2.49) becomes

$$\psi_t(t,y) + \tfrac{1}{2}|\psi_y(t,y)|_2^2 + \int_\Omega (N_C(y) - \Delta y)\psi_y(t,y)\,dx = 0,$$

$$\psi(0,y) = \tfrac{1}{2}\int_\Omega (y(x) - y^0(x))^2\,dx \qquad \forall y \in C, \tag{2.55}$$

where

$$N_C(y) = \{w \in L^2(\Omega);\ w(x) = 0 \text{ a.e. in } [x \in \Omega;\ y(x) > 0],$$
$$w(x) \le 0 \qquad \text{a.e. in } [x \in \Omega;\ y(x) = 0]\}.$$

By Theorem 2.2, we have

$$\psi(t,y) = \lim_{\varepsilon \to 0} \psi^\varepsilon(t,y) \qquad \forall y \in C,\ t \in [0,T],$$

where, in view of (2.54),

$$\psi^\varepsilon(t,y) = \inf_{p \in L^2(\Omega)} \left\{ \tfrac{1}{2}\int_{i\varepsilon}^t \int_\Omega z^2(x,s)\,ds\,dx + \psi_-^\varepsilon(i\varepsilon, w^+(t - i\varepsilon)) \right\}$$

$$\forall t \in [i\varepsilon, (i+1)\varepsilon],\ y \in L^2(\Omega), \tag{2.56}$$

where $w^+ = \max(w, 0)$ and

$$\begin{aligned}
\frac{\partial z}{\partial s} + \Delta z &= 0 && \text{in } (i\varepsilon, t) \times \Omega,\ i = 0,1,\ldots,N-1,\\
z &= 0 && \text{in } (i\varepsilon, t) \times \partial\Omega,\\
z(t,x) &= p(x) && x \in \Omega,\\
\frac{\partial w}{\partial s} - \Delta w &= z && \text{in } (i\varepsilon, t) \times \Omega,\\
w &= 0 && \text{in } (i\varepsilon, t) \times \partial\Omega,\\
w(i\varepsilon, x) &= y(x) && \text{in } \Omega.
\end{aligned} \tag{2.57}$$

Thus, at every step the calculation of ψ^ε reduces to a minimization problem on the space $L^2(\Omega)$, which can be solved numerically by standard methods.

As a second example, we shall consider the dynamic programming equation associated with the optimal control problem

$$\text{Minimize} \quad \left\{ \int_0^T g(y(t))\,dt + \varphi_0(y(T)); \right.$$

$$\left. y' + \beta(y) = u, |u(t)| \le 1, y(0) = y_0 \right\}, \tag{2.58}$$

i.e.,

$$\begin{aligned} \psi_t + |\psi_x| + \beta(x)\psi_x = g(x), \qquad & x \in \mathbf{R}, t \in (0,T), \\ \psi(0,x) = \varphi_0(x), \qquad & x \in \mathbf{R}, \end{aligned} \tag{2.59}$$

where β is a locally Lipschitz function. In this case, we have $\psi(t,x) = \lim_{\varepsilon \to 0} \psi^{\varepsilon}(t,x)\ \forall (t,x) \in [0,T] \times \mathbf{R}$, where

$$\begin{aligned} &\psi_t^{\varepsilon} + |\psi_x^{\varepsilon}| = 0 \qquad \text{in } [i\varepsilon, (i+1)\varepsilon] \times \mathbf{R}, \\ &\psi_+^{\varepsilon}(i\varepsilon, x) = \varepsilon g(x) + \psi_-^{\varepsilon}(i\varepsilon, z(\varepsilon)), \qquad i = 0,1,\ldots,N-1, \end{aligned} \tag{2.60}$$

$$\begin{aligned} z' + \beta(z) = 0 \qquad & \text{in } (0,\varepsilon), \\ z(0) = x, & \end{aligned} \tag{2.61}$$

and arguing as in the proof of (2.54) we get

$$\begin{aligned} \psi^{\varepsilon}(t,x) = \inf_{|p| \le 1} \{ & \varepsilon g(x + (t - i\varepsilon)p) + \psi_-^{\varepsilon}(i\varepsilon, w(\varepsilon)); \\ & w' + \beta(w) = 0 \quad \text{in } (0,\varepsilon), w(0) = x + (t - i\varepsilon)p \}, \\ u = -\operatorname{sign} & \psi_y^{\varepsilon}(T - t, y) \end{aligned} \tag{2.62}$$

as an approximating (suboptimal) feedback control of problem (2.58).

Bibliographical Notes and Remarks

Section 1. Most of the results of this section have been previously established in a related form in the works of Barbu and DaPrato [1] and Barbu [6, 7, 10]. For a recent treatment of dynamic programming for finite dimensional optimal control problems, we refer to the book of St. Mirica [1]. For other related results we refer to the recent works of Cannarsa and Frankowska [1], Cannarsa and DaPrato [1], Barbu, Barron, and Jensen [1], and Fattorini and Sritharan [1] (the last concerned with optimal feedback controllers for the Navier–Stokes systems).

The theory of viscosity solutions to Hamilton–Jacobi equations was developed in the works of M. G. Crandall and P. L. Lions [1–3], and M. G. Crandall *et al.* [1].

In particular, the results of D. Tătaru [1] cover the existence and uniqueness theory of viscosity solution for the dynamic programming equation associated with problem (P) or the time-optimal problem (in this context, we mention also the work of M. Bardi [1]).

Section 2. The results of Section 2.1 and 2.2 have been previously established in the author's work [13]. The semigroup approach to Hamilton–Jacobi equations was also used by Barbu and DaPrato [2] and Th. Havârneanu [1] (The latter work extends the results of Section 2.1 to Hamilton–Jacobi equations with nonconvex Hamiltonian.)

Theorem 2.2 along with other results of this type were given in the author's work [14, 15]. Related results were obtained in a more general context by C. Popa [1] and Th. Havârneanu [2]. We mention also the work of L. C. Evans [2] for a min–max type approximation formula for the solutions to Hamilton–Jacobi equations.

References

S. Agmon, A. Douglas, and L. Nirenberg

[1] Estimates near boundary for solutions of elliptic partial differential equations satisfying general boundary conditions, *Comm. Pure Appl. Math.* **12**(1959), pp. 623–727.

S. Aniţa

[1] Optimal control of a nonlinear population dynamics with diffusion, *J. Math. Anal. Appl.* **152**(1990), pp. 176–208.

V. Arnăutu

[1] Approximation of optimal distributed control problems governed by variational inequalities, *Numer. Math.* **38**(1982), pp. 393–416.

[2] On approximation of the inverse one-phase Stefan problem, *Numerical Methods for Free Boundary Problems*, pp. 69–82, P. Neittaanmäki, ed., Birkhäuser, Basel, Boston, Berlin 1991.

V. Arnăutu and V. Barbu

[1] Optimal control of the free boundary in a two-phase Stefan problem, *Preprint Series in Mathematics* 11, INCREST, Bucharest 1985.

E. Asplund

[1] Averaged norms, *Israel J. Math.* **5**(1967), pp. 227–233.

H. Attouch

[1] Familles d'opérateurs maximaux monotones et mesurabilité, *Annali Mat. Pura Appl.* **CXX**(1979), pp. 35–111.

[2] *Variational Convergence for Functions and Operators*, Pitman, Boston, London, Melbourne 1984.

H. Attouch and A. Damlamian

[1] Problèmes d'évolution dans les Hilbert et applications, *J. Math. Pures Appl.* **54**(1975), pp. 53–74.

[2] Solutions fortés d'inéquations variationnelles d'évolution, *Publication Mathématique d'Orsay* 184, Université Paris XI 1976.

P. Aubin and A. Cellina

[1] *Differential Inclusions*, Springer-Verlag, Berlin, New York, Heidelberg 1984.

A. V. Balakrishnan

[1] *Applied Functional Analysis*, Springer-Verlag, Berlin, Heidelberg, New York 1976.

H. T. Banks and K. Kunisch

[1] *Estimation Techniques for Distributed Parameter Systems*, Birkhäuser, Boston —Basel Stuttgart 1989.

P. Baras

[1] Compacité de l'opérateur $f \to u$ solution d'une équation nonlinéaire $du/dt + Au \ni f$, *C. R. Acad. Sci. Paris* **286**(1978), pp. 1113–1116.

V. Barbu

[1] Continuous perturbation of nonlinear m-accretive operators in Banach spaces, *Boll. Unione Mat. Ital.* **6**(1972), pp. 270–278.

[2] *Nonlinear Semigroups and Differential Equations in Banach Spaces*, Noordhoff International Publishing, Leyden 1976.

[3] Necessary conditions for nonconvex distributed control problems governed by elliptic variational inequalities, *J. Math. Anal. Appl.* **80**(1981), pp. 566–597.

[4] Necessary conditions for distributed control problems governed by parabolic variational inequalities, *SIAM J. Control and Optimiz.* **19**(1981), pp. 64–86.

[5] Boundary control problems with nonlinear state equations, *SIAM J. Control and Optimiz.* **20**(1982), pp. 125–143.

[6] Optimal feedback controls for a class of nonlinear distributed parameter systems, *SIAM J. Control and Optimiz.* **21**(1983), pp. 871–894.

[7] *Optimal Control of Variational Inequalities*, Research Notes in Mathematics 100, Pitman, Boston 1984.

[8] Optimal feedback controls for a class of nonlinear distributed parameter systems, *SIAM J. Control and Optimiz.* **21**(1983), pp. 871–894.

[9] The time optimal control problem for parabolic variational inequalities, *Appl. Math. Optimiz.* **116**(1984), pp. 1–22.

[10] The time optimal control of variational inequalities. Dynamic programming and the maximum principle, *Recent Mathematical Methods in Dynamic Programming*, pp. 1–19, Capuzo Dolceta, W. H. Fleming, and T. Zolezzi, eds., Lecture Notes in Mathematics 1119, Springer-Verlag, Berlin, Heidelberg, New York, Tokyo 1985.

[11] Optimal control of free boundary problems, *Confer. Sem. Mat. Univ. Bari* (1985), No. **206**.

[12] The time optimal problem for a class of nonlinear distributed systems, *Control Problems for Systems Described by Partial Differential Equations*, pp. I. Lasiecka and R. Triggiani, eds. Lecture Notes in Control and Information Sciences **97**, Springer-Verlag, Berlin, Heidelberg, New York, Tokyo 1987.

[13] A semigroup approach to Hamilton–Jacobi equation in Hilbert space, *Studia Univ. Babes–Bolyai, Mathematica* **XXXIII**(1988), pp. 63–78.

[14] A product formula approach to nonlinear optimal control problems, *SIAM J. Control and Optimiz.* **26**(1988), pp. 497–520.

[15] Approximation of the Hamilton–Jacobi equations via Lie–Trotter product formula, *Control Theory and Advanced Technology* **4**(1988), pp. 189–208.

[16] The approximate solvability of the inverse one phase Stefan problem, *Numerical Methods for Free Boundary Problems*, pp. 33–43, P. Neittaanmäki, ed., Birkhäuser 1991.

[17] Null controllability of first order quasi linear equations, *Diff. Integral Eqs.* **4**(1991), pp. 673–681.

[18] The minimal time function for the nonlinear diffusion equation, *Libertas Mathematica* **13**(1990), pp.

[19] The fractional step method for a nonlinear distributed control problem, Differential Equations and Control Theory; V. Barbu, ed., pp. 7–17 *Research Notes in Mathematics* 250, Longman, London 1991.

[20] The dynamic programming equation for the time optimal control problem in infinite dimension, *SIAM J. Control and Optimiz.* **29**(1991), pp. 445–456.

V. Barbu and N. Barron

[1] Bang-bang controllers for an optimal cooling problem, *Control and Cybernetics* **16**(1987), pp. 91–102.

V. Barbu, N. Barron, and R. Jensen

[1] The necessary conditions for optimal control in Hilbert spaces, *J. Math. Anal. Appl.* **133**(1988), pp. 151–162.

V. Barbu and G. DaPrato

[1] *Hamilton–Jacobi equations in Hilbert spaces*, Research Notes in Mathematics 86, Pitman, Boston, 1984.

[2] Hamilton–Jacobi equations in Hilbert spaces; Variational and semigroup approach, *Annalli Mat. Pura Appl.* **CXLII**(2)(1985), pp. 303–349.

V. Barbu, G. DaPrato, and R. P. Zolessio

[1] Feedback controllability of the free boundary of the one phase Stefan problem, *Diff. Integral Eqs.* **4**(1991), pp. 225–239.

V. Barbu and A. Friedman

[1] Optimal design of domains with free boundary problems, *SIAM J. Control and Optimiz.* **29**(1991), no. 3, pp. 623–637.

V. Barbu and Ph. Korman

[1] Approximating optimal controls for elliptic obstacle problem by monotone interation scheme, *Numer. Funct. Anal.* **12**(1991), pp. 429–442.

V. Barbu and T. Precupanu

[1] *Convexity and Optimization in Banach Spaces*, D. Reidel, Dordrecht 1986.

V. Barbu and S. Stojanovic

[1] Controlling the free boundary of elliptic variational inequalities on a variable domain, (to appear).

V. Barbu and D. Tiba

[1] Boundary controllability for the coincidence set in the obstacle problem, *SIAM J. Control and Optimiz.* **29**(1991), pp. 1150–1159.

M. Bardi

[1] A boundary value problem for the minimum time function, *SIAM J. Control Optimiz.* **27**(1989), pp. 776–785.

Ph. Bénilan

[1] *Equations d'évolution dans un espace de Banach quelconque et applications*, Thèse, Orsay 1972.

[2] Opérateurs accretifs et semi-groups dans les espaces L^p, $1 \leq p \leq \infty$, *Functional Analysis and Numerical Analysis*, pp. 15–51, I. Fujita, ed., Japan Society for Promotion of Science, Tokyo, 1978.

Ph. Bénilan, M. G. Crandall, and M. Pierre

[1] Solutions of the porous medium equations in $\mathbf{R}^N$ under optimal conditions on initial values, *Indiana Univ. Math. J.* **33**(1984), pp. 51–87.

Ph. Bénilan and S. Ismail

[1] Générateurs des semigroupes nonlinéaires et la formule de Lie–Trotter, *Annales Faculté de Science, Toulouse* **VII**(1985), pp. 151–160.

A. Bermudez and C. Saguez

[1] Optimal control of variational inequalities. Optimality conditions and numerical methods, *Free Boundary; Theory and Applications*, Maubuisson 1984, pp. 478–487, Research Notes in Mathematics 121, Pitman, Boston, 1985.

[2] Optimal control of a Signorini problem, *SIAM J. Control and Optimiz.* **25**(1987), pp. 576–582.

[3] Optimal control of variational inequalities, *Control and Cybernetics* **14**(1985), pp. 9–30.

[4] Optimality conditions for optimal control problems of variational inequalities, *Control Problems for Systems Described by Partial Differential Equations and Applications*. R. Lasiecka and I. Triggiani, eds. Lecture Notes in Control and Information Sciences 97, Springer-Verlag 1987.

F. Bonnans

[1] Analysis and control of a nonlinear parabolic unstable system, *J. Large Scale Systems* **6**(1984), pp. 249–262.

F. Bonnans and E. Casas

[1] Quelques méthodes pour le contrôle optimal de problèmes comportant des contraints sur l'état, *An. St. Univ. "Al. I. Cuza" Iaşi* 1986, pp. 58–62.

F. Bonnans and D. Tiba

[1] Pontryagin's principle in the control of semilinear elliptic variational inequalities, *Appl. Math. Optimiz.* **23**(1991), pp. 299–312.

H. Brézis

[1] Propriétés régularisantes de certaines semi-groupes nonlinéaires, *Isreal J. Math.* **9**(1971), pp. 513–514.

[2] Monotonicity methods in Hilbert spaces and some applications to nonlinear partial differential equations, *Contributions to Nonlinear Functional Analysis*, E. Zarantonello, ed. Academic Press, New York 1971.

[3] Problèmes unilatéraux, *J. Math. Pures Appl.* **51**(1972), pp. 1–168.

[4] *Opérateurs Maximaux Monotones et Semigroupes de Contractions dans un Espace de Hilbert*, North Holland 1973.

[5] Integrales convexes dans les espaces de Sobolev, *Isreal J. Math.* **13**(1972), pp. 9–23.

[6] New results concerning monotone operators and nonlinear semigroups, *Analysis of Nonlinear Problems*, pp. 2–27, RIMS 1974.

[7] *Analyse Fonctionnelle. Théorie et Applications*, Masson, Paris 1983.

H. Brézis, M. G. Crandall, and A. Pazy

[1] Perturbations of nonlinear maximal monotone sets, *Comm. Pure Appl. Math.* **13**(1970), pp. 123–141.

H. Brézis and F. Browder

[1] Some properties of higher order Sobolev spaces, *J. Math. Pures Appl.* **61**(1982), pp. 245–259.

H. Brézis and A. Friedman

[1] Nonlinear parabolic equations involving measures as initial conditions, *J. Math. Pures Appl.* **62**(1983), pp. 73–97.

H. Brézis and A. Pazy

[1] Semigroups of nonlinear contractions on convex sets, *J. Funct. Anal.* **6**(1970), pp. 367–383.

[2] Convergence and approximation of semigroups of nonlinear operators in Banach spaces, *J. Funct. Anal.* **9**(1971), pp. 63–74.

H. Brézis and G. Stampacchia

[1] Sur la regularité de la solution d'inéquations ellyptiques, *Bull. Soc. Math. France*, **96**(1968), pp. 153–180.

H. Brézis and W. Strauss

[1] Semilinear second order elliptic equations in L^1, *J. Math. Soc. Japan* **25**(1973), pp. 565–590.

F. E. Browder

[1] *Problèmes Nonlinéaires*, Les Presses de l'Université de Montréal 1966.

[2] *Nonlinear Operators and Nonlinear Equations of Evolution in Banach Spaces*, Nonlinear Functional Analysis, Symposia in Pure Math., vol. 18, Part 2. F. Browder ed. Amer. Math. Soc., Providence, Rhode Island 1970.

P. Cannarsa and G. DaPrato

[1] Some results on nonlinear optimal control problems and Hamilton–Jacobi equations in infinite dimensions, *J. Funct. Anal.*

P. Cannarsa and H. Frankowska

[1] Value functions and optimality conditions for semilinear control problems (to appear).

O. Cârjă

[1] On the minimal time function for distributed control systems in Banach spaces, *JOTA* **44**(1984), pp. 397–406.

C. Castaign and M. Valadier

[1] *Convex Analysis and Measurable Multifunctions*, Lecture Notes in Mathematics 580, Springer-Verlag, Berlin, Heidelberg 1977.

P. Chernoff

[1] Note on product formulas for operator semi-groups, *J. Funct. Anal.* **2**(1968), pp. 238–242.

F. Clarke

[1] Generalized gradients and applications, *Trans. Amer. Math. Soc.* **205**(1975), pp. 247–262.

[2] *Optimization and Nonsmooth Analysis*, John Wiley and Sons, New York 1983.

F. Clarke and R. Vinter

[1] The relationship between the maximum principle and dynamic programming, *SIAM J. Control and Optimiz.* **25**(1987), pp. 1291–1311.

M. G. Crandall

[1] The semigroup approach to first-order quasilinear equations in several space variables, *Isreal J. Math.* **12**(1972), pp. 108–132.

[2] Nonlinear semigroups and evolutions generated by accretive operators, *Nonlinear Functional Analysis and its Applications* pp. 305–338, F. Browder, ed., American Mathematical Society, Providence, Rhode Island 1986.

M. G. Crandall and L. C. Evans

[1] On the relation of the operator $\partial/\partial s + \partial/\partial t$ to evolution governed by accretive operators, *Isreal J. Math.* **21**(1975), pp. 261–278.

M. G. Crandall, L. C. Evans, and P. L. Lions

[1] Some properties of viscosity solutions of Hamilton–Jacobi equations, *Trans. Amer. Math. Soc.* **282**(1984), pp. 487–502.

M. G. Crandall and T. M. Liggett

[1] Generation of semigroups of nonlinear transformations in general Banach spaces, *Amer. J. Math.* **93**(1971), pp. 265–298.

M. G. Crandall and P. L. Lions

[1] Viscosity solutions of Hamilton–Jacobi equations, *Trans. Amer. Math. Soc.* **277**(1983), pp. 1–42.

[2] Hamilton–Jacobi equations in infinite dimension. I. Uniqueness of viscosity solutions, *J. Funct. Anal.* **62**(1985), pp. 379–396; II. Existence of viscosity solutions, *ibid.* **65**(1986), pp. 368–405.

[3] Viscosity solutions of Hamilton–Jacobi equations in infinite dimensions. Part V. Unbounded linear terms and B-continuous solutions. *J. Funct. Anal.* **90**(1990), pp. 273–283.

M. G. Crandall and A. Pazy

[1] Semigroups of nonlinear contractions and dissipative sets, *J. Funct. Anal.* **3**(1969), pp. 376–418.

[2] On accretive sets in Banach spaces. *J. Funct. Anal.* **5**(1970), pp. 204–217.

[3] Nonlinear evolution equations in Banach spaces, *Israel J. Math.* **11**(1972), pp. 57–92.

[4] On the range of accretive operators, *Israel J. Math.* **27**(1977), pp. 235–246.

M. G. Crandall and M. Pierre

[1] Regularizing effect for $u_t = \Delta\varphi(u)$, *Trans. Amer. Math. Soc.* **274**(1982), pp. 159–168.

C. Dafermos and M. Slemrod

[1] Asymptotic behaviour of nonlinear contraction semigroups, *J. Funct. Anal.* **3**(1973), pp. 97–106.

J. Dautray and J. L. Lions

[1] *Mathematical Analysis and Numerical Methods for Science and Technology*, Springer-Verlag, Berlin, Heidelberg, New York, Tokyo 1982.

E. DeGiorgi, M. Degiovanni, A. Marino, and M. Tosques

[1] Evolution equations for a class of non-linear operators, *Atti Acad. Naz. Lincei Rend.* **75**(1983), pp. 1–8.

M. Degiovanni, A. Marino, and M. Tosques

[1] Evolution equations with lack of convexity, *J. Nonlinear Anal. Theory and Appl.* **9**, 1985, pp. 1401–1443.

K. Deimling

[1] *Nonlinear Functional Analysis*, Springer-Verlag, Berlin, New York, Heidelberg 1985.

J. I. Diaz

[1] *Nonlinear Partial Differential Equations and Free Boundaries*. Vol. I, *Elliptic Equations*, Research Notes in Mathematics 106, Pitman, Boston, London, Melbourne 1985.

[2] Qualitative properties of solutions of some nonlinear diffusion equations via a duality argument, *Semigroup Theory and Applications*, H. Brézis, M. G. Crandall, F. Kappel, eds. Research Notes in Mathematics 141, Longman 1986.

G. Duvaut

[1] Résolution d'un problème de Stefan, *C. R. Acad. Sci. Paris*, **267**(1973), pp. 1461–1463.

G. Duvaut and J. L. Lions

[1] *Inequalities in Mechanics and Physics*, Springer-Verlag, Berlin, New York, Heidelberg 1976.

R. E. Edwards

[1] *Functional Analysis*, Holt, Rinehart, and Winston, New York 1965.

C. M. Elliott and J. R. Ockendon

[1] *Weak and Variational Methods for Moving Boundary Problems*, Research Notes in Mathematics 59, Pitman, Boston 1982.

L. C. Evans

[1] Differentiability of a nonlinear semigroup in L^1, *J. Math. Anal. Appl.* **60**(1977), pp. 703–715.

[2] Some mini-max methods for the Hamilton–Jacobi equations, *Indiana Univ. Math. J.* **33**(1985), pp. 31–50.

H. O. Fattorini

[1] The time-optimal control problem in Banach spaces, *Appl. Math. Optimiz.* **1**(1974), pp. 163–188.

[2] The time optimal problem for boundary control of heat equation, *Calculus of Variations and Control Theory*, pp. 305–320, D. Russell, ed. Academic Press, New York 1976.

[3] A unified theory of necessary conditions for nonlinear nonconvex control problems, *Appl. Math. Optimiz.* **15**(1987), pp. 141–185.

H. O. Fattorini and S. S. Sritharan

[1] Optimal control theory for viscous flow problems, *Arch. Rat. Mech. Anal.* (to appear).

A. F. Filipov

[1] *Differential Equations with Discontinuous Right Hand Side*, Kluwer Academic Publishers, Dordrecht, Boston, London 1985.

P. M. Fitzpatrick

[1] Surjectivity results for nonlinear mappings from a Banach space to its dual, *Math. Annalen* **204**(1973), pp. 177–188.

A. Friedman

[1] *Variational Principles and Free Boundary Problems*, John Wiley and Sons, New York, Chichester, Brisbane, Toronto, Singapore 1982.

[2] Optimal control for variational inequalities, *SIAM J. Control and Optimiz.* **24**(1986), pp. 439–451.

[3] Optimal control for parabolic variational inequalities, *SIAM J. Control and Optimiz.* **25**(1987), pp. 482–497.

[4] Optimal control for free boundary problems, *Control Problems For Systems Described by Partial Differential Equations*, pp. 56–63, I. Lasiecka, R. Triggiani, eds. Lecture Notes in Control and Information Sciences 97, Springer-Verlag, Berlin, Heidelberg, New York, Tokyo 1987.

A. Friedman, S. Huang, and J. Yong

[1] Bang-bang optimal control for the dam problem, *Appl. Math. Optimiz.* **15**(1987), pp. 65–85.

[2] Optimal periodic control for the two phase Stefan problem, *SIAM J. Control and Optimiz.* **26**(1988), pp. 23–41.

J. Goldstein

[1] Approximation of nonlinear semigroups and evolution equations, *J. Math. Soc. Japan* **24**(1972), pp. 558–573.

P. Grisvard

[1] *Elliptic Problems in Nonsmooth Domains*, Pitman Advanced Publishing Program, Boston, London, Melbourne 1984.

M. E. Gurtin and R. C. MacCamy

[1] Nonlinear age-dependent population dynamics, *Arch. Rat. Mech. Anal.* **54**(1974), pp. 281–300.

A. Haraux

[1] *Nonlinear Evolution Equations—Global Behaviour of Solutions*, Lecture Notes in Mathematics 841, Springer-Verlag, Berlin, New York, Heidelberg 1981.

J. Haslinger and P. Neittaanmäki

[1] *Finite Element Approximation for Optimal Shape Design. Theory and Applications*, John Wiley and Sons, Chichester, New York, Brisbane, Toronto, Singapore 1988.

J. Haslinger and P. D. Panagiotopoulos

[1] *Optimal Control by Hemivariational Inequalities*, in *Control of Boundaries and Stabilization*, pp. 128–139, J. Simon, ed. Lecture Notes in Control and Information Sciences 125, Springer-Verlag, New York 1989.

T. Havârneanu

[1] A semigroup approach to a class of Hamilton–Jacobi equations with nonconvex Hamiltonians, *Nonlinear Anal.* (to appear).

[2] An approximation scheme for the solutions of the Hamilton–Jacobi equation with max–min Hamiltonians in Hilbert spaces, *Nonlinear Anal.* (to appear).

L. I. Hedberg

[1] The approximation problems in function spaces, *Ark. Mat.* **16**(1978), pp. 51–81.

E. Hille and R. S. Phillips

[1] *Functional Analysis and Semigroups*, Amer. Math. Soc. Coll. Publ., vol. 31, Providence, Rhode Island, 1957.

I. Hlavacek, I. Bock, and J. Lovisek

[1] Optimal control of a variational inequality with applications to structural analysis. Optimal design of a beam with unilateral supports, *Appl. Math. Optimiz.* **11**(1984), pp. 111–143.

K. H. Hoffmann and J. Haslinger

[1] On identification of incidence set for elliptic free boundary value problems, *European J. Appl. Math.* (to appear).

K. H. Hoffmann and M. Niezgodka

[1] Control of parabolic systems involving free boundaries, *Free Boundary Problems, Theory and Applications*, vol. 2, pp. 431–462, A. Fasano and M. Primicerio, eds., Pitman, London, Boston 1983.

K. H. Hoffmann and J. Sprekels

[1] On the automatic control of the free boundary in an one-phase Stefan problem, *Applied Nonlinear Functional Analysis*, pp. 301–310, K. Gorenflo and K. H. Hoffmann, eds., Verlag P. Lang, Frankfurt a. M. 1983.

[2] Real time control of the free boundary in a two phase Stefan problem, *Numer. Funct. Anal. and Optimiz.* **5**(1982), pp. 47–76.

A. D. Ioffe and V. I. Levin

[1] Subdifferential of convex functions, *Trud. Mosk. Mat. Obsc.* **26**(1972), pp. 3–13.

J. W. Jerome

[1] *Approximation of Nonlinear Evolution Systems*, Academic Press, New York, London 1983.

F.Kappel and W. Schappacher

[1] Autonomous nonlinear functional differential equations and averaging approximations, *J. Nonlinear Analysis; Theory Methods Appl.* **2**(1978), pp. 391–422.

T. Kato

[1] *Perturbation Theory For Linear Operators*, Springer-Verlag, New York 1966.

[2] Nonlinear semigroups and evolution equations, *J. Math. Soc. Japan* **19**(1967), pp. 508–520.

[3] Accretive operators and nonlinear evolution equations in Banach spaces, *Nonlinear Functional Analysis*, pp. 138–161, F. Browder ed., Amer. Math. Soc., Providence, Rhode Island 1970.

[4] Differentiability of nonlinear semigroups, *Global Analysis*, Proc. Symposia Pure Math. ed. Amer. Math. Soc., Providence, Rhode Island 1970.

N. Kenmochi

[1] The semidiscretization method and nonlinear time-dependent parabolic variational inequalities, *Proc. Japan Acad.* **50**(1974), pp. 714–717.

N. Kikuchi and J. T. Oden

[1] Finite element methods for certain free boundary value problems in mechanics, *Moving Boundary Problems*, pp. 147–164, D. J. Wilson, ed. Academic Press, New York 1978.

D. Kinderlehrer and G. Stampacchia

[1] *An Introduction to Variational Inequalities and Their Applications*, Academic Press, New York 1980.

K. Kobayashi, Y. Kobayashi, and S. Oharu

[1] Nonlinear evolution operators in Banach spaces, *Osaka Math. J.* **26**(1984), pp. 281–310.

Y. Kobayashi

[1] Difference approximation of Cauchy problem for quasi-dissipative operators and generation of nonlinear semigroups *J. Math. Soc. Japan* **27**(1975), pp. 641–663.

[2] Product formula for nonlinear semigroups in Hilbert spaces, *Proc. Japan Acad.* **58**(1982), pp. 425–428.

[3] A product formula approach to first order quasilinear equations, *Hiroshima Math. J.* **14**(1984), pp. 489–509.

Y. Kobayashi and I. Miyadera

[1] Convergence and approximation of nonlinear semigroups, *Japan–France Seminar*, pp. 277–295, H. Fujita, ed. Japan Soc. Promotion Sci., Tokyo 1978.

Y. Komura

[1] Nonlinear semigroups in Hilbert spaces, *J. Math. Soc. Japan* **19**(1967), pp. 508–520.

[2] Differentiability of nonlinear semigroups, *J. Math. Soc. Japan* **21**(1969), pp. 375–402.

Y. Konishi

[1] On the nonlinear semigroups associated with $u_t = \Delta\beta(u)$ and $\rho(u_t) = \Delta u$, *J. Math. Soc. Japan* **25**(1973), pp. 622–628.

G. Köthe

[1] *Topological Vector Spaces*, Springer-Verlag, Berlin 1969.

S. N. Kružkov

[1] First order quasilinear equations in several independent variables, *Mat. Sbornik* **10**(1970), pp. 217–243.

O. A. Ladyzhenskaya, V. A. Solonnikov, and N. N. Uraltzeva

[1] *Linear and Quasilinear Equations of Parabolic Type*, AMS Translations, Providence, Rhode Island 1968.

J. M. Lasry and P. L. Lions

[1] A remark on regularization in Hilbert spaces, *Isreal J. Math.* **55**(1986), pp. 257–266.

J. L. Lions

[1] *Quelques Méthodes de Resolution des Problèmes aux Limites Nonlinéaires*, Dunod-Gauthier-Villars, Paris 1969.

[2] *Optimal Control of Systems Governed by Partial Differential Equations*, Springer-Verlag, Berlin, New York, Heidelberg 1971.

[3] *Contrôle Optimale de Systèmes Distribués Singuliers*, Dunod, Paris 1983.

[4] Various topics in the theory of optimal control of distributed parameter systems, *Optimal Control Theory and Mathematical Applications*, pp. 166–309, B. J. Kirby, ed., Lecture Notes in Economics and Mathematical Systems, Springer-Verlag, Berlin, Heidelberg 1974.

J. L. Lions and E. Magenes

[1] *Non-Homogeneous Boundary Value Problems and Applications*, T. I., Springer-Verlag, Berlin, Heidelberg, New York 1972.

P. L. Lions

[1] *Generalized Solutions of Hamilton–Jacobi Equations*, Research Notes in Mathematics 69, Pitman, Boston, London 1982.

W. B. Liu and J. E. Rubio

[1] Optimal shape design for systems governed by variational inequalities, *JOTA* **69**(1991), pp. 351–396.

E. Magenes

[1] Topics in parabolic equations; some typical free boundary problems, *Boundary Value Problems for Linear Evolution Partial Differential Equations*, pp. 239–312, H. G. Garnir, ed., D. Reidel Publishing Company, Dordrecht 1976.

[2] Remarques sur l'approximation des problèmes non linéaires paraboliques, *Analyse Mathématique et Applications*, Gauthier-Villars, Paris 1989.

E. Magenes and C. Verdi

[1] The semigroup approach to the two phase Stefan problem with nonlinear flux conditions, *Free Boundary Problems. Application and Theory*, vol. III, pp. 121–140, A. Bossavit, A. Damlamian, M. Fremond, eds., Pitman 1985.

R. H. Martin

[1] Differential equations on closed subsets of a Banach space, *Trans. Amer. Math. Soc.* **179**(1973), pp. 399–414.

[2] *Nonlinear Operators and Differential Equations in Banach Spaces*, John Wiley and Sons, New York 1976.

F. J. Massey

[1] Semilinear parabolic equations with L^1 initial data, *Indiana Univ. Math. J.* **26**(1977), pp. 399–411.

A. M. Meirmanov

[1] *The Stefan Problem* (in Russian), Izd. Nauka, Novosibirsk 1986.

F. Mignot

[1] Controle dans les inéquations variationelles elliptiques, *J. Funct. Anal.* **22**(1976), pp. 130–185.

F. Mignot and J. Puel

[1] Optimal control in some variational inequalities, *SIAM J. Control Optimiz.* **22**(1984), pp. 466–476.

G. Minty

[1] Monotone (nonlinear) operators in Hilbert spaces, *Duke Math. J.* **29**(1962), pp. 341–346.

[2] On the generalization of a direct method of the calculus of variations, *Bull. Amer. Math. Soc.* **73**(1967), pp. 315–321.

St. Mirica

[1] *Optimal Control. Sufficient Conditions and Synthesis* (in Romanian), Editura Stiinţifica, Bucharest 1990.

I. Miyadera and S. Oharu

[1] Approximation of semigroups of nonlinear operators, *Tohoku Math. J.* **22**(1970), pp. 24–47.

J. J. Moreau

[1] Proximité et dualité dans un espace hilbertien, *Bull. Soc. Math. France* **93**(1965), pp. 273–299.

[2] *Fonctionnelle Convexes*, Seminaire sur les équations aux dérivées partielles, Collège de France, Paris 1966–1967.

[3] Retraction d'une multiapplication, *Seminaire d'Analyse Convexe*, Montpelier 1972.

Ch. Moreno and Ch. Saguez

[1] Dependence par rapport aux données de la frontière libre associée à certaines inéquations variationnelles d'évolution, *Rapport de Recherche* **298**, IRIA, Rocquencourt 1978.

G. Moroşanu

[1] *Nonlinear Evolution Equations and Applications*, D. Reidel Publishing, Dordrecht, Boston, Lancaster, Tokyo 1988.

G. Moroşanu and Zheng-Xu He

[1] Optimal control of biharmonic variational inequalities, *An. St. Univ. "Al. I. Cuza" Iaşi*, **35**(1989), pp. 153–170.

P. Neittaanmäki, J. Sokoilowski, and J. P. Zolesio

[1] Optimization of the domain in elliptic variational inequalities, *Appl. Math. Optimiz.* **18**(1988), pp. 85–98.

P. Neittaanmäki and D. Tiba

[1] On the finite element approximation of the boundary control for two-phase Stefan problems, *Analysis and Optimization of Systems*, A. Bensoussan and J. L. Lions, eds., Lecture Notes in Control and Information Sciences 62, pp. 481–493, Springer-Verlag, Berlin, Heidelberg, New York, Tokyo 1984.

J. W. Neuberger

[1] Lie generators for one parametric semigroups of transformations, *J. für die reine und angewandte Math.* **258**(1973), pp. 133–136.

[2] Generation of nonlinear semigroups by a partial differential equation, *Semigroup Forum* **40**(1990), pp. 93–99.

L. Nicolaescu

[1] Optimal control for a nonlinear diffusion equation, *Richerche di Mat.* **XXXVII** (1988), pp. 3–27.

M. Niezgodka and I. Pawlow

[1] Optimal control for parabolic systems with free boundaries, *Optimization Techniques*, K. Iracki, ed., Lecture Notes in Control and Information Sciences 22, pp. 23–47, Springer-Verlag, Berlin 1980.

M. Otani

[1] Nonmonotone perturbations for nonlinear parabolic equations associated with subdifferential operators; Cauchy problem. *J. Diff. Eqs.* **46**(1982), pp. 268–299.

P. D. Panagiotopoulos

[1] *Inequality Problems in Mechanics and Applications. Convex and Nonconvex Energy Functions*, Birkäuser, Boston, Basel, Stuttgart 1985.

[2] Necessary conditions for optimal control governed by hemivariational inequalities, *Proceedings of a Conference of Distributed Parameter Systems*, J. L. Lions and A. Jay, eds., pp. 188–190, Perpignan 1989.

A. Papageorgiou

[1] Necessary and sufficient conditions for optimality in nonlinear distributed parameter systems with variable initial state, *J. Math. Soc. Japan* **42**(1990), pp. 387–396.

N. Pavel

[1] Nonlinear evolutions generated governed by f-quasi dissipative operators, *Nonlinear Analysis* **5**(1981), pp. 449–468.

[2] *Differential Equations Flow Invariance and Applications*, Research Notes in mathematics 113, Pitman, Boston, London, Melbourne 1984.

I. Pawlow

[1] Variational inequality formulation and optimal control of nonlinear evolution systems governed by free boundary problems, *Applied Nonlinear Functional Analysis*, K. Gorenflo and K. H. Hoffmann, eds., pp. 230–241, P. Lang Verlag, Frankfurt a. Main 1983.

A. Pazy

[1] The Lyapunov method for semigroups of nonlinear contractions in Banach spaces, *J. Analyse Math.* **40**(1982), pp. 239–262.

[2] *Semigroups of Linear Operators and Applications to Partial Differential Equations*, Springer-Verlag, New York, Berlin, Heidelberg 1983.

J. C. Peralba

[1] Un problème d'évolution relatif à un opérateur sous-différentiel dépendent du temps, *C.R.A.S. Paris* **275**(1972), pp. 93–96.

C. Popa

[1] Trotter product formulae for Hamilton–Jacobi equations in infinite dimension, *J. Diff. Integral Eqs.* **4**(1991), pp. 1251–1268.

[2] The relationship between the maximum principle and dynamic programming for the control of parabolic obstacle problem, *SIAM J. Control and Optimiz.* (to appear).

M. H. Porter and H. F. Weinberger

[1] *Maximum Principle in Differential Equations*, Springer-Verlag, New York, Berlin, Heidelberg, Tokyo 1984.

M. Reed and B. Simon

[1] *Methods of Modern Mathematical Physics*, T. 3, Academic Press, New York, San Francisco, London 1979.

S. Reich

[1] Product formulas, nonlinear semigroups and accretive operators in Banach spaces, *J. Funct. Anal.* 36(1980), pp. 147–168.

[2] A complement to Trotter's product formula for nonlinear semigroups generated by subdifferentials of convex functionals, *Proc. Japan Acad.* 58(1982), pp. 134–146.

R. T. Rockafellar

[1] *Convex Analysis*, Princeton University Press, Princeton, New Jersey 1969.

[2] On the maximal monotonicity of subdifferential mappings, *Pacific J. Math.* **33**(1970), pp. 209–216.

[3] Local boundedness of nonlinear monotone operators, *Michigan Math. J.* **16**(1969), pp. 397–407.

[4] On the maximality of sums of nonlinear operators, *Trans. Amer. Math. Soc.* **149**(1970), pp. 75–88.

[5] Integrals which are convex functional II, *Pacific J. Math.* **39**(1971), pp. 439–469.

[6] Integral functionals, normal integrands and measurable selections, *Nonlinear Operators and the Calculus of Variations*, pp. 157–205, J. P. Gossez, E. Dozo, J. Mawhin, and L. Waelbroeck, eds. Lecture Notes in Mathematics, Springer-Verlag 1976.

[7] The theory of subgradients and its applications to problems of optimization, *Lecture Notes, University of Montreal* 1978.

[8] Directional Lipschitzian functions and subdifferential calculus, *Proc. London Math. Soc.* **39**(1979), pp. 331–355.

P. H. Rodriguez

[1] Optimal control of unstable nonlinear evolution systems, *Ann. Faculté Sci. Toulouse* **VI**(1984), pp. 33–50.

Ch. Saguez

[1] Contrôle optimal d'inéquations variationnelles avec observation de domains, *Raport Laboria* **286**, IRIA 1978.

[2] Conditions nécessaires d'optimalité pour des problèmes de contrôl optimal associés à des inéquations variationnelles, *Rapport Laboria* **345**, IRIA 1979.

[3] *Contrôle Optimal de Systemes à Frontière Libre*, Thèse, L'Université de Technologie Compiègne 1980.

E. Schecter

[1] Stability conditions for nonlinear products and semigroups, *Pacific J. Math.* **85**(1979), pp. 179–199.

J. Schwartz

[1] *Nonlinear Functional Analysis*, Gordon and Breach, New York 1969.

Th. Seidman

[1] Some control-theoretic questions for a free boundary problem, *Control of Partial Differential Equations*, pp. 265–276, A. Bermudez, ed., Lecture Notes in Control and Information Sciences 114, Springer-Verlag, Berlin, Heidelberg, New York, Tokyo 1989.

Shuzong Shi

[1] Optimal control of strongly monotone variational inequalities, *SIAM J. Control and Optimiz.* **26**(1988), pp. 274–290.

E. Sinestrari

[1] Accretive differential operators, *Boll. UMI* **13**(1976), pp. 19–31.

M. Slemrod

[1] Existence of optimal controls for control systems governed by nonlinear partial differential equations, *Ann. Scuola Norm. Sup. Pisa* **3–4**(1974), pp. 229–246.

D. Tătaru

[1] Viscosity solutions of Hamilton–Jacobi equations with unbounded nonlinear term, *J. Math. Anal. Appl.* (to appear).

[2] Viscosity solutions for the dynamic programming equation, *Appl. Math. Optimiz.* (to appear).

[3] Viscosity solutions for Hamilton–Jacobi equations with unbounded nonlinear terms; a simplified approach (to appear).

D. Tiba

[1] Optimality conditions for distributed control problems with nonlinear state equations, *SIAM J. Control and Optimiz.* **23**(1985), pp. 85–110.

[2] Boundary control for a Stefan problem, *Optimal Control of Partial Differential Equations*, K. H. Hoffmann and J. Krabs, eds., Birkhäuser, Basel, Boston, Stuttgart 1984.

[3] Optimal control for second order semilinear hyperbolic equations, *Control Theory and Advanced Technology* **3**(1987), pp. 274–290.

[4] *Optimal Control of Nonsmooth Distributed Parameter Systems*, Lecture Notes in Mathematics, Springer-Verlag, Berlin, Heidelberg, New York, Tokyo, 1991.

D. Tiba and Z. Meike

[1] Optimal control for a Stefan problem, *Analysis and Optimization of Systems*, A. Bensoussan and J. L. Lions, eds., Lecture Notes in Control and Information Sciences 44, Springer-Verlag, Berlin, Heidelberg, New York, 1982.

L. Véron

[1] Effets régularisant de semi-groupes non linéaire dans des espaces de Banach. *Ann. Fac. Sci. Toulouse* **1**(1979), pp. 171–200.

I. Vrabie

[1] *Compactness Methods For Nonlinear Evolutions*, Longman Scientific and Technical, London 1987.

J. Watanabe

[1] On certain nonlinear evolution equations, *J. Math. Soc. Japan* **25**(1973), pp. 446–463.

G. F. Webb

[1] Continuous nonlinear perturbations of linear accretive operators, *J. Funct. Anal.* **10**(1972), pp. 181–203.

J. Yong

[1] Pontryagin maximum principle for semilinear second order elliptic partial differential equations and variational inequalities with state constraints (to appear).

K. Yosida

[1] *Functional Analysis*, Springer-Verlag, Heidelberg, New York 1978.

D. Zeidler

[1] *Nonlinear Operators*, Springer-Verlag, Berlin, New York, Heidelberg 1983.

Zheng-Xu He

[1] State constrained control problems governed by variational inequalities, *SIAM J. Control and Optimiz.* **25**(1987), pp. 1119–1144.

J. P. Zolesio

[1] Shape controllability of free boundaries, *J. Struct. Mech.* **13**(1985), pp. 354–361.

Index

Mathematics in Science and Engineering

Edited by William F. Ames, *Georgia Institute of Technology*

Recent titles

Anthony V. Fiacco, *Introduction to Sensitivity and Stability Analysis in Nonlinear Programming*
Hans Blomberg and Raimo Ylinen, *Algebraic Theory for Multivariate Linear Systems*
T. A. Burton, *Volterra Integral and Differential Equations*
C. J. Harris and J. M. E. Valenca, *The Stability of Input-Output Dynamical Systems*
George Adomian, *Stochastic Systems*
John O'Reilly, *Observers for Linear Systems*
Ram P. Kanwal, *Generalized Functions: Theory and Technique*
Marc Mangel, *Decision and Control in Uncertain Resource Systems*
K. L. Teo and Z. S. Wu, *Computational Methods for Optimizing Distributed Systems*
Yoshimasa Matsuno, *Bilinear Transportation Method*
John L. Casti, *Nonlinear System Theory*
Yoshikazu Sawaragi, Hirotaka Nakayama, and Tetsuzo Tanino, *Theory of Multiobjective Optimization*
Edward J. Haug, Kyung K. Choi, and Vadim Komkov, *Design Sensitivity Analysis of Structural Systems*
T. A. Burton, *Stability and Periodic Solutions of Ordinary and Functional Differential Equations*
Yaakov Bar-Shalom and Thomas E. Fortmann, *Tracking and Data Association*
V. B. Kolmanovskii and V. R. Nosov, *Stability of Functional Differential Equations*
V. Lakshmikantham and D. Trigiante, *Theory of Difference Equations: Applications to Numerical Analysis*
B. D. Vujanovic and S. E. Jones, *Variational Methods in Nonconservative Phenomena*
C. Rogers and W. F. Ames, *Nonlinear Boundary Value Problems in Science and Engineering*
Dragoslav D. Siljak, *Decentralized Control of Complex Systems*
W. F. Ames and C. Rogers, *Nonlinear Equations in the Applied Sciences*
Christer Bennewitz, *Differential Equations and Mathematical Physics*
Josip E. Pečarić, Frank Proschan, and Y. L. Tong, *Convex Functions, Partial Ordering, and Statistical Applications*
E. N. Chukwu, *Stability and Time-Optimal Control of Hereditary Systems*
E. Adams and U. Kulisch, *Scientific Computing with Automatic Result Verification*
Viorel Barbu, *Analysis and Control of Nonlinear Infinite Dimensional Systems*

ISBN 0-12-078145-X